FARM
MANAGEMENT

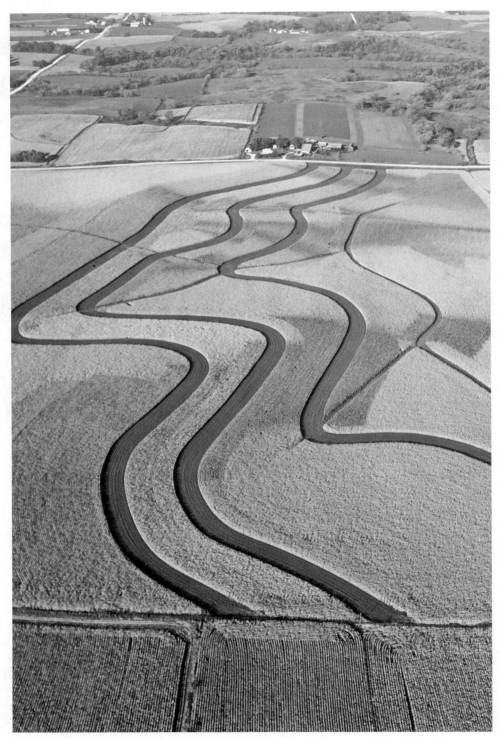

Photo by Tim McCabe, USDA Natural Resource Conservation Service

Eighth Edition

FARM
MANAGEMENT

Ronald D. Kay
Professor Emeritus,
Texas A&M University

William M. Edwards
Professor Emeritus, Iowa State University

Patricia A. Duffy
Professor, Auburn University

Mc
Graw
Hill
Education

FARM MANAGEMENT, EIGHTH EDITION

Published by McGraw-Hill Education, 2 Penn Plaza, New York, NY 10121. Copyright © 2016 by McGraw-Hill Education. All rights reserved. Printed in the United States of America. Previous editions © 2012, 2008, and 2004. No part of this publication may be reproduced or distributed in any form or by any means, or stored in a database or retrieval system, without the prior written consent of McGraw-Hill Education, including, but not limited to, in any network or other electronic storage or transmission, or broadcast for distance learning.

Some ancillaries, including electronic and print components, may not be available to customers outside the United States.

This book is printed on acid-free paper.

1 2 3 4 5 6 7 8 9 0 DOC/DOC 1 0 9 8 7 6 5

ISBN 978-0-07-340094-5
MHID 0-07-340094-7

Senior Vice President, Products & Markets: *Kurt L. Strand*
Vice President, General Manager, Products & Markets: *Marty Lange*
Vice President, Content Design & Delivery: *Kimberly Meriwether David*
Managing Director: *Michael Hackett*
Brand Manager: *Marija Magner*
Senior Director, Product Development: *Rose Koos*
Product Developer: *Mandy Clark*
Marketing Manager: *Kristine Rellihan*
Senior Digital Product Analyst: *Jake Theobald*
Director, Content Design & Delivery: *Terri Schiesl*
Content Production Manager: *Faye Schilling Herrig*
Content Project Managers: *Mary Jane Lampe, Judi David*
Buyer: *Laura M. Fuller*
Cover Design: *Studio Montage*

Content Licensing Specialist: *Beth Thole*
Cover Image Titles: *FRONT: "Case IH Early Riser© 1255 planter with Magnum™ 380 CVT tractor and producer." BACK: Tashner Dairy Farm Building, DSCN7680.JPG.*
Cover Image Credits: *FRONT: © Case IH, Early Riser 1255_B-0194_05-16-13. BACK:© Marty Tashner, Tashner Dairy Farm, 1995 Model Road, Cuba City Wisconsin.*
Compositor: *Lumina Datamatics, Inc.*
Typeface: *10/12 Times Roman*
Printer: *R.R Donnelley*

All credits appearing on page or at the end of the book are considered to be an extension of the copyright page.

Library of Congress Cataloging-in-Publication Data

Kay, Ronald D., author.
  Farm management / Ronald D. Kay, Professor Emeritus, Texas A&M University, William M. Edwards, Professor Emeritus, Iowa State University, Patricia A. Duffy, Professor, Auburn University. – Eighth edition.
    pages cm
  Includes index.
    ISBN 978-0-07-340094-5 (alk. paper) – ISBN 0-07-340094-7
1. Farm management.   I. Edwards, William M., author.   II. Duffy, Patricia Ann, 1955– author.   III. Title.
S561.K36 2014
630.68—dc23

2014035732

The Internet addresses listed in the text were accurate at the time of publication. The inclusion of a website does not indicate an endorsement by the authors or McGraw-Hill Education, and McGraw-Hill Education does not guarantee the accuracy of the information presented at these sites.

www.mhhe.com

CONTENTS

Preface xi

PART I

MANAGEMENT   *3*

CHAPTER 1

FARM MANAGEMENT NOW AND IN THE FUTURE   *7*

Chapter Outline   *7*
Chapter Objectives   *7*
Structure of Farms and Ranches   *8*
New Technology   *11*
The Information Age   *12*
Controlling Assets   *13*
Human Resources   *13*
Producing to Meet Consumer Demands   *14*
Contracting and Vertical Integration   *15*
Environmental and Health Concerns   *15*
Globalization   *16*
Summary   *17*
Questions for Review and Further Thought   *17*

CHAPTER 2

MANAGEMENT AND DECISION MAKING   *19*

Chapter Outline   *19*
Chapter Objectives   *19*
Functions of Management   *20*
Strategic Farm Management   *21*
Decision Making   *26*
Characteristics of Decisions   *29*
The Decision-Making Environment in
   Agriculture   *30*

Summary   *32*
Questions for Review and Further Thought   *32*

PART II

MEASURING MANAGEMENT
PERFORMANCE   *35*

CHAPTER 3

ACQUIRING AND ORGANIZING MANAGEMENT
INFORMATION   *39*

Chapter Outline   *39*
Chapter Objectives   *39*
Purpose and Use of Records   *40*
Farm Business Activities   *42*
Basic Accounting Terms   *43*
Options in Choosing an Accounting System   *43*
Chart of Accounts   *44*
Basics of Cash Accounting   *48*
Basics of Accrual Accounting   *49*
A Cash Versus Accrual Example   *50*
Farm Financial Standards Council
   Recommendations   *52*
Output from an Accounting System   *52*
Summary   *55*
Questions for Review and Further
   Thought   *55*

CHAPTER 4

THE BALANCE SHEET AND ITS ANALYSIS   *57*

Chapter Outline   *57*
Chapter Objectives   *57*

Purpose and Use of a Balance Sheet   *58*
Balance Sheet Format   *58*
Asset Valuation   *62*
Cost-Basis Versus Market-Basis
   Balance Sheet   *63*
Balance Sheet Example   *65*
Balance Sheet Analysis   *69*
Statement of Owner Equity   *72*
Summary   *74*
Questions for Review and Further
   Thought   *74*

CHAPTER 5

THE INCOME STATEMENT
AND ITS ANALYSIS   *77*

Chapter Outline   *77*
Chapter Objectives   *77*
Identifying Revenue and Expenses   *78*
Depreciation   *81*
Income Statement Format   *85*
Accrual Adjustments to a Cash-Basis Income
   Statement   *87*
Analysis of Net Farm Income   *89*
Change in Owner Equity   *95*
Statement of Cash Flows   *97*
Summary   *98*
Questions for Review and Further
   Thought   *99*

CHAPTER 6

FARM BUSINESS ANALYSIS   *101*

Chapter Outline   *101*
Chapter Objectives   *101*
Types of Analysis   *102*
Standards of Comparison   *103*
Diagnosing a Farm Business
   Problem   *104*
Measures of Profitability   *105*
Measures of Size   *109*
Efficiency Measures   *110*
Financial Measures   *114*
Summary   *118*
Questions for Review and Further
   Thought   *118*

PART III

APPLYING ECONOMIC PRINCIPLES   *121*

CHAPTER 7

ECONOMIC PRINCIPLES: CHOOSING PRODUCTION
LEVELS   *123*

Chapter Outline   *123*
Chapter Objectives   *123*
The Production Function   *124*
Marginal Analysis   *125*
Law of Diminishing Marginal Returns   *126*
How Much Input to Use   *127*
Using Marginal Concepts   *128*
Marginal Value Product and Marginal Input
   Cost   *132*
The Equal Marginal Principle   *133*
Summary   *136*
Questions for Review and Further
   Thought   *137*

CHAPTER 8

ECONOMIC PRINCIPLES: CHOOSING INPUT
AND OUTPUT COMBINATIONS   *139*

Chapter Outline   *139*
Chapter Objectives   *139*
Input Combinations   *140*
Output Combinations   *144*
Summary   *149*
Questions for Review and Further
   Thought   *149*

CHAPTER 9

COST CONCEPTS IN ECONOMICS   *153*

Chapter Outline   *153*
Chapter Objectives   *153*
Opportunity Cost   *154*
Cash and Noncash Expenses   *155*
Fixed, Variable, and Total Costs   *156*
Application of Cost Concepts   *159*
Economies of Size   *163*
Long-Run Average Cost Curve   *167*

Summary   *169*
Questions for Review and Further Thought   *169*
Appendix. Cost Curves   *170*

PART **IV**

BUDGETING FOR GREATER PROFIT   *175*

CHAPTER **10**

ENTERPRISE BUDGETING   *177*

Chapter Outline   *177*
Chapter Objectives   *177*
Purpose, Use, and Format of Enterprise
   Budgets   *178*
Constructing a Crop Enterprise Budget   *180*
Constructing a Livestock Enterprise Budget   *185*
General Comments on Enterprise Budgets   *187*
Interpreting and Analyzing Enterprise
   Budgets   *188*
Summary   *191*
Questions for Review and Further Thought   *191*

CHAPTER **11**

WHOLE-FARM PLANNING   *193*

Chapter Outline   *193*
Chapter Objectives   *193*
What Is a Whole-Farm Plan?   *193*
The Planning Procedure   *194*
Example of Whole-Farm Planning   *198*
Other Issues   *205*
Summary   *209*
Questions for Review and Further Thought   *209*
Appendix. Graphical Example of Linear
   Programming   *210*

CHAPTER **12**

PARTIAL BUDGETING   *215*

Chapter Outline   *215*
Chapter Objectives   *215*
Uses of a Partial Budget   *216*
Partial Budgeting Procedure   *216*

The Partial Budget Format   *217*
Partial Budgeting Examples   *219*
Factors to Consider When Computing Changes
   in Revenue and Costs   *222*
Sensitivity Analysis   *222*
Limitations of Partial Budgeting   *223*
Final Considerations   *224*
Summary   *224*
Questions for Review and Further Thought   *225*

CHAPTER **13**

CASH FLOW BUDGETING   *227*

Chapter Outline   *227*
Chapter Objectives   *227*
Features of a Cash Flow Budget   *228*
Constructing a Cash Flow Budget   *230*
Uses for a Cash Flow Budget   *238*
Monitoring Actual Cash Flows   *239*
Investment Analysis Using a Cash Flow
   Budget   *239*
Summary   *242*
Questions for Review and Further Thought   *243*

PART **V**

IMPROVING MANAGEMENT SKILLS   *245*

CHAPTER **14**

FARM BUSINESS ORGANIZATION
AND TRANSFER   *249*

Chapter Outline   *249*
Chapter Objectives   *249*
Life Cycle   *250*
Sole Proprietorship   *251*
Joint Ventures   *252*
Operating Agreements   *253*
Partnerships   *255*
Corporations   *258*
Limited Liability Companies   *261*
Cooperatives   *263*
Transferring the Farm Business   *264*
Summary   *267*
Questions for Review and Further Thought   *267*

CHAPTER **15**

MANAGING RISK AND UNCERTAINTY   *269*

Chapter Outline   *269*
Chapter Objectives   *269*
Sources of Risk and Uncertainty   *270*
Risk-Bearing Ability and Attitude   *272*
Expectations and Variability   *273*
Decision Making Under Risk   *278*
Tools for Managing Risk   *281*
Summary   *290*
Questions for Review and Further Thought   *290*

CHAPTER **16**

MANAGING INCOME TAXES   *293*

Chapter Outline   *293*
Chapter Objectives   *293*
Types of Income Taxes   *294*
Objectives of Tax Management   *295*
The Tax Year   *295*
Tax Accounting Methods   *296*
The Tax System and Tax Rates   *298*
Some Tax Management Strategies   *299*
Depreciation   *302*
Capital Gains   *306*
Summary   *309*
Questions for Review and Further Thought   *309*

CHAPTER **17**

INVESTMENT ANALYSIS   *311*

Chapter Outline   *311*
Chapter Objectives   *311*
Time Value of Money   *312*
Investment Analysis   *317*
Financial Feasibility   *322*
Income Taxes, Inflation, and Risk   *323*
Summary   *326*
Questions for Review and Further Thought   *327*
Appendix. An Example of an Investment
   Analysis   *328*
Initial Cost   *328*
Estimating Cash Expenses and Revenues   *328*

The Discount Rate   *329*
Net Cash Revenues   *330*
Net Present Value   *330*

CHAPTER **18**

ENTERPRISE ANALYSIS   *333*

Chapter Outline   *333*
Chapter Objectives   *333*
Profit and Cost Centers   *334*
The Accounting Period   *335*
Types of Enterprises   *336*
Land Costs   *341*
Verifying Production   *342*
Accounting Systems   *343*
Summary   *344*
Questions for Review and Further Thought   *345*

PART **VI**

ACQUIRING RESOURCES
FOR MANAGEMENT   *347*

CHAPTER **19**

CAPITAL AND CREDIT   *351*

Chapter Outline   *351*
Chapter Objectives   *351*
Economics of Capital Use   *352*
Sources of Capital   *353*
Types of Loans   *355*
The Cost of Borrowing   *362*
Sources of Loan Funds   *363*
Establishing and Developing Credit   *365*
Liquidity   *366*
Solvency   *368*
Summary   *370*
Questions for Review and Further Thought   *371*

CHAPTER **20**

LAND: CONTROL AND USE   *373*

Chapter Outline   *373*
Chapter Objectives   *373*

Factors that Affect Farmland Values   *374*
The Economics of Land Use and Management   *375*
Controlling Land: Own or Lease?   *377*
Buying Land   *379*
Leasing Land   *384*
Conservation and Environmental Concerns   *394*
Summary   *396*
Questions for Review and Further Thought   *396*
Cash Farm Lease   *397*

CHAPTER **21**

HUMAN RESOURCE MANAGEMENT   *403*

Chapter Outline   *403*
Chapter Objectives   *403*
Characteristics of Agricultural Labor   *405*
Planning Farm Labor Resources   *405*
Measuring the Efficiency of Labor   *410*
Improving Labor Efficiency   *411*
Improving Managerial Capacity   *412*
Obtaining and Managing Farm Employees   *413*
Agricultural Labor Regulations   *420*

Summary   *422*
Questions for Review and Further Thought   *423*

CHAPTER **22**

MACHINERY MANAGEMENT   *425*

Chapter Outline   *425*
Chapter Objectives   *425*
Estimating Machinery Costs   *426*
Examples of Machinery Cost Calculations   *431*
Factors in Machinery Selection   *433*
Alternatives for Acquiring Machinery   *436*
Improving Machinery Efficiency   *441*
Summary   *444*
Questions for Review and Further
  Thought   *445*

APPENDIX   *446*

GLOSSARY   *452*

INDEX   *460*

PREFACE

Farms and ranches, like other small businesses, require sound management to survive and prosper. The continual development of new agricultural technologies means that farm and ranch managers must stay informed of the latest advances and decide whether to adopt them. Adopting a risky, unproven technology that fails to meet expectations can cause financial difficulties or even termination of the farm business. On the other hand, failing to adopt profitable new technologies will put the farm business at a competitive disadvantage that could also prove disastrous in the long run. In addition, changing public policies regarding environmental protection, taxes, and income supports can make certain alternatives and strategies more or less profitable than they have been in the past. Finally, changes in consumer tastes, the demographic makeup of our population, and world agricultural trade policies affect the demand for agricultural products.

The continual need for farm and ranch managers to keep current and update their skills motivated us to write this eighth edition.

This book is divided into six parts. Part I begins with the chapter "Farm Management Now and in the Future." It describes some of the technological and economic forces driving the changes we see in agriculture. By reading this chapter, students will find an incentive to study farm management and an appreciation for the management skills modern farm managers must have or acquire. Part I concludes with an explanation of the concept of management and the decision-making process, with an emphasis on the importance of strategic planning and decision making.

Part II presents the basic tools needed to measure management performance, financial progress, and the financial condition of the farm business. It discusses how to collect and organize accounting data and how to construct and analyze farm financial statements. Data from an example farm is used to demonstrate the analysis process in the chapter on farm business analysis.

Part III contains three chapters on basic microeconomic principles and cost concepts. The topics in this part provide the basic tools needed to make good management decisions. Students will learn how and when economic principles can be used in management decision making, along with the importance of the different types of economic costs in both the short run and the long run. Economies and diseconomies of size and their causes are discussed.

Practical use of budgeting as a planning tool is emphasized in Part IV. The discussion includes chapters on enterprise, partial, whole farm, and cash flow budgets. The format and use for each type of budget, sources of data to use, and break-even analysis techniques are discussed in detail.

Topics necessary to further refine a manager's decision-making skills are included in Part V. Farm business organization and transfer, risk control, income tax management, investment analysis, and enterprise analysis are discussed. The chapter on income tax management has been updated with the latest changes available. The chapter on investment analysis includes a discussion of the concepts of annual equivalent and capital recovery values. The final chapter discusses how to separate the whole-farm analysis into profit centers and cost centers.

Part VI discusses strategies for acquiring the resources needed on farms and ranches, including capital and credit, land, human resources, and machinery. The human resource chapter includes sections on improving managerial capacity and bridging the cultural barriers that may be encountered in managing agricultural labor.

New materials to help instructors have been incorporated into the current edition's Web site. An electronic slide presentation covering each chapter, a test question bank, class exercises, and answers to the end-of-chapter questions can be found at www.mhhe.com/kay8e.

The authors would like to thank the instructors who have adopted the previous edition for their courses and the many students who have used it both in and out of formal classrooms. Your comments and suggestions have been carefully considered and many were incorporated in this edition. Suggestions for future improvements are always welcome. A special thanks goes to the McGraw-Hill reviewers for their many thoughtful ideas and comments provided during the preparation of this edition.

New to this edition:

- 56 new and revised tables
- 16 new and revised figures
- 14 new and revised boxes
- 12 new glossary terms

Updated material about:

- Income tax brackets and rates
- 2012 Census of Agriculture data
- Current commodity price levels for examples
- Current production costs for examples
- Multiple peril crop insurance
- Land values and rental rates
- Farm financial data and benchmark values
- Agricultural labor laws

New or expanded discussion of:

- Chart of accounts
- Treatment of forward-priced commodities on the balance sheet
- Treatment of deferred taxes and capital gains on the balance sheet
- Definitions and equations for FFSC analysis measures
- Financial repayment capacity measures
- Using calculus to find optimal input levels
- Break-even yields and prices
- Limited liability companies
- Gift and estate taxes
- Adjusting yield estimates for trends
- USDA farm commodity programs
- Amortization of balloon payment loans
- Farm lease example
- Employee benefits and bonuses

Ronald D. Kay
William M. Edwards
Patricia A. Duffy

ABOUT THE AUTHORS

Ronald D. Kay is Professor Emeritus in the Department of Agricultural Economics at Texas A&M University. Dr. Kay taught farm management at Texas A&M University for 25 years, retiring at the end of 1996. He was raised on a farm in southwest Iowa and received his B.S. in agriculture and Ph.D. in agricultural economics from Iowa State University. He has experience as both a professional farm manager and a farm management consultant, and he maintains an active interest in a farming operation. He is a member of several professional organizations, including the American Society of Farm Managers and Rural Appraisers, where he was a certified instructor in their management education program. Dr. Kay received the Society's Excellence in Education award for 2002.

William M. Edwards is Professor Emeritus of Economics at Iowa State University, from where he received his B.S., M.S., and Ph.D. degrees in agricultural economics. He grew up on a family farm in south-central Iowa, and worked as an agricultural economist with the Farmer's Home Administration and a Peace Corps volunteer with the Colombian Agrarian Reform Institute. From 1974 through 2013, he taught on-campus and distance education courses and carried out extension programs in farm management at Iowa State University. In 2013, he received the Distinguished Service to Agriculture Award from the American Society of Farm Managers and Rural Appraisers. Dr. Edwards served as president of the Extension Section of the Agricultural and Applied Economics Association in 2006–2007.

Patricia A. Duffy is Professor in the Department of Agricultural Economics and Rural Sociology at Auburn University, where she has taught farm management since 1985. She grew up in Massachusetts and received her B.A. from Boston College. After finishing this degree, she served as a Peace Corps volunteer for two years, teaching basic agriculture sciences in a vocational secondary school. She received her Ph.D. in agricultural economics from Texas A&M University. Her research papers in farm management and policy have been published in a variety of professional journals. In 1994, she received an award from the Southern Agricultural Economics Association for distinguished professional contribution in teaching programs. In 2001, she received Auburn University's College of Agriculture teaching award.

FARM
MANAGEMENT

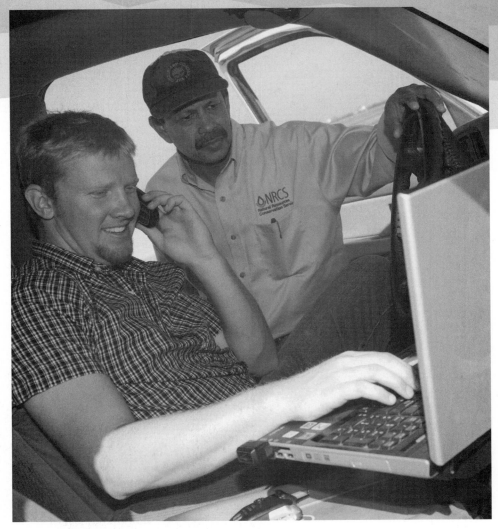

© *Photo by Jeff Vanuga, USDA Natural Resources Conservation Service*

MANAGEMENT

Good management is a crucial factor in the success of any business. Farms and ranches are no exception. To be successful, farm and ranch managers need to spend more time making management decisions and developing management skills than their parents and grandparents did.

This is because production agriculture in the United States and other countries is changing along the following lines: more mechanization, increasing farm size, continued adoption of new production technologies, growing capital investment per worker, more borrowed or leased capital, new marketing alternatives, and increased business risk. These factors create new management problems, but also present new opportunities for managers with the right skills.

These trends will likely continue throughout the rest of the twenty-first century. Farmers will make the same type of management decisions as in the past, but will be able to make them faster and more accurately. Advances in the ability to collect, transfer, and store data about growing conditions, pest and disease problems, and product quality will give managers more signals to which to respond. Moreover, future farm and ranch operators will have to balance their personal goals for an independent lifestyle, financial security, and rural living against societal concerns about food safety, environmental quality, and agrarian values.

The long-term direction of a ranch or farm is determined through a process called strategic planning. Farm families establish goals for themselves and their businesses based on their personal values, individual skills and interests, financial and physical resources, and the economic and social conditions facing agriculture in the next generation. They can choose to emphasize wider profit margins or higher volumes of production or to produce special services and products. After identifying and selecting strategies that will help them achieve their goals, farm and ranch operators employ tactical management to carry them out. Many decisions need to be made and many alternatives analyzed. Finally, the results of those decisions must be monitored and evaluated and control measures implemented where results are not acceptable.

Chapter 1 discusses factors affecting the management of farms and ranches now and in the coming decades. These factors will require a new type of manager who can absorb, organize, and use large amounts of information—particularly information related to new technologies. Resources will be a mix of owned, rented, and borrowed assets. Products will need to be more differentiated to match consumer tastes and safety standards. Industrial uses of agricultural products will increase relative to food uses. The profitability of a new technology must be determined quickly and accurately before it is or is not adopted. A modern manager will also need new human resource skills as the number and diversity of employees increase.

Chapter 2 explains the concept of management, including strategic planning and tactical decision making. What is management? What functions do managers perform? How should managers make decisions? What knowledge and skills are needed to be a successful manager? Answers to the first three questions are discussed in Chapter 2. Answers to the last question will require studying the remainder of the book.

© Comstock/PunchStock

FARM MANAGEMENT NOW AND IN THE FUTURE

CHAPTER OUTLINE

Structure of Farms and Ranches
New Technology
The Information Age
Controlling Assets
Human Resources
Producing to Meet Consumer Demands
Contracting and Vertical Integration
Environmental and Health Concerns
Globalization
Summary
Questions for Review and Further Thought

CHAPTER OBJECTIVES

1. Discuss how changes in the structure and technology of agriculture will affect the next generation of farm and ranch managers

2. Identify the management skills that future farmers and ranchers will need to respond to these changes

What will future farm managers be doing as we progress through the remaining decades of the twenty-first century? They will be doing what they are doing now, making decisions. They will still be using economic principles, budgets, record summaries, investment analyses, financial statements, and other management techniques to make those decisions. What types of decisions will managers be making in future decades?

They will still be deciding input and output levels and combinations and when and how to acquire additional resources. They will continue to analyze the risks and returns from adopting new technology, making new capital investments, adjusting farm size, changing enterprises, and seeking new markets for their products.

Will anything about management decisions in the future be different? Yes. While the broad types of decisions being made will be the same, the details and information used will change. Technology will continue to provide new inputs to employ and new, more specialized products for production and marketing. Management information systems, aided by electronic innovations, will provide more accurate and timely information for use in making management decisions. Farmers and ranchers will have to compete more aggressively with nonagricultural businesses for the use of land, labor, and capital resources. As in the past, the better managers will adapt to these changes and efficiently produce commodities that consumers and industry want.

STRUCTURE OF FARMS AND RANCHES

The number of farms in the United States has been decreasing since 1940, as shown in Figure 1-1. The amount of land in farms and ranches has been relatively constant; this means the average

production per farm has increased considerably, as shown in Figure 1-2. Several factors have contributed to this change.

First, labor-saving technology in the form of larger agricultural machinery, more efficient planting and harvesting systems, automated equipment, and specialized livestock buildings has made it possible for fewer farm workers to produce more crops and livestock. Second, employment opportunities outside agriculture have become more attractive and plentiful, encouraging labor to move out of agriculture. Also during this period of change, the cost of labor has increased faster than the cost of capital, making it profitable for farm managers to substitute capital for labor in many areas of production.

Third, farm and ranch operators have aspired to earn higher levels of income and to enjoy a standard of living comparable to that of nonfarm families. One way to achieve a higher income has been for each farm family to control more resources and produce more output while holding costs per unit level or even decreasing them. Other managers, though, have

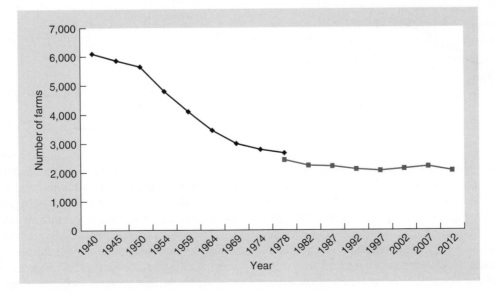

Figure 1-1 Number of farms in the United States (1000s).
Source: U.S. Census of Agriculture, USDA. Definition adjusted in 1997.

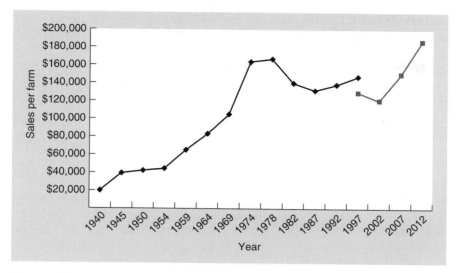

Figure 1-2 Total sales per farm in 2002 dollars.
Source: U.S. Census of Agriculture, USDA. Definition adjusted in 1997.

worked to increase profit margins per unit while keeping the size of their business the same. The desire for an improved standard of living has provided much of the motivation for increasing farm size, and new technology has provided the means for growth.

Fourth, some new technology is available only in a minimum size or scale, which encourages farmers to expand production and spread the fixed costs of the technology over enough units to be economically efficient. Examples include grain drying and handling systems, four-wheel drive tractors, large harvesting machines, confinement livestock buildings, and automated cattle feedlots. Perhaps even more important are the time and effort required for a manager to learn new skills in production, marketing, and finance. These skills also represent a fixed investment and thus generate a larger return to the operator when they are applied to more units of production. Chapter 9 contains more discussion about economies of size in agriculture.

Operators who do not wish to grow their individual businesses will look for alliances and partnerships, both formal and informal, with other producers that will allow them to achieve

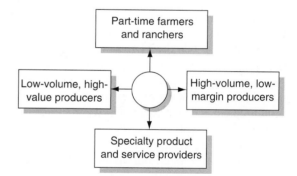

Figure 1-3 Alternative strategies for farm and ranch businesses.

the same economies as larger operations. Examples include jointly owning machinery and equipment with other producers, outsourcing some tasks such as harvesting or raising replacement breeding stock, and joining small, closed cooperatives.

As illustrated in Figure 1-3, farmers and ranchers will choose among four general business strategies: low-volume, high-value producers; high-volume, low-margin producers; specialty product and service providers; and part-time operators.

Low-Volume, High-Value Producers

Lack of access to additional land, labor, and capital effectively limits the potential of many growers for expanding their businesses. For them, the key to higher profits is producing higher valued commodities. Some look for nontraditional enterprises such as emus, bison, asparagus, or pumpkins. Promotion, quality standards, and marketing become critical to their success. Others try variations of traditional commodities, such as organically grown produce, tofu soybeans, free-range poultry, or seed crops. Margins may be increased even more through added processing and direct marketing. Such enterprises often involve high production risks, uncertain markets, and intensive management, but can be quite profitable even on a small scale.

High-Volume, Low-Margin Producers

There will always be a demand for generic feed grains, oil seeds, fruits and vegetables, cotton, and livestock products. Many producers choose to stick with familiar enterprises and expand production as a means of increasing their income. For them, squeezing every nickel out of production costs is critical. Growing the business usually involves leveraging it with borrowed or rented assets. Profit margins are thin, so it is critical to set a floor under market prices or total revenue through insurance products and marketing contracts.

Specialty Product and Service Providers

A third strategy is to specialize in just one or two skills and become one of the best at performing them. Examples are custom harvesting, custom cattle feeding, raising seed stock or replacement breeding stock, repairing and refurbishing equipment, hauling and applying manure, and applying pesticides and fertilizers. Even *agri-tourism* can be considered a special service to consumers. Often a key component of this strategy is making maximum use of expensive, highly specialized equipment and facilities. Marketing the services of the business and interacting with customers are also important ingredients for success.

Part-Time Operators

Many farmers hold other jobs in addition to farming. Part-time farmers and ranchers account for about 52 percent of the U.S. total, according to data from the U.S. Census of Agriculture. However, they produce only 13 percent of total agricultural sales. Many of these small-scale operations are *lifestyle* farms run by people who enjoy producing crops and livestock even when the potential profits are low. Their primary management concerns are to limit their financial risk and balance farm labor needs with off-farm employment. A combination of farming and non-farm employment may provide the most acceptable level of financial security and job satisfaction for many families.

Farms of all sizes will continue to find their niche in U.S. agriculture. Naturally, the largest farms contribute the highest proportion of total sales of farm products, as shown in Table 1-1. The consolidation of small- and medium-sized farms and ranches into larger units will likely continue, as older operators retire and their land is combined with existing farm units.

Management and operation of farms by family units will continue to be the norm. This is especially true for agricultural enterprises that

TABLE 1-1	Distribution of Farm Sales, United States	
Sales class	Percent of farms	Percent of sales
Less than $50,000	75.4	3.0
$50,000–$99,999	6.1	2.3
$100,000–$249,999	6.6	5.8
$250,000–$999,999	8.1	22.5
$1,000,000 or more	3.8	66.4

Source: 2012 Census of Agriculture, USDA.

cannot concentrate production into a small geographic area, such as crop production or extensive grazing of cattle or sheep. Enterprises that can centralize production, such as poultry and hog production or cattle feeding, can be more easily organized into large-scale business entities. Management of these farms will be segregated into several layers, and areas of responsibility will be more specialized. Most managers of centralized production enterprises will be salaried employees rather than owner-operators.

Some family farm businesses will find that by cooperating with their neighbors and relatives they can achieve many of the same advantages that larger-scale operations enjoy. Decades ago, farmers formed grain threshing or haying crews to take advantage of new harvesting technology. Today, several farmers join together to guarantee a constant, uniform supply of livestock or crops in a quantity that can be transported and processed efficiently. As the number of input suppliers and processing firms diminishes, producers must collaborate to maintain their bargaining position. This is one example of how a cooperative effort or *strategic alliance* can provide economic benefits. Another example is several operators forming an input purchasing group to achieve quantity discounts or purchasing large equipment jointly. A small amount of managerial independence must be sacrificed to conform to the needs of the group. However, personal ownership and operation of each business is preserved.

NEW TECHNOLOGY

Agricultural technology has been evolving for many decades and will continue to do so. The field of biotechnology offers possible gains in production efficiency, which may include crop varieties engineered to fit growing conditions at particular locations, resistant to herbicide damage or to certain insects and diseases, or having a more highly valued chemical composition such as higher protein or oil content. Livestock performance may be improved by introducing new genetic characteristics or by improving nutrient use. New nonfood uses for agricultural products, such as biodiesel and ethanol, will open new markets, but may also cause changes in the desired characteristics or composition of products grown specifically for these uses.

One example of a recent technology is the use of global positioning systems (GPS) to pinpoint the exact location of equipment in a field. By combining satellite reception with a yield monitor on harvesting equipment, the crop yield can be measured and recorded continuously for every point in the field. Variations in yield due to soil type, previous crops, different tillage methods, and fertilizer rates can be identified quickly and recommendations made to correct problems. This technology is now being used to automatically adjust the application rates of fertilizer and chemicals as the applicator moves across the field. Fertilizer and chemicals are applied only at the rates and locations needed, which improves efficiency and lowers costs.

Automated GPS can also keep crop production machinery on a consistent course, when used with automatic guidance systems on tractors, harvesters, and sprayers. Field time and operator fatigue are reduced, and more efficient use of crop inputs results from less overlapping of applications. Operator errors while using equipment at night are reduced as well.

These technologies and others yet to be developed will provide the farm manager with a continual challenge. Should this or any new technology be adopted? The cost of any new technology must be weighed against its benefits, which may come in several forms. There may be increased yields, an improvement in product quality, less variation in yield, or a reduced impact on the environment. Decisions about if and when to adopt a new technology will affect the profitability and long-term viability of a farm or ranch business.

THE INFORMATION AGE

Many decision-making principles and budgeting tools have been underused in the past. Individual farm data needed to use them were not available, or the process for analyzing the data was too complex. Recent years have seen rapid changes in methods of data collection, analysis, and interpretation. Electronic sensors and processors used in large-scale industries are now accessible and affordable to farms and ranches, as well as to purchasers of agricultural products.

Not only will more whole-farm data be available, but data specific to small land areas or to individual animals will also become more common. These specific data will help managers customize the treatment of each acre of land or each head of livestock. Yields can be monitored and recorded as harvesting machines move across the field. GPS can use satellite signals to identify the exact position of harvesting units when the data are collected. Automated machines may be able to take a soil sample every few yards, analyze it instantly, and record the results by field location. Satellite photographs and other techniques may provide information on the specific location of weed and insect infestations or moisture, permitting a limited, pinpoint application of pesticide or irrigation water.

Miniature electronic sensors will be able to collect and record information from livestock by continuously monitoring individual animal performance levels, feed intake, and health status. When undesirable changes are detected, there could be automatic adjustment in environmental conditions and feed rations. This information could also be related back to genetic background, physical facilities, feed rations, health programs, and other management factors to improve and fine-tune animal performance. Ear tags, electronic implants, and detailed production records can provide *identity preservation* of both crops and livestock from the original producer to the final consumer.

Financial transactions may be recorded and automatically transferred to accounts through

---

Box 1-1   Meeting New Challenges: Berilli Farms

Berilli Farms, Inc. consists of only a few hundred acres. These acres have been transformed from growing common field grains to producing high-value specialty crops. Fresh vegetables are sold to a local wholesale grocer. High-protein alfalfa has been contracted to a dairy in the next county. High-quality turf grass seed goes to a chain of nurseries.

Keeping a stable work crew of 25 machinery operators, truck drivers, sorters, and crop scouts is a real test of the Berilli family's human relations skills. All of their employees are trained to gather data on crop growth and yields from monitors mounted on machinery or in fields, and to download it into their handheld computers. Each morning before chemicals or fertilizers are applied, a variable-rate application plan is read into the control units of the applicators.

The Berillis use sophisticated crop simulation computer models to formulate these recommendations, taking into account current input prices and the selling prices for their products that they have contracted or hedged. Each week they review their cash-flow position and electronically transfer operating funds into their business account. All their crops are protected by multiple peril crop insurance and are committed to delivery according to a detailed production contract.

The grocers, dairies, and nurseries they supply send them real-time data about the results of quality tests performed on their products and the varieties selling the fastest. At the end of the year, the Berillis analyze the costs and returns from each crop, field, and buyer and replace the least profitable ventures with more promising ones.

the use of debit cards and bar-code symbols whenever purchases and sales occur. Smaller purchases may be made with preloaded cash cards. These transactions can also be posted automatically to the accounting system for an individual farm and classified by enterprise, production period, vendor, or business unit. These technological advances mean that the information in a farmer's accounting system can be accurate and up to date at the end of each day.

Personal computers have greatly enhanced capacities to receive, process, and store information and to communicate with outside data sources. Portable computers and personal data recorders allow precise decisions to be made in the pickup or on the tractor, as well as in the office. The first computers were used primarily to sort data and do calculations, but increasingly computers are being designed and used as communication tools. Wireless transmission technology and global computer networks are increasing the availability, speed, and accuracy of information sharing about weather, markets, and other critical events.

Managers in the past century often found the lack of accurate, timely, and complete information to be frustrating. Modern managers may still be frustrated by information; only the cause of their frustration will be the large quantity and continual flow of information available to them. A vital task for managers will be to determine which information is critical to their decision making, which is useful, and which is irrelevant. Even when this is done, the critical and useful information must be analyzed and stored in an easily accessible manner for future reference.

CONTROLLING ASSETS

Outside capital will continue to be needed to finance large-scale operations. Management of traditional sources of farm credit, such as rural banks, is becoming more vertically integrated, and funds will come from national money markets. Credit will also be available from nontraditional sources such as input suppliers and processors. Farm managers will increasingly have to compete with nonfarm businesses for access to capital, as the rural and urban financial markets become more closely tied together. This competition will necessitate more detailed documentation of financial performance and credit needs, and more conformity to generally accepted accounting principles and performance measures. Farmers will need to use standard accounting methods and principles and perhaps even have audited financial statements to gain access to commercial capital markets.

Standardized records and online databases will help make comparative analysis with similar farms more meaningful. The farm manager will have to decide whether to train an employee to carry out the required accounting and analysis or hire this expertise from outside the business. Even if outside help is used, the manager must have the skills and knowledge to read, interpret, and use this accounting information.

Controlling assets is becoming more important than owning them. Farmers have long gained access to land by renting it. Leasing machinery, buildings, and livestock has been less common, but will likely increase in use. Custom farming and contract livestock production are other means by which a good manager can apply his or her expertise without taking the financial risks of ownership. When other parties supply much of the capital, the operator can produce a larger volume at less risk, although the profit margin may be smaller.

HUMAN RESOURCES

Farm managers are currently depending more on a team of employees or partners to carry out specific duties in the operation. Working with other people will become a more important factor in the success of the operation. Motivation, communication, evaluation, and training of personnel will become essential skills.

Farm businesses will have to offer wages, benefits, and working conditions competitive with nonfarm employment opportunities. They will likely have to follow more regulations regarding worker safety in handling farm chemicals and equipment and see that employees are properly trained in the use of new technologies. Many of the most efficient farms and ranches will be those with a small number of operators or employees who have specialized responsibilities. They will have mastered the communication and teamwork skills needed in such operations.

Modern managers will need to take advantage of the expertise of paid consultants and advisors. For some very technical decisions, such as diagnosing animal and plant diseases, developing legal contracts, or executing commodity pricing strategies, the manager may pay a consultant to make recommendations. In other cases, the farm manager will obtain information from outside sources but do the analysis and decision making. Examples include formulating livestock rations or crop fertility programs based on the results of laboratory tests. In either case, the successful manager must learn to communicate clearly and efficiently with the consultant. This means understanding the terminology and principles involved and summarizing information in a concise form before submitting it.

PRODUCING TO MEET CONSUMER DEMANDS

Agriculture has long been characterized by the production of *undifferentiated* commodities. Historically, grain and livestock products from different farms have been treated alike by buyers if these products met basic quality standards and grades. The trend is to offer more highly specialized and processed food products to the consumer, so buyers are beginning to implement stricter product standards for producers.

For example, livestock processors want uniform animals with specific size and leanness characteristics to fit their processing equipment, packaging standards, and quality levels. Improved measuring devices, product identification, and data processing will make it easier to pay differential prices to producers based on product characteristics and to trace each lot to its point of origin. As processors invest in larger-scale plants, they must operate them at full capacity to reduce costs and remain competitive. Producers who can assure the packer of a continuous supply of high-quality, uniform animals will receive a premium price. Those who cannot may find themselves shut out of many markets or forced to accept a lower price.

In crop production, the protein and oil content of grain and forages is becoming easier to

---

Box 1-2   Custom Pork Production: Producing for the Market

Howard Berkmann continues to produce traditional cross-bred, uniform lean hogs for the local packing plant. One morning each week, he delivers a load of hogs, and by evening, he receives by electronic mail a summary of the carcass data and pricing formula from the packer. He downloads the information onto his swine production computer software and prints out a current summary for the facility from which the hogs came and the genetic group they represented.

A few years ago, Howard started a specialty group of Berkshire hogs designed for the Japanese market. The particular coloring and marbling of the meat earns him a premium price. He negotiated an agreement with a Berkshire breeder in a neighboring state to supply him with a regular stream of replacement gilts. Several times a week, he checks the Japanese livestock markets for forward pricing opportunities, and he has visited his marketing contact in Tokyo.

measure, making differential pricing possible. Biotechnology research will allow plant characteristics to be altered and genetically engineered varieties to be produced for specific uses, regions, and production technologies.

More agricultural products will be used for industrial purposes, such as biofuels, renewable energy, pharmaceutical products, and biodegradable packaging. This will require increased attention to product quality, segregation of production, record keeping, and marketing contracts. Traditional marketing channels and price patterns will change.

So-called niche markets will also become more important. Organic produce, extra-lean meat, specialty fruits and vegetables, and custom-grown products for restaurants and food services will be in greater demand. As international trade barriers continue to fall, foreign markets will also be more important. These markets may require products with special characteristics. Farm managers who seek out these markets and learn the production techniques necessary to meet their specifications can realize a higher return from their resources. The manager will have to evaluate the additional costs and increased risks associated with specialty markets and compare them with the potentially higher returns.

CONTRACTING AND VERTICAL INTEGRATION

Just as some farmers and ranchers will produce specific products, others will specialize in a particular phase of producing more generic products. Examples include raising dairy replacement heifers, harvesting crops on a custom hire basis, or producing bedding plants for home gardeners. Such operators can develop a high degree of expertise in their particular area and apply it to a high volume of production.

Many of these managers produce an intermediate product or service so there may not be a widespread market at an established market price. To ensure that they can sell their product,

they may enter into a marketing contract with a processor, wholesale distributor, or other farmers. The contract may guarantee that a constant supply of product of a minimum quality and type will be delivered. In some cases the buyer may supply some of the inputs and management, such as when pigs or broilers are finished in contract facilities on independent farms. Such arrangements are called *vertical integration*.

ENVIRONMENTAL AND HEALTH CONCERNS

As the availability of an adequate quantity of food becomes ever more taken for granted, concerns about food quality and food safety as well as the present and future condition of our soil, water, and air will continue to receive high priority from the nonfarm population. Farmers and ranchers have always had a strong interest in maintaining the productivity of natural resources under their control. However, the off-farm and long-term effects that new production technologies have on the environment have not always been well quantified or understood. As more people decide to live in rural areas, the contact between farm and nonfarm residents will increase. This will lead to increased concern about agricultural wastes and their effects on air and water quality. Pressure from nonfarm rural residents may even cause some production systems such as concentrated livestock feeding to shift to less-populated regions. Farm managers will have to choose between discontinuing those enterprises and moving their businesses.

As research and experience improve, the understanding of the interactions among various biological systems, education, and regulation will be used to increase the margin of safety for preserving resources for future generations. Top agricultural managers of today recognize the need to keep abreast of the environmental implications of their production practices and are often leaders in developing sustainable production systems. All farm managers must be aware of the effects their production practices have on the environment, both on and off the farm, and take the steps necessary to keep their agricultural resources productive and environmentally safe.

The value of agricultural assets, particularly farmland, will be affected by environmental conditions and regulations. When farms are sold or appraised, environmental audits become routine to warn potential buyers of any costs that might be incurred to clean up environmental hazards. The crop production combinations and practices allowed by a farm's conservation plan also affect its value. Farm managers will have to evaluate every decision for profitability and for how it affects the environment. The successful managers will be those who can generate a profit while sustaining resources on the farm and minimizing environmental problems off the farm.

GLOBALIZATION

Agricultural producers all over the world are finding that their success or failure is increasingly tied to weather, public policies, and consumer tastes that exist thousands of miles away. Expansion of markets through international trade has long been an avenue by which farmers have sought to enhance the prices of their products and channel increased production to consumers. However, the governments of many countries, including the United States, have tried to protect their farmers from foreign competition through the use of trade barriers such as tariffs, quotas, and sanitary regulations.

In recent years many of these barriers have been lowered or eliminated. The World Trade Organization (WTO) is an international organization dedicated to negotiating freer trade throughout the world to increase the efficiency of food production and improve standards of living for millions of people. Other cooperative arrangements such as the North American Free Trade Agreement (NAFTA) have been able to achieve similar objectives among smaller groups of nations.

One long-term effect of such efforts is for countries and regions to specialize in products for which they have a *comparative advantage*, that is, those that their particular climate, soil, or labor supply allows them to produce more efficiently than other regions. Those countries can then exchange commodities with each other, and citizens in both countries end up with a higher and more varied standard of living. For example, since the implementation of NAFTA began in 1994, the United States and Canada have sold increasing quantities of feed grains to Mexico, allowing Mexico to increase its livestock production and the quantity of meat in the diets of its citizens. Likewise, Mexico has been able to supply more fresh fruits and vegetables to U.S. and Canadian markets. These are examples of a much larger set of changes known as *globalization*.

Along with the lowering of trade barriers, the WTO is working to reduce subsidies and other favorable treatments to farmers by national governments that would encourage them to produce more of a certain product than they would produce based solely on competitive market prices. This is to prevent policies in some countries from driving down international commodity prices for producers in other countries. Losing price supports or input subsidies will cause short-term financial losses for some farmers, but it will increase the efficiency of world agriculture in the long run.

Opportunity or Threat?

Some producers and commodity groups recognize globalization as an opportunity to expand the markets for their products. Others see the trends as a threat, especially if they are unable to produce as efficiently as farmers in other countries and no longer enjoy the protection of trade barriers. They may need to develop a strategic plan that involves reducing production costs, looking for new enterprises, or finding alternative markets in which they can better compete.

Besides changing the flow of international trade, globalization can affect consumer tastes and preferences. Improved communication and transportation can introduce consumers to products and types of food they were not familiar with previously. A decade ago, bananas and other tropical fruits were not common in eastern European countries. Likewise, consumers in the United States were not familiar with kiwifruit or some types of imported cheeses.

Globalization also means that farmers and other producers around the world will increasingly compete for the same raw materials. Petroleum and other forms of energy are becoming increasingly scarce and expensive. Higher transportation costs will alter trade patterns. Agricultural labor will move across borders to fill the demand for workers, regardless of immigration laws. Investment capital will flow to where the highest returns are available. All of these changes will force successful farmers and ranchers to continually assess their external environments and internal resources to meet their long-term goals.

SUMMARY

Farmers and ranchers in the twenty-first century are making most of the same basic decisions that they made in the past century. The difference is they are making them faster and with more accurate information. Farm businesses will continue to become larger, and their operators will have to acquire specialized skills in managing personnel, interpreting data, competing for resources with nonfarm businesses, and customizing products to meet the demands of new markets. Changes in world trade policies and globalization of agriculture will have both positive and negative effects to which farmers must respond. All this must be done while balancing the need to earn a profit in the short run with the need to preserve agricultural resources and environmental quality into the future. While some farm managers will look at these trends as threats to the way they have traditionally operated their businesses, others will see them as new opportunities to gain a competitive advantage and to prosper.

QUESTIONS FOR REVIEW AND FURTHER THOUGHT

1. What forces have caused farms and ranches to become larger? Which of them are likely to continue? How can smaller businesses compete successfully?

2. How will quick access to more information help farm managers in the twenty-first century make better decisions?

3. List two examples of specialty agricultural markets and the changes a conventional producer might have to make to fill them.

4. What agricultural products from other countries do you consume? Do any of these compete with products produced by farmers in your country?

5. List other new challenges not discussed in this chapter that you think farm and ranch managers may have to face in the future.

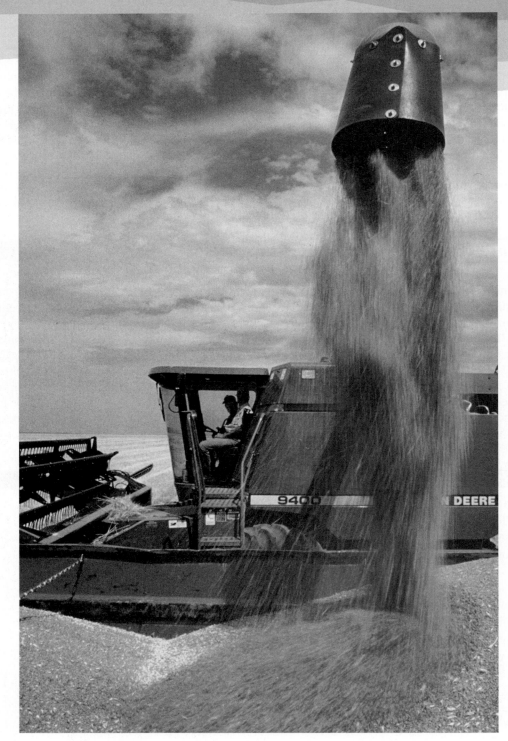

Photo by, Scott Bauer, USDA-ARS

MANAGEMENT AND DECISION MAKING

CHAPTER OUTLINE

Functions of Management
Strategic Farm Management
Decision Making
Characteristics of Decisions
The Decision-Making Environment
    in Agriculture
Summary
Questions for Review and Further Thought

CHAPTER OBJECTIVES

1. Understand the functions of management

2. Present the steps in developing a strategic management plan for a farm or ranch

3. Identify some common goals of farm and ranch managers and show how they affect decision making

4. Explain the steps in the decision-making process

5. Describe some unique characteristics of the decision-making environment for agriculture

Successful managers cannot simply memorize answers to problems, nor can they do exactly as their parents did. Some managers make decisions by habit. What worked in the past year will also work this year, and maybe again next year. But good managers learn to continually rethink their decisions as economic, technological, and environmental conditions change. Farmers and ranchers are continually bombarded by new information about prices, weather, technology, public regulations, and consumers' tastes. This information affects the organization of their businesses; what commodities to produce; how to produce them; what inputs to use; how much

19

of each input to use; how to finance their businesses; and how, where, and when to market their products. New information is vital for making new decisions and will often cause old management strategies to be reconsidered.

Important changes can occur in climate, weather, government programs and policies, imports and exports, international events, and many other factors that affect the supply and demand situation for agricultural commodities. Long-term trends must be recognized and taken into account.

Technology is also a constant source of change. Examples include the development of new seed varieties; new methods for weed and insect control; new animal health products and feed ingredients; and new designs, controls, and monitors for machinery. Other changes occur in income tax rules, environmental regulations, and farm commodity programs. These factors are all sources of new information that the manager must take into account when formulating strategies and making decisions.

Some managers achieve better results than others, even when faced with the same economic conditions, climate, and technology choices. Table 2-1 contains some evidence of this difference in results from a group of farms in a farm business association. Farms in the top one-third of the group had an average return to management and net farm income ratio many times higher than those of farms in the lowest one-third. However, the high-profit farms had slightly less land and only slightly more labor than the low-profit farms. Therefore, the wide range in net farm income and return on assets cannot totally be explained by the different quantities of resources available. The explanation must lie in the management ability of the farm operators.

FUNCTIONS OF MANAGEMENT

Farm and ranch managers perform many functions. Much of their time is spent doing routine jobs and chores. However, the functions that distinguish a manager from a mere worker are those that involve a considerable amount of thought and judgment. They can be summarized under the general categories of *planning*, *implementation*, *control*, and *adjustment*.

Planning

The most fundamental and important of the functions is planning. It means choosing a course of action, policy, or procedure. Not much will happen without a plan. To formulate a plan, managers must first establish *goals*, or be sure they clearly understand the business owner's goals. Second, they must identify the quantity and quality of *resources* available to meet the goals. In agriculture, such resources include land, water, machinery, livestock, buildings, and labor. Third, the resources must be allocated among several competing uses. The manager must identify all possible *alternatives*, analyze them, and select those that will come closest to meeting the goals of the business. All these steps require the manager to make careful long- and short-run decisions.

Implementation

Once a plan is developed, it must be implemented. This includes acquiring the resources and materials necessary to put the plan into effect, plus overseeing the entire process. Coordinating, staffing, purchasing, and supervising are steps that fit under the implementation function.

TABLE 2-1	Comparison of Mid-Size Grain Farms in Kentucky	
Item	Highest third (average)	Lowest third (average)
Gross farm returns	$1,272,835	$1,055,700
Return to management	$ 342,517	$ 833
Net income as % of gross	34.8%	10.6%
Crop acres farmed	1,408	1,461
Months of labor utilized	43.4	43.0

Source: Kentucky Farm Business Management Program, Annual Summary Data 2012, University of Kentucky.

Control

The control function includes monitoring results, recording information, and comparing results to a standard. It ensures that the plan is being followed and producing the desired results, or provides an early warning so adjustments can be made if it is not. Outcomes and other related data become a source of new information to use for improving future plans.

Adjustment

If the information gathered during the control process shows that outcomes are not meeting the manager's objectives, adjustments need to be made. This may involve fine-tuning the technology being used, or it may require changing enterprises. In some cases, more detailed production and cost data will have to be collected to identify specific problems.

Figure 2-1 illustrates the flow of action from planning through implementation and control to adjustment. It also shows that information obtained from the control function can be used for revising future plans. This circular process of constant improvement and refinement of decisions can continue through many cycles. But first, some basic decisions must be made about exactly why the business even exists and where it is headed.

STRATEGIC FARM MANAGEMENT

Management of a farm or ranch can be divided into two broad categories: strategic and tactical. *Strategic management* consists of charting the overall long-term course of the business. *Tactical management* consists of taking short-run actions that keep the business moving along the chosen course until the destination is reached.

Always doing things right in farming is not enough to ensure success. Farmers and ranchers must also do the right things. Strategic management seeks to discover what the right things are for a particular business at a particular time. Simply doing what the previous generation did will not keep the farm competitive in the long run.

Strategic management is an ongoing process. However, this process can be broken down into a series of logical steps:

1. Define the mission of the business.
2. Formulate the goals of the business.
3. Assess the resources of the business (internal scanning).
4. Survey the business environment (external scanning).
5. Identify and select strategies that will reach the goals.
6. Implement and refine the selected strategies.

Defining the Mission of the Business

A *mission statement* is a short description of why a business exists. For some farms and ranches, the mission statement includes strictly business considerations. For a family-owned and -operated business, the mission of the farm may be only one component of the overall family mission, which may reflect social, religious, and cultural values as well as economic considerations. Mission statements should emphasize the special talents and concerns of each farm business and its managers.

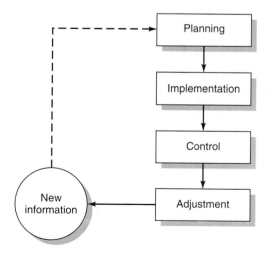

Figure 2-1 Management flow chart based on four functions of management.

Box 2-1   A Mission Statement

George and Connie Altman have been milking cows and growing crops since their early twenties. At age 35, they decided to assess where their farming operation was and where they wanted it to go. They chose the following mission statement for their business: "Our mission is to produce safe and nutritious milk at a reasonable cost, to maintain and enhance the quality of the natural resources under our control, and to contribute toward making our community a satisfying place to live."

Formulating the Goals of the Business

Goals provide a reference point for making decisions and measuring progress. For a family-owned and -operated farm, the goals of the business may be a subset of the overall family goals. For larger farms where managers are hired, the owners may define the goals while the manager strives to achieve them.

Not all farm managers will have the same goals, even when their resources are similar. This is because people have different *values*. Values influence the goals people set and the priorities they put on them. Table 2-2 lists some typical values held by farmers and ranchers. How strongly they feel about each of them will affect their business and family goals. When more than one person is involved in setting goals, it is important to recognize differences in values and to be willing to compromise, if necessary, to arrive at a mutually acceptable set of goals.

When goals are being established, keep in mind the following important points:

1. Goals should be written. This allows everyone involved to see and agree on them and provides a record for review at later dates.
2. Goals should be specific. "To own 240 acres of class I farmland in Washington County" is a more useful goal than "to own land." It helps the manager determine whether a goal has been reached and provides a sense of accomplishment

and an opportunity to think about defining new goals.
3. Goals should be measurable. The goal of owning 240 acres is measurable, and each year the manager can gauge progress toward the goal.
4. Goals should have a timetable. "To own 240 acres within five years" is more useful than a goal with an open-ended or vague completion date. The deadline helps keep the manager focused on achieving the goal.

TABLE 2-2   Common Values Among Farmers and Ranchers

Do you agree or disagree?

1. A farm or ranch is a good place to raise a family.
2. A farm or ranch should be run as a business.
3. It is acceptable for farmers to borrow money.
4. A farmer should have at least two weeks of vacation each year.
5. It is better to be self-employed than to work for someone else.
6. It is acceptable for a farmer to also work off the farm.
7. It is more enjoyable to work alone than with other people.
8. Farmers should strive to conserve soil and keep water and air resources clean.
9. A family farm should be passed on to the next generation.
10. All family members should be involved in the operation.

A farm operated by a family unit often has more than one set of goals because of the close and direct involvement of family members with the farm business. There can be personal goals as well as business goals, and each individual within the family may have different goals. In these situations, it is important to use a family conference or discussion to agree on at least the business goals. Without an agreement, everyone may go in different directions, and none of the business goals will be reached. This also applies to farms and ranches with multiple partners or shareholders.

Individuals and the businesses they manage differ, so many potential goals exist. Surveys of farm operators have identified the following common farm and ranch goals:

- Survive; stay in business; do not go broke; avoid foreclosure.
- Maximize profits; get the best return on investment.
- Maintain or increase standard of living; attain a desirable family income.
- Own land; accumulate assets.
- Reduce debt; become free of debt.
- Avoid years of low profit; maintain a stable income.
- Pass the entire farm on to the next generation.
- Increase leisure and free time.
- Increase farm size; expand; add acres.
- Maintain or improve the quality of soil, water, and air resources.
- Own and manage my own business.

These goals are stated in a general manner and would need to be made more specific before they would be useful to an individual business. Rarely does a single goal exist; farm operators usually have multiple goals. When this occurs, the manager must decide which goals are most important. Some combinations of goals may be impossible to achieve simultaneously, which makes the ranking process even more important. Another job for the manager is to balance the tradeoffs among conflicting goals.

Prioritizing Goals

Any of the goals listed may rank first for a certain individual, depending on the time and circumstances. Goals can and do change with changes in age, financial condition, family status, and experience. Also, long-run goals may differ from short-run goals. Profit maximization is often assumed to be the major goal of all business. However, farm operators often rank survival or staying in business above profit maximization. Achieving a profit plays a direct, or at least an indirect, role in meeting many other possible goals, including business survival.

Profit is needed to pay family living expenses and taxes, increase owner equity, decrease debt, and expand production. However, several possible goals on the list imply minimization or avoidance of risk, which may conflict with profit maximization. The most profitable long-run production plans and strategies are often among the most risky as well. Highly variable profits from year to year may greatly reduce the chances for survival and conflict with the desire for a stable income. For these and other reasons, profit maximization is not always the most important goal for all farm operators. Profit may be maximized subject to achieving minimum acceptable levels of other goals, such as security, leisure, and environmental stewardship. Nevertheless, profit maximization has the advantage of being easily measured, quantified, and compared across different businesses.

Assessing the Resources of the Business

Farms and ranches vary widely in the quantity and quality of physical, human, and financial resources available to them. An honest and thorough assessment of these resources will help the manager choose realistic strategies for achieving the goals of the business. This process is often called *internal scanning*.

Physical Resources   The land base is probably the most critical physical resource. Productivity, topography, drainage, and fertility are just a few of the qualities that determine the potential of land for agricultural use. The number of acres available and their location are also important. In many states, detailed databases exist that describe the important characteristics of a particular tract of land.

Other physical resources that should be evaluated include breeding livestock, buildings and fences, machinery and equipment, irrigation installations, and established perennial crops such as orchards, vineyards, and pasture.

Human Resources The skills of the operator(s) and other employees often determine the success or failure of certain enterprises. Some workers are talented with machinery, while others do better with livestock. Still others excel at marketing or accounting. Equally important is the degree to which each person in the operation likes or dislikes doing certain jobs. It is a good idea to conduct a thorough audit of personal skills and preferences before identifying competitive strategies for a farm business.

Financial Resources   Even when the physical and human resources are present to carry out certain enterprises, capital may be a limiting factor. Financial resources can be evaluated by completing a set of financial statements and by exploring the possibility of obtaining additional capital from lenders or outside investors. These tools and strategies will be discussed in detail in later chapters.

An honest and thorough appraisal of the farm's physical, human, and financial strengths and weaknesses will steer it toward realistic strategies. Particular attention should be given to identifying resources that will give the farm or ranch a competitive advantage over other firms. If certain key resources are found to be in short supply, strategies to fill these gaps need to be formulated.

Surveying the Business Environment

Critically analyzing the business environment in which a ranch or farm operates is called *external scanning*. Although the major types of livestock and crops grown in various parts of the world do not change rapidly, many of their characteristics

have changed. Changing consumer tastes and expanded international markets have led some customers to pay premiums for lean meat or high-protein grains, for example.

Other trends also affect the availability of new resources and the choices of technology. Changes in government regulations may create new constraints, or even remove some. The prudent manager must be aware of all these changes in the external environment and react to them early. If new production practices that lower costs per unit are adopted by most producers, then the operation that does not change will soon be at a competitive disadvantage.

Prices of some key inputs such as fuel and fertilizer may rise faster than others. This can affect crop and livestock production practices used, the choice of products, and the marketing channels used.

Some trends may represent threats to the farm or ranch, which could decrease profits if no corrective action is taken. For example, decreasing consumption of a crop such as tobacco may require alternative crops to be considered. Other trends, such as a desire for low-fat diets, may present opportunities for a farm that can help it reach its goals faster.

Whether a trend represents an opportunity or a threat will sometimes depend on the particular nature and location of the farm. Lowering of international trade barriers may expose farmers to foreign competition that they have been protected from in the past. By the same token, freer trade may open up new markets for products for which producers have a comparative advantage.

Identifying and Selecting Strategies

Everyone connected with the farm should brainstorm about possible plans for the future. By matching up the most promising opportunities with the strong points of the particular farm or ranch, an overall business strategy with a high chance of success can be formulated. Changes may have to be made, but they will be part of a

---

Box 2-2   **Internal and External Scanning**

June and Carl Washington have raised corn and hay on their rolling 460-acre farm for nearly 18 years. They have also run 50 stock cows on their rough pastureland, and farrowed and sold feeder pigs from 35 sows each year. Through hard work and careful budgeting, they have managed to pay down their mortgage and send their children off to school.

It is getting harder to sell the feeder pigs through local sale barns. They would like to sell pigs directly to one of the finishing operations in the area, but all of them want a larger volume of pigs, delivered at regular intervals. Without their children around to help, June and Carl don't think they can handle increased hog chores. Besides, they would have to buy extra corn.

Their county beef producers association is negotiating a contract to sell high-quality feeder calves to an out-of-state feedlot, by pooling calves from all their members. Carl has always enjoyed working with cattle. After comparing a series of whole farm budgets developed with the help of their farm business association consultant, the Washingtons decide to liquidate their swine operation and purchase 30 first-calf heifers. They also plan to gradually renovate their pastures and subdivide them and increase their hay acres. By selecting their best heifer calves for replacements, they hope to build their herd up to 100 females in five years. They will keep supervised performance analysis (SPA) records to measure the production and financial success of their venture.

---

deliberate, integrated plan, not just haphazard reactions.

Four general business strategies were identified in Chapter 1: a low-volume, high-value producer; a high-volume, low-margin producer; a special service provider; and a part-time operator. Some businesses can expand their options by forming *strategic alliances* with other farms or ranches that have complementary skills, such as a feeder pig producer and a custom hog finisher. Alliances can also be formed with processors and wholesalers.

Some businesses have more possible strategies for reaching their goals than others. In the arid regions of the western United States, for example, the land resource is such that the only alternative may be to use it as pasture for livestock production. But even in this situation, the manager must still decide whether to use the pasture for cow/calf production, for grazing stocker steers during the summer, or for sheep and goat production. Other regions have land suitable for both crop and livestock production, so more alternatives exist. As the number of

alternatives for the farm's limited resources increases, so does the complexity of the manager's decisions.

Implementing and Refining the Selected Strategies

Even the best strategy does not happen by itself. The manager must formulate action steps, place them in a timetable, and execute them promptly. In some cases, a formal *business plan* will be developed and presented to potential lenders or partners. Some common elements found in farm and ranch business plans are outlined in Box 2-3. Concrete, short-term objectives need to be set so that progress toward long-term goals can be measured. The manager then needs to decide what information will be needed to evaluate the success or failure of the strategy and how to collect and analyze the data.

Above all, strategic management should not be a one-time, limited process. It is an ongoing activity in which the manager is constantly alert

---

Box 2-3   Creating a Business Plan

Once you have carried out the strategic planning process, you may want to organize your conclusions into a business plan. A well-written business plan can be useful for justifying a loan application to obtain the capital you need to follow your plan, for convincing possible partners and land owners that your farm or ranch has a viable chance for success in the long run, and for guiding your own decisions in the future.

A farm or ranch business plan can include the following elements:

- **Executive summary:** a brief overview of the current situation and your aspirations for the future
- **Mission statement:** why the business will exist
- **Description of the business:** location, major enterprises, history, legal structure
- **Products and services:** what the business will produce and sell

- **Resources available:** owned and rented land, machinery line, useful buildings, breeding livestock
- **Potential markets:** where products can be sold, customer base for services or specialty products
- **Personnel:** who will be involved in the operation, experience, special skills and training; availability
- **Financial statements:** net worth and net income statements, cash flow projections, sources of capital
- **Risk management strategies:** how risk will be limited to a level consistent with financial resources
- **Letters of reference:** from people familiar with the farming operation and the managers

Online resources for developing a farm or ranch business plan are available from Purdue University and the University of Minnesota.

---

for new threats or opportunities, ready to take advantage of new resources, and willing to adapt the farm's strategies to changes in the values and goals of the individuals involved.

Tactical Management

Once an overall strategy for the farm or ranch has been developed, the manager must make *tactical decisions* about how to implement it. Tactical decisions include when and where to market crops, what rations to feed livestock, when to trade machinery, and whom to hire for the milking parlor. They may be as minute as which field to till on a given day or which telephone service to buy.

Many different tactics are available to carry out the same business strategy. Later we will examine some budgeting tools useful for making tactical decisions.

Decision Making

Without decisions, nothing will happen. Even allowing things to drift along as they are implies a decision, perhaps not a good decision, but a decision nevertheless.

The decision-making process can be broken down into several logical and orderly steps:

1. Identify and define the problem or opportunity.
2. Identify alternative solutions.
3. Collect data and information.
4. Analyze the alternatives and make a decision.
5. Implement the decision.
6. Monitor and evaluate the results.
7. Accept responsibility.

Following these steps will not make every decision perfect. It will, however, help a manager act in a logical and organized manner when confronted with choices.

Identifying and Defining the Problem or Opportunity

Many problems confront a farm or ranch manager. Most are tactical decisions such as choosing what seed to use, selecting a livestock ration, deciding how to market production, and deciding how to obtain access to land.

Problems may be identified by comparing results from the business to the levels that could be attained or that similar farms are achieving. For example, a farm may have a cotton yield 100 pounds per acre lower than the average for other farms in the same county on the same soil type. This difference between what is (the farm yield) and what should be (the county average yield or better) identifies a condition that needs attention. What appears to be a problem is often a symptom of a deeper problem, however. The low cotton yields could be caused by low fertility or inadequate pest control. These, in turn, could be caused by even more fundamental production problems.

A manager must constantly be on the alert to identify problems as quickly as possible. Most problems will not go away by themselves. Once a problem area is identified, it should be defined as specifically as possible. Good problem definition will minimize the time required to complete the remainder of the decision-making steps.

Identifying Alternative Solutions

Step two is to begin listing potential solutions to the problem. Some may be obvious once the problem is defined, while some may require time and research. Still others may become apparent during the process of collecting data and information. This is the time to brainstorm and list any idea that comes to mind. Custom, tradition, or habit should not restrict the number or types of alternatives considered. The most feasible ones can be sorted out later.

Collecting Data and Information

The next step is to gather data and information about the alternatives. Data may be obtained from many sources, including university extension services, bulletins and pamphlets from agricultural experiment stations, electronic data services, farm input dealers, salespersons of agricultural inputs, radio and television, computer networks, farm magazines and newsletters, and neighbors. Perhaps the most useful source of data and information is an accurate and complete set of past records for the manager's own farm or ranch. New technology for collecting and analyzing data has made it much easier to have current and complete information available. Whatever the source, the accuracy, usefulness, and cost of the information obtained should be carefully considered.

Decision making typically requires information about future events, because plans for producing crops and livestock must be made long before the final products are ready to market. The decision maker may have to formulate some estimates or expectations about future prices and yields. Past observations provide a starting point, but will often need to be adjusted for current and projected conditions. Later we will study *risk management* techniques that farm and ranch managers use to soften the effects of forecasts of future conditions that turn out to be wrong.

Gathering data and facts and transforming them into useful information can be a never-ending task. A manager may never be satisfied with the accuracy and reliability of the data and the resulting information. However, this step must end at some point. Gathering data has a cost in terms of time and money. Too much time spent gathering and analyzing data may result in a higher cost than can be justified by the extra benefit received. Good judgment and practical experience may have to substitute for information that is unavailable or available only at a cost greater than the additional return from its use.

Analyzing the Alternatives and Making a Decision

Each alternative should be analyzed in a logical and organized manner. The principles and procedures discussed in Part Three provide the basis for sound analytical methods.

Choosing the best solution to a problem is not always easy, nor is the best solution always obvious. Sometimes the best solution is to change nothing, or to go back, redefine the problem, and go through the decision-making steps again. These are legitimate actions, but they should not be used to avoid making a decision when a promising alternative is available.

After carefully analyzing each alternative, the one that will best meet the established goals is normally selected. Some managers create a list of desired outcomes and assign a score to each alternative strategy based on how well it meets each goal. The scores for all the alternatives can then be summed and used to rank them.

Sometimes none of the alternatives appears to be definitely better than any other. If profit maximization is the primary goal, the alternative resulting in the largest profit or increase in profit would be chosen. However, the selection is often complicated by uncertainty about the future, particularly future prices. If several alternatives have nearly the same profit potential, the manager must then assess the probability that each will achieve the expected outcome and the potential problems that could arise if it doesn't.

Making decisions is never easy, but it is what people must do when they become managers. Most decisions will be made with less than the desired level of information. An alternative must be selected from a set of possible actions, all of which have some disadvantages and carry some risk. Just because a decision is difficult is no reason to postpone making it, though. Many opportunities have been lost by delay and hesitation.

Box 2-4   The Decision-Making Process: An Example

S tep 1. Identify the problem
Soil erosion rates on the more sloping parts of the farm are above acceptable rates.

Step 2. Identify alternatives
Some of the neighbors use terraces or strip cropping on similar slopes. Many farmers are experimenting with reduced tillage or no-till practices.

Step 3. Collect information
Study research results from similar soils, comparing reduced tillage, terraces, and strip cropping. Obtain prices for different equipment and for building terraces. Visit with neighbors about their results. Consult extension experts about what changes in crop production practices and fertility would be needed.

Step 4. Analyze the alternatives and select one
Taking into account long-term costs and effects on yield, reduced tillage seems to be the most profitable way to bring soil erosion down to an acceptable level.

Step 5. Implement the decision
Purchase a new planter and modify the tillage implements.

Step 6. Monitor the results
Compare yields, calculate machinery and chemical costs, and measure erosion rates for several years.

Step 7. Accept responsibility for results
Yields and erosion are acceptable, but costs have increased. Fine-tune fertilizer and pesticide applications.

Implementing the Decision

Nothing will happen and no goals will be met by simply making a decision. That decision must be correctly and promptly implemented, which means taking some action. Resources need to be acquired, financing arranged, a timetable constructed, and expectations communicated to partners and employees. This takes organizational skills. Remember that *not* implementing a decision has the same result as not making any decision at all.

Monitoring and Evaluating the Results

Managers must know the results of their decisions. The longer it takes for the results of a decision to become known, the more likely it is that the results will be different than expected. Sometimes even a good decision will have bad results. Good managers will monitor the results of a decision with an eye toward modifying or changing it.

The more frequently a decision is repeated, the more useful it is to evaluate it. Deciding where to market crops or what genetic lines to choose for livestock is done over and over. Monitoring prices received or performance traits achieved allows better alternatives to be identified over time. Decisions that can be easily reversed also deserve to be evaluated more closely than those that cannot be changed.

Managers must set up a system to assess the results of any decision, so that any deviation from the expected outcome can be quickly identified. This is part of the control function of management. Profit and loss statements summarize the economic impact of a decision, yield records measure the impact on crop production, and daily milk or feed efficiency logs monitor livestock performance. Careful observation and good records will provide new data to be analyzed. The results of this analysis provide new information to use in modifying or correcting the original decision and making future decisions. Evaluating decisions is a way to learn from your past mistakes.

Accepting Responsibility

Responsibility for the outcome of a decision rests with the decision maker. A reluctance to bear responsibility may explain why some individuals find it so difficult to make decisions. Sometimes even good decisions bring bad results, due to uncertainties of markets and production. Blaming the government, the weather, or suppliers and processors when a decision turns out bad will not improve the results of the next decision. The manager must try to control the damage and then turn attention to the future.

CHARACTERISTICS OF DECISIONS

The amount of time and effort a manager devotes to making a decision will not be the same in every case. Some decisions can be made almost instantly, while others may take months or years of investigation and thought. Some of the characteristics that affect how the steps in the decision-making process are applied to a specific problem include the following:

1. Importance
2. Frequency
3. Imminence
4. Revocability
5. Number of alternatives

Importance

Some farm and ranch decisions are more important than others. Importance can be measured in several ways, such as the number of dollars involved in the decision or the size of the potential gain or loss. Decisions risking only a few dollars, such as sorting hogs into pens or buying small tools, may be made rather routinely. For these, little time needs to be spent gathering data and proceeding through the steps in the decision-making process.

On the other hand, decisions involving a large amount of capital or potential profits and losses need to be analyzed more carefully.

Decisions about purchasing land, installing an irrigation system, and constructing a new total-confinement hog building easily justify more time spent on gathering data and analyzing the alternatives.

Frequency

Some decisions may be made only once in a lifetime, such as choosing farming or ranching as a vocation or buying a farm. Other decisions are made almost daily, such as scheduling field work activities, balancing livestock rations, and setting breeding schedules. Frequent decisions are often made based on some rule of thumb or the operator's intuitive judgment. Nevertheless, small errors in frequent decisions can accumulate into a substantial problem over time.

Imminence

A manager is often faced with making a decision quickly or before a deadline to avoid a potential loss. Grain prices moving up or down rapidly call for quick action. Other decisions have no deadline, and there is little or no penalty for delaying the decision until more information is available and more time can be spent analyzing the alternatives. The thoroughness with which any decision is made will depend on the time available to make it.

Revocability

Some decisions can be easily reversed or changed if later observations indicate that the first decision was not the best. Examples are calibrating a seeder or adjusting a feeder, which can be changed rather quickly and easily. Managers may spend less time making the initial decision in these situations, because corrections can be made quickly and at little cost.

Other decisions may not be reversible or can be changed only at a high cost. Examples would be the decision to drill a new irrigation well or to construct a new livestock building.

Once the decision is made to go ahead with these projects and they are completed, the choice is to either use them or abandon them. It may be difficult or impossible to recover the money invested. These nonreversible decisions justify much more of the manager's time.

Number of Alternatives

Some decisions have only two possible alternatives. They are of the *yes or no* or *either, or* type. The manager may find these decisions to be easier and less time consuming than others that have many alternative solutions or courses of action. Where many alternatives exist, such as selecting seed varieties, the manager may be forced to spend considerable time identifying the alternatives and analyzing each one.

THE DECISION-MAKING ENVIRONMENT IN AGRICULTURE

Is managing a farm or ranch greatly different from managing other types of businesses? The basic functions, principles, and techniques of management are the same everywhere, but a typical farm or ranch business has some unique characteristics that affect the way decisions are made.

Biological Processes and Weather

One distinguishing characteristic of agriculture is the limitation placed on a manager's decisions by the biological and physical laws of nature. Managers soon find there are some things they cannot control. Nothing can be done to shorten the gestation period in livestock production, there is a limit on how much feed a pig can consume in a day, and crops require some minimum time to reach maturity. Even attempts to control the effects of climate with irrigation equipment and confinement buildings can be thwarted by sudden rainstorms and blizzards.

The unpredictability of the production process is unique to agriculture. Not even the best manager can forecast with certainty the effects of variations in rainfall, temperature, disease, or genetic combinations. This introduces an element of risk that most nonfarm businesses do not face.

Fixed Supply of Land

In most industries, a business can buy more raw materials or replicate production facilities when demand for their products increases. However, the supply of the most valuable resource in agricultural production, land, is essentially fixed. Farm managers can only try to increase the productivity of the existing land base or outbid other farms for the limited amount of land available for sale or rent. This makes decisions about buying, selling, and renting land particularly crucial. It also causes the sale and rental prices of farmland to be especially sensitive to changes in the prices of agricultural commodities and inputs.

Small Size

Fifty-six percent of the farms in the United States have only one operator (Table 2-3), and only 1.5 percent have more than three. No other sector of the economy is so dominated by small-scale operations. By contrast, in a large corporation, the stockholders own the business, while

TABLE 2-3	Number of Operators per Farm in the United States

Number of operators	Percentage
One	56.0
Two	37.2
Three	5.3
Four or more	1.5

Source: 2012 Census of Agriculture, USDA.

the board of directors sets goals, defines policies, and hires managers to follow them. It is generally easy to identify three distinct groups: owners, management, and labor. These distinct groups do not exist on the typical family farm or ranch, where one individual or a small group owns the business, provides the management, and contributes most or all of the labor. This makes it difficult to separate the management activity from labor, because the same individuals are involved. It also causes the constant need for labor *to get a job done*, placing management in a secondary role, with decisions constantly being delayed or ignored. Moreover, when the manager's residence and home life are located on the farm, personal and business goals and activities become more intertwined.

Perfect Competition

Production agriculture is often used as an example of a perfectly competitive industry. This means that each individual farm or ranch is only one of many and represents a small part of the total industry. Most agricultural products are homogeneous; that is, grain, fruits, and vegetables or livestock from one farm are virtually identical to those from another farm. Therefore, the individual manager usually cannot affect either the prices paid for resources or the prices received for products sold. Prices are determined by national and international supply and demand factors, over which an individual manager has little control except possibly through some type of collective action. Some managers attempt to overcome this by finding niche markets or localized markets where they are the only suppliers. Examples include growers of heirloom produce or nontraditional livestock.

All these factors taken together create a unique environment for making management decisions in agriculture. The remaining chapters will explain in more detail the principles and tools used by farm and ranch managers and the particular types of problems to which they apply them.

SUMMARY

Good management usually means the difference between earning a profit or suffering a loss in a commercial farm or ranch business. Managers must make plans for the farm business, implement the plans, monitor their success, and make adjustments where needed.

The overall direction in which the farm is headed is defined through the process of strategic planning. A strategic plan begins with a vision statement of why the business exists. Goals provide the direction and focus for the process and reflect the values of the managers. After assessing the internal resources and the external environment of the business, strategies can be identified and selected. Finally, the strategies must be implemented and the results monitored. The strategic plan is carried out by making a large number of short-run tactical decisions. Tactical decisions are made by defining the problem, gathering information about and analyzing alternative solutions, choosing and implementing an alternative, and evaluating the results.

Farm and ranch managers operate in a different environment than managers of other businesses. Agriculture relies heavily on biological processes; the supply of farmland is essentially fixed; many of the same people combine ownership, labor, and management; and the firms usually operate in a perfectly competitive economic environment.

QUESTIONS FOR REVIEW AND FURTHER THOUGHT

1. What is your definition of management? Of a farm manager?

2. Do farm and ranch managers need different skills than managers of other businesses? If so, which skills are different? Which are the same?

3. How do strategic management and tactical management differ?

4. Would you classify the following decisions as strategic or tactical?

    a. deciding whether a field is too wet to till

    b. deciding whether to specialize in beef or dairy production

    c. deciding whether to take a partner into the business

    d. deciding whether to sell barley today or wait until later

5. Why are goals important? List some examples of long-term goals for a farm or ranch business. Make them specific and measurable and include a time line.

6. What are some common goals of farm and ranch families that might be in conflict with each other?

7. What are your personal goals for the next week? For next year? For the next five years?

8. What internal characteristics of the farm should a manager consider when developing a strategic plan?

9. Identify several trends in technology or consumer tastes that a farm manager should consider when developing a strategic plan.

10. List the steps in the decision-making process. Which steps are part of the planning function of management? Implementation? Control? Evaluation?

11. What characteristics of a decision affect how much time and effort a manager devotes to making it?

12. What are some characteristics of agriculture that make managing a farm or ranch different from managing other businesses?

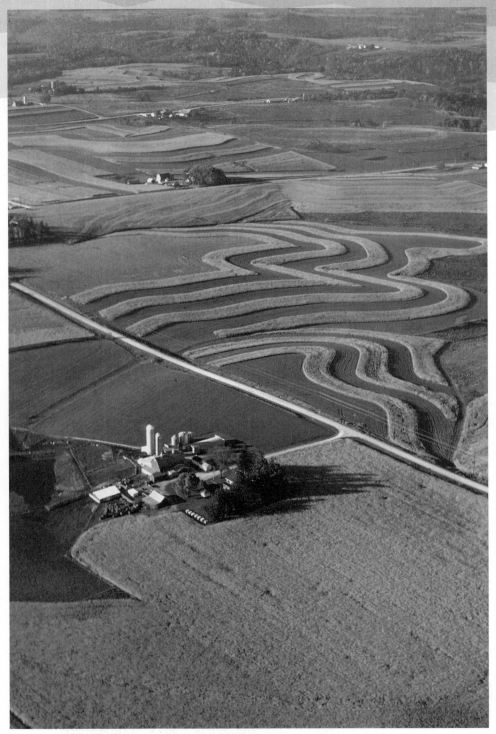

Photo by Lynn Betts, USDA Natural Resources Conservation Service

MEASURING MANAGEMENT PERFORMANCE

T wo points were made in Chapter 2: (1) setting goals is important and (2) control is one of the functions of management. In Part Two we will describe how to measure and analyze profit and other financial characteristics of a farm or ranch business. Results of the analyses will allow the manager to determine how well and to what degree the financial goals are being met.

This discussion is also related to the control function of management. Control is a monitoring system used to assess whether the business plan is being followed and how well the farm is meeting the goals of the plan. Many of the same records needed to measure profit and the financial status of the business are also needed to perform the control function of management. The records provide a method of measuring not only how well the business is doing but also how well the manager is doing.

This discussion introduces the need for an information management system for the farm or ranch business. Many choices and options are available in the design and implementation of a system, ranging from the simple to the complex. The best system for any business will depend on many factors, including the size of the business, the form of business organization, the amount of capital borrowed, lender requirements, and what specific financial reports are needed and in what detail.

Chapter 3 discusses the purposes and components of an information management system. The next two chapters cover the most common financial statements. Chapter 4 emphasizes that a balance sheet is intended to measure the financial condition of a business at a point in time. The statement of owner equity shows the sources of a farm's net worth.

Chapter 5 introduces the income statement, which provides an estimate of the value of products and services produced during an accounting period and the costs of the resources used to produce them. The accuracy of the reported profit depends on many factors, including the type of record system employed and the effort put forth to keep good records. The proper recording of cash transactions and how they are summarized in the statement of cash flows are discussed, and the difference between cash flows and income/expenses is explained. The emphasis will be on understanding what it takes to accurately measure profit or net farm income. Without an accurate measurement, the effects of past management decisions will be distorted and unreliable information will be used to make future decisions.

The tools developed in Chapters 4 and 5 then can be used in the control functions of management. These are consolidated in Chapter 6 and combined with some additional analytical measurements to perform a whole-farm business analysis. Measures of the financial strength and profitability of the business can be compared to goals and standards. Areas of strength and weakness can be identified, and specific problems addressed to improve the overall business performance and better meet the operator's goals.

© Comstock/PunchStock

ACQUIRING AND ORGANIZING MANAGEMENT INFORMATION

CHAPTER OUTLINE

Purpose and Use of Records

Farm Business Activities

Basic Accounting Terms

Options in Choosing an Accounting System

Chart of Accounts

Basics of Cash Accounting

Basics of Accrual Accounting

A Cash Versus Accrual Example

Farm Financial Standards Council
   Recommendations

Output from an Accounting System

Summary

Questions for Review and Further Thought

CHAPTER OBJECTIVES

1. Appreciate the importance and value of establishing a good farm or ranch accounting system

2. Discuss some choices that must be made when selecting an accounting system

3. Outline the basic concepts of the cash accounting method

4. Present the basic concepts of the accrual accounting method and compare them with cash accounting

5. Review some recommendations of the Farm Financial Standards Council related to choice of accounting method

6. Introduce some financial records that can be obtained from a good accounting system

A business with inadequate records can be likened to a ship in the middle of the ocean that has lost the use of its rudder and navigational aids. It does not know where it has been, where it is going, or how long it will take to get there. Records tell the manager where the business has been and whether it is now on the path to making profits and creating financial stability. Records are, in one respect, the manager's *report card*, because they show the results of management decisions over past periods. Records may not directly show where a business is going, but they can provide considerable information that can be used to correct or amend past decisions and to improve future decision making. In that way, they at least influence the future direction of the business.

For a number of reasons, including the small scale and diverse production systems of most farm businesses, farm and ranch records have not always been totally consistent with the standards that the accounting profession follows for other types of businesses. Financial problems on farms and ranches during the 1980s focused attention on the many different styles and formats of financial reports being used, differences in terminology, and inconsistent treatment of some accounting transactions unique to agriculture.

PURPOSE AND USE OF RECORDS

Several uses for records have already been mentioned. Following is a more detailed and expanded list of the purpose and use of farm records:

1. Measure profit and assess financial condition.
2. Provide data for business analysis.
3. Assist in preparing reports for partners, lenders, landlords, input providers, and government agencies.
4. Measure the profitability of individual enterprises.
5. Assist in the analysis of new investments.
6. Prepare income tax returns.

This is not a complete list of all possible reasons for keeping and using farm records. Other possible uses include demonstrating compliance with environmental regulations, establishing insurance needs, planning and valuing estates, monitoring inventories, dividing landlord/tenant expenses, reporting to partners and shareholders, and developing marketing plans. Good records are also essential for splitting income and expenses in multiple-owner businesses, such as those with absentee landowners, and to assist with profit-sharing distributions and share-lease arrangements. However, the six uses in the list are the more common and will be discussed in more detail.

Box 3-1   Farm Financial Standards Council

The *Farm Debt Crisis* from 1983 to 1987 provided evidence that the farm record-keeping methods and financial analyses of that time were often inadequate or underused. Following the debt crisis, farm financial education increased, leading to growth in the number of available books, farm record systems, and services, but the new methods were generally not standardized. In 1989, the Farm Financial Standards Task Force (FFSTF) was formed to address accounting and record-keeping problems on farms and ranches. Subsequently, they changed their name to the Farm Financial Standards Council (FFSC).

According to the FFSC Web site, the mission of the FFSC is "to create and promote uniformity and integrity in financial reporting and analysis for agricultural producers."

The first report of this group was issued in 1991, with the goal of making farm financial reporting more uniform and technically correct. Since then agricultural educators, accountants, and software vendors have made a significant effort to bring more consistency to formats, nomenclature, and definitions in farm accounting and analysis. In 2006, the FFSC developed a report concerning management accounting guidelines for agricultural producers. This chapter and the subsequent chapters of this text generally follow the financial accounting recommendations of the FFSC.

Measure Profit and Assess Financial Condition

These two reasons for keeping and using farm records are among the more important. Profit is estimated by developing an income statement, the topic for Chapter 5. The financial condition of the business as shown on a balance sheet will be covered in detail in Chapter 4.

Provide Data for Business Analysis

After the income statement and balance sheet are prepared, the next logical step is to use this information to do an in-depth business analysis. There is a difference between just making *a profit* and having *a profitable* business. Is the business profitable? How profitable? Just *how* sound is the financial condition of the business? The answers to these and related questions require more than just preparing an income statement and balance sheet. A financial analysis of the business can provide information on the results of past decisions, and this information can be useful when making current and future decisions.

Assist in Obtaining Credit

Lenders need and require financial information about the farm business to assist them in their lending decisions. After the financial difficulties of the 1980s, many agricultural lenders and bank examiners are requiring more and better farm records. Good records can greatly increase the odds of getting a loan approved and receiving the full amount requested.

Measure the Profitability of Individual Enterprises

A farm or ranch showing a profit may include several different enterprises. It is possible that one or two of the enterprises are producing all or most of the profit, and one or more of the other enterprises are losing money. A record system can be designed that will show revenue and expense not only for the entire business but also for each enterprise. With this information, the unprofitable or least profitable enterprises can be eliminated, and resources can be redirected for use in the more profitable ones.

Box 3-2   Biosecurity and Farm Records

Following a number of high-profile food contamination problems in recent years, interest in food safety has grown considerably. Of particular concern is product traceability, or the ability to determine where food items were grown, how they were grown, and how and where they were subsequently processed. To gain market strength, some producers, especially those growing produce, are voluntarily engaging in certification processes that require extensive record keeping. Legislation has been proposed that could make such record keeping mandatory for many types of operations. Under the proposed legislation, farmers would need to keep detailed records for a minimum of two years. Exempt from the requirements of this proposed legislation would be animal food products regulated under the Federal Meat Inspection Act, the Poultry Products Inspection Act, or the Egg Products Inspection Act.

Because of disease concerns, especially following the discovery of the *mad cow* disease in Washington state in 2003, there is also increased interest in tracking livestock and poultry. The National Animal Identification System (NAIS) was implemented by the USDA in 2004. The long-term goal is to allow identification within 48 hours of all livestock and premises that have had contact with a disease of concern. NAIS is currently a voluntary program, but it is possible that government could require mandatory tracking of animals at some point in the future.

Assist in the Analysis of New Investments

A decision to commit a large amount of capital to a new investment can be difficult and may require a large amount of information to do a proper analysis. The records from the past operation of the business can be an excellent source of information to assist in analyzing the potential investment. For example, records on the same or similar investments can provide data on expected profitability, expected life, and typical repairs over its life.

Prepare Income Tax Returns

Internal Revenue Service (IRS) regulations *require* keeping records that permit the proper reporting of taxable income and expenses. This type of record keeping can often be done with a minimal set of records not adequate for management purposes. A more detailed management accounting system can produce income tax benefits. It may identify additional deductions and exemptions, for example, and allow better management of taxable income from year to year, reducing income taxes paid over time. In case of an IRS audit, good records are invaluable for proving and documenting all income and expenses. Chapter 16 expands on managing information for income tax purposes, while the present chapter will focus on accounting practices for managerial purposes.

FARM BUSINESS ACTIVITIES

In designing a farm accounting system, it is useful to think of the three types of business activities that must be incorporated into the system. Figure 3-1 indicates that an accounting system must be able to handle transactions relating not only to the *production* activities of the business but also to the *investment* and *financing* activities.

Production Activities

Accounting transactions for production activities are those related to the production of crops and livestock. Revenue from their sale, and other farm revenue such as government program payments and custom work done for others, would

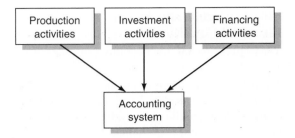

Figure 3-1 Farm business activities to be included in an accounting system.

be included here. Expenses incurred in producing that revenue—such as feed, fertilizer, chemicals, fuel, interest, and depreciation—are also part of the production activities that need to be recorded in the accounting system.

Investment Activities

Investment activities are related to the purchase, depreciation, and sale of long-lived assets. Examples would be land, buildings, machinery, orchards, vineyards, and breeding livestock. Records kept on each asset should include purchase date, purchase price, annual depreciation amount, book value, current market value, sale date, sale price, and gain or loss when sold.

Financing Activities

Financing activities are all transactions related to borrowing money and paying interest and principal on debt of all kinds. This includes money borrowed to finance new investments, operating money borrowed to finance production activities for the year, and accounts payable at farm supply stores.

Dividing the farm business activities into these three types illustrates the broad range of transactions that should be recorded in any accounting system. It also shows some interrelations among these activities. Interest expense comes from financing activities, but it is a production or operating expense. Depreciation results from investment in a depreciable asset, but it is also a production or operating expense. Therefore, a good accounting system must be able not only to record all the various types of transactions but also to assign them to the appropriate activity and enterprise of the operation.

BASIC ACCOUNTING TERMS

A person does not need an accounting degree to keep and analyze a set of farm or ranch records. However, some knowledge of basic accounting and accounting terminology is useful. One must fully understand and use any accounting system and accurately communicate accounting information to others. The following terms and definitions provide the foundation to understand the material in the remainder of this chapter and those that follow. Other terms will be defined as they are introduced.

Account payable—an expense that has been incurred but not yet paid. Typical accounts payable are for items charged at farm supply stores where the purchaser is given 30 to 90 days to pay the amount due.

Account receivable—revenue for a product that has been sold or a service provided but for which no payment has yet been received. Examples would be custom work done for a neighbor who has agreed to make payment by the end of next month or grain sold on a deferred payment contract.

Accrued expense—an expense that accrues or accumulates daily but has not yet been paid. An accrued expense has typically not been paid because the due date or payment date is in the future. Examples are interest on loans and property taxes.

Asset—any item of value, tangible or financial. On a farm or ranch, examples would be machinery, land, bank accounts, buildings, grain, and livestock.

Credit—in accounting, an entry on the right-hand side of a double-entry ledger. A credit entry is used to record a decrease in the value of an asset or an increase in a liability, an owner equity, or an income account.

Debit—an entry on the left-hand side of a double-entry ledger. A debit entry is used to record an increase in an asset or expense account and a decrease in a liability or an owner equity account.

Expense—a cost or expenditure incurred in the production of revenue.

Inventory—the physical quantity and financial value of products produced for sale that have not yet been sold. Farm or ranch examples would be grain in storage, or livestock ready for sale or that could be sold at the time the inventory is taken.

Liability—a debt or other financial obligation that must be paid in the future. Examples would include loans from a bank or other lending institutions, accounts payable, and accrued expenses.

Net farm income—revenue minus expenses. It is also the return to the owner's equity capital, unpaid labor, and management.

Owner equity—the difference between business assets and business liabilities. It represents the net value of the business to the owner(s) of the business.

Prepaid expense—a payment made for a product or service in an accounting period before the one in which it will be used to produce revenue.

Profit—revenue minus expenses minus opportunity costs. Also equal to net farm income minus opportunity costs.

Revenue—the value of products and services produced by a business during an accounting period. Revenue may be either cash or noncash.

OPTIONS IN CHOOSING AN ACCOUNTING SYSTEM

Before anyone can begin entering transactions into an accounting system, several decisions must be made about the type of system to be used. A number of options are available, which generally fall into the following areas:

1. What accounting period should be used?
2. Should it be a cash or an accrual system?
3. Should it be a single- or double-entry system?
4. Should it be a basic or complete accounting system?
5. Should analysis be done only for the whole farm, or for each enterprise?

It is often difficult to make certain types of changes in an accounting system once one is established and users are familiar with it. Therefore, considerable thought should be given and advice obtained when making the initial selection of an accounting system.

Accounting Period

An accounting period is a period of time, such as a quarter or a year, for which a financial statement is produced. Under a *calendar-year* accounting period, all transactions occurring between January 1 and December 31 of each year are organized and summarized into financial reports. By contrast, *fiscal-year* accounting uses a 12-month period that may begin on any date. Accounting can be done on a fiscal-year basis for management as well as income tax purposes. When a shorter period is chosen, such as quarterly reporting, everything should still be consolidated into annual reports.

It is generally recommended that a firm's accounting period follow the production cycle of the major enterprises and end at a time when business activities are slow. For most crop production activities and some livestock, a December 31 ending date fits this recommendation, so most farmers and ranchers use a calendar-year accounting period. However, winter wheat, citrus crops, and winter vegetables are examples of crops where intensive production or harvesting activities may be under way around December 31. These producers may want to

consider a fiscal-year accounting period that ends after harvesting is completed. Large dairies and commercial feedlots with continuous feeding activities would have a difficult time finding a month when business transactions are much slower than any other month. They could just as well use any convenient accounting period.

CHART OF ACCOUNTS

A chart of accounts lists and organizes all accounts used by the accounting system. It includes the broad categories of assets, liabilities, equity, revenue, and expenses, each with subaccounts and perhaps other subaccounts under those. For example, repairs would be a subaccount under expenses and could have its own subaccounts for building and machinery repairs. The number of accounts will vary from business to business depending on size, number of enterprises, needs of the manager, and many other factors. More accounts allow for a more detailed analysis but require more time and accounting knowledge.

Each account is typically assigned a number to assist in tracking accounts. These numbers are assigned from a range of numbers allotted to each broad category. For example, all assets may be assigned numbers between 100 and 199, liabilities 200 to 299, and so forth. This type of numbering system allows the person making entries to quickly find, organize, and track accounts. When choosing account names, it may be useful to look at IRS Schedule F, Form 1040, used to report farm income and expenses for income tax purposes. If the chart of accounts contains the same names as Schedule F, it is easy to transfer amounts from the accounting system to Schedule F.

Table 3-1 is an example of a basic chart of accounts for a farm business. Many more accounts would be necessary for a large business with multiple enterprises, because such a business would need a detailed financial management analysis.

TABLE 3-1	Example Chart of Accounts

Account	Description	Type
Asset-related accounts		
1010	Cash on hand	Cash
1020	Checking account	Cash
1100	Accounts receivable	Accounts receivable
1200	Inventory—crops	Inventory
1300	Inventory—market livestock	Inventory
1400	Prepaid expense	Other current assets
1470	Other current assets	Other current assets
1500	Equipment	Fixed assets
1510	Building	Fixed assets
1600	Inventory—breeding livestock	Fixed assets
1690	Land	Fixed assets
1710	Accumulated depreciation	Accumulated depreciation
Liability-related accounts		
2000	Accounts payable	Accounts payable
2420	Current portion long-term debt	Other current liabilities
2480	Other current liabilities	Other current liabilities
2600	Noncurrent liabilities—land	Noncurrent liabilities
2740	Other noncurrent liabilities	Noncurrent liabilities
Equity-related accounts		
3000	Retained earnings	Equity
3010	Contributed capital	Equity
3020	Valuation equity	Equity
Revenue-related accounts		
4020	Sales crops	Income
4080	Sales livestock	Income
4100	Government payments	Income
4200	Other income	Income
5000	Gain/loss on sale of assets	Income
Expense-related accounts		
6400	Crop expense	Expenses
6500	Livestock expense	Expenses
6600	Depreciation expense	Expenses
6750	Feed and grain expense	Expenses
6760	Repairs	Expenses
7010	Property taxes	Expenses
7050	Insurance expense	Expenses
7100	Interest expenses	Expenses
7900	Other expense	Expenses

Cash Versus Accrual Accounting

This topic will be covered again when discussing income taxes in Chapter 16. However, the discussion here will be restricted to accounting for management purposes and not for income taxes. While the concepts are the same in either case, the advantages and disadvantages of each accounting method may be different, depending on the use for management or income tax purposes. The basics of cash and accrual accounting will be discussed in later sections of this chapter.

Single Versus Double Entry

With a single-entry cash system, only one entry is made in the books to record a receipt or an expenditure. Sale of wheat would have the dollar amount recorded under the *Grain Sales* column in the ledger. A check written to pay for feed would have the amount entered under the *Feed Expense* column. The other side of the transaction is always assumed to be cash, which changes the balance in the checking account. In practice, the checkbook register might be thought of as the other entry, but one not included in the ledger.

A double-entry system records changes in the values of assets and liabilities as well as revenue and expenses. There must be equal and offsetting entries for each transaction. This system will result in more transactions being recorded during an accounting period, but it has two important advantages:

1. Improved accuracy, because the accounts can be kept in balance more easily
2. The ability to produce complete financial statements, including a balance sheet, at any time, directly from data already recorded in the system

The improved accuracy of double-entry accounting comes from the two offsetting entries, which means that *debits* must equal *credits* for each transaction recorded. It also means that the basic accounting equation of

$$\text{Assets} = \text{Liabilities} + \text{Owner Equity}$$

will be maintained. The double-entry system maintains the current values of assets and liabilities within the accounting system, allowing financial

---

Box 3-3   Debits and Credits

It is easy to fall into the trap of thinking all debits are *bad* or decreases and all credits are *good* or increases. Accounting procedures make no such distinction. A debit is an entry in the left column of a two-column ledger account and a credit is an entry in the right column. Whether a debit or credit is good or bad or an increase or a decrease depends on the type of account. For example, a checking account is an asset account where a debit is an increase and a credit a decrease in the account balance. The receipt of cash for a sale would be recorded in the debit column of the checking account and the same amount in the credit column of a sales account,

a revenue account. In this example, both the debit and credit are good, representing an increase in both sales and cash. When a check is written to pay a bill, the appropriate expense account has a debit entry and the checking account a credit entry. A debit entry to an expense account increases it and the credit entry to the checking account decreases it. Both debit and credit entries have bad results, an increase in expenses and a decrease in cash.

REMEMBER, debit means left column and credit means right column. Nothing more can be inferred until something is known about which accounts are affected by the transaction.

statements to be generated directly from the accounting system without any need for outside information.

Whole Farm Versus Enterprise Accounting

Most farm and ranch record systems summarize income, expenses, and profits for the entire farm. Managers can then evaluate the performance of the business compared to past years and other farms. However, if the farm produces multiple products, it is helpful to know which ones contribute most to the overall profitability. Dividing the business into individual enterprises or profit centers allows the manager to identify areas that are generating a satisfactory return and those that need to be improved or discontinued. Chapter 18 discusses enterprise analysis in more detail.

Basic Versus Complete System

The most basic and simple accounting system would be one that is manual and uses cash accounting only. A complete system would be computerized with capabilities for both cash accounting for tax purposes and accrual accounting for management purposes. It would also be able to track inventories, compute depreciation, track loans, perform enterprise analysis, and handle all employee payroll accounting.

Between these two extremes are many possibilities. For example, a simple basic system can be maintained on a computer with any one of a number of personal finance software programs available. The next step up would be any of several small business accounting programs that can be used for farm accounting by changing the account names. These programs can do either cash or basic accrual accounting. Most are inexpensive but relatively powerful programs that often do a good job of *disguising* the fact that the user is dealing with debits and credits. Very little accounting knowledge is needed to use these programs and to get useful and accurate output.

Most farmers and ranchers desiring a complete system use computer software accounting programs written specifically for agricultural use. These programs use agricultural terminology and are designed to handle many of the situations unique to agriculture such as accounting for raised breeding livestock, government farm program payments, operating loans, and quantities of product sold or in inventory. Many also allow the user to maintain both cash and accrual records for each year within the same program. This makes it easy to pay income taxes on a cash basis while still having complete accrual records for making management decisions. Some companies sell a basic program along with a number of optional add-on modules. This allows the user to become familiar with the program and its basic output without being overwhelmed by its complexity. Later, modules for inventory, depreciation, payroll, and production records can be added as needed or desired.

How complete the accounting system should be for any given farm business will depend primarily on the answers to three questions:

1. How much accounting knowledge does the user have?
2. How large and complex is the farm business and its financial activities?
3. How much and what type of information is needed or desired for management decision making?

The lack of accounting training should not deter farmers and ranchers from using one of the more complete software accounting programs. Many of them require only a limited knowledge of accounting. Additional accounting information is available in the form of self-help manuals, adult education courses, and community college courses. The price of the more complete and therefore more expensive farm accounting programs may include some training on using that specific software program as well as free technical assistance for a time.

The larger the farm business, the more the enterprises involved, the more the employees hired, the more the depreciable assets owned, and the more the money borrowed, the more the need for a complete accounting system. Accounting programs that compute depreciation, track inventories, generate budgets, reconcile checks, and complete employee payrolls as part of the accounting system become increasingly necessary and useful. Users who begin with a basic accounting system often find additional output useful and perhaps needed. This is one advantage of beginning with a complete system or one that can be easily upgraded.

BASICS OF CASH ACCOUNTING

The term *cash* in the name is perhaps the best description of this accounting method. With only a few exceptions, no transaction is recorded unless cash is spent or received.

Revenue

Revenue is recorded only when cash is received for the sale of products produced or services provided. The accounting period during which the products were produced or the services provided is not considered when recording revenue. Revenue is recorded in the accounting period when cash is received regardless of when the product was produced or the service provided.

Cash accounting can and often does result in revenue being recorded in an accounting period other than the one in which the product was produced. A common example is a crop produced in one year, placed in storage, and sold the following year. Any accounts receivable at the end of an accounting period also result in cash being received in an accounting period after the product was produced or the service provided.

Expenses

Cash accounting records expenses in the accounting period during which they are paid, that is, when the cash is expended. The accounting period in which the product or service was purchased is not considered. Expenses can therefore be recorded in an entirely different year or accounting period than the one in which the purchased product or service generated a product and related revenue. Items may be purchased and paid for late in one year but not used until the next, and items used in one year may not be paid for until the following year. The first case is an example of a prepaid expense, and the latter of an account payable. An accrued expense such as interest is another example. Here, an item (borrowed money) is being used in one year, but the cost of that item (interest) will not be paid until the next year when the annual payment is due.

There is one major exception to the rule that only cash expenses are recorded in a cash accounting system. Although depreciation is a noncash expense, it is generally considered an expense when using cash accounting.

Advantages and Disadvantages

Cash accounting is a relatively simple, easy-to-use system that requires very little knowledge of accounting. It also has some definite advantages for many farmers and ranchers when computing taxable income for income tax purposes.

However, these advantages are offset by one major disadvantage. As noted, it is common to have revenue and expenses recorded in a year other than the year the product was produced or the expense was used to produce a product. Therefore, neither the revenue nor the expenses may have any direct relation to the actual production activities for a given year. The result is an estimated profit that may not truly represent the profit from the year's production activities.

This inability of cash accounting to properly match revenue and expenses within the same year that the related production took place is a major disadvantage of this method. Compared to the true profit, the profit shown by cash accounting can be greatly distorted. It may be overestimated in some years and underestimated in others. If this estimate of profit is then used to make management decisions for the future, the result is often poor decisions.

BASICS OF ACCRUAL ACCOUNTING

Accrual accounting is the standard of the accounting profession. It requires more entries and accounting knowledge than cash accounting. However, it provides a much more accurate estimate of annual profit than does cash accounting.

Revenue

Accrual accounting records as revenue the value of all products produced and all services provided during a year. Whenever the products are produced and sold and the cash received all in the same year, this is no different than cash accounting. The difference occurs when the product is produced in one year and sold in the next, or when cash for a service is not received until a later year. Accrual accounting emphasizes that the value of a product or service should be counted as revenue in the year it was produced, no matter when the cash is received.

The handling of inventories is a major difference between cash and accrual accounting on farms and ranches. Accrual accounting records inventories as revenue. A simple example will show why this is important. Assume a farmer produces a crop but places it all in storage for sale the following year, when prices are expected to be higher. In the year of production, there would be no cash sales but presumably some cash expenses. Under cash accounting, there would be a negative profit (loss) for the year. This result is a poor indicator of the results of the production activities for the year and completely ignores the value of the crop in storage.

Accrual accounting includes an estimate of the value of the crop in storage as revenue in the year it was produced. This is done by adding an inventory *increase* from the beginning to the end of the year to other revenue. An inventory *decrease* is deducted from other revenue. The result is an estimated profit that more accurately describes the financial results of the production activities for the year. In the same manner and for the same reason, any uncollected amounts for services provided (accounts receivable) are also recorded as revenue.

Expenses

One principle of accounting is *matching*. This principle states that once revenue for a year is determined, all expenses incurred in the production of that revenue should be recorded in the same year. The results for items purchased and paid for in the same year are the same for cash and accrual accounting. Differences arise when items are purchased in the year before the one in which they produce revenue (prepaid expenses) or payment is not made until the year after the items are used (accounts payable and accrued expenses).

To match expenses with revenue in the proper year, an accounting entry must cause (1) prepaid expenses to show up as expenses in the year after the item or service was purchased and paid for; (2) accounts payable to be entered as expenses, although no cash has yet been expended to pay for the items; and (3) accrued expenses at year-end to be entered as expenses, although no cash has been expended. A typical example of the latter is interest accrued from the last interest payment to the end of the year. This accrued expense recognizes that the borrowed capital was used to produce revenue in one year, but the next cash interest payment may not be due for several months into the following year.

Advantages and Disadvantages

A major advantage of accrual accounting is that it produces a more accurate estimate of profit than can be obtained with cash accounting. Related to this is the accurate information it provides for financial analysis and management decision making. These advantages make accrual accounting the standard of the accounting profession and generally required for all corporations selling stock to the public. The latter requirement ensures that potential investors can base their investment decision on the best financial information possible.

The disadvantages of accrual accounting are primarily the additional time and knowledge required to properly use this method. Also, accrual accounting may not be the best choice for all farmers and ranchers to use when calculating their taxable income.

A CASH VERSUS ACCRUAL EXAMPLE

The differences between cash and accrual accounting and the resulting effect on annual profit may be best explained with an example. Assume that the information in the following table contains most of the relevant transactions related to producing a crop in the year 2017. Note, however, that transactions related to the 2017 crop year occur in three years. How would these transactions be handled with both cash and accrual accounting, and what would the estimated profit be with each method? Each transaction will be looked at individually, and then the 2017 profit for each method will be calculated.

(a) November 2016

> **Cash:** Increase fertilizer expense by $16,000. The result is a 2016 fertilizer expense $16,000 higher than it should be, because this fertilizer will not be used to produce a crop or revenue until 2017.
>
> **Accrual:** Decrease cash by $16,000, and increase prepaid expense by the same

amount. The result is an exchange of one asset (cash) for another (prepaid expense), with no effect on fertilizer expense for 2016.

(b) May 2017

> **Cash:** Increase seed, chemicals, and fuel expenses each by their appropriate share of $50,000.
>
> **Accrual:** Decrease cash by $50,000, and increase seed, chemical, and fuel expenses each by their appropriate share of $50,000.

(c) May 2017

> Sometime during 2017, the prepaid expense must be converted to fertilizer expense.
>
> **Cash:** No entry. It has already been counted as an expense.
>
> **Accrual:** Decrease prepaid expense, and increase fertilizer expense each by $16,000. The prepaid expense is eliminated, and the $16,000 fertilizer expense now shows up as a 2017 expense, as it should.

(d) October 2017

> **Cash:** No entry for the drying fuel, because no cash has been expended.
>
> **Accrual:** Increase fuel expense and increase account payable each by $6,000. The result places the fuel expense in the proper year (it was used to dry 2017 grain) and establishes a liability, an account payable in the amount of $6,000.

Month/Year	Transaction
November 2016	Purchased, paid for, and applied fertilizer to be used by the 2017 grain crop, $16,000
May 2017	Purchased and paid for seed, chemicals, fuel, etc., $50,000
October 2017	Purchased and charged to account fuel for drying, $6,000
November 2017	One-half of grain sold and payment received, $100,000. The remaining one-half is placed in storage and has an estimated value of $100,000
January 2018	Paid bill for fuel used to dry grain, $6,000
May 2018	Remaining 2017 grain sold, $120,000

(e) November 2017

Cash: Increase grain sold by $100,000.

Accrual: Increase cash by $100,000 and grain revenue by $100,000. As a part of this entry, or as a separate entry, grain revenue should be increased by another $100,000, and inventory (a new asset) also should be increased by $100,000. These entries result in all $200,000 of the 2017 grain being included in 2017 revenue, even though cash has been received for only one-half of it.

(f) January 2018

Cash: Increase fuel expense by $6,000.

Accrual: Decrease cash by $6,000, and decrease account payable by $6,000. This eliminates the account payable but does not increase fuel expense, because that was done in October 2017.

(g) May 2018

Cash: Increase grain revenue by $120,000.

Accrual: Increase cash by $120,000, increase grain revenue by $20,000, and decrease inventory by $100,000. This sale indicates there was more grain in the bin than was estimated in November 2017, or that the price increased since then. To adjust for either or both of these pleasant outcomes,

grain revenue must be increased by $20,000. Inventory is decreased by the original amount to give it a $0 balance. It must have a $0 balance, because all grain has now been sold.

It can be seen from this example that accrual accounting requires more entries and more knowledge of accounting than does cash accounting. However, there are several benefits from this extra work and knowledge. The most important is a more accurate estimate of profit.

Review the transactions again. This business produced grain with a value of $200,000 during 2017 but sold only one-half of it. The other half was in storage at the end of the year. Expenses to produce this grain totaled $72,000. However, $16,000 of this amount was paid in 2016, and $6,000 was not paid until 2018. A comparison of the 2017 profit under both cash and accrual accounting will show how this distribution of cash revenue and cash expenditures affects profit.

Cash accounting includes only cash receipts and cash expenditures during 2017, so there are only two entries. The calculated profit is $50,000. Accrual accounting, by the use of an inventory change, includes the value of all grain produced in 2017, even though not all was sold. Similarly, adjustments through the use of a prepaid expense and an account payable record all expenses incurred in producing this grain as 2017 expenses. The result is a profit of $128,000. This is a much

	2017 Income and Expenses			
	Cash accounting		**Accrual accounting**	
Cash grain sales	100,000 (e)		100,000 (e)	
Grain inventory increase	N/A		100,000 (e)	
Total revenue		$100,000		$200,000
Fertilizer	0		16,000 (c)	
Seed, chemicals, fuel	50,000 (b)		50,000 (b)	
Drying fuel	0		6,000 (d)	
Total expenses		50,000		72,000
Net farm income (profit)		$50,000		$128,000

more accurate estimate of what the 2017 production activities contributed to the financial condition of this business than the $50,000 estimated by cash accounting.

While cash accounting shows a lower profit than accrual accounting in 2017, the opposite may occur in 2018. Assume all grain produced in 2018 is sold at harvest. The result is cash receipts for both one-half the 2017 grain and all the 2018 crop being received in the same year. Cash accounting would show large cash receipts, only one year's cash expenses, and therefore, a large profit. Accrual accounting would show the same cash receipts, but this would be offset by an inventory decrease of $100,000, because no grain is in storage at the end of 2018. The net total revenue would then be only the value of the 2018 grain plus the $20,000 increase in value from the 2017 grain. Again, accrual accounting would result in a more accurate estimate of what the 2018 production activities contributed to the financial position of the business than would cash accounting.

An alternative to accrual accounting is to follow cash accounting procedures throughout the year, and then make accrual adjustments at the end of the accounting year. Adjustments include adding or subtracting inventories of inputs and products, accounts receivable and payable, and other accrued expenses. The end results for annual gross income, total expenses, and profit will be the same as for accrual accounting implemented throughout the year. This hybrid approach to accounting will be explained in more detail in Chapter 5.

FARM FINANCIAL STANDARDS COUNCIL RECOMMENDATIONS

The FFSC report assumes an accrual-based system throughout its discussion and recommendations on financial analysis measures. But it also recognizes that a large majority of farmers and ranchers currently use cash accounting and will continue to do so for some time. Simplicity, ease of use, and often income tax advantages account for the popularity of cash accounting. Therefore, the FFSC accepts the use of cash accounting during the year but strongly recommends that end-of-year adjustments be made to convert the cash accounting profit to an *accrual adjusted* profit. The latter should then be used for analysis and management decision making.

The type and nature of some of these accrual adjustments may be evident in the cash versus accrual example given. A full discussion of these adjustments and how they should be done will be delayed until Chapter 5.

OUTPUT FROM AN ACCOUNTING SYSTEM

Any accounting system should be able to produce some basic financial reports. Computerized accrual systems can generate many different reports. Figure 3-2 expands on Figure 3-1 to show the possible products from an accounting system. The balance sheet and income statement are shown first, for two reasons. First, they are the two most common reports to come out of an accounting

---

Box 3-4   Annual Versus Lifetime Profit

The different results from using cash or accrual accounting only show up in *annual* estimates of profit. Over the full lifetime of the farm business, the *total lifetime* before-tax profit would be the same using either method. The two methods only serve to divide the total lifetime profit in a different manner among individual years in the life of the business. The total lifetime, after-tax profit could be different because of differences in timing and amounts of annual income tax.

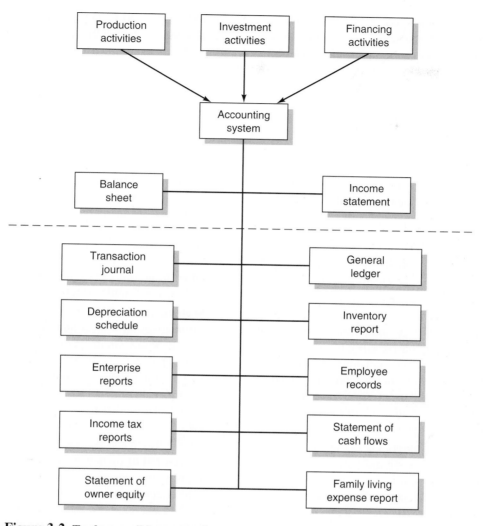

Figure 3-2 Twelve possible reports from an accounting system.

system, and second, they are the subjects of the next two chapters. Some other possible reports are often necessary and useful but may not be available from all systems nor widely used.

Balance sheet The balance sheet is the report that shows the financial condition of the business at a point in time. A detailed discussion of this report and its use will be covered in Chapter 4.

Income statement An income statement is a report of revenue and expenses ending with

an estimate of net farm income. This report will be discussed in detail in Chapter 5.

Transaction journal This is a record of all financial transactions, including check and deposit numbers, dates, payees and payers, amounts, and descriptions. A check register is a form of a transaction journal but does not contain all of the above information. This journal is used to make entries into the general ledger and to provide an audit trail.

General ledger The general ledger contains the different financial accounts for the business and the balances in these accounts. Balances in the revenue and expense accounts are used to prepare an income statement, and balances in the asset, liability, and owner equity accounts are used to prepare a balance sheet.

Depreciation schedule A depreciation schedule is a necessary part of any accounting system. Annual depreciation on all depreciable assets must be computed and recorded as an expense before an income statement can be produced. This is true whether a cash or an accrual system is used. Depreciation and the information contained in a depreciation schedule are discussed in detail in Chapter 5.

Inventory report This is a useful report, particularly for large crop farms and livestock operations. It tracks the quantity and value of crops and livestock on hand by recording purchases, sales, births, deaths, amounts harvested, and amounts fed. This report is useful for monitoring feed availability and usage, for developing a marketing program, and for monitoring any inventory pledged as collateral for a loan.

Enterprise reports These look like income statements for each individual enterprise. They are useful for determining which enterprises are contributing the most profit to the business and are therefore candidates for expansion. Enterprises identified as unprofitable become candidates for elimination.

Employee records Any business with employees must keep considerable data related to each employee. This includes not only information such as hours worked but financial data on gross pay, deductions for income taxes and social security, and so forth. Several payroll reports must be filed in a timely manner with both state and federal agencies. All of this payroll-related work can be done by hand, but many computer programs are designed especially to compute and record gross pay, deductions,

and net pay. If the payroll program is part of a general accounting program, all of this information can be automatically entered into the general ledger.

Income tax reports The information from any farm accounting system must be sufficient to prepare the farm tax return. In some systems, it will be necessary to take values from the accounting reports and enter them on IRS Schedule F, Form 1040. It is helpful if the general ledger accounts are named and organized in the same way as the categories on this schedule. Some computer accounting systems can compile and print the tax information in the same format, making it easy to transfer the data. Other programs can duplicate a Schedule F and print a completed return.

Statement of cash flows This statement summarizes all sources and uses of cash during the accounting period and is useful when analyzing the business activities during that period. When prepared monthly, it allows comparison of actual cash flows with budgeted cash flows. It is also important as a source of data when completing a cash flow budget for the next accounting period.

Statement of owner equity Financial transactions during the accounting period will affect the owner equity or net worth of the business. This statement identifies and summarizes the sources of these changes.

Family living expense report While not really a part of the business financial activities, it is desirable to keep detailed records of family living expenses. This is particularly so for any expenditures that may be deductible on income tax returns. Again, this can be done manually or included as part of the farm accounting system, provided care is taken to be sure business and personal records are not mixed. Rather inexpensive computer programs are designed solely for recording, summarizing, and analyzing personal expenses and investments.

SUMMARY

This chapter discussed the importance, purpose, and use of a good information management system as a management tool. Records provide the information needed to measure how well the business is doing in terms of meeting its goals. They also provide feedback so the results of past decisions can be evaluated, as can be the decision-making ability of the manager. Finally, individual farm records are perhaps the single best source of information needed to make current and future decisions.

Any accounting or record system must be able to handle transactions from the production, investment, and financing activities of the farm business. The choices of accounting period, cash or accrual accounting, single- or double-entry accounting, whole farm or enterprise accounting, and a basic or complete system are important. They affect the quantity, quality, and accuracy of the information provided by the accounting system and the time required to maintain the records. The output required or desired from an accounting system also must be considered when making these choices.

QUESTIONS FOR REVIEW AND FURTHER THOUGHT

1. What factors affect the choice of accounting period for farmers and ranchers?

2. How would one construct a balance sheet if the accounting was done using a single-entry cash system?

3. Is it possible to use double entry with a cash accounting system? If so, what are the advantages and disadvantages?

4. Is it possible to use single entry with an accrual system? Why or why not?

5. Check advertising material for several farm accounting software programs. Are they cash or accrual systems? Single or double entry? How many of the 12 reports from an accounting system discussed in this chapter are available from each program? Are any additional reports available?

6. Place an X under the column(s) to indicate whether each business event is an operating, investment, or financing activity.

Event	Operating	Investment	Financing
Pay cash for tractor repairs	_____	_____	_____
Borrow $40,000	_____	_____	_____
Pay interest on the loan	_____	_____	_____
Charge $12,000 of feed	_____	_____	_____
Equipment depreciates	_____	_____	_____
Sell $35,000 of corn	_____	_____	_____
Purchase pickup	_____	_____	_____
Pay principal on a loan	_____	_____	_____

7. Explain the difference between an account payable and an account receivable.

8. What products might a typical farm or ranch have in inventory at the end of a year?

9. Why are the results from an accrual accounting system recommended for use when making management decisions?

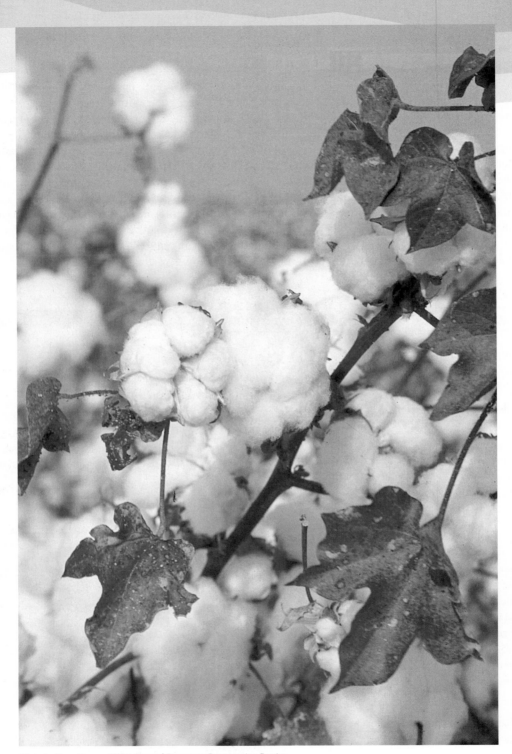

Photo by Gary Kramer, USDA Natural Resources Conservation Service

THE BALANCE SHEET AND ITS ANALYSIS

CHAPTER OUTLINE

Purpose and Use of a Balance Sheet
Balance Sheet Format
Asset Valuation
Cost-Basis Versus Market-Basis Balance Sheet
Balance Sheet Example
Balance Sheet Analysis
Statement of Owner Equity
Summary
Questions for Review and Further Thought

CHAPTER OBJECTIVES

1. Discuss the purpose of a balance sheet

2. Illustrate the format and structure of a balance sheet

3. Discuss the different methods of valuing assets and the recommended valuation methods for different types of assets

4. Show the differences between a cost-basis and a market-basis balance sheet

5. Define owner equity or net worth and show its importance

6. Analyze a firm's solvency and liquidity through the use of a number of financial ratios derived from the balance sheet

7. Introduce the statement of owner equity and explain its construction

Chapter 3 introduced the balance sheet and the income statement as two of the products or outputs from an accounting system. They are part of a complete set of financial statements but are meant to serve different purposes. The balance sheet, also called the net worth statement,

57

summarizes the financial condition of the business at a point in time, while an income statement summarizes those financial transactions that affected revenue and expenses over a period of time. The purpose of an income statement is to provide an estimate of net farm income or profit, while the balance sheet concentrates on estimating net worth or owner equity by valuing and organizing assets and liabilities.

Many transactions affect both the balance sheet and the income statement. Transactions can occur daily, so the balance sheet can change daily. That is why the point-in-time concept is emphasized when discussing the balance sheet. The connection between these two financial statements will be discussed at the end of this chapter.

PURPOSE AND USE OF A BALANCE SHEET

A balance sheet is a systematic organization of everything *owned* and *owed* by a business or an individual at a given time. Anything of value owned by a business or an individual is called an asset, and any debt or other financial obligation owed to someone else is referred to as a liability. Therefore, a balance sheet is a list of assets and liabilities that concludes with an estimate of net worth, calculated by taking the difference between the value of assets and liabilities. It is the amount the owner(s) have invested in the business, and is often referred to as *owner equity*. The *balance* in a balance sheet comes from the requirement that the ledger be in balance through the basic accounting equation of

$$\text{Assets} = \text{liabilities} + \text{owner equity.}$$

Rearranging this equation allows finding owner equity once assets and liabilities are known:

$$\text{Owner equity} = \text{assets} - \text{liabilities}$$

A balance sheet can be completed any time during an accounting period. One of the advantages of a computerized accounting system is the ease with which an interim balance sheet can be prepared, such as might be needed for a loan application. However, most balance sheets are prepared at the end of the accounting period, December 31 on most farms and ranches. This procedure allows a single balance sheet to be an end-of-the-year statement for one accounting period and a beginning-of-the-year statement for the next accounting period. For purposes of comparison and analysis, it is necessary to have a balance sheet available for the beginning and end of each year.

The financial position of a business at a point in time is measured primarily through the use of two concepts.

1. *Solvency*, which measures the liabilities of the business relative to the amount of owner equity invested in the business. It also provides an indication of the ability to pay off all financial obligations or liabilities if all assets were sold; that is, it measures the degree to which assets are greater than liabilities. If assets are not greater than liabilities, the business is insolvent and a possible candidate for liquidation.

2. *Liquidity*, which measures the ability of the business to meet financial obligations as they come due without disrupting the normal operations of the business. Liquidity measures the ability to generate cash in the amounts needed at a given time. These cash requirements and possible sources of cash are generally measured only over the next accounting period, making liquidity a short-run concept.

BALANCE SHEET FORMAT

A condensed and general format for a balance sheet is shown in Table 4-1. Assets are shown on the left side or top part of a balance sheet, and liabilities are placed to the right of or below assets.

TABLE 4-1	General Format of a Balance Sheet		

Assets		Liabilities	
Current assets	$100	Current liabilities	$ 60
Noncurrent assets	$400	Noncurrent liabilities	$200
		Total liabilities	$260
		Owner equity	$240
Total assets	$500	Total liabilities and owner equity	$500

Assets

An asset has value for one of two reasons. First, it can be sold to generate cash, and second, it can be used to produce other goods that in turn can be sold for cash at some future time. Goods that have already been produced, such as grain and feeder livestock, can be sold quickly and easily without disrupting future production activities. They are called liquid assets. Assets such as machinery, breeding livestock, and land are owned primarily to produce agricultural commodities that can then be sold to produce cash income. Selling income-producing assets to generate cash would affect the firm's ability to produce future income, so they are less liquid, or illiquid. These assets are also more difficult to sell quickly and easily at their full market value.

Current Assets

Accounting principles require current assets to be separated from other assets on a balance sheet. Current assets are the more liquid assets, which will be either used up or sold within the next year as part of normal business activities. Cash on hand and checking and savings account balances are current assets and are the most liquid of all assets. Other current assets include accounts and notes receivable (which represent money owed to the business because of loans granted, products sold, or services rendered), investments in growing crops and prepaid expenses, and inventories of feed, grain, supplies, and feeder livestock. The latter are livestock held primarily for sale and not for breeding purposes.

Noncurrent Assets

Any asset not classified as a current asset is, by default, a noncurrent asset. On a farm or ranch, these assets include primarily machinery and equipment, breeding livestock, fences and corrals, buildings, and land. Shares of stock owned in cooperatives or other businesses closely related to the farming or ranching operation can also be included in noncurrent assets.

Liabilities

A liability is an obligation or debt owed to someone else. It represents an outsider's claim against one or more of the business assets.

Box 4-1   Valuing Farm Inventories

Accounting principles allow grain, forages, fruits and vegetables, and certain other commodities to be valued at current market value on a balance sheet, provided they meet the following conditions: (1) have a reliable, determinable, and realizable market price; (2) have relatively small and known selling expenses; and (3) are ready for immediate delivery for sale; that is, they have been harvested. As the FFSC suggests, most market livestock would also meet these conditions.

Current Liabilities

Current liabilities must be separated from all other liabilities for the balance sheet to follow basic accounting principles. Current liabilities are financial obligations that will become due and payable within one year from the date of the balance sheet and will therefore require that cash be available in these amounts within the next year. Examples would be accounts payable at farm supply stores for goods and services received but not yet paid for and the full amount of principal and accumulated interest on any short-term loans or notes payable. Short-term loans are those requiring complete payment of the principal in one year or less. These would typically be loans used to purchase crop production inputs, feeder livestock, and feed for the feeder livestock.

Loans obtained for the purchase of machinery, breeding livestock, and land are typically for a period longer than one year. Principal payments may extend for three to five years for machinery and for 20 years or more for land. However, a principal payment is typically due annually or semiannually, and these payments will require cash within the next year. Therefore, all principal payments due within the next year, whether they are for short-term loans or for noncurrent loans, are included as current liabilities.

The point-in-time concept requires identifying all liabilities that exist as of the date of the balance sheet. In other words, what obligations would have to be met if the business sold out and ceased to exist on this date? Some expenses tend to accrue daily but are paid only once or twice a year. Interest and property taxes are examples. To properly account for these, accrued expenses are included as current liabilities. Included in this category would be interest that has accumulated from the last interest payment on each loan to the date of the balance sheet. Accrued property taxes would be handled in a similar manner, and there may be other accrued expenses, such as wages and employee tax withholdings, that have been incurred but not yet paid. Income taxes on farm income are typically paid several months after the close of an accounting period. Therefore, a balance sheet for the end of an accounting period should also show accrued income taxes, or income taxes payable, as a current liability.

Payments that must be made in the coming year for leased assets are generally not shown on the balance sheet. These could include rental payments for land, machinery, or breeding livestock. Because leased assets are not actually owned by the business, they are not included under the asset section of the balance sheet, nor are the lease payments shown as liabilities, unless they are past due.

Noncurrent Liabilities

These include all obligations that do not have to be paid in full within the next year. As discussed, any principal due within the next year would be shown as a current liability, and the remaining balance on the debt would be listed as a noncurrent liability. Care must be taken to be sure the current portion of these liabilities has been deducted, and only the amount remaining to be paid after the next year's principal payment is recorded as a noncurrent liability.

For example, suppose a rancher owed $24,000 to an auto dealer for the purchase of pickup, at 6 percent annual interest rate. A principal payment of $8,000 is due in the coming year, which is listed as a current liability. The remaining balance of $16,000 is shown as a noncurrent liability. The rancher made a payment of principal and interest last April 1. Assuming the balance sheet is being developed on December 31, interest has been accruing for nine months or 0.75 years. The total amount of interest owed would be $24,000 \times 6\% \times 0.75$ years $= \$1,080$, which would also be included under current liabilities. Note that the amount shown is the interest that has accrued as of the date of the statement, not the amount that will be owed when the next payment is due.

Owner Equity

Owner equity or net worth represents the amount of money left for the owner of the business should the assets be sold and all liabilities paid as of the date of the balance sheet. It is also called net worth, especially when the balance sheet contains financial information for the family as well as the business. Equity can be found by subtracting total liabilities from total assets and is therefore the *balancing* amount, which causes total assets to exactly equal total liabilities plus owner equity. Owner equity is the owner's current investment or equity in the business.

Owner equity can and does change for a number of reasons. One common and periodic change comes from using assets to produce crops and livestock, with the profit from these production activities then used to purchase additional assets or to reduce liabilities. This production process takes time, and one of the reasons for comparing a balance sheet for the beginning of the year with one for the end of the year is to study the effects of the year's production on owner equity and the composition of assets and liabilities. Owner equity will also change if there is a change in an asset's value, a gift or inheritance is received, cash or property is contributed or withdrawn from the business, or an asset is sold for more or less than its balance sheet value.

However, changes in the composition of assets and liabilities may not cause a change in owner equity. For example, if $10,000 cash is used to purchase a new machine, owner equity does not change. There is now $10,000 less current assets (cash) but an additional $10,000 of noncurrent assets (the machine). Total assets remain the same and, therefore, so does owner equity. If $10,000 is borrowed to purchase this machine, both assets and liabilities will increase by the same amount, leaving the difference or owner equity the same as before. The purchase will not affect owner equity. However, over time the loss of value in the item recorded by depreciation will affect owner equity. Using the $10,000 to make a principal payment on a loan will also have no effect on equity. Assets have been reduced by $10,000, but so have liabilities. Equity will remain the same.

These examples illustrate an important point. Owner equity in a business changes only when

the owner puts additional personal capital into the business (including gifts or inheritance), withdraws capital from the business, or when the business shows a profit or loss. Changes in asset values due to changes in market prices also affect equity if assets are valued at market value. However, many business transactions only change the mix or composition of assets and liabilities and do not affect owner equity.

Alternative Format

Farm and ranch balance sheets have traditionally included three categories of both assets and liabilities, while the accounting profession uses two, as shown in Table 4-1. The FFSC restates the accounting principle of separating current from noncurrent items on a balance sheet. However, it does not restrict all noncurrent items to a single category. A three-category balance sheet is permitted if the preparer believes such a format would add information and the definition used for intermediate assets and liabilities is disclosed.

The usual categories for the traditional farm or ranch balance sheet are shown in Table 4-2. Both current assets and current liabilities are separated from other assets and liabilities as required by accounting principles. These categories would each contain exactly the same items as mentioned in the discussion of Table 4-1. The difference in the two formats comes in

TABLE 4-2	Format of a Three-Category Balance Sheet

Assets		Liabilities	
Current assets	$100	Current liabilities	$ 60
Intermediate assets	$120	Intermediate liabilities	$ 75
Fixed assets	$280	Long-term liabilities	$125
		Total liabilities	$260
		Owner equity	$240
		Total liabilities and	
Total assets	$500	owner equity	$500

the division of noncurrent assets and liabilities into two categories, intermediate and fixed or long term.

Intermediate assets are generally defined as less liquid than current assets and with a life greater than one year but less than about 10 years. Machinery, equipment, perennial crops, and breeding livestock would be the usual intermediate assets found on farm and ranch balance sheets. Fixed assets are the least liquid and have a life greater than 10 years. Land and buildings are the usual fixed assets.

Intermediate liabilities are debt obligations where repayment of principal occurs over a period of more than one year and up to as long as 10 years. Some principal and interest would typically be due each year, and the current year's principal payment would be shown as a current liability, as discussed earlier. Most intermediate liabilities would be loans where the money was used to purchase machinery, breeding livestock, and other intermediate assets. Long-term liabilities are debt obligations where the repayment period is for a length of time exceeding 10 years. Farm mortgages and contracts for the purchase of land are the usual long-term liabilities, and the repayment period may be 20 years or longer.

Intermediate assets and liabilities are often a substantial and important part of total assets and total liabilities on farms and ranches. Also, land values may be volatile in some areas, affected by economic and market conditions unrelated to agricultural production. These factors may explain why the three-category balance sheet came into use in agriculture. However, the FFSC encourages the use of the two-category format and predicts a movement away from the use of a three-category balance sheet.

ASSET VALUATION

Before constructing a balance sheet, it is often necessary to estimate the value of business assets. These values may be needed for existing or newly acquired assets or for those that have been produced in the normal course of business. Several valuation methods can be used, and the choice depends on the type of asset and the purpose of the valuation.

Market Value

This method values an asset using its current market price. It can sometimes be called the fair market value or net market price method. Any normal marketing charges, such as transportation, selling commissions, and other fees, are subtracted to find the net market value. This method can be used for many types of property, but it works particularly well for items that could or will be sold in a relatively short time as a normal part of the business activities and for which current market prices are available. Examples are hay, grain, feeder livestock, stocks, and bonds.

Sometimes the selling price for grain or livestock is set in advance through the use of a *forward price contract*. The seller promises to deliver a fixed quantity for a specified price by a certain date. Current assets that have been forward contracted on or before the date of the balance sheet should be valued at the contract price, minus potential selling costs, regardless of whether the current market price is higher or lower than the contract price.

Commodities for which a future selling price has been established through the use of futures contracts on a commodity exchange through a process known as *hedging* can be valued at current market price, and any potential gain or loss on the futures contracts can be shown as a separate entry under current assets.

Cost Value

Items that have been purchased can be valued at their original cost. This method works well for items that have been purchased recently and for which cost records are still available. Land is generally valued at cost for a conservative valuation. Feed, fertilizer and supplies are often valued at cost. Items such as buildings and machinery, which normally lose value or depreciate over time, should not be valued with this method. Livestock and crops that have been raised cannot be valued at cost, because there is no purchase price to use.

Lower of Cost or Market Value

This valuation method requires valuing an item at both its cost and its market value and then using whichever value is lower. This is a conservative method, because it minimizes the chance of placing an overly high value on any item. Using this method, property increasing in value because of inflation will have a value equal to its original cost. Valuing such property at cost eliminates any increase in value over time caused solely by inflation or a general increase in prices. When prices have decreased since the item was purchased, this method results in valuation at market value.

An exception is made for feeder livestock, which are expected to increase in value over time due to gains in weight. They should be valued at their respective current market price times their current weight, rather than their original cost.

Farm Production Cost

Items produced on the farm can be valued at their farm production cost. This cost is equal to the accumulated costs of producing the item but should not include profit or any opportunity costs associated with the production. Established but immature crops growing in a field are generally valued this way, with the value equal to the actual, direct production expenses incurred to date. It is not appropriate to value a growing crop using expected yield and selling price, because poor weather, a hailstorm, or lower prices could drastically change the value before harvest and final sale. This procedure is another example of conservatism in valuation. However, detailed enterprise accounts are needed to use this method.

Cost Less Accumulated Depreciation

Depreciation is a method of accounting for the loss in value of certain durable assets over their expected years of use in the business. The cost-less-depreciation method of valuation is thus used only for assets such as machinery, buildings, and purchased breeding livestock, which are expected to decline in value over their years of use. Book value is defined as cost less accumulated depreciation. Therefore, this method results in an asset's estimated value being the same as its current book value. Depreciation methods are discussed in detail in Chapter 5.

Whatever valuation method is used for any asset, the accounting concepts of *conservatism* and *consistency* should be kept in mind. Conservatism cautions against placing too high a value on any asset, while consistency stresses using the same valuation method or methods over time. Use of these concepts makes financial statements directly comparable from year to year and prevents an overly optimistic portrayal of the firm's financial condition.

COST-BASIS VERSUS MARKET-BASIS BALANCE SHEET

The general approach to asset valuation in agriculture and the proper method to use for specific assets have long been debated. Much of the discussion involves whether agriculture should use a cost-basis or a market-basis balance sheet.

Cost Basis

The cost-basis balance sheet values assets using the cost, cost-less-depreciation, or farm-production cost methods. The one general exception would be inventories of grain and market livestock. Stored grain can be valued at market value less selling costs, provided it meets several conditions, which it normally would. While market livestock are not specifically mentioned in accounting principles, the FFSC suggests that there is little difference between grain and market livestock, particularly when the livestock are nearly ready for market. They recommend market valuation for raised market livestock even on a cost-based balance sheet. For purchased market livestock, market valuation is acceptable.

Raised breeding livestock can pose an especially difficult valuation problem for a cost-basis balance sheet. The FFSC recommends that an agricultural producer should use either the *full-cost absorption* method or the *base value* approach for reporting values for raised breeding livestock. Under full-cost absorption, all the costs of raising the breeding livestock are accumulated over time until the animal reaches its current status (e.g., a bred heifer). Both the direct and the indirect cost required to bring breeding livestock into production should be included in the accumulated costs when using this method. These costs are not taken as expenses on the income statement in the period in which they are incurred, but instead are capitalized into the animal's value. After the animal reaches maturity, its accumulated value is subsequently depreciated over the useful life, as would be the case for purchased breeding animals. Clearly, this method of valuing raised breeding livestock requires extensive records.

The base value method is considerably simpler. Under this valuation plan, the producer may use: (1) the actual or estimated cost of raising the animal to its current status, (2) the market value of similar animals when the base value is established, (3) values provided by the Internal Revenue Service (IRS), or (4) a standardized value specific to the business. The important thing is that the chosen value remains relatively fixed over time, so that changes in the value of the breeding stock would result only from changes in the number of animals. Raised breeding livestock are not depreciated when using the base value method. Instead, costs associated with raising the animals are taken as expenses in the year they are incurred.

Advantages of cost-basis balance sheets include conformity to generally accepted accounting principles, conservatism, and their direct comparability with balance sheets from other types of businesses using cost basis. Also, changes in equity come only from net income that has been earned and retained in the business or from putting more personal assets into the business, as through inheritance, but not from asset price changes.

Market Basis

A market-basis balance sheet would have all assets valued at market value less estimated selling costs. Long-run inflation could cause land owned for a number of years to have a market value higher than its cost. Inflation and fast depreciation methods also would result in machinery and breeding livestock market values being higher than their book values. Therefore, a market-basis balance sheet typically would show a higher total asset value, and consequently a higher equity, than one using cost valuation. An exception would be during periods of falling prices and corresponding decreases in asset values.

The main advantage of market-basis balance sheets is the more accurate indication of the current financial condition of the business and the value of collateral available to secure loans. A primary use for farm and ranch balance sheets is to show an operation's financial situation when the operator is applying for a loan, so market-basis valuation, which shows the current value of available collateral, has been in general use.

Evaluating the Farm's Financial Condition

The FFSC considered both types of balance sheets and concluded that information on both cost and market values is needed to properly analyze the financial condition of a farm or ranch business. The FFSC guidelines state that acceptable formats are: (1) a market-basis balance sheet with cost information included as footnotes or shown in supporting schedules or (2) a double-column balance sheet, with one column containing cost values and the other market values.

Table 4-3 contains FFSC-recommended or acceptable methods for valuing assets on both types of balance sheets. Cost-basis valuation, as used in this table, represents one of three valuation methods: cost, cost less depreciation, or farm production cost. Cost less depreciation (or book value) would be used for all depreciable assets, such as machinery, purchased breeding livestock, and buildings. Farm production cost,

TABLE 4-3	**Valuation Methods for Cost-Basis and Market-Basis Balance Sheets**	

Asset	Cost basis	Market basis
Inventories of grain and market livestock	Market[*]	Market
Accounts receivable	Cost	Cost
Prepaid expenses	Cost	Cost
Investment in growing crops	Production cost	Production cost
Purchased breeding livestock	Cost less depreciation	Market
Raised breeding livestock	Production cost or a base value	Market
Machinery and equipment	Cost less depreciation	Market
Buildings and improvements	Cost less depreciation	Market
Land	Cost	Market

[*]Market is an acceptable method for raised grain and market livestock. Lower of cost or market is the preferred method for purchased grain and supplies.

or the accumulated direct expenses incurred to date, would be the value used for the investment in growing crops.

Even on a market-basis balance sheet, not every asset is valued at market. Accounts receivable and prepaid expenses are valued at cost, their actual dollar amount. Investment in growing crops is valued under both valuation methods at an amount equal to the direct expenses incurred for that crop to date. The crop is not yet harvested, not ready for market, and still subject to production risks. Valuing it at expected market value would be an optimistic approach and not in compliance with conservative accounting principles.

BALANCE SHEET EXAMPLE

Table 4-4 is an example of a balance sheet with a format and headings that follow the FFSC recommendations. It uses the double-column format to

present cost and market values on a single balance sheet. For simplicity, only farm business assets and liabilities are included. Many farm and ranch balance sheets include personal as well as business assets and liabilities. It is often difficult to separate them, and lenders may prefer they be combined for their analysis. This presentation of a complete balance sheet with both cost and market values includes some new headings and concepts that need to be discussed along with a review of the asset valuation process.

Asset Section

With inventories valued at market in both cases, there is often little or no difference between the total value of current assets under the two valuation methods.

Most of the difference in asset values between cost- and market-based balance sheets will show up in the noncurrent asset section. A combination of inflation and rapid depreciation can result in the book values for machinery, equipment, purchased breeding livestock, and buildings being much less than their market values. Land that has been owned for a number of years during which there was only moderate inflation can still have a market value considerably higher than its original cost. These factors can combine to make the noncurrent asset value much higher with market valuation than with cost valuation. It is also possible for market values to be less than cost during periods when asset values are declining.

Liability Section

There is little difference in the valuation of the usual liabilities on a cost- or market-based balance sheet. However, there are several entries related to income taxes in the liability section of this balance sheet that have not been discussed before. Income taxes payable under current liabilities represent taxes due on any taxable net farm income for the past year. Taxes on farm income are generally paid several months after the year ends, but they must be paid. Caused by the past year's activity but not yet paid, they are like an account payable,

TABLE 4-4	Balance Sheet for I. M. Farmer, December 31, 2017

Assets				Liabilities		
Current assets:	Cost in ($)	Market in ($)		**Current liabilities:**	Cost in ($)	Market in ($)
Cash/checking acct.	7,000	7,000		Accounts payable	3,500	3,500
Inventories				Notes payable within one year	6,000	6,000
Crops	40,640	40,640		Current portion of term debt	22,250	22,250
Livestock	22,400	22,400		Accrued interest	12,300	12,300
Supplies	860	860		Income taxes payable	2,300	2,300
Accounts receivable	3,570	3,570		Other accrued expenses	1,800	1,800
Prepaid expenses	3,780	3,780		Current portion—deferred taxes	14,448	14,448
Investment in growing crops	5,000	5,000				
Other current assets	2,500	2,500		Total Current Liabilities	62,598	62,598
Total Current Assets	$85,750	$85,750		**Noncurrent liabilities:**		
Noncurrent assets:				Notes payable		
				Machinery	18,000	18,000
Machines and equipment	108,000	120,000		Breeding livestock	—	—
Breeding livestock (purchased)	8,550	9,000		Real estate debt	177,500	177,500
Breeding livestock (raised)	113,400	144,000		Noncurrent portion—deferred taxes	17,010	74,318
Buildings and improvements	56,000	80,000		Total Noncurrent Liabilities	212,510	269,818
Land	315,000	630,000		Total Liabilities	275,108	332,416
Total Noncurrent Assets	600,950	983,000		**Owner equity:**		
Total Assets	686,700	1,068,750				
				Contributed capital	160,000	160,000
				Retained earnings	251,592	251,592
				Valuation adjustment		324,742
				Total Owner Equity	411,592	736,334
				Total Liabilities and Owner Equity	686,700	1,068,750

which would still have to be paid even if the business ceased to operate as of December 31. Accounting for income taxes payable is necessary regardless of the valuation method being used.

Deferred Taxes on Current Assets

The current portion of deferred income taxes represents taxes that would be paid on revenue from the sale of current assets less any current liabilities that would be tax-deductible expenses.

They are called *deferred* or *contingent taxes* because the assets have not been sold nor the expenses paid. Therefore, no taxes are payable at this time. If cash accounting is used for tax purposes, taxes are deferred into a future accounting period when the assets are converted into cash and the expenses paid. However, because the assets exist and the expenses have been incurred, there is no question that there will be taxable events in the future.

Generally income-increasing items would include: (1) inventories of crops, feed, feeder livestock, and livestock products; (2) accounts receivable; (3) cash investment in growing crops; and (4) prepaid expenses. The differences between the balance sheet values and the tax bases (generally zero under cash accounting) for these items are totaled and then the accrued expenses that would decrease tax liability when paid are subtracted. Estimated federal and state income taxes and Social Security taxes are then calculated using an estimated tax rate.

For the example farm, the values of the crop and livestock inventories, the accounts receivable, the investment in growing crops, and the prepaid expenses total $75,390. From this amount, $17,600 is subtracted to account for the accounts payable, the accrued interest, and the other accrued expenses. The net of $57,790 is then multiplied by an estimated 25 percent tax rate to get the current portion of deferred taxes of $14,448.

Deferred Taxes on Noncurrent Assets

Raised breeding livestock may be another source of deferred taxes, a noncurrent source. If the costs of raising the breeding livestock are taken as expenses on the tax return in the period incurred, then the tax basis of the raised breeding livestock will be zero. All the income from its sale (minus selling costs) is taxed as capital gain and is not subject to self-employment tax. To qualify as breeding stock, livestock must have been held for breeding or dairy purposes for at least 24 months for cattle and horses and for at least 12 months for other species. (If a full-cost absorption method is used, then only the portion of the value that exceeds the tax basis will be subject to taxation.)

In our example farm, raised breeding livestock were assumed to have a cost basis value of $113,400, determined using the *base value* method. Assuming a tax basis of zero for the raised breeding livestock, and ignoring any selling expenses, deferred taxes associated with the breeding livestock under the cost basis would be

$113,400 × 0.15, or $17,010. Because the raised breeding livestock have a market value of $144,000, deferred taxes associated with the breeding livestock under the market basis would be $144,000 × 0.15, or $21,600.

On the market-basis balance sheet, noncurrent deferred taxes will also arise from the difference between cost and market values for noncurrent assets. Market values are often higher than cost and result in a market-basis balance sheet presenting a stronger financial position than one done on a cost basis. If the assets were sold for market value, the business would have to pay capital gains taxes on the difference between market value and each asset's cost or tax basis. Ignoring these taxes on a market balance sheet results in owner equity higher than would actually result from a complete liquidation of the business. Therefore, these noncurrent deferred taxes are included on a market-basis balance sheet and are an estimate of the taxes that would result from a liquidation of the assets at their market value.

For the market-value balance sheet in the example, the difference between market-basis values and cost-basis values for all noncurrent assets other than the raised breeding livestock is calculated to be $351,450, as shown in Table 4-5. This potential gain would be subjected to an

TABLE 4-5	Potential Capital Gain on Noncurrent Assets		
	Market value (in $)	**Cost value (in $)**	**Potential gain (in $)**
Machinery and equipment	120,000	108,000	12,000
Breeding livestock (purchased)	9,000	8,550	450
Buildings and improvements	80,000	56,000	24,000
Land	630,000	315,000	315,000
Total			351,450

estimated 15 percent tax rate, for an additional $52,718 in deferred taxes. When this amount is added to the deferred taxes associated with the raised breeding livestock, noncurrent deferred taxes total $74,318 for the market-basis balance sheet.

Owner Equity Section

Owner equity has three basic sources: (1) capital contributed to the business by its owner(s), (2) earnings or business profit that has been left in the business rather than withdrawn, and (3) any change caused by fluctuating market values when market valuation, rather than cost, is used. The FFSC recommends showing all three sources of equity separately, rather than combining them into one value. This breakdown provides additional information for anyone analyzing the balance sheet.

The example in Table 4-4 shows $160,000 of contributed capital. This is the value of any personal cash or property the owner used to start the business and any that might have been contributed since that time. In the absence of any further contributions or withdrawals charged against contributed capital, this value will remain the same on all future balance sheets.

Any before-tax net farm income not used for family living expenses, income taxes, or withdrawals for other purposes remains in the business. These retained earnings will be used to increase assets (not necessarily cash), decrease liabilities, or some combination of the two. Retained earnings, and therefore owner equity, will increase during any year in which net farm income is greater than the combined total of income taxes paid and net withdrawals. If these latter two items are greater than net farm income, retained earnings and cost-basis owner equity will decrease. In the example, a total of $251,592 has been retained since the business began. Retained earnings will be discussed again in further detail in Chapter 5.

There is no valuation adjustment on a cost-basis balance sheet. Owner equity will change

---

Box 4-2   Notes to a Balance Sheet

The numbers shown on a balance sheet provide little information about the nature of the business, its accounting procedures, and the computations done to arrive at a balance sheet value. Accounting principles require this and other information be disclosed in notes included with the balance sheet.

The FFSC recommends that, at a minimum, three types of information be included in notes to the balance sheet.

1. *Basis of Accounting*—a brief description of the accounting methods, procedures, and policies used and whether the balance sheet is based on market or cost values
2. *Nature of the Operation*—a brief description of the operation including acres owned and rented, types and number of crops and

livestock raised, and the form of business organization used
3. *Depreciation Methods*—information on the years of depreciable life and depreciation methods used for each type of depreciable asset

Additional notes may be desirable or necessary in some cases to help the user (often a lender) to understand and trust the values shown. Examples would be details on the valuation methods used; details about inventory assets (bushels, head, weights, etc.); a listing of machinery, equipment, and buildings; the calculation of deferred income taxes; details about leased land and other leased assets controlled by the business; and details about debt and debt payments, including names of creditors.

only when there are changes in contributed capital or retained earnings. Whenever a market-based balance sheet includes an asset valued at more than its cost, it creates equity that is neither contributed capital nor retained earnings. For example, during inflationary periods, the market value of land may increase substantially. A balance sheet with market values would include an increase in the land value each year, which would cause an equal increase in equity. However, this increase is due only to the ownership of the land and not from its direct use in producing agricultural products. Any differences between cost and market values, which can be either positive or negative, should be shown as a valuation adjustment in the equity section of the balance sheet. An entry for valuation adjustment makes it easy for an analyst to determine what part of total equity on a market-based balance sheet is a result of valuation differences. On the market-basis balance sheet, total owner equity equals the sum of contributed capital, retained earnings, and valuation adjustment.

BALANCE SHEET ANALYSIS

A balance sheet is used to measure the financial condition of a business and, more specifically, its liquidity and solvency. Analysts often want to compare the relative financial condition of different businesses and of the same business over time. Differences in business size cause potentially large differences in the dollar values on a balance sheet, and therefore cause problems comparing their relative financial condition. A large business can have serious liquidity and solvency problems, as can a small business, but the difficulty is measuring the size of the problem relative to the size of the business.

To get around this problem, ratios are often used in balance sheet analysis. They provide a standard procedure for analysis and permit comparison over time and between businesses of different sizes. A large business and a small one would have substantial differences in the dollar values on their balance sheets, but the same ratio

value would indicate the same relative degree of financial strength or weakness. Ratio values can be used as goals and easily compared against the same values for other businesses. In addition, many lending institutions use ratio analysis from balance sheet information to make lending decisions and to monitor the financial progress of their customers.

Most farm and ranch balance sheets are used for loan purposes, where market values are of the most interest to the lender. Therefore, the analysis of the balance sheet will use values from the *market* column.

Analyzing Liquidity

An analysis of liquidity concentrates on current assets and current liabilities. The latter represent the need for cash over the next 12 months, and the former, the sources of cash. Liquidity is a relative rather than an absolute concept, because it is difficult to state that a business is or is not liquid. Based on an analysis, however, it is possible to say that one business is more or less liquid than another.

Current Ratio
The current ratio is one of the more common measures of liquidity and is computed from the equation

$$\text{Current ratio} = \frac{\text{current asset value}}{\text{current liability value}}$$

The current ratio for the balance sheet example in Table 4-4 would be

$$\text{Current ratio} = \frac{\$85,750}{\$62,598} = 1.37$$

This ratio measures the amount of current assets relative to current liabilities. A value of 1.0 means current liabilities are equal to current assets, and while there are sufficient current assets to cover current liabilities, there is no safety margin. Asset values can and do change with changes in market prices, so values larger than 2.0 are preferred to provide a safety margin for price changes and other factors. The larger the ratio value, the more liquid the business, and vice versa. There is no hard and fast rule for how high the current ratio should be. Whether a particular current ratio indicates sufficient liquidity will depend on many factors, including the planned sales times for inventories and the schedule of loan repayments. A cash flow budget, which will be discussed in detail in Chapter 13, provides much more information about potential liquidity problems.

Working Capital

Working capital is the difference between current assets and current liabilities:

Working = current assets − current liabilities
capital

This equation computes the dollars that would remain after selling all current assets and paying all current liabilities. It is an indication of the margin of safety for liquidity measured in dollars. I. M. Farmer's balance sheet in Table 4-4 shows working capital of $23,152, calculated by subtracting $62,598 from $85,750.

Working capital measures the dollars that could, in theory, be available to purchase new inputs or other items after the sale of current assets and the payment of the current liabilities. Working capital is also available to pay family living expenses if the business and family are not treated separately.

Working capital is a dollar value, not a ratio. This makes it difficult to use working capital to compare the liquidity of businesses of different sizes. A larger business would be expected to have larger current assets and liabilities and would need more working capital to have the same relative liquidity as a smaller business. Therefore, it is important to relate the amount of working capital to the size of the business.

Dividing working capital by the total revenue generated by the farm in the most recent year provides a ratio that can be used to compare farms of different scales. I. M. Farmer's total income in 2017 is $406,548, so the ratio of working capital to gross revenues is $23,152/$406,548 = 0.06.

Analyzing Solvency

Solvency measures the relations among assets, liabilities, and equity. It is a way to analyze the business debt and to see if all liabilities could be paid off by the sale of all assets. The latter requires assets to be greater than liabilities, indicating a solvent business. However, solvency is generally discussed in relative terms by measuring the degree to which assets exceed liabilities.

Three ratios are commonly used to measure solvency and are recommended by the FFSC. Each of them uses two of the three items mentioned earlier, making them all related to one another. Any one of these ratios will, when properly computed and analyzed, provide full information about solvency. However, all three have been in common use, and some individuals prefer one over another.

There are no uniform standards for these solvency measures. According to the FFSC, the range of acceptable values will vary from lender to lender and will depend on the income variability and other production risks of the farm enterprises, the types of assets owned by the business, and potential fluctuations in farm asset values.

Debt/Asset Ratio

The debt/asset ratio is computed from the equation

$$\text{Debt/asset ratio} = \frac{\text{total liabilities}}{\text{total assets}}$$

and measures what part of total assets is owed to lenders. This ratio should have a value less than 1.0; even smaller values are preferred. A debt/asset ratio of 1.0 means debt or liabilities equal assets, and therefore, equity is zero. Ratios greater than 1.0 would be obtained for an insolvent business. I. M. Farmer has a debt/asset ratio of $332,416/$1,068,750, or 0.31, using market values.

Equity/Asset Ratio

This ratio is computed from the equation

$$\text{Equity/asset ratio} = \frac{\text{total equity}}{\text{total assets}}$$

and measures what part of total assets is financed by the owner's equity capital. Higher values are preferred, but the equity/asset ratio cannot exceed 1.0. A value of 1.0 is obtained when equity equals assets, which means liabilities are zero. An insolvent business would have a negative equity/asset ratio, because equity would be negative. For the example in Table 4-4, the equity/asset ratio is $736,334/$1,068,750, or 0.69, using the market values.

Debt/Equity Ratio

The debt/equity ratio is also called the leverage ratio by some analysts and is computed from the equation

$$\text{Debt/equity ratio} = \frac{\text{total liabilities}}{\text{total equity}}$$

This ratio compares the proportion of financing provided by lenders with that provided by the business owner. When the debt/equity ratio is equal to 1.0, lenders and the owner are providing an equal portion of the financing. Smaller values are preferred, and the debt/equity ratio will approach zero as liabilities approach zero. Large values result from small equity, which means an increasing chance of insolvency. I. M. Farmer has a debt/equity ratio of $332,416/$736,334, or 0.45.

All three solvency provide essentially the same information, and are linked mathematically by the following formulas:

$$\text{Debt/asset} =$$
$$(\text{Debt/equity}) \times (\text{equity/asset})$$

and

$$\text{Equity/asset} = 1 - (\text{debt/asset})$$

Summary of Analysis

Table 4-6 summarizes the values obtained in the analysis of the I. M. Farmer balance sheet. These values show current assets to be 38 percent higher than current liabilities. Using market values, the business financing is just under one-third financed from debt, and the rest from equity capital.

Performing an analysis of this balance sheet using the cost values would show two general results. First, there would be little difference in the liquidity measures, because there is little difference in the values of current assets and current liabilities under the two valuation methods. Second, cost values, being lower, would show a weaker solvency position. These results would be typical. Market valuation will have little effect on liquidity measures, but it has the potential for a large impact on solvency measures. This example illustrates the importance of knowing how assets were valued on any balance sheet being analyzed and shows why the FFSC recommends including information on both cost and market values. Providing complete information allows a more thorough analysis and eliminates any possible confusion about which valuation method was used.

TABLE 4-6	**Summary of I. M. Farmer's Financial Condition**

Measure	Market basis
Liquidity:	
Current ratio	1.37
Working capital	$23,152
Working capital to gross revenues	0.06
Solvency:	
Debt/asset ratio	0.31
Equity/asset ratio	0.69
Debt/equity ratio	0.45
Other:	
Debt structure ratio	0.19

Other Measures

The FFSC recommends the use of the five analysis measures previously discussed. However, other analyses have also been widely used and continue to be used. One is the debt structure ratio, computed as

$$\text{Debt structure ratio} = \frac{\text{current liabilities}}{\text{total liabilities}}$$

This ratio shows what proportion current liabilities are to total liabilities and can be converted to a percentage by multiplying by 100. It cannot be greater than 1, or 100 percent, which would result when all liabilities are current liabilities. All current liabilities must be paid within the next year, so smaller numbers are preferred. Higher values mean a large proportion of total liabilities must be paid within the next year, which may require more cash than will be available. This could indicate a need to convert some current liabilities into noncurrent ones, which would reduce current payments by spreading payments over more years. However, a relatively high debt structure ratio may not indicate a problem if both current and total liabilities are small amounts. For the example farm, the debt structure ratio is $62,598/332,416 = 0.19$, or 19 percent.

STATEMENT OF OWNER EQUITY

The balance sheet shows the amount of owner equity at a point in time, but not what caused changes in this value over time. The FFSC therefore recommends that a statement of owner equity be part of a complete set of financial records. This statement shows the sources of changes in owner equity and the amount that came from each source. It also serves to reconcile the beginning and ending owner equity.

An Example

The balance sheet for I. M. Farmer in Table 4-4 shows an owner equity of $736,334 using market-basis valuation. However, it does not show how much this total changed over the past year. With a balance sheet from a year earlier, the change could be calculated by subtraction, but the sources or causes of any change would still not be evident. To show the sources and amounts of any changes, a statement of owner equity is needed.

Table 4-7 is an example of a statement of owner equity for I. M. Farmer. It shows an owner equity of $714,237 at the beginning of Year 2017, which increased to $736,334 at the end of the year. Where did this increase of $22,097 come from? An item that always affects equity is net farm income for the year, which in this case was $46,286. However, cash income taxes were paid during the year and a change in taxes payable amounted to $12,800. Therefore, the net effect of the year's after-tax profit was a $33,486 increase in equity.

Any money I. M. Farmer takes from the business for personal or other use reduces farm equity. A $1,600 increase in the current portion of deferred income taxes, associated with higher values of current assets and/or lower levels of debts for current expenses over the accounting period, reduces net worth and is entered as a negative number. This statement shows that $36,000 was withdrawn from the business during the year. However, nonfarm income of $9,500 was contributed to the business, leaving a net withdrawal of $26,500. This net withdrawal is shown as a negative amount because it reduces equity. Possible changes in equity can also come from other contributions to or from the business. Contributions to the business may be in the form of gifts or inheritances of either cash or property. In a similar manner, property may be removed from the business and gifted to someone else or converted to personal use rather than business use. Contributions to the business would increase equity, and contributions from the business would decrease equity.

A statement of owner equity for someone using cost-basis balance sheets would end at this point, showing an increase in equity of $5,386. The only things affecting equity on a

TABLE 4-7	Statement of Owner Equity for I. M. Farmer for Year Ending December 31, 2017 (Market Valuation)

Owner equity, January 1, 2017		$714,237
Net farm income for 2017	46,286	
Less adjustment for income taxes paid and payable	(12,800)	
Net after-tax farm income		33,486
Less increase in current portion—deferred income taxes		(1,600)
Owner withdrawals from farm business	(36,000)	
Nonfarm income contributed to farm business	9,500	
Net owner withdrawals from farm business		(26,500)
Other capital contributions to farm business		0
Other capital distributions from farm business		0
Increase in market value of farm assets	19,660	
Less increase in noncurrent portion of deferred income taxes	(2,949)	
Net increase in valuation equity		16,711
Owner equity, December 31, 2017		$736,334

cost-basis balance sheet are net farm income, owner withdrawals, and other contributions to and from the business. However, changes in the market value of assets will affect equity when market values are used.

This balance sheet assumes that the increase in market values for the year totaled $19,660. An increase in land values, an increase in the difference between market value and book value for depreciable assets, and other factors could contribute to this increase. However, should these assets be sold, income taxes would be due on this increase in value, reducing the net effect on equity. Additional income taxes due were estimated to be $2,949. This amount should reflect the increase from January 1 to December 31 in the noncurrent portion of deferred taxes.

Therefore, the net after-tax effect of the increase in market values for the year is an increase of $16,711. This amount, plus the $33,486 after-tax net farm income, less the $26,500 net owner withdrawals, less the $1,600 increase in current portion-deferred taxes, accounts for and reconciles the $22,097 increase in market-value owner equity during 2017.

A statement of owner equity pulls together accounting information from a number of sources to document and reconcile the reasons behind any change in owner equity. Any failure to completely explain the changes indicates a less-than-adequate accounting system. It takes an accounting system that is complete, accurate, detailed, and consistent to provide the information needed.

SUMMARY

A balance sheet shows the financial position of a business at a point in time. It does this by presenting an organized listing of all assets and liabilities belonging to the business. The difference between assets and liabilities is owner equity, which represents the investment the owner has in the business.

An important consideration when constructing and analyzing balance sheets is the method used to value assets. They can be valued using cost methods, which generally reflect the original investment value or cost, or by using current market valuations. The latter generally results in higher asset values, and therefore higher owner equity, but more accurately reflects the current collateral value of the assets. There are advantages to each method, and the FFSC recommends providing both cost and market values on a farm or ranch balance sheet. This provides full information to the user of the balance sheet.

Two factors, liquidity and solvency, are used to analyze the financial position of a business as shown on a balance sheet. Liquidity considers the ability to generate the cash needed to meet cash requirements over the next year without disrupting the production activities of the business. Solvency measures the debt structure of the business and whether all liabilities could be paid by selling all assets; that is, it measures whether assets are greater than liabilities. Several ratios are used to measure liquidity and solvency and to analyze the relative strength of the business in these areas.

A statement of owner equity completes the analysis of a balance sheet by showing the sources and amounts of changes in owner equity during the accounting period. Without the detail shown on this statement, it is difficult to identify and explain what caused owner equity to change and the dollar amount of each change factor.

QUESTIONS FOR REVIEW AND FURTHER THOUGHT

1. True or false? If the debt/equity ratio increases, the debt/asset ratio will also increase. Why?

2. True or false? A business with a higher working capital will also have a higher current ratio. Why?

3. Use your knowledge of balance sheets and ratio analysis to complete the following abbreviated balance sheet. The current ratio = 2.0, and the debt/equity ratio = 1.0.

Assets		Liabilities	
Current assets	$80,000	Current liabilities	_____
Noncurrent assets	_____	Noncurrent liabilities	_____
	_____	Total liabilities	_____
	_____	Owner equity	$100,000
Total assets	_____	Total liabilities and owner equity	_____

4. Can a business be solvent but not liquid? Liquid but not solvent? How?

5. Does a balance sheet show the annual net farm income for a farm business? Why or why not?

6. True or false? Assets + liabilities = Equity.

7. Assume you are an agricultural loan officer for a bank, and a customer requests a loan based on the following balance sheet. Conduct a ratio analysis and give your reasons for granting or denying an additional loan. What is the weakest part of this customer's financial condition?

Assets		Liabilities	
Current assets	$ 40,000	Current liabilities	$ 60,000
Noncurrent assets	$240,000	Noncurrent liabilities	$ 50,000
		Total liabilities	$110,000
		Owner equity	$170,000
Total assets	$280,000	Total liabilities plus equity	$280,000

8. Why is there no valuation adjustment on a cost-basis balance sheet?

9. Assume that a mistake was made and the value of market livestock on a balance sheet is $10,000 higher than it should be. How does this error affect the measures of liquidity and solvency? Would results be the same if land had been overvalued by $10,000?

10. Would the following entries to a farm balance sheet be classified as assets or liabilities? As current or noncurrent?

   a. Machine shed

   b. Feed bill at local feed store

   c. A 20-year farm mortgage contract

   d. A 36-month certificate of deposit

   e. Newborn calves

David Hughes/Getty Images

THE INCOME STATEMENT AND ITS ANALYSIS

CHAPTER OUTLINE

Identifying Revenue and Expenses
Depreciation
Income Statement Format
Accrual Adjustments to a Cash-Basis Income
    Statement
Analysis of Net Farm Income
Change in Owner Equity
Statement of Cash Flows
Summary
Questions for Review and Further Thought

CHAPTER OBJECTIVES

1. Discuss the purpose and use of an income statement

2. Illustrate the structure and format of an income statement

3. Define depreciation and illustrate the different methods of computing depreciation

4. Define the sources and types of revenue and expenses that should be included on an income statement

5. Show how net farm income is computed from an income statement, what it means and what it measures

6. Analyze farm profitability by computing returns to assets and equity and by studying other measures of profitability

An income statement is a summary of revenue and expenses for a given accounting period. It is sometimes called an operating statement or a profit-and-loss statement. However, income statement is the preferred term and is used by the Farm Financial Standards Council (FFSC).

77

Its purpose is to measure the difference between revenue and expenses. A positive difference indicates a profit, or a positive net farm income, and a negative value indicates a loss, or a negative net farm income, for the accounting period. Therefore, an income statement answers the question: Did the farm or ranch business have a profit or loss during the last accounting period, and how large was it?

A balance sheet and an income statement are two different, yet related, financial statements. A balance sheet shows financial position at a point in time, while an income statement is a summary of the revenue and expenses as recorded over a *period of time*. This distinction is shown in Figure 5-1. Using a calendar-year accounting period, a balance sheet is prepared at the end of each year, and it can also serve as the balance sheet for the beginning of the following year. The result is a record of financial position at the beginning and end of each accounting period. However, comparing these balance sheets does not permit a direct calculation of the net farm income for the year. This is the purpose of the income statement.

Even though the balance sheet and the income statement contain different information and have different purposes and uses, they are both financial statements for the same business. It seems only logical that revenue, expenses, and the resulting profit or loss affect the financial position of the business, and indeed they do. However, an explanation of this relationship will be delayed until the end of this chapter, after a complete discussion of the structure and components of an income statement.

IDENTIFYING REVENUE AND EXPENSES

To construct an income statement showing the difference between revenue and expenses, a necessary first step is to identify all revenue and expenses that should be included in the computation of net farm income. The discussion of cash versus accrual accounting in Chapter 3 introduced some of the difficulties that might be encountered in this seemingly simple task. Chapter 4 introduced the problem of asset valuation and introduced the concept of depreciation. This chapter will expand on these earlier discussions before introducing the format and construction of an income statement.

Revenue

An income statement should include all business revenue earned during the accounting period but no other revenue. The problem is one of determining when revenue should be recognized, that is, in what accounting period it was earned. This problem is further compounded because revenue can be either cash or noncash.

When revenue is received in the form of cash for a commodity produced and sold within the same accounting period, recognition is easy and straightforward. However, revenue should also be recognized whenever an agricultural commodity is ready for sale. Inventories of grain and market livestock fit this classification, and changes in these inventories (ending value minus beginning value) are included on an accrual

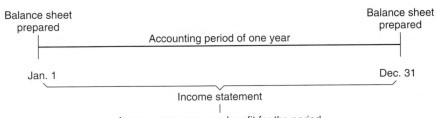

Figure 5-1 **Relationship between balance sheet and income statement.**

Box 5-1   Are All Cash Receipts Counted as Revenue?

The answer is NO! Not all receipts of cash result in revenue. Cash received from a gift or an inheritance is obviously not business revenue. A new loan from the bank results in the receipt of cash but is also not revenue. Nonfarm income such as salary and investment income would not be counted as farm business revenue. Sales of land or depreciable assets result in revenue only to the extent of any gain or loss realized. Farm business revenue results only from the production of agricultural commodities, services performed, and the gain or loss on the sale of assets used in that production.

---

income statement. Accounts receivable represent earned revenue, which should be recognized, but they are revenue for which a cash payment has not yet been received. Any change in their value from the beginning to the end of the year must be included as revenue. These two items represent sources of revenue that may be recognized in one accounting period even though cash will not be received until a later accounting period.

Payment may sometimes be received in the form of goods or services instead of cash, such as feed or livestock received for performing custom work or other services. This noncash payment will eventually show up as income when the products received are sold, or indirectly as reduced expenses for feed or other inputs.

Gain or Loss on Sale of Capital Assets

The gain or loss on the sale of a capital asset is an entry that often shows up in the revenue section of an income statement. It is the difference between the sale price and the cost of a capital asset such as land. For depreciable assets, such as machinery, orchards, and purchased breeding livestock, it is the difference between the selling price and the asset's book value. Gain or loss is recognized only when an asset is actually sold. Before then, the market value or selling price is subject to considerable uncertainty.

Any gain or loss on the sale of a nondepreciable asset such as land is the direct result of an increase or a decrease in the market price since its purchase. For depreciable assets, changes in market value affect gains and losses, as well as how accurately the depreciation was estimated. If an asset is sold for exactly its book value, there is no gain or loss. The depreciation over its life has perfectly matched its decline in market value. A sale price higher than book value implies that too much total depreciation has been deducted (perhaps due to unanticipated strength in market values) and that past annual depreciation expenses have been too high. Recognizing the gain as revenue adjusts for the fact that depreciation expense was not correct in past years. Selling a depreciable asset for less than its book value means that it lost market value faster than what was accounted for by the annual depreciation. There should have been a higher depreciation expense in the past, and an adjustment is made in the form of a *loss on sale* entry on the income statement.

Given that (1) useful lives and salvage values are only estimates made at the time of purchase and (2) the choice of depreciation method will affect the amount of annual depreciation, there will generally be a gain or loss to be recognized when a depreciable asset is sold.

When a capital asset that is already owned is traded for another similar asset, such as a new equipment item, no gain or loss is recognized. The book value of the asset that is traded is added to the dollars paid to complete the trade to calculate the initial value of the new item. For example, if a farmer trades a used sprayer with a book value of $60,000 for a new sprayer, and needs to pay the dealer an additional $120,000 in cash, the beginning value of the new item is $180,000. A capital gain or loss is recognized only when a capital asset is sold outright.

Expenses

Once all revenue for an accounting period has been identified, the next step is to identify all expenses incurred in producing that revenue. They may be either cash or noncash expenses. For example, cash expenses would include purchases of and payment for feed, fertilizer, seed, market livestock, and fuel. Noncash expenses would include depreciation, accounts payable, accrued interest, and other accrued expenses. There is also an adjustment for prepaid expenses.

Depreciation is a noncash expense that reflects decreases in the value of assets used to produce the revenue. Depreciation will be discussed in detail later in this chapter. Accounts payable, accrued interest, and other accrued expenses such as property taxes are expenses incurred during the past accounting period but not yet paid. To properly match expenses with the revenue they helped produce, these expenses must be included on the current year's income statement. They must also be subtracted from next year's income statement, when the cash will be expended for their payment. The difference in timing between the year in which the expense was incurred and the one in which it will be paid for creates the need for these entries.

Accounts payable and accrued expenses are paid for in an accounting period later than the one in which the products or services were used. This is opposite from prepaid expenses, goods and services paid for in one year but not used to produce revenue until next year. Examples of prepaid expenses are seed, fertilizer, pesticides, and feed purchased and paid for in December to take advantage of price discounts and income tax deductions or to assure availability. However, because they will not be used until the next calendar year, the expense should be deferred until then to properly match expenses with the revenue produced by this expense. The timing of paying the expense and using the product is just opposite of an account payable, so the accounting procedure is opposite. Prepaid expenses should not be included in the current year's expenses as they did not produce any revenue in the current year. They should be included with next year's expenses, ensuring that the business records properly match expenses with their associated revenue in the same period.

Income taxes can be included on an income statement, making the final result *after-tax* net farm income. However, taxes due on farm income

---

Box 5-2   Is Every Expenditure of Cash an Expense?

The answer is NO! Not all cash expenditures are business expenses. Cash expenditures for food, clothing, gifts, and other personal items are not business expenses. Business expenses are only those items required to produce agricultural commodities and services and, hence, revenue. Principal paid on loans is not an expense, because it represents only the return of borrowed property. Interest is an expense, however, because it is the *rent* paid for the use of the borrowed property. Cash paid to purchase depreciable assets is not an expense in the year of purchase. However, it is converted into an expense over time through an annual depreciation expense.

can be difficult to estimate, particularly when there is also off-farm income to consider. For simplicity and other reasons, the examples used in this text will omit income taxes from the income statement and concentrate on estimating a *before-tax* net farm income. All discussion and use of the term *net farm income* refers to before-tax net farm income, unless stated otherwise.

Payments of personal income taxes on total family income, including farm income, are sometimes included with family living expenses as cash withdrawals from the farm business.

DEPRECIATION

Machinery, buildings, and similar assets are purchased because they are required or helpful in the production of farm products, which, in turn, produce revenue. Their use in the production process over time causes them to grow old, wear out, and become less valuable. This loss in value is considered a business expense, because it is a direct result of the asset's use in producing revenue and profit.

Depreciation is often defined as the annual loss in value due to use, wear, tear, age, and technical obsolescence. It is both a business expense that reduces annual profit and a reduction in the value of the asset. What types of assets would be depreciated? To be depreciable, an asset must have the following characteristics:

1. A useful life of more than one year
2. A determinable useful life but not an unlimited life
3. Use in a business for the depreciation to be a *business* expense (loss in value on a personal automobile or personal residence is not a business expense)

Examples of depreciable assets on a farm or ranch are vehicles, machinery, equipment, buildings, fences, livestock and irrigation wells, and purchased breeding livestock. Land is not a depreciable asset, because it has an unlimited life. However, some improvements to land, such as drainage tile, can be depreciated.

How much does an asset depreciate each year? Several methods or mathematical equations

---

Box 5-3 Quick Estimate of Economic Depreciation

A quick and easy way to estimate economic depreciation, used by the Iowa Farm Business Association, is to add together all machinery and equipment values at the beginning of the year; adjust for purchases, trades, or sales; and then take 10 percent of the value. An advantage of using this method is that it is not necessary to track the value of each item.

For machinery and equipment, the formula is:

Economic depreciation = (beginning value + purchases or trades – sales) × 10%

For example, if the total value of machinery and equipment on January 1 is $345,000, one item valued at $80,000 is acquired, and another item valued at $50,000 is sold, economic depreciation

for all machinery and equipment would be approximated as

($345,000 + $80,000 – $50,000) × 10% = $37,500.

This method approximates a 10-year life 100 percent declining balance method.

After accounting for depreciation, the cost value of machinery and equipment at the end of the year would be

($345,000 + $80,000 – $50,000 – $37,500) = $337,500.

For buildings, the Association uses the same type of formula, but with a 5 percent factor to approximate a 20-year declining balance calculation.

can be used to compute annual depreciation. However, it should always be remembered that they are only estimates of the actual loss in value. The true depreciation can be determined only by finding the asset's current market value and comparing it to the original cost. This would require an appraisal or the sale of the asset.

Several terms used with depreciation must be defined before reviewing the depreciation methods. *Cost* is the price paid for the asset, including taxes, delivery fees, installation, and any other expenses directly related to placing the asset into use. *Useful life* is the number of years the asset is expected to be used in the business. This may be something less than the total potential life of the asset if it will be traded or sold before it is completely worn out. *Salvage value* is the expected market value of the asset at the end of its assigned useful life. Therefore, the difference between cost and salvage value is the total depreciation or loss in value expected over the useful life.

A salvage value will generally be some positive value. However, it may be zero if the asset will be used until it is completely worn out and will have no scrap or junk value at that time. There should be a relation between useful life and salvage value. The shorter the useful life, the higher the salvage value, and vice versa.

Book value is another term related to depreciation. It is equal to the asset's cost less accumulated depreciation. The latter is all depreciation from purchase date to the current date. Book value will always be somewhere between cost and salvage value. It will never be less than salvage value, and it will equal salvage value at the end of the asset's useful life. Although book value is one way to determine the value of an asset, it should not be confused with market value. Both useful life and salvage value are only estimates made at the time the asset is purchased. It is unlikely that either or both will turn out to be the actual values. Therefore, book value and market value will be equal only by chance.

Depreciation Methods

Several methods or equations can be used to compute annual depreciation. No single correct choice exists for every business or every asset. The correct choice will depend on the type of asset, its pattern of use over time, how rapidly or slowly its market value declines, and other factors. The two most common depreciation methods, which may be used for managerial purposes, are illustrated in this section. These methods also form the basis of the depreciation methods used for income tax purposes, to be discussed in Chapter 16.

Straight Line

The straight-line method of calculating depreciation is widely used. This easy-to-use method gives the same annual depreciation for each full year of an item's life.

Annual depreciation can be computed from the equation

$$\frac{\text{Annual}}{\text{depreciation}} = \frac{\text{cost} - \text{salvage value}}{\text{useful life}}$$

Straight-line depreciation can also be computed by an alternative method using the equation

$$\frac{\text{Annual}}{\text{depreciation}} = (\text{cost} - \text{salvage value}) \times R$$

where R is the annual straight-line percentage rate found by dividing 100 percent by the useful life. For example, assume the purchase of a machine for \$100,000 that is assigned a \$20,000 salvage value and a 10-year useful life. The annual depreciation using the first equation would be

$$\frac{\$100,000 - \$20,000}{10 \text{ years}} = \$8,000$$

Using the second equation, the percentage rate would be 100 percent divided by 10, or 10 percent, and the annual depreciation is

$$(\$100,000 - \$20,000) \times 10\% = \$8,000$$

The result is the same for either procedure, and the total depreciation over 10 years would be $8,000 × 10 years = $80,000, reducing the machine's book value to its salvage value of $20,000.

Declining Balance

There are a number of variations or types of declining-balance depreciation. The basic equation for all types is

$$\text{Annual depreciation} = \text{(book value at beginning of year)} \times R$$

where R is a constant percentage value or rate.

The same R value is used for each year of the item's life and is multiplied by the book value, which declines each year by an amount equal to the previous year's depreciation. Therefore, annual depreciation declines each year with this method. The percentage rate is multiplied by each year's book value, *not* cost minus salvage value as was done with the straight-line method.

The various types of declining balance come from the determination of the R value. For all types, the first step is to compute the straight-line percentage rate, 10 percent in our previous example. The declining balance method then uses a multiple of the straight-line rate, such as 200 (or double), 150, or 100 percent, as the R value. If double declining balance is chosen, R would be 200 percent, or two times the straight-line rate. The R value would be determined in a similar manner for the other variations of declining-balance depreciation.

Using the previous example, the double declining balance rate would be two times 10 percent, or 20 percent, and the annual depreciation would be computed in the following manner:

Year 1: $100,000 × 20% = $20,000
Year 2: $80,000 × 20% = $16,000
Year 3: $64,000 × 20% = $12,800
Year 4: $51,200 × 20% = $10,240
Year 5: $40,960 × 20% = $8,192

Year 6: $32,768 × 20% = $6,554
Year 7: $26,214 × 20% = $5,243
Year 8: $20,972 × 20% = $4,194

In Year 8, if $4,194 in depreciation were taken, the book value would drop to $16,777, $3,223 less than the salvage value of $20,000. Accordingly, depreciation in Year 8 must be adjusted to $972, the amount of depreciation remaining before the salvage value is reached. In Years 9 and 10, there is no remaining depreciation.

This example is not unusual, as double declining balance, with a nonzero salvage value, will often result in the total allowable depreciation being taken before the end of the useful life, and depreciation must stop when the book value equals salvage value.

Notice also that the declining-balance method will never reduce the book value to zero. With a zero salvage value, it is necessary to switch to straight-line depreciation at some point to get all the allowable depreciation or to take all remaining depreciation in the past year. The typical practice is to switch to straight line on the remaining value for the remaining life when equal or higher depreciation results compared to continuing to use double declining balance. If the $100,000 machine in the earlier example had a zero salvage value, then in Year 7, applying the straight-line method to the remaining value ($26,214) for the remaining life (four years) would result in depreciation of $6,554, the same as double declining balance, and the switch would be made then. Under this method, for Years 7, 8, 9, and 10, annual depreciation would then be $6,554 and a zero salvage value would be reached.

If a 150 percent declining balance is used, R is one and one-half times the straight-line rate, or 15 percent for this example. The annual depreciation for the first three years would be

Year 1: $100,000 × 15% = $15,000
Year 2: $85,000 × 15% = $12,750
Year 3: $72,250 × 15% = $10,838

Each year's depreciation is smaller than if the double declining balance is used, and therefore, the book value declines at a slower rate.

Comparing Depreciation Methods

Figure 5-2 graphs the annual depreciation for both depreciation methods illustrated based on a $100,000 asset with a $20,000 salvage value and a 10-year life. The annual depreciation is different for the two methods. Double declining balance has a higher annual depreciation in the early years than straight line, with the reverse being true in the later years.

The choice of depreciation method does not change the *total* depreciation over the useful life. That is determined by the cost and salvage value ($100,000 − $20,000 = $80,000 in this example) and is the same regardless of the depreciation method. The different methods only spread or allocate the $80,000 in a different pattern over the 10-year life. The most appropriate

depreciation method will depend on the type of asset and the use to be made of the resulting book value. For example, the market value of vehicles, tractors, and other motorized machinery tends to decline most rapidly during the first few years of life and more slowly in the later years. If it is important for depreciation on these items to approximate their decline in value as closely as possible, declining balance should probably be used. Assets such as fences and buildings have little or no market value without the land they are attached to, and they provide a rather uniform flow of productive services over time. Straight-line depreciation might be the more appropriate depreciation method to use. Since the amount of annual depreciation depends on the depreciation method used, so does the book value at the end of each year. With straight line, book value declines by the constant annual depreciation each year, that is, from $100,000 to $20,000 over the useful life, as shown in Figure 5-3. Using double declining

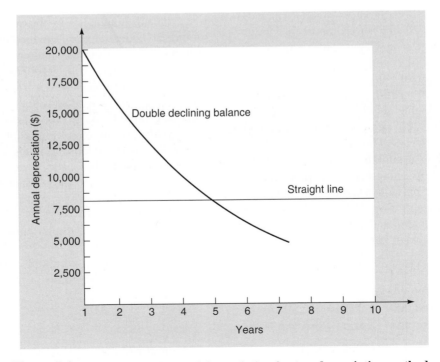

Figure 5-2 Comparison of annual depreciation for two depreciation methods.

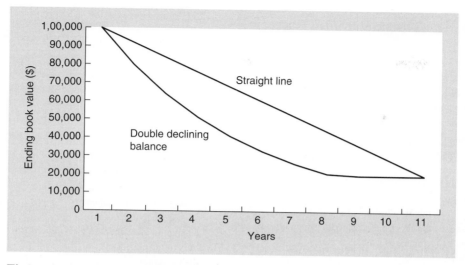

Figure 5-3 **Comparison of ending book value for two depreciation methods.**

balance causes book value to decrease faster in the early years because annual depreciation is higher. Book value will decline more slowly in the later years of useful life than for straight line. Regardless of the method used, book value will be equal to salvage value at the end of the useful life. It should also be noted that the Internal Revenue Service (IRS) has specific methods of deprecation that must be used, which may not reflect economic deprecation. These methods will be discussed in Chapter 16.

Partial-Year Depreciation

The example used here assumed that the asset was purchased at the beginning of the year, with a full year's depreciation that year. An asset purchased during the year should have the first year's depreciation prorated according to the length of time it was owned. For example, a tractor purchased on April 1 would be eligible for 9/12 of a full year's depreciation, and a pickup purchased on October 1 would get only 3/12 of a year's depreciation in the year of purchase. Any time there is a partial-year depreciation the first year, there will be less than a full year of depreciation in the final year of depreciation as

well. The IRS has specific rules for depreciation of items based on the date of purchase, which will be discussed in Chapter 16.

INCOME STATEMENT FORMAT

Most of the entries on an income statement have already been discussed, but several items need additional discussion. The outline of an income statement shown in Table 5-1 follows one of the formats recommended by the FFSC. In a very condensed form, the basic structure is

Gross revenue

Less total expenses

Equals net farm income from operations

Plus or *minus* gain/loss on sale of capital assets

Equals net farm income

Any gain/loss on the sale of culled breeding livestock is shown in the revenue section, but other gains/losses are included in a separate section at the end of the income statement. The reason for including the former in revenue is that the sale of culled breeding livestock is a normal,

TABLE 5-1	Income Statement Format

Revenue:

Cash crop sales
Cash livestock sales
Livestock product sales
Government program payments
Other farm income
Adjustments
    Change in value of crops inventory
    Change in value of market livestock inventory
    Change in value of raised breeding stock
    Gain/loss from sale of culled breeding stock
    Change in accounts receivable
Gross revenue

Expenses:

Purchased feed and grain
Purchased market livestock
Other cash operating expenses:
    Crop expenses
    Livestock expenses
    Fuel, oil
    Labor
    Repairs, maintenance

Property taxes
Insurance
Cash land rent
Other: _____
     _____
     _____
Adjustments
    Change in accounts payable
    Change in accrued expenses
    Change in prepaid expenses
    Change in unused supplies
    Change in investment in growing crops
Total operating expenses
    Depreciation
Cash interest paid
    Change in accrued interest
Total interest expense
Total expenses
Net farm income from operations
Gain/loss on sale of capital assets:
    Machinery
    Land
    Other
Net farm income

The "Line" (handwritten annotation)

expected, and usual part of the ongoing production activities of the business. While the FFSC specifies including all gains/losses in net farm income, it recommends that those for machinery and land—along with improvements such as buildings and permanent crops—be shown separately. Any gains/losses from the sale of these assets are less frequent, result from investment rather than production activities, and may be large. Therefore, net farm income from operations is the profit from normal, ongoing production activities. Net farm income includes this amount, plus any gains/losses on the sale of land and its improvements, buildings, and machinery. Showing these items separately toward the end of the income statement allows the

user to determine easily whether they had an unusually large effect on net farm income.

Cash interest paid, accrued interest, and total interest are shown separately from other expenses. While often categorized as an operating expense, the position of the FFSC is that interest is a result of financing activities rather than production activities. Therefore, it should not be included with the direct operating expenses associated with the production of crops and livestock. This separation also makes it easy for the user to find and note the amount of interest expense when conducting an analysis of the income statement.

Some income statement formats compute *value of farm production* as an intermediate step.

In a condensed form, this format has the following structure:

> Gross revenue
>
> *Less* cost of purchased feed and grain
>
> *Less* cost of purchased market livestock
>
> *Equals* value of farm production
>
> *Less* all other expenses
>
> *Equals* net farm income from operations
>
> *Plus* or *minus* gain/loss on sale of capital assets
>
> *Equals* net farm income

The value of farm production measures the dollar value of all goods and services produced by the farm business. Any purchased feed, grain, and market livestock were produced by some other farm or ranch and should not be credited to this business. Subtracting the cost of these items from the gross revenue results in a *net* production value. Any increase in value obtained by feeding purchased feed and grain to purchased livestock will end up being credited to the business being analyzed through sales revenue or an inventory increase.

The value of farm production is a useful measure of farm size when analyzing a farm business and is used to compute other measures as well. Although the FFSC does not require use of the format showing value of farm production, it does recognize its usefulness. Therefore, for the format shown in Table 5-1, it recommends that the cost of purchased feed and grain and purchased market livestock be shown first, separate from other operating expenses. This makes it easy for the user to find the values needed to compute the value of farm production.

ACCRUAL ADJUSTMENTS TO A CASH-BASIS INCOME STATEMENT

While the FFSC encourages the use of accrual accounting, it recognizes that cash accounting is used currently and will probably continue to be used by a majority of farmers and ranchers for some time. Simplicity and advantages for income tax purposes explain its popularity. However, net farm income from a cash accounting system can be misleading and can result in poor decisions when used for management purposes.

The FFSC recommends that anyone using cash accounting convert the resulting net farm income to an *accrual-adjusted* net farm income at the end of each year. This value can then be used for analysis and decision making. A cash-basis income statement includes only cash receipts, cash expenses, and depreciation, so a number of adjustments have to be made. Figure 5-4 shows the necessary adjustments and the recommended procedure.

Adjustments to Receipts

Two adjustments are made to cash receipts: change in inventory values and accounts receivable. Adjustments should be made for inventories of grain, market livestock, and raised breeding livestock. The beginning inventory value is subtracted from cash receipts, because this inventory has been sold (or otherwise consumed) during the year, and the sales have been included as cash receipts. The ending inventory value is added to cash receipts, because this is the current year's production that has not yet been sold.

The process for accounts receivable is the same. Beginning accounts receivable are subtracted from cash receipts, because this amount was collected during the year and recorded as cash receipts. Ending accounts receivable are added, because they represent production during the past year that has not yet shown up in cash receipts. An alternative procedure for inventories and accounts receivable is to subtract beginning values from ending values and enter only the difference or change. This change can be either a positive or a negative adjustment to cash receipts.

Adjustments to Expenses

Several adjustments are made to cash disbursements or expenses. Some of these involve expenses that have been incurred but not paid

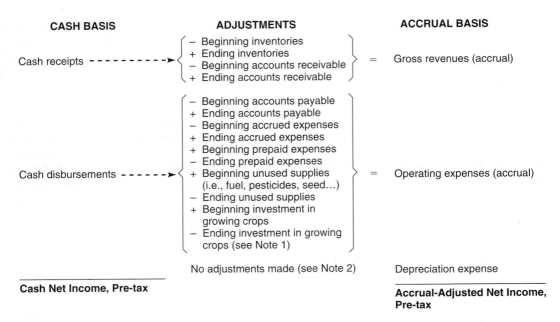

CASH BASIS   **ADJUSTMENTS**   **ACCRUAL BASIS**

Cash receipts - - - - - - - - - ▸ { − Beginning inventories
                                    + Ending inventories
                                    − Beginning accounts receivable } =   Gross revenues (accrual)
                                    + Ending accounts receivable

Cash disbursements - - - - - - ▸ { − Beginning accounts payable
                                    + Ending accounts payable
                                    − Beginning accrued expenses
                                    + Ending accrued expenses
                                    + Beginning prepaid expenses
                                    − Ending prepaid expenses
                                    + Beginning unused supplies   } =   Operating expenses (accrual)
                                      (i.e., fuel, pesticides, seed…)
                                    − Ending unused supplies
                                    + Beginning investment in
                                      growing crops
                                    − Ending investment in growing
                                      crops (see Note 1)

                                    No adjustments made (see Note 2)   Depreciation expense
_____   _____
Cash Net Income, Pre-tax   **Accrual-Adjusted Net Income,
                                                                          Pre-tax**

Note 1: Remember to avoid double-counting items included in prepaid expenses and unused supplies when determining the investment in growing crops.

Note 2: Because depreciation is a noncash expense, it is not reflected on a cash-basis income statement. However, the IRS allows capital assets to be depreciated even for cash-basis taxpayers.

Figure 5-4 **Adjustments to get accrual-adjusted net farm income from a cash-basis income statement.**
Source: Adapted from *Financial Guidelines for Agricultural Producers*, Recommendations of the Farm Financial Standards Council (Revised), 2011.

yet, including accounts payable and accrued expenses. They are often the largest expense adjustments, and accrued interest typically will be the largest accrued expense. Both accounts payable and accrued expenses are expenses that helped produce income during the accounting year but for which no cash has yet been spent. The cash will be spent in the next year. Making the adjustments, as shown in Figure 5-4, will get these expenses into the proper year, the year they helped produce revenue.

Adjustments are also made to expenses that are paid in advance, such as prepaid expenses, supplies, and investment in growing crops. For expenses that are paid in advance, there is a different type of adjustment. The beginning values are added to cash disbursements, while the ending values are subtracted. This procedure is

different from the unpaid expense adjustments, because the timing of the cash expenditures is different. For prepaid expenses, inventory of supplies, and investment in growing crops, cash was spent during the past accounting period, but it will not produce revenue until the next period. The different signs on the beginning and ending values correctly account for the difference in timing and result in the correct accrual-adjusted operating expenses.

This adjustment process is rather straightforward but does require some effort and attention to detail. It is necessary to know the dollar value of each item at the beginning of the year and again at the end of the year. These values are not part of a cash accounting system, so they must be measured and recorded in some other way. This adjustment process results in the correct accrual-adjusted net farm income only if all measurements recorded have the same values as they would have had if an accrual accounting system had been used. The most accurate method, and one that will keep all financial records consistent and correlated, is to use values from the beginning- and end-of-the-year balance sheets.

Table 5-2 shows the calculation of the accrual adjustments that I. M. Farmer will need to make before completing his income statement. To complete an accrual adjustment table, values are taken from the two balance sheets corresponding to the first and last day of the accounting period (in this case a year) and the difference in these values is found. For example, crop inventories were valued at $32,000 at the beginning of the year and at $40,640 at the end of the year, an $8,640 increase. This increase is shown as a positive number in the column reflecting the change in value. Similar calculations are made for all other accrual adjustments. For accounts payable, for example, the beginning value is subtracted and the ending value is added, as indicated in

Figure 5-4. For prepaid expenses, by contrast, the beginning value is added and the ending value is subtracted to find the accrual adjustment, also as indicated in Figure 5-4.

ANALYSIS OF NET FARM INCOME

Table 5-3 contains a complete accrual-basis income statement for I. M. Farmer. It shows a net farm income from operations of $45,486 and a net farm income of $46,286. The business shows a profit for the year, but is it a *profitable* business? Profitability is concerned with the size of the profit relative to the size of the business. Size is measured by the value of the resources used to produce the profit. A business can show a profit but can have a poor profitability rating if this profit is small relative to the size of the business. For example, two farms with the same net farm income are not equally profitable if one used twice as much land, labor, and capital as the other to produce that profit.

Profitability is a measure of the efficiency of the business in using its resources to produce

TABLE 5-2	Accrual Adjustments for Income Statement for I. M. Farmer			
	Beginning of the year value ($)	End of the year value ($)		Change in value ($)
Revenue items				
Crop inventories	−32,000	+40,640	=	8,640
Market livestock inventories	−22,000	+22,400	=	400
Accounts receivable	−2,120	+3,570	=	1,450
Expense items accrued but not paid				
Accounts payable	−3,620	+3,500	=	−120
Accrued expenses	−1,290	+1,800	=	510
Accrued interest	−12,080	+12,300	=	220
Expense items paid but not accrued				
Prepaid expenses	+3,646	−3,780	=	−134
Unused supplies	+520	−860	=	−340
Investment in growing crops	+5,000	−5,000	=	0

TABLE 5-3	Income Statement for I. M. Farmer for Year Ending December 31, 2017

Revenue:

Cash crop sales	$284,576
Cash livestock sales	88,112
Livestock product sales	0
Government program payments	22,500
Other farm income	0
Inventory changes:	
Crops	8,640
Market livestock	400
Change in value of raised breeding stock	0
Gain/loss from sale of culled breeding stock	870
Change in accounts receivable	1,450
Gross revenue	$406,548

Expenses:

Purchased feed and grain	11,550
Purchased market livestock	22,720
Other cash operating expenses:	
Crop expenses	190,940
Livestock expenses	43,845
Fuel, oil	31,840
Labor	0

Repairs, maintenance	7,960	
Property taxes	2,700	
Insurance	3,240	
Cash land rent	8,500	
Other: utilities	3,780	
Adjustments		
Change in accounts payable	(120)	
Change in accrued expenses	510	
Change in prepaid expenses	(134)	
Change in unused supplies	(340)	
Change in investment in growing crops	0	
Depreciation	16,000	
Total operating expenses		$342,991
Cash interest paid	17,851	
Change in accrued interest	220	
Total interest expense		$ 18,071
Total expenses		$361,062
Net farm income from operations		$ 45,486
Gain/loss on sale of capital assets		
Machinery	800	
Land	0	
Other	0	
Net farm income		$ 46,286

profit or net farm income. The FFSC recommends five measures of profitability: (1) net farm income, (2) rate of return on farm assets, (3) rate of return on farm equity, (4) operating profit margin ratio, and (5) earnings before interest, taxes, depreciation, and amortization (EBITDA). It is also possible to compute a return to unpaid operator labor and to management alone, as other measures of profitability.

Net Farm Income

As shown on the income statement, net farm income is the amount by which revenue exceeds expenses, plus any gain or loss on the sale of capital assets. It can also be thought of as the

amount available to provide a return to the operator for the unpaid labor, management, and equity capital used to produce that net farm income. As discussed earlier, net farm income is an absolute dollar amount, making it difficult to use it by itself as a measure of profitability. It should be considered more as a starting point for analyzing profitability than as a good measure of profitability itself.

Rate of Return on Farm Assets

Return on assets (ROA) is the term used by the FFSC, but the same concept has been called return to capital, or return on investment (ROI). It measures profitability with a ratio obtained by dividing the dollar return to assets by the average farm asset value for the year. The latter value is found by averaging the beginning and ending total asset values from the farm's balance sheets. Expressing ROA as a percentage allows an easy comparison with the same values from other farms, over time for the same farm, and with returns from other investments. The equation is

$$\text{Rate of return on assets } (\%) = \frac{\text{return to assets } (\$)}{\text{average assets } (\$)}$$

The ROA is the dollar return to debt and equity capital, so net farm income from operations must be adjusted. Interest on debt capital was deducted as an expense when calculating net farm income from operations. Any such interest must be added back to net farm income from operations when computing the return to assets. This step eliminates any expense that comes from the type and amount of financing used by the business. The result is equal to what net farm income from operations *would have been* if there had been no debt capital and therefore no interest expense. For I. M. Farmer, the computations are

Net farm income from operations	$45,486
Plus interest expense	$18,071
Equals adjusted net farm income from operations	$63,557

This is what Farmer's net farm income from operations would have been if there had been no borrowed capital, or 100 percent equity in the business.

It is also necessary to make an adjustment for the unpaid labor and management provided by the farm operator and the farm family. Without this adjustment, that part of net farm income from operations credited to assets would also include the contribution of labor and management toward earning that income. Therefore, a charge for unpaid labor and management must be subtracted from adjusted net farm income from operations to find the actual return to

assets in dollars. There is no way of knowing exactly what part of that income was earned by labor and management, so opportunity costs are used as estimates. Assuming an opportunity cost of $40,000 for the unpaid labor provided by I. M. Farmer and other family members, and $5,000 for management, the calculations are

Adjusted net farm income from operations	$63,557
Less opportunity cost of unpaid labor	–40,000
Less opportunity cost of management	–5,000
Equals return to assets	$18,557

The final step is to convert this dollar return to assets into a percentage of total assets, the dollar value of all capital invested in the business. An immediate problem is which total asset value to use, cost or market basis? Although either can be used, the market-basis value is generally used, because it represents the current investment in the business and allows a better comparison of ROAs between farms. It also makes the resulting ROA comparable to the return that these assets could earn in other investments should the assets be converted into cash at current market values and invested elsewhere. However, cost basis provides a better indicator of the return on the actual funds invested and is the better indicator for trend analysis.

As shown in Chapter 4, I. M. Farmer's total asset value, on a market basis, was $1,068,750 on December 31, 2017. Assuming a value of $1,030,755 on January 1, 2017, the average asset value over this period was $1,049,753. Therefore, the rate of return on I. M. Farmer's assets for 2017 was

$$\text{ROA} = \frac{\$18,557}{\$1,049,753} = 1.8\%$$

This ROA of 1.8 percent represents the return on capital invested in the business after adjusting net farm income from operations for the opportunity cost of labor and management. The *profitability* of the farm can now be judged by comparing this ROA to those from similar farms, the returns from other possible investments, the opportunity cost of the farm's capital, and past ROAs for the same farm.

Other Considerations

Several things should be noted whenever ROA is computed and analyzed. First, the FFSC recommends using net farm income from operations rather than net farm income. Any gains or losses on sale of capital assets included in the latter value are sporadic in nature, can be very large, and do not represent income generated by the use of assets in the normal production activities of the business. Therefore, they should not be included in any calculation of how well assets are used in generating profit. Second, the opportunity costs of labor and management are estimates, and changing them will affect ROA. Third, any comparisons of ROAs should be done only after making sure they were all computed the same way and that the same method was used to value assets. Market valuation is recommended for comparison purposes; cost valuation is recommended for checking trends on the same farm. Fourth, the value of any personal or nonfarm assets that might be included on the balance sheet should be subtracted from total assets. Likewise, any nonfarm earnings, such as interest on personal savings accounts, should not be included in the farm income. The computed ROA should be for farm earnings generated by farm assets. Finally, ROA is an average return and not a marginal return. It should not be used when making decisions about investing in additional assets where the marginal return is the important value.

Rate of Return on Farm Equity

The ROA is the return on all assets or capital invested in the business. On most farms and ranches, there is a mixture of debt and equity capital. Another important measure of profitability is the return on equity (ROE), the return on the

owner's share of the capital invested. Should the business be liquidated and the liabilities paid off, only the equity capital would be available for alternative investments.

The calculation of ROE begins directly with net farm income from operations. No adjustment is needed for any interest expense. Interest is the payment for the use of borrowed capital, and it must be deducted as an expense before the ROE is computed. Interest has already been deducted when computing net farm income from operations. However, the opportunity cost of operator and family labor and management again must be subtracted, so the ROE does not include their contribution toward earning income.

Continuing with the I. M. Farmer example, the dollar ROE would be

Net farm income from operations	$45,486
Less opportunity cost of unpaid labor	–40,000
Less opportunity cost of management	–5,000
Equals return on equity	$486

Note that the return to equity can also be computed by subtracting interest expense from the return to assets, $18,557 - 18,071 = \$486$. The rate of ROE is computed from the equation

$$\frac{\text{Rate of return}}{\text{on equity (\%)}} = \frac{\text{return to equity (\$)}}{\text{average equity (\$)}}$$

Average equity for the year is used in the divisor and is the average of beginning and ending market-basis equity for the year. ROE is also expressed as a percent to allow easy comparison among farms and with other investments.

In Table 4.7, I. M. Farmer's ending market basis equity was $736,334. The beginning equity was $714,237, so the average equity value for the year was $725,285. I. M. Farmer's ROE would be

$$\text{ROE} = \frac{\$486}{\$725,285} = 0.1\%$$

The ROE can be either greater or less than the ROA, depending on the ROA's relation to the average interest rate on borrowed capital. If the ROA is greater than the interest rate paid on borrowed capital, this extra margin or return above interest cost accrues to equity capital. Its return becomes greater than the average return on total assets. Conversely, if the ROA is less than the interest rate on borrowed capital, the ROE will be less than the ROA. Some of the equity capital's earnings had to be used to make up the difference when the interest was paid, thereby lowering the ROE. With i being the interest rate on debt, these relations can be summarized as follows:

If ROA $>$ i, then ROE $>$ ROA
If ROA $<$ i, then ROE $<$ ROA

Given either of these relations, the absolute difference between the two values depends on another factor: the amount of debt compared to equity. If ROA $>$ i, and debt is large relative to equity, ROE can be much greater than ROA. The relatively large amount of debt generates many dollars of return above interest cost, which in turn accrues to the return earned by the relatively small amount of equity capital itself. This can result in a large ROE. However, a combination of large relative debt and ROA $<$ i has the opposite effect. All or more of the ROE may be needed to make up the difference between the return earned by the debt capital and the interest paid for it. The ROE can become negative, meaning equity must be used to pay part of the interest. Several years of this relationship can lead to insolvency.

Although, ROE is positive in this example, it is very close to zero, and lower than ROA. Thus, although net farm income for this farm was positive in 2017, the farm does not seem very profitable. In the next chapter, the farm profitability will be compared to other, similar farms, and possible problems will be investigated.

Operating Profit Margin Ratio

This ratio computes operating profit as a percentage of gross revenue. A higher value means the business is making more profit per dollar of revenue. The first step in computing the operating profit margin ratio is to find the absolute dollar value for operating profit. The process is

> Net farm income from operations
> *Plus* interest expense
> *Less* opportunity cost of unpaid labor
> *Less* opportunity cost of management
> *Equals* operating profit

Box 5-4   Sources of Comparative Data

It is often difficult to judge the strength or weakness of the results of a net farm income analysis without comparing the results to those from similar farms. One source of comparative data is farm business management associations, which exist in many states that have a large number of farms. These associations assist farmers and ranchers with their record keeping during the year and then do a complete farm financial analysis for them at year end. Average values and those for the highest and lowest one-quarter or one-third of all association members are generally available to the public. Average values of farms sorted according to size, region, or major enterprise are often available as well. Information on any associations in your state and published information may be obtained from local county extension directors or from the Web site of the agricultural economics department of your state land grant university.

Interest is added back to net farm income from operations to eliminate the effect of debt on operating profit. This allows the operating profit margin ratio to focus strictly on the profit made from producing agricultural commodities without regard to the amount of debt, which can vary substantially from farm to farm. Eliminating this variable permits a valid comparison of this ratio across different farms. To recognize that unpaid labor and management contributed to earning the profit, their opportunity costs are subtracted. This makes the results comparable to those from businesses where all labor and management is hired, as these expenses have already been deducted in the computation of net farm income from operations. Note that the value for operating profit is the same as for the dollar return to assets.

The equation for the operating profit margin ratio is

$$\frac{\text{Operating profit}}{\text{margin ratio}} = \frac{\text{operating profit}}{\text{gross revenue}}$$

I. M. Farmer's operating profit margin ratio would be

$$\frac{\$45,486 + 18,071 - 40,000 - 5,000}{\$406,548} = \frac{0.046 \text{ or}}{4.6\%}$$

This means that, on the average, for every dollar of revenue, 4.6¢ remained as profit after paying the operating expense necessary to generate that dollar. Farms with a low operating profit margin ratio should concentrate on improving this ratio before expanding production. It does little good to increase gross revenue if there is little or no profit per dollar of revenue. Net farm income is related to both the operating profit margin ratio, a measure of profitability, and the gross revenue, which measures volume of business.

Earnings Before Interest, Taxes, Depreciation, and Amortization

The FFSC notes that commercial analysts often begin with earnings before interest, taxes, depreciation, and amortization (EBITDA) as a measure of potential repayment capacity and will compare this figure to total interest payments or to principal and interest payments.[1] The basic formula is

Net farm income from operations	$45,486
Plus interest expense	18,071
Plus depreciation and amortization expense	16,000
Equals earnings before interest, taxes, depreciation, and amortization (EBITDA)	$79,557

Our sample farm does not have any amortization expense. Amortization expenses can arise from accounting treatment of intangible assets, such as patents on inventions, to spread the cost over the expected life of the associated benefit.

It is important to note that EBITDA ignores capital expenditures, which can be large in some operations, and thus it may significantly overstate cash available for repayment capacity. We will learn more about measures of repayment capacity in Chapter 6, and in Chapter 13 we will learn about the cash flow budget, the most accurate way of estimating cash flow over the next accounting period.

Return to Labor and Management

Net farm income was described as the amount available to provide a return to unpaid labor, management, and equity capital. A rate of ROE was computed in an earlier section, and similarly, it is possible to compute a return to labor and management. Return to labor and management is a dollar amount that represents the part of net farm income from operations that remains to pay for operator labor and management after equity capital (net worth) is paid a return equal to its opportunity cost.

The procedure is similar to that used to compute returns on assets and equity, except the result is expressed in dollars and not as a ratio or

---

[1]There is no legal requirement under GAAP for publically traded companies to disclose EBITDA.

percentage. Return to labor and management is computed as follows, using net farm income from operations as the starting point:

Net farm income from operations

Less opportunity cost of equity capital

Equals return to labor and management

The average of I. M. Farmer's beginning net worth and ending net worth, as shown in Table 4.7, is

($714,237 + $736,334) ÷ 2 = $725,286

If the opportunity cost of I. M. Farmer's equity capital is assumed to be 5 percent, the opportunity cost on equity capital (average net worth) is $725,286 × 5% = $36,264. Therefore, after equity capital is assigned a return equal to its opportunity cost, I. M. Farmer's labor and management earned $45,486 − $36,264 = $9,222.

Assuming that the opportunity costs of I. M. Farmer's labor and management are $40,000 and $5,000, respectively, the $9,222 indicates that capital, labor, management, or some combination of the three did not receive a return equal to their opportunity cost. Unfortunately, there is no way to determine which did or did not have a return equal to or higher than its opportunity cost.

Return to Management

Management is often considered the residual claimant to net farm income, for several reasons. It is difficult to estimate, and in some ways, it measures how well the manager organized the other resources to generate a profit. Return to management for average or typical farms is often reported in farm business summaries. It is computed by subtracting the opportunity cost of labor from the return to labor and management.

The opportunity cost for I. M. Farmer's labor is assumed to be $40,000, about what full-time farm employees in the region are earning annually.

Return to labor and management	$9,222
Less opportunity cost of labor	−40,000
Equals return to management	−30,778

The return to management is highly variable from year to year. A negative return is rather common on many farms and ranches, assuming as this process does that capital and labor earned their opportunity costs. However, a negative return means that net farm income was not sufficient to provide a return to capital, labor, and management equal to or higher than their opportunity costs. The net farm income may have been a substantial amount, particularly on a large farm or ranch. It just should have been better to provide labor, management, and capital a return equal to their individual opportunity costs.

CHANGE IN OWNER EQUITY

After computing and analyzing net farm income, several questions may come to mind. What was this profit used for? Where is this money now? Did this profit affect the balance sheet? If so, how? The answers to these questions are related and illustrate the relation between the income statement and the beginning and ending balance sheets.

Any net farm income must end up in one of four uses: (1) owner withdrawals for family living expenses and other uses, (2) payment of income and Social Security taxes (another reason for withdrawals), (3) increases in cash or other farm assets, or (4) a reduction in liabilities through principal payments on loans or payment of other liabilities. Table 4.6 provided an example of how owner equity changed for I. M. Farmer from January 1 to December 31, 2017. Figure 5-5 provides a diagram that shows what happens to net farm income and how it affects the balance sheet. Withdrawals from the business to pay for family living expenses, income and Social Security taxes, and other purposes reduce the amount of net farm income available for use in the farm business. What remains is called *retained farm earnings*.

As the name implies, retained farm earnings is that part of the farm earnings, after personal withdrawals and taxes, retained for use in the farm business. Asset and liability values will have changed, but if retained farm earnings is positive, equity must have increased. Retained farm earnings represents increased assets or decreased liabilities, or some other combination of changes, which will increase the difference between assets and liabilities and therefore increase equity. This change in equity may not be positive. If living expenses, taxes, and other withdrawals are greater than net farm income, retained farm earnings will be negative. In this case, assets had to be withdrawn from the farm business, or additional borrowing was needed to meet living expenses, income taxes, and other withdrawals.

The direct relation between retained farm earnings and change in equity, shown in Figure 5-5, applies only to a cost-basis balance sheet. When depreciable assets are valued at book value, their decrease in value matches the depreciation expense on the income statement. Land value is unchanged, and if there are no new farm assets from gifts, inheritance, or nonfarm income, retained farm earnings will equal the change in owner equity for the year. Under these conditions, equity on a cost-basis balance sheet will increase only if retained farm earnings is positive; that is, net farm income is greater than the sum of living expenses, taxes, and other withdrawals. This further emphasizes the necessity of making a substantial profit if the farm operator/manager wants to live well and increase business equity at the same time.

On a market-basis balance sheet, changes in equity will come not only from retained farm earnings but also from any changes in the value of assets due to changes in their market value. As shown in Figure 5-5, the change in cost-basis equity must be adjusted for any changes in market valuation of assets and related deferred income taxes. Farms or ranches with a large proportion of their assets in land can experience rapid gains or losses in equity due to fluctuations in land values. Without this adjustment, it is impossible to correlate and account for all changes in equity.

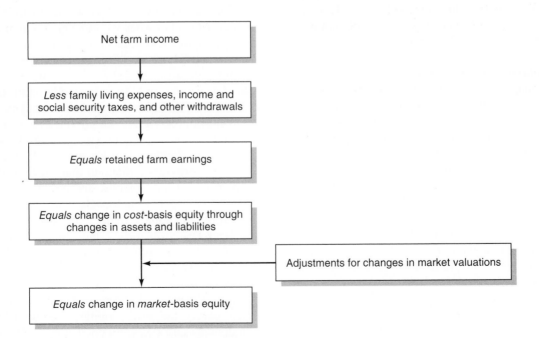

Figure 5-5 Relation between net farm income and change in equity.

STATEMENT OF CASH FLOWS

The FFSC recommends that the set of financial statements for a farm or ranch operation should include a statement of cash flows along with the balance sheet, the statement of owner equity, and the income statement. The statement of cash flows, as the name implies, is a summary of the actual cash inflows and cash outflows experienced by a business during an accounting period. Table 5-4 is I. M. Farmer's Statement of Cash Flows for the year ending December 31, 2017. This statement is organized around five broad categories: operating, for example, cash farm income and expenses; investing, for example income and expenses; investing, for example

TABLE 5-4	Statement of Cash Flows for I. M. Farmer for Year Ending December 31, 2017		
	Cash in	**Cash out**	**Net cash flow**
Operating (cash farm income and expenses)			
Cash received from operations	$395,188	NA	
Cash paid for feeder livestock, purchased feed, and other items for resale	NA	$ 34,270	
Cash paid for operating expenses	NA	292,805	
Cash paid for interest	NA	17,851	
Net cash—income and Social Security taxes	NA	10,500	
Net cash—other miscellaneous income	0	NA	
Net cash provided by operating activities			$39,762
Investing (capital assets)			
Cash received from sale of breeding livestock	19,140		
Cash received from sale of machinery and equipment	8,600	NA	
Cash received from sale of real estate	0	NA	
Cash received from sale of marketable securities	0	NA	
Cash paid to purchase breeding livestock	NA	1,500	
Cash paid to purchase machinery and equipment	NA	53,400	
Cash paid to purchase real estate	NA	0	
Cash paid to purchase marketable securities	NA	0	
Net cash provided by investing activities			(27,160)
Financing (loans)			
New loans received	238,148		
Principal paid		222,250	
Net cash provided from financing activities			15,898
Nonfarm			
Nonfarm income (wages, rents, interest, etc.)	9,500	NA	
Nonfarm expenditures (family living, etc.)	NA	36,000	
Cash on hand (farm cash, checking, savings)			(26,500)
Beginning of year	5,000	NA	
End of year	NA	7,000	(2,000)
			Discrepancy
Total	$675,576	$675,576	$ 0

capital assets; financing, for example loans and repayments; nonfarm items; and a balancing section for cash on hand. Noncash transactions, even though they affect farm profit or net worth, are not included in a statement of cash flows.

I. M. Farmer recorded cash inflows of $395,188 from operations, $372,688 in crop and cash livestock sales, and $22,500 in government program payment. He paid $34,270 in cash for purchased market livestock and purchased feed and grain, $17,851 in cash interest, and $292,805 in cash for other operating expenses. In addition, he paid out $10,500 in income and Social Security taxes on the previous year's taxable income. (Because of the timing of tax payments, these cash expenditures will not match exactly with the period in which the taxes were incurred.) When cash expenses for operations were subtracted from cash receipts, there was a net cash flow from operating, adjusted by the taxes paid, of $39,762.

The next section, on investing, shows that I. M. Farmer bought and sold some capital items during this period. Cash expenditures for capital items purchased exceeded cash receipts for capital items sold, so the net cash flow for the capital assets activities was negative $27,160. The total of new loans received ($238,148) was higher than the amount of principal paid ($222,250), yielding a positive net cash flow for financing activities of $15,898. Finally, the nonfarm cash income of $9,500 and nonfarm expenditures of $36,000 yield a negative net cash flow for nonfarm activities; that is, a net amount of $26,500 was withdrawn from the farm account. If the items in the net cash flow column are totaled, they yield an increase in cash of $2,000, exactly the difference in beginning- and end-of-year cash on hand. The total cash in, including the beginning balance of $5,000, exactly equals the total cash out, including the ending balance of $7,000. Hence, there is no discrepancy to reconcile.

SUMMARY

An income statement organizes and summarizes revenue and expenses for an accounting period and computes net farm income for that period. First, all revenue earned during the period should be identified and recorded. Next, the same thing is done for all expenses incurred in producing that revenue. Depreciation is a noncash expense that can be important in agriculture, so accurate records of depreciation must be kept. Net farm income is the amount by which revenue exceeds expenses. If records are kept using cash-basis accounting, accrual adjustments should be made at the end of each accounting period to derive the accrual net farm income. A cash-basis net farm income can be misleading, and management decisions should be made only on the information obtained from an accrual net farm income.

Net farm income or profit is an actual dollar amount, whereas profitability refers to the size of that profit relative to the resources used to produce it. An analysis of profitability should be done each year to provide a means of comparing results against previous years and against other farm businesses. Net farm income, rate of ROA, rate of ROE, and operating profit margin ratio are recommended measures of profitability.

Part of net farm income is withdrawn from the business to pay for family living expenses, income and Social Security taxes, and other items. The remaining net farm income is called retained farm earnings, because it has been retained for use in the farm business. A positive retained farm earnings ends up as an increase in assets, a decrease in liabilities, or some other combination of changes in assets and liabilities that causes equity to increase. Conversely, a negative retained farm earnings means some farm equity has been used for living expenses, taxes, and other personal withdrawals. Equity based on market values for assets also changes by the amount these market values change during the year. A statement of cash flows should also be developed as part of a complete set of records.

QUESTIONS FOR REVIEW AND FURTHER THOUGHT

1. Assume that a new tractor is purchased on January 1 for $124,000 and given a salvage value of $20,000 and a useful life of eight years. What would the annual depreciation be for the first two years under each depreciation method?

	Year 1	Year 2
Straight line	_____	_____
Double declining balance	_____	_____

2. For the previous problem, what would the tractor's book value be at the end of Year 2 under each depreciation method?

3. Rework the answers to question 1 assuming the purchase date was May 1.

4. Why is inventory change included on an accrual income statement? What effect does an increase in inventory value have on net farm income? A decrease?

5. What are the differences between net farm income computed on a cash basis versus an accrual basis? Which is the better measure of net farm income? Why?

6. What factors determine the change in cost-basis equity? In market-basis equity?

7. Why are changes in land values not included with the inventory changes shown on an income statement?

8. Why are there no entries for the purchase price of new machinery on an income statement? How does the purchase of a new machine affect the income statement?

9. Use the following information to compute values for each of the following items.

Net farm income	$ 36,000	Opportunity cost of labor	$16,000
Average equity	$220,000	Opportunity cost of management	$ 8,000
Average asset value	$360,000	Family living expenses	$20,000
Interest expense	$ 11,000	Income and social security taxes	$ 4,000
Gross revenue	$109,500	Opportunity cost of capital	10%

   a. Rate of return on assets _____%
   b. Rate of return on equity _____%
   c. Operating profit margin ratio _____%
   d. Change in equity $ _____
   e. Return to management $_____

10. Use the following information to compute the values:

Cash expenses	$110,000	Beginning inventory value	$42,000
Cash revenue	$167,000	Ending inventory value	$28,000
Depreciation	$ 8,500	Cost of new tractor	$48,000
Beginning accounts receivable	$ 2,200	Ending accounts receivable	$ 0
Beginning accounts payable	$ 7,700	Ending accounts payable	$ 1,500

   a. Cash-basis net farm income $ _____
   b. Gross revenue on accrual basis $ _____
   c. Accrual-basis net farm income $ _____

S. Meltzer/PhotoLink/Getty Images

FARM BUSINESS ANALYSIS

CHAPTER OUTLINE

Types of Analysis
Standards of Comparison
Diagnosing a Farm Business Problem
Measures of Profitability
Measures of Size
Efficiency Measures
Financial Measures
Summary
Questions for Review and Further Thought

CHAPTER OBJECTIVES

1. Show how farm business analysis contributes to the control function of management

2. Suggest standards of comparison to use in analyzing the farm business

3. Identify measures of business size

4. Outline a procedure for locating economic or financial problem areas in the farm or ranch business

5. Review measures that can be used to analyze the solvency, liquidity, and profitability of the business

6. Illustrate the concept of economic efficiency and show how it is affected by physical efficiency, product prices, and input prices

Farm managers and economists have always been interested in the reasons some farms have higher net incomes than others. The observation and study of these differences and their causes began in the early 1900s and marked the beginning of farm management and farm business analysis. Differences in management are a common explanation for different net farm incomes, but this explanation is not complete. Unless the specific differences in management can be identified, there can be no precise recommendations for improving net income on farms with *poor management*.

The differences in profitability of similar farms can be illustrated in several ways. Table 6-1 shows some observed differences per farm on a group of Alabama cattle farms. Similar differences can be found in other states and for other types of farms and ranches. The average net operating income per cow was –$41 for low-profit farms and $137 in the most profitable group. Several questions beg an answer: What accounts for the differences in operations? What can be done to improve returns on the less profitable farms? Differences in the operators' goals or the quality of resources available may be partial answers.

Other information in this table sheds light on the differences in profitability. The number of cows was roughly similar across all three groups, as was the average weight of the feeder calves produced. However, the high-profit group produced more pounds of feeder calf per brood cow, with less hay fed and less money spent on other purchased inputs. More detailed answers can be found only by performing a complete farm business analysis. This is part of the control function of management.

Control is critical for the entire farm business as well as for individual enterprises. Three steps in the control function are: (1) establishing standards for comparing results, (2) measuring the actual performance of the business, and (3) taking corrective action to improve the performance once problem areas have been identified. Improvement in performance is the payoff for the time spent controlling the business.

TYPES OF ANALYSIS

A farm business analysis can be divided into four areas of investigation:

1. *Profitability*: Profitability is analyzed by comparing income and expenses. High net farm income is usually an important goal of the farm manager, though not necessarily the only one.
2. *Farm size*: Not having adequate resources is often a cause for low profits. Growing too rapidly or exceeding the size that the operator can manage effectively can also reduce profits.
3. *Financial*: Financial analysis concentrates on the capital position of the business, including solvency, liquidity, and changes in owner's equity.

Table 6-1	**Comparisons of Alabama Cattle Farms by Net Operating Income**		
Item	Low third	Middle third	High third
Net operating income ($/cow)	–$ 41	$ 47	$ 137
Average calf weight (lbs)	657	642	650
Pounds of feeder calf per cow	520	528	576
Hay fed (tons/cow)	2.3	1.8	1.6
Seed, fertilizer, chemicals ($/cow)	$ 119	$ 91	$ 78
Purchased feed ($/cow)	$ 116	$ 91	$ 64
Breakeven price ($/lb)	$1.29	$0.99	$0.82
Number of brood cows	262	272	278

Source: Alabama Farm Analysis Association.

4. *Efficiency*: Low profitability can often be traced to inefficient use of resources in one or more areas of the business. Both economic and physical efficiency measures should be examined.

The efficiency and profitability of each enterprise should also be examined. Enterprise analysis will be discussed in Chapter 18.

STANDARDS OF COMPARISON

The remainder of this chapter will examine some measures and ratios that can be used to conduct a complete farm analysis. Once a measure has been calculated, the problem becomes one of evaluating the result. Is the value good, bad, or average? Compared with what? Can it be improved? These questions emphasize the need to have some standards against which the measures and ratios can be compared. There are three basic standards to use in analyzing farm business results: budgets, comparative farms, and historical trends.

Budgets

In budget analysis, the measures or ratios are compared against budgeted goals or objectives identified during the planning process. When results in an area consistently fall short of the

---

Box 6-1   Farm Business Analysis: An Ancient Art

The following conclusions about farm business analysis are reprinted from *Iowa Farm Management Surveys*, a study of 965 farms in north central Iowa summarized by H. B. Munger in 1913:

CONCLUSIONS

"From a business standpoint a farm cannot be called successful that does not pay operating expenses, a current mortgage rate of interest on capital and a fair return as pay for the farmer's labor and management. By analyzing the business of large numbers of farms in one area, the reasons why some farms are more profitable than others stand out clearly.

"The first consideration is to have a type of farming adapted to the region. Farmers doing the wrong thing cannot expect to make labor incomes. Change in the demands of the market, increase in value of farm land, and other factors make necessary a readjustment of the crops and livestock raised. A farmer in this region for the period of this investigation has better chances of success who had more corn and hogs than the average, and less pasture and beef cattle. Most dairy farmers were doing well.

"For efficient farming in Iowa, where the usual crops and livestock are raised, at least 160 acres is necessary in order to use labor, horses, and machinery to good advantage. Many farmers who own less than 160 acres rent additional land in order to utilize efficiently labor, teams, and tools.

"Good crop yields are fundamental to successful farming. Regardless of other factors, most farms with poor crops were not highly profitable. One should aim to get at least one-fifth better yields than the average of other farms on similar soil.

"In a region where most of the farm receipts come from the sale of livestock, the returns per animal unit are of highest importance. No matter how good the crops may be, if they are fed to unproductive livestock the chances for success are small.

"The highest profits come from a well-balanced business. One should have a farm large enough to provide for the economical use of labor, horses, and up-to-date machinery. He should raise better crops than the average and feed most of them to livestock that can use them profitably. Each of these factors is important, and the combination of all on an individual farm brings the highest success."

budgeted objectives, either that area needs additional managerial effort or the budgets need to be more realistic.

Comparative Farms

Unfortunately, poor weather or low prices may prevent a farm from reaching its goals, even in years when no serious management problems exist. Therefore, a second source of useful standards for comparison are results from other farms of similar size and type, for the same year. These may not represent ideal standards but will indicate if the farm being analyzed was above or below average given the weather and market conditions that existed. Organized record-keeping services are available through lending agencies, accounting firms, farm cooperatives, and university extension services. One of the advantages of participating in such a service is the annual record summary provided for all farms using the service.

These summaries contain averages and ranges of values for different measures for all farms, as well as for groups of farms sorted by size and type. Such results provide an excellent set of comparative standards as long as the farms being compared are similar and in the same general geographic area.

Historical Trends

When using trend analysis, the manager records the various measures and ratios for the same farm for a number of years and observes any trends in the results. The manager looks for improvement or deterioration in each measure over time. Those areas showing a decline deserve further analysis to determine the causes. This method does not compare results against any particular standards, only against previous performance. The objective becomes one of showing improvement over the results of the past. Trend analysis must take into account year-to-year changes in weather, prices, and other random variables.

Any of the analytical measures discussed in the remainder of this chapter can be used with budgets, comparative farms, and historical trend analysis. Many of these measures are shown in tables that contain results from an example farm as well as average values for a group of similar farms.[1]

DIAGNOSING A FARM BUSINESS PROBLEM

A complete whole-farm business analysis can be carried out by using a systematic procedure to identify the source of a problem. Figure 6-1 illustrates a procedure that can make the process more efficient.

Profitability is generally the first area of concern. Low net farm income or poor returns to management can have many causes. The farm or ranch may not be large enough to generate the level of production needed for an adequate income. If inadequate resources are a problem, the operator should look for ways to farm more acres, increase the labor supply, expand livestock, or obtain more capital. If farm size cannot be increased, then fixed costs such as machinery and building depreciation, interest, and general farm overhead costs should be evaluated carefully. Steps should be taken to reduce the costs that have the least effect on the level of production. Off-farm employment or other types of self-employment may also have to be considered as a way to improve family income.

If adequate resources are available but production levels are low, then resources are not being used efficiently. Computing several *economic efficiency* measures can help pinpoint the problem. Poor economic efficiency may be due to low physical efficiency, low selling prices, and/or high input costs. If efficiency for the current enterprises cannot be improved, it may mean the wrong enterprises are in the current

---

[1]The analysis measures discussed in the remainder of this chapter include those recommended by the Farm Financial Standards Council as well as a number of others. Those recommended by the Council are identified by the abbreviation FFSC shown after their name, and are summarized in Table 6-2.

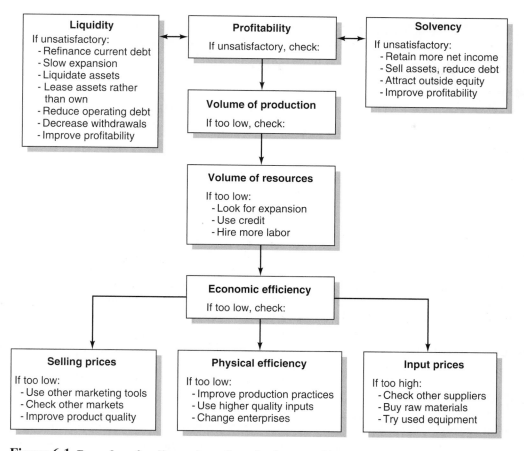

Figure 6-1 **Procedure for diagnosing a farm business problem.**

farm plan. Enterprise analysis and enterprise budgeting can be used to identify more profitable enterprises and develop a new whole-farm plan. This procedure should help managers isolate and identify the causes of a profitability problem quickly and systematically. However, even profitable farms need to be concerned about *liquidity* and *solvency*. If cash flow always seems to be tight, even when net farm income is satisfactory, the operation may need to refinance some current debt, slow down expansion, sell off some assets, reduce nonfarm withdrawals, or pay down operating debt. If solvency is not as strong as the operator would like, more of the

net farm income may need to be retained in the business each year, or some assets can be sold to reduce debt levels. The three areas of farm business management—profitability, solvency, and liquidity—are all closely related (Table 6-2). Outstanding farm managers do an above-average job in all of them.

MEASURES OF PROFITABILITY

Profitability analysis starts with the income statement, as was explained in Chapter 5, where several measures of profitability were discussed. These measures for our sample farm are shown

TABLE 6-2	**Farm Financial Ratios**

Measure	Calculation
Liquidity	
Current ratio	Total current farm assets ÷ total current farm liabilities
Working capital	Total current farm assets − total current farm liabilities
Working capital to gross revenue	Working capital ÷ gross revenue
Solvency	
Debt-to-asset ratio	Total farm liabilities ÷ total farm assets
Equity-to-asset ratio	Total farm net worth ÷ total farm assets
Debt-to-equity ratio	Total farm liabilities ÷ total farm net worth
Profitability	
Net farm income from operations	Gross revenue (accrual) − total expenses (accrual)
Net farm income	Net farm income from operations +/− gains or losses on sales of capital assets
Rate of return on farm assets	(Net farm income from operations + interest expense − value of unpaid labor and management) ÷ average of beginning and ending total farm assets
Rate of return on farm equity	(Net farm income from operations − value of unpaid labor and management) ÷ average of beginning and ending farm net worth
Operating profit margin ratio	(Net farm income from operations + interest expense − value of unpaid labor and management) ÷ gross revenue
Earnings before interest, taxes, depreciation, and amortization (EBITDA)	Net farm income from operations + interest expense + depreciation and amortization
Repayment capacity	
Capital debt repayment capacity	Net farm income from operations + depreciation + nonfarm income − nonfarm withdrawals + interest expense on term loans
Capital debt repayment margin	Capital debt repayment capacity − scheduled principal and interest payments on term loans
Term debt coverage ratio	Capital debt repayment capacity ÷ scheduled principal and interest payments on term loans
Financial efficiency	
Asset turnover ratio	Gross revenue (or value of farm production) ÷ average of beginning and ending total farm assets
Operating expense ratio	(Total expenses − interest expense − depreciation) ÷ gross revenue
Depreciation expense ratio	Depreciation expense ÷ gross revenue
Interest expense ratio	Interest expense ÷ gross revenue
Net farm income ratio	Net farm income from operations ÷ gross revenue

Source: Farm Financial Standards Council, *Financial Guidelines for Agricultural Producers*, 2013.

TABLE 6-3	Measures of Farm Profitability*		
Item	Average of comparison group	Our farm	Comments
1. Net farm income from operations	$130,230	$45,486	Much lower
2. Return to labor and management	$ 87,320	$ 9,222	Much lower
3. Return to management	$ 64,654	–$30,778	Much lower
4. Rate of return on farm assets	5.5%	1.8%	Much lower
5. Rate of return on farm equity	5.6%	0.1%	Much lower
6. Operating profit margin ratio	20.1%	4.6%	Much lower

*Comparison group figures based on Alabama Farm Analysis Association information.

in Table 6-3, along with average values based on a group of Alabama cattle and row crop operations, which can be used for comparison.

Net Farm Income (FFSC)

One method of establishing a goal for net farm income is to estimate the income that the owner's labor, management, and capital could earn in nonfarm uses. In other words, the total opportunity cost for these factors of production becomes the minimum goal or standard for net farm income. In any year for which the income statement shows a lower net farm income, one or more of these factors did not earn its opportunity cost. Another goal would be to reach a specific income level at some time in the future, such as at the end of five years. Progress toward this goal can be measured over time.

Although net farm income is influenced heavily by the profitability of the farm, it also depends on what proportion of the total resources used are contributed by the operator. Replacing borrowed capital with equity capital, rented land with owned land, or hired labor with operator labor can all improve net farm income. Figure 6-2 shows how the gross farm revenue *pie* is divided among the parties who supply resources to the farm or ranch business. The size of the farmer's slice of the pie depends on how

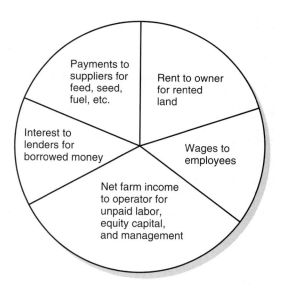

Figure 6-2 How the gross revenue pie is divided?

many of the resources he or she contributes, as well as the size of the pie. Other resource contributors, such as lenders, landlords, employees, and input suppliers, must be paid first. What is left over is net farm income. Farms with high net incomes are often those that have accumulated a large equity over time or that depend heavily on operator and family labor.

Return to Labor and Management

Some farm businesses have more net worth than others. As shown in Chapter 5, a fair comparison requires that the interest expense incurred by the business be added back to net farm income from operations to get adjusted net farm income. Next, an opportunity cost for the investment in farm net worth is subtracted from net farm income from operations. What remains is the return to the unpaid labor and management used in the business. One goal or standard for return to labor and management is the opportunity cost of the operator's labor and management in another use.

Return to Management

The return to management is the portion of net farm income from operations that remains after the opportunity costs of both unpaid labor and equity capital have been subtracted. It represents the residual return to the owner for the management input, and it can be highly variable from year to year. Negative returns to management in years of low prices or production are not unusual, but the goal should be to earn a positive return over the long run.

Rate of Return on Farm Assets (FFSC)

The rate of return on farm assets (or return on assets [ROA]) was discussed in Chapter 5. This value can be compared to rates of return on other long-term investments, although differences in risk and potential gains in asset values must also be considered. It also allows comparison among farms of different sizes and types. ROA includes a return to both equity capital and debt capital (interest on loans). It includes only net income from operations, though, and not capital gains or losses.

Rate of Return on Farm Equity (FFSC)

The return on equity (ROE) is perhaps more indicative of financial progress than ROA, as it measures the percent return to the owner's business equity. If the farm business has no debts, the rate of return on farm equity will be equal to the rate of return on farm assets, because they are one and the same. When debt capital is used, though, interest must be paid. The dollars available as a ROE then are only that part of net farm income remaining after interest is paid. Opportunity costs for labor and management are also subtracted before computing the rate of return on equity. See Chapter 5 for details on how to compute ROA and ROE.

Operating Profit Margin Ratio (FFSC)

As explained in Chapter 5, the operating profit margin ratio (OPMR) measures the proportion of gross revenue left after paying all expenses. It is calculated by dividing the dollar return to farm assets (adjusted net farm income minus the opportunity cost of labor and management) by the gross revenue of the farm.

Farms with large investments in fixed assets such as land and buildings, and few operating expenses such as cash rent, will generally show higher operating profit margin ratios. Conversely, farms with more rented assets will often have a higher ROA but a lower OPMR. Rented and owned assets are substitutes for each other. As more of one is used, the rate of return to the other will generally increase. Thus, these two measures of profitability must be considered together. An operation may be high in one ratio and low in the other because of the mix of owned and rented assets used. If both measures are below the average or the goal, then profitability problems are evident.

General Comments

Two cautions should be exercised regarding the procedures used to calculate returns to labor, management, assets, and equity. The first is the somewhat arbitrary nature of estimating the opportunity costs used in the calculations. No rigid rules exist, and individuals may have different opinions about the appropriate values to use. The calculated return to any factor will change if different opportunity costs are used for the others.

The second caution is that these values are *average* returns to the factors, not marginal returns. For example, the rate of return to assets is the average rate of return on all capital invested in the business, not the marginal return on the last dollar invested. The marginal rates of return would be much more useful for planning purposes, especially for new investments. Nevertheless, if the opportunity costs are estimated carefully and used consistently from year to year, and the average nature of the result is kept in mind, these measures can provide a satisfactory means for making historical and between-farm comparisons of profitability. They are, however, less useful for evaluating marginal changes in the organization and operation of the farm business.

The measures of profitability included in Table 6-3 can be used to identify the existence of a problem. However, these measures do not identify the exact source of the problem. Further analysis is needed to clarify the cause of the problem and suggest the corrective action needed. The next three sections proceed with a more in-depth analysis of the farm business to pinpoint more specific problem areas.

MEASURES OF SIZE

An income or profitability problem can exist in any single year due to low selling prices or low crop yields. If the problem persists even in years with above-average prices and yields, it may be caused by insufficient farm size. Historically, net farm income has been highly correlated with farm size. Small farms sometimes make inefficient use of fixed capital investments and labor. Small farms that do make efficient use of resources may still fall short of generating enough net income to support the operator and family. Several common measures of farm size are shown in Table 6-4. Some capture the quantity or value of production generated, while others measure the volume of resources used.

Gross Revenue

The volume of production from a farm producing several different products can be measured by gross revenue. Table 5-3 showed the calculations where *gross revenue* is defined as total sales and other cash income, adjusted for changes in inventories, accounts receivable, and any gains/losses on the sale of breeding livestock.

Total Farm Assets

Another measure of size is the total capital invested in land, buildings, machinery, livestock, crops, and other assets. The market value of total farm assets from the balance sheet is the easiest place to obtain this value. With all values in dollar terms, this measure allows an easy comparison of farm size across different farm types. However, it does not take into account the value of rented or leased assets.

TABLE 6-4	**Measures of Farm Size**		
Item	Average of comparison group	Our farm	Comments
1. Gross revenue	$ 565,179	$ 406,548	Lower
2. Total farm assets	$1,427,154	$1,068,750	Lower
3. Total crop acres farmed	930	720	Lower
4. Number of brood cows	180	150	Lower
5. Total labor (person-years)	2.2	1.5	Lower

Total Acres Farmed

The number of acres farmed includes both owned and rented land. It is useful for comparing similar types of farms but is not particularly good for comparing farms with different types of land. A California farm with 200 acres of irrigated vegetables is not comparable to an operation with 200 acres of dryland wheat in western Kansas or to a 200-acre cattle feedlot in Texas. Number of acres is best used for comparing the size of crop farms of the same general type in an area with similar soil resources and climate.

Livestock Numbers

It is common to speak of a 500-cow ranch or a 250-cow dairy farm or a 500-sow hog farm. These measures of size are useful for comparing size among farms with the same class of livestock.

Total Labor Used

Labor is a resource common to all farms. Terms such as a one-person or two-person farm are frequently used to describe farm size. A full-time worker equivalent can be calculated by summing the months of labor provided by operator, family, and hired labor and dividing the result by 12. For example, if the operator works 12 months, family members provide 4.5 months of labor, and 5 months of labor is hired, the full-time worker equivalent is 21.5 months divided by 12, or 1.8 years of labor. This measure of size is affected by the amount of labor-saving technology used and should therefore be interpreted carefully when comparing farm sizes.

Quantity of Sales

For specialized farms, the number of units of output sold annually is a convenient measure of size. Examples are 5,000 head of farrow-to-finish hogs or 25,000 boxes of apples. On a diversified farm, it may be necessary to record the units sold from several different enterprises to more accurately reflect farm size. Where a crop share or livestock share lease is being used, both the owner's and tenant's shares of production should be included when making comparisons with other farms.

Value of Farm Production

Some livestock operations buy feeder livestock and feed, while others produce all their own feed and animals. To make a fair comparison of business size, the value of feed and livestock purchases is subtracted from gross revenue to arrive at the *value of farm production*. In this measure, only the value added to purchased feed and livestock is credited to the value of production. Details of the calculation were given in Chapter 5.

The value of farm production for a given farm may vary substantially from year to year, depending on weather, disease, pests, and product prices. However, it is a convenient way to compare the size of different types of farms.

EFFICIENCY MEASURES

Some managers are able to generate more production or use fewer resources than their neighbors because they use their resources more efficiently. A general definition for efficiency is the quantity or value of production achieved per unit of resource employed.

$$\text{Efficiency} = \frac{\text{production}}{\text{resources used}}$$

If a comparison with other farms or with a budget goal shows that an operation has an adequate volume of resources but is not reaching its production goals, then some resources are not being used efficiently. A farm or ranch business may use many types of resources, so there are many ways to measure both economic and physical efficiency. Some of the more useful and common ones will be discussed here. They are summarized in Table 6-5.

TABLE 6-5	Measures of Economic Efficiency*		

Item	Average of comparison group	Our farm	Comments
Capital efficiency			
1. Asset turnover ratio	0.40	0.39	About the same
2. Operating expense ratio	0.67	0.81	Much higher
3. Depreciation ratio	0.07	0.04	Lower
4. Interest expense ratio	0.03	0.04	Somewhat higher
5. Net farm income from operations ratio	0.23	0.11	Much lower
Labor efficiency			
6. Gross revenue per person	$256,899	$271,032	Higher
Marketing			
7. Price received for calves ($/cwt)	$ 106.00	$ 95.00	Lower
8. Price received for cotton lint ($/lb)	$ 0.55	$ 0.56	About the same
Livestock efficiency			
9. Purchased feed ($/cow)	$ 68	$ 77	Higher
10. Production per $100 of feed fed	$ 178	$ 142	Lower

*Comparison group figures based on Alabama Farm Analysis Association information.

Economic Efficiency

Economic efficiency measures are computed either as dollar values per unit of resource or as some rate or percentage relating to capital use. Measures of economic efficiency must be interpreted carefully. They measure average values, not the marginal values or the effect that small changes would have on overall profit.

Asset Turnover Ratio (FFSC)

This ratio measures how efficiently capital invested in farm assets is being used. It is found by dividing the gross revenue generated by the market value of total farm assets. For example, an asset turnover ratio equal to 0.25, or 25 percent, indicates that gross revenue for one year was equal to 25 percent of the total capital invested in the business. At this rate, it would take four years to produce agricultural products with a value equal to the total assets.

The asset turnover ratio will vary by farm type. Dairy, hog, and poultry farms will generally have higher rates, beef cow farms tend to be lower, and crop farms usually have intermediate values. Farms or ranches that own most of their resources will generally have lower asset turnover ratios than those that rent land and other assets. Therefore, the asset turnover ratio should be compared only among farms of the same general type.

Sometimes the value of farm production is used to calculate the asset turnover ratio instead of gross revenue. As was explained earlier, the value of farm production is equal to gross revenue minus the cost of purchased feed and livestock, and is a more accurate measure of the value of the crops and livestock actually produced on the farm.

The asset turnover ratio (ATR) is an overall measure of financial efficiency. The operating profit margin ratio is a measure of profitability per unit produced. When the two are multiplied, they equal the rate of return on farm assets:

$$ATR \times OPMR = ROA$$

For I. M. Farmer:

$$0.39 \times 4.6\% = 1.8\%$$

So, overall profitability can be improved by producing more units of output while maintaining the same profit margin per unit, or by producing the same output but with a larger profit margin per unit. Of course, increasing both the asset turnover ratio and the operating profit margin ratio will increase the return on assets even more!

Operating Expense Ratio (FFSC)
Four operational ratios are recommended to show what percent of gross revenue went for operating expenses, depreciation, interest, and net income. The operating expense ratio is computed by dividing total operating expenses (excluding depreciation) by gross revenue. Farms with a high proportion of rented land and machinery or hired labor will tend to have higher operating expense ratios.

Depreciation Expense Ratio (FFSC)
This ratio is computed by dividing total depreciation expense by gross revenue. Farms with a relatively large investment in newer machinery, equipment, and buildings will have higher depreciation expense ratios. A higher-than-average ratio may indicate underused capital assets.

Interest Expense Ratio (FFSC)
Total farm interest expense (adjusted for accrued interest payable at the beginning and end of the year) is divided by gross revenue to find this ratio. Ratios higher than average, or higher than desired, may indicate too much dependence on borrowed capital or high interest rates on existing debt.

Net Farm Income from Operations Ratio (FFSC)
Dividing net farm income from operations by gross revenue measures what percent of gross revenue is left after paying all expenses (but before subtracting any opportunity costs). These four operational ratios will sum to 1.00, or 100 percent. They can also be calculated using the value of farm production as a base instead of gross revenue. In that case, the cost of purchased feed and livestock should not be included in operating expenses. Using the value of farm production as a base is more accurate when comparing operations that purchase large quantities of feed and livestock against those that do not.

Gross Revenue per Person
The efficiency with which labor is being used can be measured by dividing gross revenue for the year by the number of full-time equivalents of labor used, including hired workers. Capital and labor can be substituted for each other, so capital and labor efficiency measures should be evaluated jointly.

Livestock Production per $100 of Feed Fed
This is a common measure of economic efficiency in livestock production. It is calculated by dividing the value of livestock production during a period by the total value of all the feed fed in the same period, and multiplying the result by 100. The value of livestock production is equal to:

Total cash income from livestock
Plus or *minus* livestock inventory changes
Minus value of livestock purchased
Equals value of livestock production

Livestock production per $100 of feed fed is computed from the following formula:

$$\frac{\text{Value of livestock production}}{\text{Value of feed fed}} \times 100$$

It is important to include the value of any raised feed in the denominator, as well as feed purchased. A ratio equal to $100 indicates that the livestock just paid for the feed consumed but no other expenses. The return per $100 of feed fed also depends on the type of livestock, as shown in the following table:

Livestock enterprise	Production per $100 of feed fed (10-year average, 2003–2012)
Beef cow	$159
Dairy	$211
Feeding cattle	$152
Hogs (farrow to finish)	$156

Source: Iowa Farm Costs and Returns, Iowa State University, 2003–2012.

Higher values for some livestock enterprises do not necessarily mean that these enterprises are more profitable. Enterprises with higher building, labor or other nonfeed costs, such as dairy and feeder pig production, need higher feed returns to be as profitable as enterprises with low nonfeed costs, such as finishing feeder animals. Because of these differences, the production per $100 of feed fed should be calculated for each individual livestock enterprise and compared only with values for the same enterprise.

Feed Cost per 100 Pounds of Production

The feed cost per 100 pounds of weight gain or per 100 pounds of milk for a dairy enterprise is found by dividing the total feed cost for each enterprise by the total pounds of production and multiplying the result by 100. The total pounds of production for the year should be equal to the pounds sold and consumed minus the pounds purchased and plus or minus any inventory changes. This measure is affected by feed prices and feed conversion efficiency, and values should be compared only among farms with the same livestock enterprise.

Crop Value per Acre

This value measures the intensity of crop production and whether high-value crops are included in the crop plan. It is calculated by dividing the total value of all crops produced during the year by the number of crop acres. The result does not take into account production costs and therefore does not measure profit.

Machinery Cost per Crop Acre

To calculate this measure, the total of all annual costs related to machinery is divided by the number of crop acres. Total annual machinery costs include all ownership and operating costs, plus the cost of custom work hired and machinery lease and rental payments. Costs for any machinery used mainly for livestock should be excluded. This is another measure that can be either too high or too low. A high value indicates that machinery investment may be excessive, while low values may indicate that the machinery is too old and unreliable or too small for the acres being farmed. There is an important interaction between machinery investment, labor requirements, timeliness of field operations, and the amount of custom work hired. These factors are examined more thoroughly in Chapter 22.

Physical Efficiency

Poor economic efficiency can result from several types of problems. *Physical efficiency*, such as the rate at which seed, fertilizer, and water are converted into crops, or the rate at which feed is turned into livestock products, may be too low. Physical efficiency measures include bushels harvested per acre, pigs weaned per sow, and pounds of milk sold per cow. Examples of physical efficiency measures are shown in Table 6-6. Just like the measures of economic efficiency, they are average values and must be interpreted as such. An average value should not be used in place of a marginal value when determining the profit-maximizing input or output level.

Selling Prices

The average price received for a product may be lower than desired, for several reasons. In some years, supply and demand conditions cause prices in general to be low. However, if the average price

TABLE 6-6	Measures of Physical Efficiency[*]			
Item	Average of comparison group	Our farm	Comments	
1. Feeder calves produced (lbs/calf)	650	540	Lower	
2. Feeder calves produced (lbs/cow)	576	473	Lower	
3. Hay fed (tons/cow)	1.6	2.1	Higher	
4. Cotton yields (lbs/acre)	762	750	Slightly lower	
5. Corn yields (bushels/acre)	118	120	About the same	

[*]Comparison group figures based on Alabama Farm Analysis Association information.

received is lower than for other similar farms in the same year, the operator may want to look for different marketing outlets or spend some time improving marketing skills. Poor product quality also can cause selling prices to be below average.

Purchase Prices

Sometimes physical efficiency is good and selling prices are acceptable, but profits are low because resources were acquired at high prices. Examples are buying land at a price above what it will support, paying high cash rents, paying high prices for seed and chemical products, buying overly expensive machinery, purchasing feeder livestock at high prices, borrowing money at high interest rates, and so forth. Input costs can be reduced by comparing prices among different suppliers, buying used or smaller equipment, buying large quantities when possible, and substituting raw materials for more refined or more convenient products.

Physical efficiency, selling price, and purchase price combine to determine economic efficiency, as shown in the following equations:

$$\frac{\text{Economic}}{\text{efficiency}} = \frac{\text{value of product produced}}{\text{cost of resources used}}$$

$$\text{or} = \frac{\text{quantity of product} \times \text{selling price}}{\text{quantity of resource} \times \text{purchase price}}$$

FINANCIAL MEASURES

Even profitable farms may suffer from tight cash flows or high debt loads. The financial portion of a complete farm analysis is designed to measure the solvency and liquidity of the business and to identify weaknesses in the structure or mix of the various types of assets and liabilities. The balance sheet and the income statement are the primary sources of data for calculating the measures related to the financial position of the business. These financial measures and ratios were presented in Chapters 4 and 5 but will be reviewed briefly to show how they fit into a complete farm analysis. It is convenient to compare the actual financial performance to values from a comparison group such as shown in Table 6-7. Actual values should be compared to historical values as well, to identify any trends that may be developing.

Solvency

Solvency refers to the value of assets owned by the business compared to the amount of liabilities, or the relation between debt and equity capital.

Debt/Asset Ratio (FFSC)

The debt/asset ratio is a common measure of business solvency and is calculated by dividing total farm liabilities by total farm assets, using current market values for assets. Smaller values are usually preferred to larger ones, as they

TABLE 6-7	Analyzing the Financial Condition of the Business*			
Item	**Average of comparison group**	**Our farm**	**Comments**	
1. Net worth	$1,013,210	$736,836	Lower	
2. Debt/asset ratio	0.29	0.31	Higher	
3. Current ratio	7.18	1.37	Much lower	
4. Working capital	$ 140,187	$ 23,654	Much lower	
5. Working capital to gross revenues ratio	24.8%	5.8%	Much lower	

*Comparison group figures based on Alabama Farm Analysis Association information.

indicate a better chance of maintaining the solvency of the business should it ever be faced with a period of adverse economic conditions. However, low debt/asset ratios may also indicate that a manager is reluctant to use debt capital to take advantage of profitable investment opportunities. In this case, the low ratio would be at the expense of developing a potentially larger business and generating a higher net farm income. A more complete discussion of the use of debt and equity capital is found in Chapter 19.

Two other common measures of solvency are the equity/asset ratio (FFSC) and the debt/equity ratio (FFSC). These were discussed in Chapter 4. They are mathematical transformations of the debt/asset ratio and provide the same information.

Change in Equity

An important indicator of financial progress for a business is an owner's equity increasing over time. This is an indication that the business is earning more net income than is being withdrawn. Reinvesting profits allows for acquiring more assets or reducing debt. As discussed in Chapter 4, equity should be measured both with and without the benefit of inflation and its effect on the value of assets such as land. This requires a balance sheet prepared on both a cost basis and a market basis, so that any increase in equity can be separated into the result of profits generated by the production activities during the accounting period and the effect of changes in asset values.

Liquidity

The ability of a business to meet its cash flow obligations as they come due is called *liquidity*. Maintaining liquidity is important to keep the financial transactions of the business running smoothly.

Current Ratio (FFSC)

This ratio is a quick indicator of a farm's liquidity. The current ratio is calculated by dividing current farm assets by current farm liabilities, as shown in Chapter 4. The current (or short-term) assets will be sold or turned into marketable products in the near future. They will generate cash to pay the current liabilities, that is, the debt obligations that come due in the next 12 months.

Working Capital (FFSC)

Working capital is the difference between current assets and current liabilities. It represents the excess dollars that would be available from current assets after current liabilities have been paid, as shown in Chapter 4. Because working capital needs may differ with the size of the operation, the FFSC also recommends calculation of the working capital to gross revenue ratio. It is found by dividing working capital by gross revenue earned in the most recent accounting year. Higher ratios indicate higher liquidity.

These liquidity measures are easy to calculate but fail to take into account upcoming operating costs, overhead expenses, capital replacement needs, and nonbusiness expenditures. They also do not include sales from products yet to be produced and do not consider the timing of cash receipts and expenditures throughout the year. An operation that concentrates its production during one or two periods of the year, such as a cash grain farm, needs to have a high current ratio at the beginning of the year, because no new production will be available for sale until near the end of the year. On the other hand, dairy producers or other livestock operations with continuous sales throughout the year can operate safely with lower current ratios and working capital margins.

Chapter 13 discusses how to construct a cash flow budget and how to use it to manage liquidity. Although developing a cash flow budget takes more time than computing a current ratio, it takes into account all sources and uses of cash, as well as the timing of their occurrence throughout the year. Every farm or ranch that has any doubts about being able to meet cash flow needs should routinely construct cash flow budgets and monitor them against actual cash flows throughout the year. This allows early detection of liquidity problems.

Tests for Liquidity Problems

When overall cash flow becomes tight, several simple tests can be carried out to try to identify the source of the problem.

1. Analyze the debt structure by calculating the percent of total farm liabilities classified as current and noncurrent. Do the same for total farm assets. If the debt structure is *top heavy*, that is, a larger proportion of the farm's liabilities are in the current category than is true for its assets, it may be advisable to convert some of the current debt to a longer-term liability. This will reduce the annual principal payments and improve liquidity.
2. Compare the changes in the current assets over time, particularly crop and livestock inventories. This can be done by looking at the annual year-end balance sheets. Building up inventories can cause temporary cash flow shortfalls. On the other hand, if cash flow has been met by reducing current inventories for several successive years, then the liquidity problem has only been postponed.
3. Compare purchases of capital assets to sales and depreciation. Continually increasing the inventory of capital assets may not leave enough cash to meet other obligations. On the other hand, liquidating capital assets may help meet cash flow needs in the short run, but could affect profitability in the future.
4. Compare the amount of operating debt carried over from one year to the next for a period of several years. Borrowing more than is repaid each year will eventually reduce borrowing capacity and increase interest costs, and make future repayment even more difficult. This only hides the real problem of low profitability.
5. Compare the cash flow needed for land (principal and interest payments, property taxes, and cash rent) per acre to a typical cash rental rate for similar land. Taking on too much land debt, trying to repay it too quickly, or paying high cash rents can cause severe liquidity problems.
6. Compare withdrawals for family living and personal income taxes to the opportunity cost of unpaid labor. The business may be trying to support more people than are fully employed or at a higher level than is possible from net farm income. Any of these symptoms can cause poor liquidity, even for a profitable farm, and can eventually lead to severe financial problems.

Measures of Repayment Capacity

Repayment capacity measures the ability of the business to generate enough cash to meet all debt commitments each year. Several measures are recommended. The first is the *capital debt repayment capacity* (FFSC). This measure provides an estimate of the funds the operation can generate for debt repayment and/or asset replacement.

It is computed using this formula:

Net farm income from operations
Plus total nonfarm income
Plus depreciation/amortization expense
Plus interest expense on term debt
Less total withdrawals for family living and income taxes
Equals capital debt repayment capacity

The *capital debt repayment margin* (FFSC) is the dollar difference between the capital debt repayment capacity and the scheduled principal and interest payments on term loans.

The *debt and capital lease coverage ratio* (FFSC) is calculated by dividing the capital debt repayment capacity by the term debt payments due in the next year, including principal and interest, on amortized loans. The greater the ratio over 1.0, the greater the margin to cover payments. It is important to note that this measure does not indicate whether payments can be made in a timely fashion within the year. A cash flow budget is needed to assess possible problems with on-time payments.

Box 6-2   Farm Business Analysis Case Study

In Tables 6-3 through 6-7, our sample farm was compared to a group of similar farms in a variety of measures of profitability, efficiency, and financial strength. Our sample farm's net farm income was substantially lower than the average of the comparison group. Return to management was negative on our sample farm, and the rate of return on farm equity was very close to zero. In addition, because the rate of return on farm assets was higher than the rate of return on farm equity, borrowed money was operating at an average loss.

Our sample farm had fewer resources than the average of comparison farms, so some of the causes for the lower net farm income can be explained by size. However, the difference in size alone would probably not explain the negative return to management or the extremely low value of the rate of return on farm equity.

Physical efficiency appears to be a problem in the cow-calf enterprise. Calves produced are considerably lighter than those in the comparison group, leading to low production in terms of pounds of calf per cow. At the same time, hay fed per cow is higher than in the comparison group, perhaps because of poor pasture quality or wasting feed. Prices received for the calves are also low, indicating either a quality problem or poor marketing strategies. Cotton yields, while somewhat lower than the average of the comparison group, are not badly out of line, and the cotton price received is about the same as that of the comparison group.

Examination of the economic efficiency measures indicates that this farm appears to be spending too much on operating expenses relative to what is being produced. The lower-than-average depreciation ratio may indicate that the farm has older or smaller machinery than the comparison group, which may contribute to the somewhat lower-than-average cotton yields.

The overall financial strength of the business, as measured by the debt/asset ratio, is about the same as that of the comparison group. The current ratio is much lower than the average of the comparison group, however, and working capital is also considerably lower.

In summary, the cow-calf enterprise appears to be operating less efficiently than it should be, contributing to low overall profits. A more detailed analysis of this enterprise would be warranted. The operator may also need to look carefully at the machinery complement, to see if it is adequate for the farm.

SUMMARY

A whole-farm business analysis is much like a complete medical examination. It should be conducted periodically to see whether symptoms exist that indicate the business is not functioning as it should. Standards for comparison can be budgeted goals, results from other similar farms, or past results from the same farm. A systematic approach can be used to compare these standards to actual results.

Profitability can be measured by net farm income and returns to labor, management, total assets, and equity. If an income or profitability problem is found, the first area to look at is farm size. An analysis of size checks to see if there are enough resources to generate an adequate net income, and how many of the resources are being contributed by the operator. Next, various measures relating to efficient use of machinery, labor, capital, and other inputs can be computed. Economic efficiency depends on physical efficiency, the selling price of products, and the purchase price of inputs.

Financial analysis uses balance-sheet values to evaluate business solvency and liquidity. The various measures calculated for a whole-farm analysis are part of the control function of management. As such, they should be used to identify and isolate a problem before it has a serious negative impact on the business.

QUESTIONS FOR REVIEW AND FURTHER THOUGHT

1. What are the steps in the control function of farm management?

2. What types of standards or values can be used to compare and evaluate efficiency and profitability measures for an individual farm? What are the advantages and disadvantages of each?

3. What are the differences between *gross revenue* and *value of farm production*? Between *net farm income* and *return to management*?

4. Use the following data to calculate the profitability and efficiency measures listed:

Gross revenue	$185,000
Value of farm production	$167,000
Net farm income from operations	$ 48,000
Interest expense	$ 18,000
Value of unpaid labor and management	$ 31,000
Total asset value: beginning	$400,000
ending	$430,000
Farm equity: beginning	$340,000
ending	$352,000

a. Rate of return on assets _____%

b. Rate of return on equity _____%

c. Asset turnover ratio _____

d. Operating profit margin ratio _____%

e. Net farm income from operations ratio _____

5. What three general factors determine economic efficiency?

6. Tell whether each of the following business analysis measures relates to volume of business, profitability, economic efficiency, or physical efficiency.

    a. Pounds of cotton harvested per acre

    b. Return to management

    c. Livestock return per $100 of feed fed

    d. Gross revenue

    e. Total farm assets

    f. Asset turnover ratio

7. Would a crop farm that cash rents 1,200 crop acres and owns 240 acres be more likely to have a better-than-average *asset turnover ratio* or *operating expense ratio*? Explain.

8. Explain the difference between solvency and liquidity. List three tests that can be used to diagnose liquidity problems.

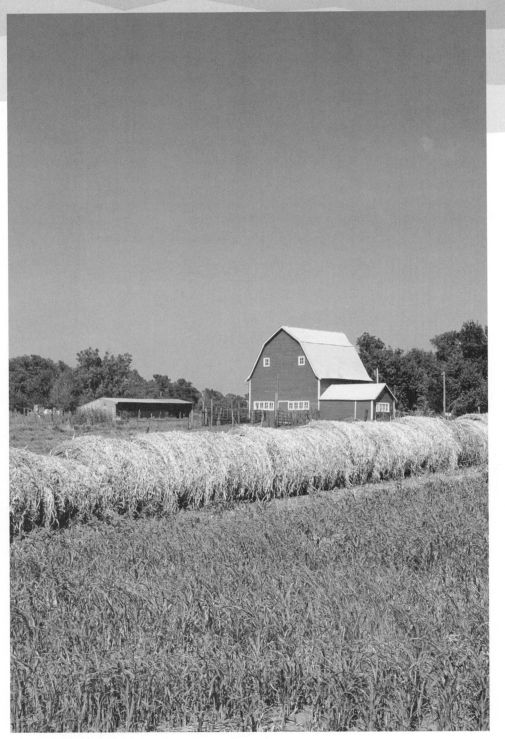

Philip Coblentz/AGE Fotostock

III

APPLYING ECONOMIC PRINCIPLES

Chapters 7, 8, and 9 discuss economic principles and cost concepts and their application to decision making in agriculture. Economic principles are important because they provide a systematic and organized procedure to identify the best alternative when profit maximization is the goal. The decision rules derived from these principles require an understanding of some marginal concepts and different types of costs. However, once they are learned, they can be applied to many different types of management problems.

Chapter 7 introduces the concept of a production function and the law of diminishing marginal returns. Marginal analysis is used to determine the profit-maximizing levels of input and output. In Chapter 8 marginal analysis is extended to answer the questions of what combination of inputs will maximize profits for a given level of output, and what combination of enterprises will maximize profits for a given set of resources. Various cost concepts are explored in Chapter 9, including total, marginal, and average costs, and fixed and variable costs. Finally, the cost concepts are used to explain how increasing farm size can result in both economies and diseconomies of size at different levels of output.

Glow Images

ECONOMIC PRINCIPLES: CHOOSING PRODUCTION LEVELS

CHAPTER OUTLINE

The Production Function
Marginal Analysis
Law of Diminishing Marginal Returns
How Much Input to Use
Using Marginal Concepts
Marginal Value Product and Marginal Input Cost
The Equal Marginal Principle
Summary
Questions for Review and Further Thought

CHAPTER OBJECTIVES

1. Explain the concept of marginal analysis

2. Show the relation between a variable input and an output by use of a production function

3. Describe the concepts of average physical product and marginal physical product

4. Illustrate the law of diminishing marginal returns and its importance in agriculture

5. Show how to find the profit-maximizing amount of a variable input using the concepts of marginal revenue and marginal cost

6. Explain the use of the equal marginal principle for allocating limited resources

Knowledge of economic principles provides the manager with a set of procedures and rules for decision making. This knowledge is useful when making plans for organizing and operating a farm or ranch business. It also provides some help and direction when moving through the steps in the decision-making process. Once a problem has been identified and defined, the next step is to acquire data and information. The information needed to apply economic principles provides a focus and direction to this search and prevents time being wasted searching for unnecessary data. Once the data are acquired, economic principles provide some guidelines for transforming the data into useful information and for analyzing the potential alternatives.

Economic principles consist of a set of rules for making a choice or decision that will result in maximum profit. Several such principles will be introduced in the next two chapters. The application of these rules follows three steps: (1) acquire physical and biological data and observe how certain resources create marketable products, (2) acquire price data for both resources and products, and (3) apply the appropriate economic decision-making rule to maximize profit. The reader should look for each of these steps as each new economic principle is discussed.

THE PRODUCTION FUNCTION

The fundamental basis for agriculture is the biological process of combining various resources to produce a useful product. Figure 7-1 illustrates one example of this production process. A systematic way of showing the relation between the resources or inputs that can be used to produce a product and the corresponding output of that product is called a *production function*. In

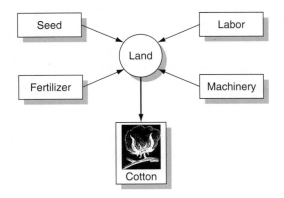

Figure 7-1 Example of the production process in agriculture.

TABLE 7-1	**Production Function for Nitrogen Fertilizer and Corn—One Acre**

Input level	Nitrogen applied (lbs)	Yield (bushels)	Total physical product (TPP) (bushels)	Average physical product (APP) (bushels)	Marginal physical product (MPP) (bushels)
0	0	130	0	—	
					18
1	25	148	18	18.00	
					14
2	50	162	32	16.00	
					8
3	75	170	40	13.33	
					7
4	100	177	47	11.75	
					3
5	125	180	50	10.00	
					2
6	150	182	52	8.67	
					1
7	175	183	53	7.57	
					0
8	200	183	53	6.62	

Source: Department of Agronomy, Iowa State University.

some agricultural disciplines, the same relation may be called a response curve, yield curve, or input/output relation. A production function can be presented in the form of a table, graph, or mathematical equation.

The first three columns of Table 7-1 are a tabular presentation of a production function. Nitrogen fertilizer is the input and corn is the product. All other inputs are assumed to be fixed in supply. Different levels of nitrogen are shown in the second column. To simplify the analysis, nitrogen input is measured in units of 25 pounds per acre. The amount of corn production expected (based on yield trials) from using each input level of nitrogen is shown in the third column. In the fourth column the added yield from applying nitrogen is labeled *total physical product (TPP)*. A yield of 130 bushels of corn per acre is expected without adding any nitrogen. This is because a certain level of fertility is already present in the soil. Applying additional nitrogen allows an even higher yield to be achieved, though.

It is possible to calculate the average amount of output produced per unit of input at each input level. This value is called *average physical product (APP)* and is shown in the fifth column of Table 7-1. APP is calculated by the formula

$$APP = \frac{total\ physical\ product}{input\ level}$$

In this example, 148 bushels of corn can be produced by adding one unit (25 pounds) of nitrogen. One hundred and thirty bushels are produced without applying any nitrogen, so 18 bushels is the physical product that can be attributed to the first unit of added nitrogen. The average physical product from adding the first unit of nitrogen, then, is $18 \div 1 = 18.00$. Likewise, applying 4 units of nitrogen produces $177 - 130 = 47$ more bushels of corn, so the APP per 25-pound unit of N is $47 \div 4 = 11.75$. The production function in Table 7-1 has an APP that declines continuously after the first unit of input, a common occurrence in agriculture.

MARGINAL ANALYSIS

Much of the economics of agricultural production is related to the concept of *marginal analysis*. Economists and managers are often interested in what changes will result from altering one or more factors under their control. For example, they may be interested in how cotton yield changes from using an additional 50 pounds of fertilizer or how 2 additional pounds of grain in the daily feed ration affects milk production from a dairy cow or how profit changes from raising an additional 20 acres of corn and reducing soybean production by 20 acres.

The term *marginal* will be used extensively throughout this chapter. It refers to incremental changes, increases or decreases that occur at the edge or the margin. It may be useful to mentally substitute *extra* or *additional* whenever the word *marginal* is used, remembering that the *extra* can be a negative amount or even zero. Also, any marginal change being measured or calculated is a result of a marginal change in some other factor.

To calculate a marginal change of any kind, it is necessary to find the difference between an original value and the new value that resulted from the change in the controlling factor. In other words, the change in some value caused by the marginal change in another factor is needed. Throughout this text, a small triangle (the Greek letter delta) will be used as shorthand for *the change in*. For example, Δ corn yield would be read as *the change in corn yield* and would be the difference in corn yield before and after some change in an input affecting yield, such as seed, fertilizer, or irrigation water. Although other inputs may also be necessary, they are assumed to be present in a fixed or constant amount in the short run. This does not mean they are unimportant, but this assumption serves to simplify the analysis.

Marginal Physical Product

Marginal analysis and production function data can be used to derive additional information about the relation between the input and TPP.

The first marginal concept to be introduced is *marginal physical product (MPP)*, shown in the sixth column of Table 7-1. Remembering that marginal means additional or extra, MPP is the additional or extra TPP produced by using an additional unit of input. It requires measuring changes in both output and input.

Marginal physical product is calculated as

$$MPP = \frac{\Delta\ total\ physical\ product}{\Delta\ input\ level}$$

The numerator is the change in TPP caused by a change in the variable input, and the denominator is the actual amount of change in the input. For example, Table 7-1 indicates that using 4 units of nitrogen instead of 3 causes TPP to increase by 7 bushels. Thus, the value for MPP at this level is $7 \div 1 = 7$. Marginal physical product can be either positive or negative. It can also be zero if changing the input level causes no change in TPP. In the example in Table 7-1, the input increases by 1-unit increments, which simplifies the calculation of MPP. Other examples and problems may show the input increasing by increments of varying amounts.

The result for MPP is not an exact determination for the last pound of nitrogen but an average MPP for the last 25-pound unit added. Many times, this calculation will provide sufficient information for decision making, unless the change in input between two possible input levels is fairly large. In this case, either more information should be obtained about expected output levels for intermediate levels of the variable input or other mathematical techniques such as graphs or calculus should be used.[1]

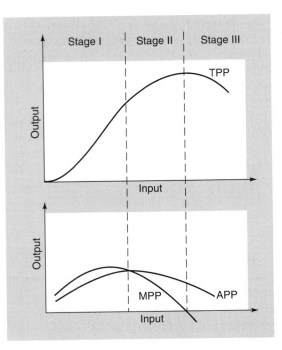

Figure 7-2 Graphical illustration of a production function.

LAW OF DIMINISHING MARGINAL RETURNS

Table 7-1 and Figure 7-2 can be used to illustrate a principle important from both a theoretical and a practical standpoint. The term *diminishing marginal returns* is used to describe what happens to MPP as additional input is used. The law of diminishing marginal returns states that as additional units of a variable input are used in combination with one or more fixed inputs, marginal physical product will eventually decline. Where MPP begins to decline depends entirely on the biological characteristics of the production process being analyzed. Three properties of this law and its definition need to be emphasized. First, for diminishing returns to exist, one or more

fixed inputs must be used in the production process in addition to the variable input. One acre of land or one head of livestock is often the fixed input used to define the production function and, therefore, to illustrate diminishing marginal returns. Second, the definition does not preclude diminishing marginal returns from beginning with the first unit of the variable input. This is often the case in agricultural applications of this law. Third, this law is based on the biological processes found in agricultural production. It results from the inability of plants and animals to provide the same response indefinitely to successive increases in nutrients or some other input.

Numerous examples of diminishing marginal returns exist in agricultural production. As additional units of seed, fertilizer, or irrigation water are applied to a fixed acreage of some crop, the additional output or MPP will eventually begin to decline. The MPP gets smaller and smaller as the crop nears its biological capacity to use the input. A similar result is obtained with additional daily feed fed to dairy cows, hogs, or cattle on feed. The importance and practical significance of these examples will become apparent in the next two sections.

A Graphical Analysis

A production function (TPP) and its corresponding APP and MPP can also be shown in a graphical form, as in Figure 7-2. The top portion of Figure 7-2 shows TPP, or output, increasing at an increasing rate as the input level is increased from zero. As the input level is increased further, TPP continues to increase, but at a decreasing rate, and eventually peaks and begins to decline.

Figure 7-2 illustrates several important relations among TPP, APP, and MPP. As long as TPP is increasing at an increasing rate, both MPP and APP are increasing as well. At the point where TPP changes from increasing at an increasing rate to increasing at a decreasing rate, MPP reaches its maximum and then continuously declines. Where TPP reaches its maximum, MPP has a value of zero, and where TPP is declining, MPP is negative.

Average physical product increases over a slightly longer range than MPP before beginning to decline, which demonstrates an interesting relation between APP and MPP. Whenever MPP is above APP, APP is increasing, and whenever MPP is below APP, APP is decreasing. This relation can be explained by remembering that to raise (lower) any average value, the additional or marginal value used in calculating the new average must be above (below) the old average. The only way baseball players can raise their batting averages is to have a daily batting average (the marginal or additional value) above their current season average. The APP curve is at its maximum where the MPP curve crosses it.

The relation among TPP, APP, and MPP often is used to divide this particular type of production function into three stages, as shown in Figure 7-2. The importance of these stages in determining the proper amount of input to use

will be discussed later. Stage I begins at the zero input level and continues to the point where APP is maximum and equal to MPP. Stage II begins where APP is maximum and ends where MPP is zero (or TPP is maximum). Stage III is the range of input levels where MPP is negative and TPP is declining absolutely. The data in Table 7-1 show only Stage II of the nitrogen and corn production function, that is, the stage where MPP is declining but not yet negative.

HOW MUCH INPUT TO USE

An important use of the information derived from a production function is to determine how much of the variable input to use. Given a goal of maximizing profit, the manager must select from all possible input levels the one that will result in the greatest profit.

Some help in this selection process can be found by looking at the three stages in Figure 7-2. Any input level in Stage III can be eliminated from consideration, because additional input causes MPP to be negative and TPP to decrease.

Stage I covers the area where adding additional units of input causes the average physical product to increase. In this stage, adding another unit of input increases the productivity of all previous inputs, as measured by average productivity or APP. If any input is to be used, it seems reasonable that a manager would want to use at least that input level that results in the greatest average physical product per unit of input. This point is the boundary of Stage I and Stage II and represents the greatest efficiency in the use of the variable input. However, profit usually can be increased further by using more input even though APP is declining along with MPP. Using only the physical information available from TPP, APP, and MPP, though, it is not possible to determine which input level in Stage II will maximize profit. More information is needed, specifically information about input and output prices plus some way to incorporate it into the decision-making process.

Box 7-1   Other Diminishing Returns

The law is specifically described as the law of diminishing marginal returns and is defined in terms of MPP. However, after some input level, both TPP and APP may also begin to decline, and the term *diminishing returns* is sometimes used to describe these results. However, the diminishing MPP is the more interesting and useful concept.

Total Cost, Total Revenue, and Profit

If we assume that the manager's goal is to maximize profit, at least in regard to fertilizer use, then we need to know how much the fertilizer will cost and for what price the corn can be sold. In Table 7-2 the expected level of corn production (TPP) for each level of nitrogen application is shown, just as in Table 7-1. However, columns for total cost (TC), total revenue (TR), and profit have been added.

At the zero nitrogen level the total cost is $750. This represents all the other costs of producing corn besides nitrogen. These will be assumed to stay constant as we add more nitrogen. If we multiply each level of nitrogen by the price of a pound of nitrogen (assume $0.50) and add it to the other costs of $750, we can show the TC of corn production for each level of nitrogen use in the fourth column of Table 7-2.

Likewise, if we multiply the bushels produced at each level of nitrogen applied by the price of corn (assume $5.00 per bushel), we can show the TR at each level in the fifth column. Finally, by subtracting TC from TR at each level, we arrive at the profit, shown in the sixth column. A quick comparison shows that the maximum profit level is $87.50, which results when 125 pounds of nitrogen is applied to produce 180 bushels of corn per acre.

USING MARGINAL CONCEPTS

The profit-maximizing level of use of an input can also be found by examining the marginal or incremental changes in costs and revenue as more input is added.

Table 7-3 is similar to the previous table, except the last two columns are now called marginal revenue and marginal cost. *Marginal revenue (MR)* is defined as the change in total revenue or the additional income received from selling one more unit of output. It is calculated from the equation

$$MR = \frac{\Delta \text{ total revenue}}{\Delta \text{ total physical product (MPP)}}$$

TABLE 7-2	Total Cost, Total Revenue, and Profit (Nitrogen Price = $0.50 per lb; Corn Price = $5.00 per bushel)

Input level	Nitrogen applied (lbs)	Total physical product (TPP) (bushels)	Total cost (TC) ($)	Total revenue (TR) ($)	Profit ($) (TR – TC)
0	0	130	750.00	650.00	(100.00)
1	25	148	762.50	740.00	(22.50)
2	50	162	775.00	810.00	35.00
3	75	170	787.50	850.00	62.50
4	100	177	800.00	885.00	85.00
5	**125**	**180**	**812.50**	**900.00**	**87.50**
6	150	182	825.00	910.00	85.00
7	175	183	837.50	915.00	77.50
8	200	183	850.00	915.00	65.00

TABLE 7-3	**Marginal Revenue, Marginal Cost, and the Optimum Output Levels (Nitrogen Price = $.50 per lb; Corn Price = $5.00 per bushel)**						

Input level	Nitrogen applied (lbs)	Total physical product (TPP) (bushels)	Marginal physical product (MPP)	Total revenue (TR) ($)	Total cost (TC) ($)	Marginal revenue (MR) ($) (ΔTR/ΔTPP)	Marginal cost (MC) ($) (ΔTC/ΔTPP)
0	0	130		650.00	750.00		
			18			5.00 >	0.69
1	25	148		740.00	762.50		
			14			5.00 >	0.89
2	50	162		810.00	775.00		
			8			5.00 >	1.56
3	75	170		850.00	787.50		
			7			5.00 >	1.79
4	100	177		885.00	800.00		
			3			5.00 >	4.17
5	**125**	**180**		**900.00**	**812.50**		
			2			5.00 <	6.25
6	150	182		910.00	825.00		
			1			5.00 <	12.50
7	175	183		915.00	837.50		
			0			5.00 <	Infinite
8	200	183		915.00	850.00		

For example, when the first 25-pound unit of nitrogen is added, TR increases by $90.00 and MPP is 18 bushels, so MR = $90.00 ÷ 18 = $5.00. It is apparent from Table 7-3 that MR is constant for all input levels and equal to the output selling price, that is, $5.00 per bushel of corn. This should not be surprising given the definition of MR. Provided the output price does not change as more or less output is sold, typical for an individual farmer or rancher, MR will always equal the price of the output. However, if the selling price varies with changes in the quantity of output sold, MR must be calculated, using the equation previously provided, at each output level. An example is when hogs or cattle are fed to higher and higher weights and eventually the price received per pound decreases due to discounts.

Marginal cost (MC) is defined as the change in cost, or the additional cost incurred, from producing another unit of output. It is calculated from the equation

$$MC = \frac{\Delta \text{ total cost}}{\Delta \text{ total physical product (MPP)}}$$

where total cost is the same as previously defined in Table 7-2. For example, when the level of nitrogen applied increases from zero to 1 unit, total cost increases by $0.50 × 25 pounds = $12.50, and MPP is 18 bushels, so MC = $12.50 ÷ 18 = $0.69. In Table 7-3, marginal cost begins to increase as additional input is used. Notice the inverse relation between MPP and MC. When MPP is declining (diminishing marginal returns), MC is increasing, because it takes relatively more input to produce an additional unit of output. Therefore, the additional cost of another unit of output is increasing.

Decision Rule

Next, MR and MC are compared to find the profit-maximizing input and output levels. As long as MR is greater than MC, an additional unit of output increases profit, because the additional income exceeds the additional cost of producing it. Conversely, if MR is less than MC, producing the additional unit of output will decrease profit. The profit-maximizing output level is therefore where marginal revenue equals marginal cost.

In Table 7-3, when the level of nitrogen applied is increased from 100 to 125 pounds, MR is $5.00 and MC is only $4.17, so it clearly pays to add more nitrogen. However, increasing the rate of nitrogen application from 125 to 150 pounds has an MC of $6.25, which exceeds the MR of $5.00. Therefore, profit is maximized by applying 5 units (125 pounds) of nitrogen per acre. Not surprisingly, this is the same conclusion reached by comparing the expected profit at each level of input use. In this example it is not possible to determine at what level of nitrogen

application MC exactly equals MR, so we must look at where MC goes from being less than MR to being greater than MR.

What would happen to the profit-maximizing level if one or more of the prices changed? A change in the price of nitrogen would cause MC to change, and a change in the price of corn would cause MR to change. Such changes are likely to change the profit-maximizing solution. Assuming that the corn price is constant at $5.00, and the price of nitrogen increases to $0.70 per pound, the marginal cost of increasing nitrogen application from 100 to 125 pounds per acre is now $5.83 per bushel of corn (Table 7-4), more than the price of corn (MR). Thus, the profit-maximizing level now is only 100 pounds of nitrogen, where MC equals $2.50, applied to produce 177 bushels of corn.

Similarly, if the price of nitrogen stayed at $0.50 per pound but the price of corn increased to $7.00 per bushel, we could apply up to 150 pounds of nitrogen and produce 182 bushels of corn before MC exceeded MR. This is shown in the last two columns of Table 7-4. There must be a change in the *relative* prices. Doubling both prices or decreasing both by 50 percent will change the numbers in Table 7-4 but will not change the choice of input and output levels. One price must move proportionately more than the

---

Box 7-2   New Technology and Production Decisions

In the past, farmers had little choice but to optimize input and output levels using marginal values computed from a field's *average* response to changes in input levels. Yet most fields have areas with different soil types, fertility levels, slope, drainage, pH, weed problems, and so forth. New technology is now available, including global positioning systems (GPS) to find exact locations in a field, yield monitors that use GPS to record yield by location in a field, and precision application

equipment that can adjust seed, fertilizer, and pesticide application rates *on the go* based on field location. Farmers now have new information and the equipment needed to apply the marginal principles and optimize input and output levels on each different area in a field. This can be done by soil type, fertility level, soil pH, or any other factor that affects productivity. Optimizing the application of crop inputs on small areas of land is known as *precision agriculture*.

| TABLE 7-4 | | Marginal Revenue and Marginal Cost Under Varying Prices | | | | | |

Input level	Nitrogen applied (lbs)	Total physical product (TPP) (bushels)	Marginal revenue (MR) ($) (corn price = $5.00)		Marginal cost (MC) ($)* (nitrogen price = $0.70)	Marginal revenue (MR) ($) (corn price = $7.00)		Marginal cost (MC) ($)* (nitrogen price = $0.50)
0	0	130						
			5.00	>	0.97	7.00	>	0.69
1	25	148						
			5.00	>	1.25	7.00	>	0.89
2	50	162						
			5.00	>	2.19	7.00	>	1.56
3	75	170						
			5.00	>	2.50	7.00	>	1.79
4	**100**	**177**						
			5.00	<	5.83	7.00	>	4.17
5	125	180						
			5.00	<	8.75	7.00	>	6.25
6	**150**	**182**						
			5.00	<	17.50	7.00	<	12.50
7	175	183						
			5.00	<	Infinite	7.00	<	Infinite
8	200	183						

*Marginal cost is calculated by dividing the cost of 25 pounds of nitrogen by the MPP.

other, remain constant, or move in the opposite direction to cause a change in the relative prices.

Price Ratios and Profit Maximization

The relation between prices and the production function can be viewed another way. Given constant prices, MR = P_o, where P_o is the price of the output, and MC = P_i/MPP, where P_i is the price of the input. Substituting these equalities for MR and MC in the profit-maximizing rule MR = MC results in P_o = P_i/MPP. Next, divide each side of this equation by P_o and multiply by MPP. The result is

$$MPP = \frac{P_i}{P_o}$$

which says that MPP will equal the ratio of the input price to the output price at the profit-maximizing input level.

What happens if this ratio changes? If it decreases (P_i decreases or P_o increases), MPP must be smaller to maximize profit. More input should be used as MPP decreases with higher input levels. Conversely, if the price ratio increases (P_i increases or P_o decreases), less input should be used to reach a higher MPP, needed to maintain the equality. These results show the importance of prices, specifically the relationship between input and output prices, in determining the profit-maximizing input level.

Take another look at Table 7-3. The price ratio in this example is $12.50 per unit of nitrogen (25 pounds @ $0.50) divided by $5.00, the price of corn, or 2.5. As long as MPP is greater than 2.5 it pays to add more nitrogen. But as soon as MPP falls below 2.5 (moving from level 5 to level 6), adding more nitrogen reduces profits.

MARGINAL VALUE PRODUCT AND MARGINAL INPUT COST

For some decisions it is convenient to compare the added revenue and added cost measured in terms of dollars per unit of input instead of output. The value of the revenue produced by adding one more unit of input is called the *marginal value product (MVP)*. It is calculated by dividing the change in revenue as we move from one input

level to the next by the units of added input. The MVP can be compared to the *marginal input cost (MIC)*, the unit price of the input being added. MIC is similar to marginal cost, but is measured in dollars per unit of input instead of unit of output. MVP and MIC for the corn and nitrogen example are shown in Table 7-5.

Theoretically, the MR = MC rule should be used to find the profit-maximizing output level, and the MVP = MIC rule should be used to find the profit-maximizing input level. However, once either of these levels has been found, the other can be found by referring to the corresponding value of the production function. What is important is to be consistent when comparing them. Always use MVP and MIC together,

TABLE 7-5	Marginal Value Product, Marginal Input Cost, and the Optimum Output Levels (Nitrogen Price = $0.50 per lb; Corn Price = $5.00 per bushel)

Input level	Nitrogen applied (lbs)	Yield (bushels)	Marginal physical product (MPP)	Total revenue (TR) ($)	Total cost (TC) ($)	Marginal value product (MVP) ($) ($\Delta TR/\Delta IP$)		Marginal input cost (MIC) ($) ($\Delta TC/\Delta IP$)
0	0	130		650.00	750.00			
			18			3.60	>	0.50
1	25	148		740.00	762.50			
			14			2.80	>	0.50
2	50	162		810.00	775.00			
			8			1.60	>	0.50
3	75	170		850.00	787.50			
			7			1.40	>	0.50
4	100	177		885.00	800.00			
			3			0.60	>	0.50
5	**125**	**180**		**900.00**	**812.50**			
			2			0.40	<	0.50
6	150	182		910.00	825.00			
			1			0.20	<	0.50
7	175	183		915.00	837.50			
			0			0.00	<	0.50
8	200	183		915.00	850.00			

because both are computed per unit of *input*. Similarly, always use MR and MC together, because both are computed per unit of *output*.

THE EQUAL MARGINAL PRINCIPLE

The discussion so far in this chapter has assumed that sufficient input is available or can be purchased to set MR equal to MC for every acre or

head of livestock. Another, possibly more likely, situation is when only a limited amount of input is available, which prevents reaching the MR = MC point for all possible uses. Now the manager must decide how the limited input should be allocated or divided among several possible uses or alternatives. Decisions must be made on the best allocation of labor among many acres, fields, and different crops; irrigation water among fields and crops; and pasture among different types and

---

| *Box 7-3* | **Using Calculus to Find the Profit-Maximizing Input and Output Levels** |

If the production function for transforming units of input into units of output can be expressed in equation form, and if the function is twice differentiable, calculus can be used to find the optimal amount of input to use or output to produce.

Let Π be the profit function, P be the price of the output (Y), and W the price of the input (X). The problem is expressed as

$$\text{Maximize } \Pi = P*Y(X) - WX$$

We take the derivative with respect to X:

$$d\Pi/dX = P(dY/dX) - W = 0$$

To be a maximum, second-order conditions must be met. That is, the second derivative with respect to X must be negative: $F''(X) < 0$.

For example, let the production function be $Y = 6X^{1/2}$, the price of the output be $20, and the price of the input be $12. Thus, we would solve the following equation for X:

$$d\Pi/dX = 20(3X^{-1/2}) - 12 = 0$$

By rearranging terms, we have

$$60X^{-1/2} = 12$$

Multiplying both sides by $X^{1/2}$ and dividing both sides by 12 yields

$$5 = X^{1/2}$$

Squaring both sides, we find that $X = 25$ is the solution.

To check that this is truly a maximum point, we take the second derivative of the profit function with respect to X. In this case, the second derivative with respect to X ($-30X^{-3/2}$) is always negative for any positive value of X; hence, the profit-maximizing amount of input to use is 25 units, giving an output of 30.

$$Y = 6(25^{1/2}) = 30$$

If the cost function is known, rather than the production function, the problem would be expressed in this fashion:

$\Pi = P*Y - C(Y)$, where C is the cost function.

In our example, letting fixed costs be zero, the cost function would be

$$C = 12X$$

Since, $Y = 6X^{1/2}$, by squaring both sides and rearranging, we get

$$X = Y^2/36$$

Thus, cost can be expressed in terms of Y as $12(Y^2/36)$, and the problem becomes

$$\text{Maximize } \Pi = 20*Y - 12*(Y^2/36)$$

Taking the derivative, setting it to zero, and solving for Y, we get $Y = 30$, as before. Adding fixed costs to the cost function will not change the solution, as fixed costs do not change when output changes.

weights of livestock. If capital is limited, it first must be allocated somehow among the purchases of livestock, machinery, land, or other inputs.

The *equal marginal principle* provides the guidelines and rules to ensure that the allocation is done in such a way that overall profit is maximized from the use of any limited input. This principle can be stated as follows: *A limited input should be allocated among alternative uses in such a way that the marginal value products of the last unit used on each alternative are equal.* Table 7-6 shows an application of this principle, in which irrigation water must be allocated among three crops in three fields of equal size. The MVPs are obtained from the production functions relating water use to the yield of each crop and from the respective crop prices. Assume that a maximum of 2,400 acre-inches of water is available and can be applied only in increments of 4 acre-inches. The limited supply of water

would be allocated among the three crops in the following manner, using the MVPs to make the decisions. The first 400 acre-inches (4 acre-inches on 100 acres) would be allocated to cotton, because it has the highest MVP, at $1,800. The second 400 acre-inches would be allocated to grain sorghum, because it has the second highest MVP, $1,600. In a similar manner, the third 400 acre-inches would be used on cotton (MVP = $1,500), and the fourth, fifth, and sixth 400-acre-inch increments on wheat, grain sorghum, and cotton, respectively ($1,200 each). Each successive 400-acre-inch increment is allocated to the field that has the highest MVP remaining after the previous allocations.

The final allocation is 4 acre-inches on wheat, 8 on grain sorghum, and 12 on cotton. Each final 4-acre-inch increment on each crop has an MVP of $1,200, which satisfies the equal marginal principle. If more water were available, the final allocation could be different. For example, if 3,600 acre-inches were available, it would be allocated 8 acre-inches to wheat, 12 to grain sorghum, and 16 to cotton. This equates the MVPs of the last 4-acre-inch increment on each crop at $800, which again satisfies the equal marginal principle. When inputs must be applied in fixed increments, it may not be possible to exactly equate the MVPs of the last units applied to all alternatives. However, the MVP of the last unit allocated should always be equal to or greater than the MVP available from any other alternative use.

When allocating an input thought to be limited, care must be taken not to use it past the point where MVP equals MIC for any alternative. This would result in less-than-maximum profit. The input is not really limited if a sufficient amount is available to reach or exceed the point where MVP equals MIC.

The profit-maximizing property of this principle can be demonstrated for the 2,400-acre-inch example given. If the 4 acre-inches used on wheat were allocated to grain sorghum or cotton, instead, $1,200 of income would be lost and $800 gained, for a net loss of $400. The same loss would be incurred if the last 4-acre-inch increment was

Table 7-6	**Application of the Equal Marginal Principle to the Allocation of Irrigation Water[*]**

	Marginal value products ($)		
Irrigation water (acre-inch)	Wheat (100 acres)	Grain sorghum (100 acres)	Cotton (100 acres)
0			
	1,200[4th]	1,600[2d]	1,800[1st]
4			
	800	1,200[5th]	1,500[3d]
8			
	600	800	1,200[6th]
12			
	300	500	800
16			
	50	200	400
20			

[*]Each application of 4 acre-inches on a crop is a total use of 400 acre-inches (4 inches times 100 acres).

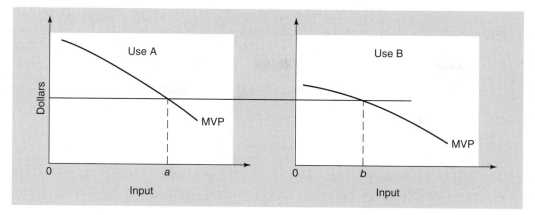

Figure 7-3 **Illustration of the equal marginal principle.**

removed from either grain sorghum or cotton and reallocated to another crop. When the MVPs are equal, profit cannot be increased by a different allocation of the limited input.

The equal marginal principle can also be presented in a graphical form, as in Figure 7-3, where there are only two alternative uses for the limited input. The problem is to allocate the input between the two uses, keeping the MVPs equal, so that input quantity 0a plus quantity 0b just equals the total amount of input available. If 0a plus 0b is less than the total input available, more should be allocated to each use, again keeping the MVPs equal, until the input is fully used. There

would have to be a decrease in the input used on both alternatives if 0a plus 0b exceeded the total input available.

The equal marginal principle applies not only to purchased inputs but also to those already owned or available, such as land, the manager's time, and machinery time. Its use also prevents making the mistake of maximizing profit from one enterprise and not having enough of an input left to use on other enterprises. Maximizing profit from the *total* business requires the proper allocation of limited inputs among competing enterprises, which will not necessarily result in maximizing profit from any *single* enterprise.

Box 7-4   Special Considerations in Livestock Production

The law of diminishing marginal returns applies to backgrounding feeder calves as well as to crop production. Backgrounding feeder calves involves taking weaned beef calves and putting them on pasture and/or a growing ration for the purpose of adding weight before the feeders go to

the feedlot for finishing. As the feeder calf gets heavier, more nutrients are used for maintenance, so the conversion rate of feed to weight gain becomes less efficient. In addition, a complicating factor in feeder calf production is that the selling price in dollars per pound usually decreases as

(Continued)

the feeder calf gets heavier, so that marginal revenue is not a constant value, but decreases with increased weight.

In the example in Figure 7-4, the feeder steer in this backgrounding system should be sold to a feedlot for finishing at 800 pounds. Beyond that weight, the additional cost of adding weight becomes more than the additional revenue from selling a heavier feeder calf. Thus, the producer in a feeder calf backgrounding system must monitor the rate of gain, the cost of the additional gain, and changes in revenue (due to changes in market prices) carefully to avoid holding cattle beyond their economic optimum selling weight.

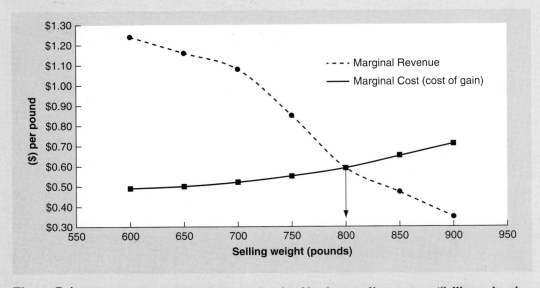

Figure 7-4 **Optimal selling weight for steers in a beef backgrounding system. (Selling prices in dollars per pound: 550 pounds, $1.72; 600 lbs, $1.68; 650 lbs, $1.64; 700 pounds, $1.60; 750 pounds, $1.55, 800 pounds, $1.49; 850 pounds, $1.43; 900 pounds, $1.37.)**

SUMMARY

Economic principles that use the concept of marginal analysis provide useful guidelines for managerial decision making. They have direct application to the basic decisions of how much to produce, how to produce, and what to produce. The production function that describes the relation between input levels and corresponding output levels provides some basic technical information. When this information is combined with price information, the profit-maximizing input and output levels can be found.

Marginal revenue and marginal cost are equated to find the profit-maximizing input/output level. These are marginal concepts that measure the changes in revenue and cost that result from small changes in input levels. When a limited amount of input is available and there are several alternative uses for it, the equal marginal principle provides the rule for allocating the input and maximizing profit under these conditions.

A manager often will not have sufficient information to fully use the economic principles discussed in this chapter. This does not detract from the importance of these principles, but their application and use are often hindered by insufficient physical and biological data. Prices must also be estimated before the product is available for sale, which adds more uncertainty to the decision-making process. However, a complete understanding of the economic principles permits making changes in the right direction in response to price and other changes. A manager should continually search for better information to use in refining the decisions made through the use of these basic economic principles.

QUESTIONS FOR REVIEW AND FURTHER THOUGHT

1. In Table 7-2, assume that the prices of both the input and the output have doubled. Calculate the new TCs and TRs, and determine the profit-maximizing input level for the new prices. Now assume that both prices have been cut in half and repeat the process. Explain your results.

2. Use several other price combinations for the production function data in Table 7-3, and find the profit-maximizing input and output for each price combination. Pay close attention to what happens to these levels when the input price increases or decreases or when the output price increases or decreases.

3. In Table 7-3, what would the profit-maximizing input level be if the input price was $0? What would be true about TPP and MPP at this point?

4. How does the law of diminishing marginal returns cause MC to increase?

5. Find an extension service or experiment station publication for your state that shows the results of a seeding, fertilizer, or irrigation rate experiment. Do the results exhibit diminishing returns? Use the concepts of this chapter and current prices to find the profit-maximizing amount of the input to use.

6. Does the equal marginal principle apply to personal decisions when you have limited income and time? How do you allocate a limited amount of study time when faced with three exams on the same day?

7. Freda Farmer can invest capital in $100 increments and has three alternative uses for the capital, as shown in the following table. The values in the table are the marginal value products for each successive $100 of capital invested.

Capital invested	Fertilizer ($)	Seed ($)	Chemicals ($)
First $100	400	250	350
Second $100	300	200	300
Third $100	250	150	250
Fourth $100	150	105	200
Fifth $100	100	90	150

a. If Freda has an unlimited amount of capital available and no other alternative uses for it, how much should she allocate to each alternative?

b. If Freda can borrow all the capital she needs for one year at 10 percent interest, how much should she borrow and how should it be used?

c. Assume that Freda has only $700 available. How should this limited amount of capital be allocated among the three uses? Does your answer satisfy the equal marginal principle?

d. Assume that Freda has only $1,200 available. How should this amount be allocated? What is the total income from using the $1,200 this way? Would a different allocation increase the total income?

ECONOMIC PRINCIPLES: CHOOSING INPUT AND OUTPUT COMBINATIONS

CHAPTER OUTLINE

Input Combinations
Output Combinations
Summary
Questions for Review and Further Thought

CHAPTER OBJECTIVES

1. Explain the concept of substitution as it is used in marginal analysis and decision making

2. Demonstrate how to compute a substitution ratio and a price ratio for two inputs

3. Show how to use input substitution and ratios to find the least-cost combination of two inputs

4. Describe the characteristics of competitive, supplementary, and complementary enterprises

5. Show the use of output substitution and profit ratios to find the profit-maximizing combination of two outputs

Substitution takes place in the daily lives of most people. It occurs whenever one product is purchased or used in place of another or whenever personal income is spent for one type or class of product rather than another. Steak replaces hamburger on the dinner table, a new

139

car replaces the old model in the garage, and a new brand of soap or toothpaste is purchased instead of the old brand. Some substitutions or replacements are made due to an increase (or decrease) in one's personal income. The motivation for others comes from changes in relative prices or perceived differences in quality.

Substitution also takes place in the production of goods and services. Usually there is more than one way to produce a product or provide a service. Machinery, computers, and robotics can substitute for labor, and one livestock ration ingredient can be replaced by another. Prices, specifically relative prices, play a major role in determining whether and how much substitution should take place. In many production activities, including those on farms and ranches, the substitution is not an all-or-nothing decision. It is often a matter of making some relatively small changes in the ratio or mix of two or more inputs being used. Substitution is therefore another type of marginal analysis that considers the changes in costs and/or revenue when making the substitution decision.

INPUT COMBINATIONS

One basic production decision that a farm or ranch manager has to make, how much of a product to produce, was discussed in Chapter 7.

A second basic decision is what resources to use to produce a given amount of some product. Most products require two or more inputs in the production process, but the manager can often choose the input combination or ratio to be used. The problem is one of determining whether more of one input can be substituted for less of another and thereby reduce total input cost. This leads to the least-cost combination of inputs to produce a given amount of output.

Substitution of one input for another occurs frequently in agricultural production. One type of grain can be substituted for another in a livestock ration. Herbicides substitute for mechanical cultivation, and computers can replace labor. The manager must select the combination of inputs that will produce a given amount of output or perform a certain task for the least cost. In other words, the problem is to find the least-cost combination of inputs, because this combination will maximize the profit from producing a given amount of output. The alert manager will always be looking for a different input combination that will do the same job for less cost.

The least-cost input combination is not always the same. Changes in the price of one or more inputs may make it profitable to substitute one resource for another or at least change the proportion in which they are used.

Input Substitution Ratio

The first step in analyzing a substitution problem is to determine whether it is physically possible to make a substitution and at what rate. Figure 8-1 illustrates the more common types of substitution between two inputs. In Figure 8-1a, the line PP' is an *isoquant* (from isoquantity, meaning the same quantity) and shows a number of possible combinations of corn and barley in a livestock feed ration. This line is called an isoquant because any of these combinations will produce the same quantity of output or weight gain in this example.

The input substitution ratio, or the rate at which one input will substitute for another, is determined from the equation

$$\text{Input substitution ratio} = \frac{\text{amount of input replaced}}{\text{amount of input added}}$$

where both the numerator and the denominator are the differences or changes in the amount of inputs being used between two different points on the isoquant PP', measured in physical units.

Constant Substitution Ratio

In Figure 8-1a, moving from point A to point B means 4 pounds of corn are being replaced by 5 additional pounds of barley to produce the same weight gain or output. The input substitution ratio

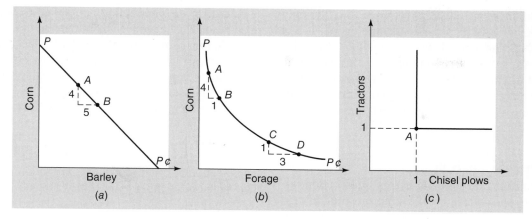

Figure 8-1 Three possible types of substitution.

is 4 ÷ 5 = 0.8, which means 1 pound of barley will replace 0.8 pounds of corn. PP' is a straight line, so the input substitution ratio will always be 0.8 between any two points on this isoquant. This is an example of a constant rate of substitution between two inputs. Whenever the input substitution ratio is equal to the same numerical value over the full range of possible input combinations, the inputs exhibit a constant rate of substitution. This occurs most often when the two inputs contribute the same or nearly the same factor to the production process. Corn and barley, for example, both contribute energy to a feed ration.

Decreasing Substitution Ratio

Another, and perhaps more common, example of physical substitution is shown in Figure 8-1b. Here the isoquant PP' shows the different combinations of corn and forage that might produce the same weight gain on a steer or the same milk production from a dairy cow. The amount of corn that can replace a given quantity of forage changes, depending on whether more corn or more forage is being used. The input substitution ratio is 4 ÷ 1 = 4 when moving from point A to point B on isoquant PP', but it is 1 ÷ 3 = 0.33 when moving from point C to point D. In this example, the input substitution ratio depends on the location on the isoquant and will decline

with any movement down and to the right on the curve. This is an illustration of a decreasing rate of substitution.

Many agricultural substitution problems have a decreasing input substitution ratio, which occurs when the two inputs are dissimilar or contribute different factors to the production process. As more of one input is substituted for another, it becomes increasingly difficult to make any further substitution and still maintain the same level of output. More and more of the added input is needed to substitute for a unit of the input being replaced, which causes the input substitution ratio to decrease. This is an indication that the inputs work together best when used at some combination containing a relatively large proportion of each. At low levels of one, there is a near surplus of the other, and it is usually not an efficient or productive combination.

No Substitution Possible

Another possible substitution situation is where no substitution is possible. Figure 8-1c illustrates one example with tractors and chisel plows. It takes one of each to make a working combination as shown by point A. The right-angled isoquant indicates that more than one tractor with just one chisel plow does not increase production but would increase costs. The same is true for more than one chisel plow. Therefore, the only efficient, and the least-cost combination, is one tractor and one chisel plow. Other examples might be the input combinations of fence posts and wire for a given fence type and many chemical reactions that require a fixed proportion of chemicals to obtain the desired chemical reaction and output.

The Decision Rule

Identifying the type of physical substitution that exists and calculating the input substitution ratio are necessary steps, but they alone do not permit a determination of the least-cost input combination. Input *prices* are also needed, and the ratio of the input prices must be compared to the input substitution ratio. The input price ratio is computed from the equation

$$\text{Input price ratio} = \frac{\text{price of input being added}}{\text{price of input being replaced}}$$

It is sometimes called the *inverse* price ratio, because it is the ratio of the *added* input price divided by the *replaced* input price. This is inverse to the input substitution ratio, the amount of *replaced* input divided by the amount of *added* input.

With this ratio, the least-cost combination of inputs can be found. The decision rule for finding this combination is where the

$$\text{Input substitution ratio} = \text{input price ratio}$$

Table 8-1 is an application of this procedure. Each of the feed rations is a combination of grain and hay that will put the same pounds of gain on a feeder steer. The problem is to select the ration that produces this gain for the least cost. The fourth and fifth columns of Table 8-1 contain the input substitution ratio and input price ratio for each feed ration. The input substitution ratio declines as more grain and less hay are used, while the input price ratio is constant for the given prices. As long as the substitution ratio is

greater than the price ratio, the total cost of the feed ration can be reduced by moving to the next lower ration combination in the table. The converse is true if the substitution ratio is less than the price ratio. This can be verified by calculating the total cost of each ration, as shown in the last column of Table 8-1. The least-cost ration combination is 1,125 pounds of grain and 625 pounds of hay, which has a cost of $138.75.

The decision rule states that the least-cost combination is where the input substitution ratio just equals the input price ratio. However, in Table 8-1 and many other problems where the choices involve a discrete rather than an infinite number of combinations, there is no combination where the two ratios are exactly equal. In these cases, select the combination where the input substitution ratio changes from being greater than the input price ratio to being less than the input price ratio. Ration E is therefore the least-cost combination for this problem. Going from Ration E to the next choice, Ration F, the substitution ratio is less than the price ratio and therefore represents a move to a higher-cost ration—$.75 higher, $139.50 versus $138.75 for Ration E.

Using the input substitution and price ratios is a convenient method of comparing the additional cost of using more of one input versus the reduction in cost from using less of the other. Whenever the additional cost is less than the cost reduction, the total cost is lower and the substitution should be made. This is the case whenever the input substitution ratio is greater than the input price ratio.

Changes in Prices

In any input substitution problem, the least-cost combination depends on both the substitution ratio and the price ratio. The substitution ratio will remain the same over time provided the underlying physical or biological relations do not change. However, the price ratio will change whenever the relative input prices change, which will result in a different input combination becoming the new least-cost combination.

TABLE 8-1	Selecting a Least-Cost Feed Ration*				

Feed ration	Grain (lb)	Hay (lb)	Input substitution ratio		Input price ratio**	Total cost of ration**
A	825	1,350				$155.25
			2.93	>	1.50	
B	900	1,130				$148.80
			2.60	>	1.50	
C	975	935				$143.85
			2.20	>	1.50	
D	1,050	770				$140.70
			1.93	>	1.50	
E	**1,125**	**625**				**$138.75**
			1.33	<	1.50	
F	1,200	525				$139.50
			1.07	<	1.50	
G	1,275	445				$141.45

*Each ration is assumed to put the same weight gain on a feeder steer with a given beginning weight.
**The price of grain is $0.09 per pound, and the price of hay is $0.06 per pound.

As the price of one input increases relative to the other, the new least-cost input combination will tend to have less of the higher-priced input and more of the now relatively less expensive input.

When two inputs exhibit a decreasing input substitution ratio, such as in Table 8-1, the least-cost combination will generally include at least some of both inputs. However, different results are obtained when the input substitution ratio is constant.

Figure 8-1a is an example of a constant input substitution ratio, in this case 0.8 pounds of corn substitutes for 1.0 pound of barley between any two points on the isoquant. Assume that the price of barley is $0.06 per pound and the price of corn is $0.10 per pound, for an input price ratio of 0.6. Beginning at the top of the isoquant and moving downward, any point would have a substitution ratio greater than the price ratio. The rule says to make this substitution and move to the next point, where the result is exactly the same. Once the horizontal axis is reached, the result is a ration of all barley, no corn. If the input price ratio should be greater than 0.8, say 1.0 (prices per pound of corn and barley are equal), all points on the isoquant have a substitution ratio less than the price ratio. Every move down the isoquant represents a higher-cost ration than the one before. Therefore, the least-cost solution is all corn, no barley.

If the substitution ratio and the price ratio are exactly the same, then any combination of corn and barley would have the same cost. This would happen if the price of barley was $0.08 per pound and the price of corn was $0.10 per pound, for example.

These results, and those for a decreasing input substitution ratio, can be summarized as follows:

1. Given a constant rate of substitution between two inputs, the least-cost combination will be all of one input or all of the other. The

---

Box 8-1   Input Prices, Price Ratio, and Profit

The least-cost input combination always depends on the *relative* prices, as shown in the input price *ratio*, not the absolute prices. Doubling or halving both prices results in the same price ratio and the same least-cost combination. However, doubling input prices will reduce profit, and halving prices will increase profit, even though the least-cost input combination is unchanged.

---

only exception occurs when the input substitution and price ratios happen to be equal. Then any combination is least-cost, because they all have the same total cost.

2. With a decreasing rate of substitution between two inputs, the least-cost combination will generally include some of each input. First, the types of input that have a decreasing rate of substitution generally must be used in some combination to produce any output. All of one or the other may not even be an option. Second, even if it is physically and biologically possible to produce the output with only one of the inputs, the input substitution ratio at that point will be either so high or so low that an input price ratio probably would never reach these extremes.

OUTPUT COMBINATIONS

The third basic decision to be made by a farm or ranch manager is what to produce, or what combination of products will maximize profit. A choice must be made from among all possible enterprises, which may include vegetables, wheat, soybeans, cotton, beef cattle, hogs, poultry, and others. Climate, soil, range vegetation, and other fixed inputs may restrict the list of possible enterprises to only a few on some farms and ranches. On others, the manager may have a large number of possible enterprises from which to select the profit-maximizing combination. In all cases, it is assumed that one or more inputs

are limited, which places an upper limit on how much can be produced of a single product or any combination of products.

Competitive Enterprises

The first step in determining a profit-maximizing enterprise combination is to determine the physical relationship among the enterprises being considered. Given a limited amount of land, capital, or some other input, the production from one enterprise can generally be increased only by decreasing the production from another enterprise. Some of the limited input must be switched from use in one enterprise to use in another, causing the changes in production levels. There is a tradeoff, or substitution, to be considered when changing the enterprise combination. These are called *competitive enterprises*, because they are competing for the use of the same limited input at the same time.

Figure 8-2 illustrates two types of competitive enterprises. In the first graph, corn and soybeans are competing for the use of the same 100 acres of land. Planting all corn would result in the production of 15,000 bushels of corn, and planting all soybeans would produce a total of 5,000 bushels of soybeans. Other combinations of corn and soybeans totaling 100 acres would produce the combinations of corn and soybeans shown on the line connecting these two points. This line is called a *production possibility curve (PPC)*; it shows all possible combinations of corn and soybeans that can be produced from the given 100 acres.

Beginning with producing all corn, replacing an acre of corn with an acre of soybeans

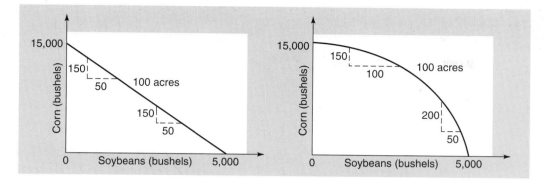

Figure 8-2 Production possibility curves for competitive enterprises.

results in a loss of 150 bushels of corn and a gain of 50 bushels of soybeans. The tradeoff, or output substitution ratio, is 3.0, because three bushels of corn must be given up to gain one bushel of soybeans. With a straight-line production possibility curve, this substitution ratio is the same between any two combinations. This is an example of competitive enterprises with a constant substitution ratio.

Over time, a combination of crop enterprises may benefit each other because of better pest control, soil fertility, erosion control, and improved timeliness in planting and harvesting large acreages. This situation is shown in the second graph in Figure 8-2. The curved production possibility curve indicates that total corn production increases at a slower rate, as a higher proportion of the land is used for corn production. The inverse situation holds true for soybeans. This causes the output substitution ratio to be different for different combinations of the two enterprises. The substitution ratio is 150 ÷ 100 = 1.5 near the top of the curve and increases to 200 ÷ 50 = 4.0 near the bottom of the curve. These enterprises are still competitive, but they have an increasing output substitution ratio.

Enterprise Combination Example

The data in Table 8-2 present the results of various combinations of two competitive crops, alfalfa and grain sorghum. The number of acres that can be grown of each crop may be limited by the availability of several fixed resources, such as suitable crop land, labor during harvest season, and irrigation water. If only one crop is grown, the maximum area possible is 1,000 acres of grain sorghum or 800 acres of alfalfa. As resources are diverted from grain sorghum production to alfalfa production, a larger and larger decrease in grain sorghum acres is needed to increase alfalfa by 100 acres.

The third and fourth columns of Table 8-2 show the total profit from each crop, assuming that alfalfa produces a profit of $50 per acre and grain sorghum produces a profit of $30 per acre. The last column shows the total profit from both crops.

Comparing the total profit for each combination of acres shows that profit is maximized when 600 acres are devoted to alfalfa and 470 acres are devoted to grain sorghum. All other combinations produce a smaller total profit.

Substitution and Profit Ratios

The most profitable combination of two competitive enterprises also can be determined by comparing the output substitution ratio and the output profit ratio. The output substitution ratio is calculated from the equation

$$\text{Output substitution ratio} = \frac{\text{quantity of output lost}}{\text{quantity of output gained}}$$

TABLE 8-2	Profit from Various Combinations of Alfalfa and Grain Sorghum*			
Acres of alfalfa	Acres of grain sorghum	Profit from alfalfa ($)	Profit from grain sorghum ($)	Total profit ($)
0	1,000	0	30,000	30,000
100	975	5,000	29,250	34,250
200	925	10,000	27,750	37,750
300	845	15,000	25,350	40,350
400	745	20,000	22,350	42,350
500	620	25,000	18,600	43,600
600	**470**	**30,000**	**14,100**	**44,100**
700	270	35,000	8,100	43,100
800	0	40,000	0	40,000

*Profit is $50 per acre for alfalfa and $30 per acre for grain sorghum.

where the quantities gained and lost are the *changes* in production between two points on the PPC. The output profit ratio is found from the equation

$$\text{Output profit ratio} = \frac{\text{profit of output gained}}{\text{profit of output lost}}$$

The decision rule for finding the profit-maximizing combination of two competitive enterprises is the point where the

output substitution ratio = output profit ratio

Table 8-3 shows the same example of two competitive enterprises, alfalfa and grain sorghum, that have an increasing output substitution ratio. The method for determining the profit-maximizing enterprise combination is basically the same as for determining the least-cost input combination but with one important exception. For enterprise combinations, when the output profit ratio is greater than the output substitution ratio, substitution should continue by moving downward and to the right on the PPC and down to the next combination in Table 8-3. Conversely, a profit ratio smaller than the substitution ratio means that too much substitution has taken place,

TABLE 8-3	Profit-Maximizing Combination of Two Competitive Enterprises		
Acres of alfalfa	Acres of grain sorghum	Output substitution ratio*	Profit ratio**
0	1,000		
		0.25 <	1.67
100	975		
		0.50 <	1.67
200	925		
		0.80 <	1.67
300	845		
		1.00 <	1.67
400	745		
		1.25 <	1.67
500	620		
		1.50 <	1.67
600	**470**		
		2.00 >	1.67
700	270		
		2.70 >	1.67
800	0		

*Decrease in sorghum acres divided by increase in alfalfa acres.
**Profit is $50 per acre for alfalfa and $30 per acre for grain sorghum.

> ### Box 8-2   Price Ratio or Profit Ratio?
>
> Traditionally, examples of two competitive enterprises have assumed that various combinations of product from two enterprises can be produced from the same set of fixed resources. That is, total production costs are the same for all combinations. In this manner the profit-maximizing combination can be found by comparing the substitution ratio of the products (such as bushels) to the inverse selling price ratio of the same products. In real life, however, two enterprises almost always use different variable inputs, and the total cost changes as the number of units of output produced changes. For the example in Table 8-2, alfalfa and grain sorghum need different amounts of seed, fertilizer, and other inputs. Moreover, the total cost of production is more directly related to the number of acres planted to each crop than to the number of tons or bushels of product obtained.
>
> For these reasons, the example in Table 8-2 assumes that different combinations of acres of two crops can be produced from a fixed set of resources, such as labor and machinery, and the profit per acre for each crop is compared. Production costs and gross revenue per acre are constant, a more realistic assumption. Finding the combination for which the substitution ratio (using acres as the units) is equal to the inverse ratio of profit per acre will lead to the maximum profit. If the two enterprises use the same fixed resources, the ratio of their respective gross margins can be used instead of the profit ratio.
>
> Chapter 11 contains a discussion of how linear programming can be used to find profit-maximizing combinations of more than two enterprises, using the same approach. For the crop enterprises, 1 acre is used as the budgeting unit.

and the adjustment should be upward and to the left on the PPC and up to the next combination in the table.

As long as the profit ratio is greater than the substitution ratio, profit will be increasing. The last combination where this relation changes from greater than when moving from the last combination, to less than when moving to the next combination, will be the profit-maximizing combination. Never move to a new combination when the output substitution ratio is greater than the output profit ratio. Even though the ratio values may be closer to being equal, the profit will be less than that in the previous combination.

Given the nine possible combinations in Table 8-3, the profit-maximizing combination is again found to be 600 acres of alfalfa and 470 acres of grain sorghum. The profit ratio is equal to the alfalfa profit per acre ($50) divided by the grain sorghum profit per acre ($30), or 1.67. As less grain sorghum and more alfalfa are produced, the output substitution ratio increases from 0.25 to 1.50 for 600 acres of alfalfa, still less than the profit ratio. Moving to 700 acres of alfalfa changes the substitution ratio to 2.00, however, indicating a decrease in total profit.

An increasing substitution ratio generally will result in the production of a combination of the enterprises, with the combination dependent on the current profit ratio. Any change in the prices or costs of the enterprises that change the profit ratio will affect the profit-maximizing enterprise combination when there is an increasing output substitution ratio. As with input substitution, it is the profit *ratio* that is important, not just the profit level.

When the enterprises have a *constant* output substitution ratio, the profit-maximizing solution will be to produce all of one or all of the other enterprise, not a combination. This is because the profit ratio will be either greater than or less than the substitution ratio for all combinations.

In Chapter 11 a technique for choosing from among more than two competitive enterprises is discussed. *Linear programming* uses the same principle of comparing output substitution ratios and profit ratios, but can compare a large number of possible enterprises at one time. The profit-maximizing solution takes into account any limits on the supply of key resources.

Supplementary Enterprises

While competitive enterprises are the most common, other types of enterprise relationships do exist. One of these is *supplementary enterprises*, and an example is shown in the left diagram of Figure 8-3. Two enterprises are supplementary if the production from one can be increased without affecting the production level of the other.

The example in Figure 8-3 shows that beef production from stocker steers run on winter wheat pasture can be increased over a range without affecting the amount of wheat produced. The PPC shows a relationship that eventually becomes competitive, however, as beef production cannot be increased indefinitely without affecting wheat production. An example of two purely supplemental enterprises could occur in states where landowners can lease hunting rights to hunters. This leasing enterprise could be supplementary with either livestock or crop production.

A manager should take advantage of supplementary relationships by increasing production from the supplementary enterprise. This should continue at least up to the point where the enterprises become competitive. If the supplementary enterprise shows any profit, however small, total profit will be increased by producing some of it, because production from the primary enterprise does not change. Two general conclusions can be drawn about the profit-maximizing combination of supplementary enterprises. First, this combination will not be within the supplementary range. It will be at least at the point where the relationship changes from supplementary to competitive. Second, it will in all likelihood be in the competitive portion of the production possibility curve. The exact location will depend, as it always does with competitive enterprises, on the output substitution and profit ratios.

Complementary Enterprises

Another possible type of enterprise relationship is *complementary enterprises*. This type of relationship exists whenever increasing the production from one enterprise causes the production from the other to increase at the same time. The right-hand graph in Figure 8-3 illustrates a possible complementary relationship between wheat production and land left fallow.

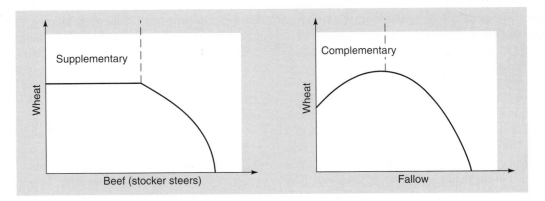

Figure 8-3 **Supplementary and complementary enterprise relationships.**

In many dryland wheat production areas with limited rainfall, some land is left fallow, or idle, each year as a way to store part of one year's rainfall to be used by a wheat crop the following year. Leaving some acres fallow reduces the acres in wheat, but the per-acre yield may increase because of the additional moisture available. This yield increase may be enough that total wheat production is greater than from planting all acres to wheat every year.

A complementary enterprise should be increased to at least the point where production from the primary enterprise (wheat in this example) is at a maximum. This is true even if the complementary enterprise has no value because production from the primary enterprise is increasing at the same time. Enterprises will generally be complementary only over some limited range, after which they will become competitive.

As with supplementary enterprises, two general conclusions can be drawn about the profit-maximizing combination of complementary enterprises. First, this combination will not be within the complementary range. It will be at least at the point where the relationship changes from complementary to competitive. Second, assuming that both enterprises produce some revenue, it will be somewhere in the competitive range. The correct combination can be found using the substitution ratio and profit ratio decision rule for competitive enterprises.

SUMMARY

This chapter continues the discussion of economic principles begun in Chapter 7. Here the emphasis is on the use of substitution principles to provide a manager with a procedure to answer the questions of how and what to produce. The question of how to produce concerns finding the least-cost combination of inputs to produce a given amount of output. Calculation and use of the input substitution and price ratios was shown, as well as the decision rule of finding the point where these two values are equal. This decision rule determines the least-cost combination of two inputs. Marginality is still an important concept here, because input substitution ratios are computed from small or marginal changes in the quantities of the two inputs used.

The question of what to produce is one of finding the profit-maximizing combination of enterprises when the quantity of one or more inputs is limited. That combination will depend first on the type of enterprise relationship that exists—competitive, supplementary, or complementary. The profit-maximizing combination for competitive enterprises is found by computing output substitution ratios and the output profit ratio. Finding the point where they are equal determines the correct combination. Supplementary and complementary enterprises have unique properties over a limited range, but they will eventually become competitive when they begin competing for the use of some limited input. The profit-maximizing combination for these two types of enterprise relationships will not be within these ranges, but generally will be within their competitive range.

QUESTIONS FOR REVIEW AND FURTHER THOUGHT

1. First double and then halve both prices used in Table 8-1, and find the new least-cost input combination in each case. Why is there no change? But what would happen to profit in each case?

2. Is there always one *best* livestock ration? Or does it depend on prices? Would you guess that feed manufacturers know about input substitution and least-cost input combinations? Explain how they might use this knowledge.

3. Explain how the slope of the isoquant affects the input substitution ratio.

4. Carefully explain the difference between an isoquant and a production possibility curve.

5. Reduce the profit from alfalfa in Table 8-2 to $36 per acre, calculate the new output profit ratio, and find the new profit-maximizing combination. Do you produce more or less of the enterprise with the now lower profit? Why? What would be the effect of doubling both profits per acre? Halving both profits per acre?

6. Why is the profit-maximizing combination of even supplementary and complementary enterprises generally in the competitive range?

7. Where climate and rainfall permit, most farms produce two or more crops. Explain the reason for this production practice in terms of the shape of the PPC and the output profit ratio.

8. For two similar inputs, such as soybean oil meal and cottonseed oil meal in a livestock feed ration, would you expect the substitution ratio to be nearly constant or to decline sharply as one input is substituted for another? Why?

U.S. Department of Agriculture

COST CONCEPTS IN ECONOMICS

CHAPTER OUTLINE

Opportunity Cost
Cash and Noncash Expenses
Fixed, Variable, and Total Costs
Application of Cost Concepts
Economies of Size
Long-Run Average Cost Curve
Summary
Questions for Review and Further Thought
Appendix. Cost Curves

CHAPTER OBJECTIVES

1. Explain the importance of opportunity cost and its use in managerial decision making

2. Clarify the difference between short run and long run

3. Discuss the difference between fixed and variable costs

4. Identify fixed costs and show how to compute them

5. Demonstrate the use of fixed and variable costs in making short-run and long-run production decisions

6. Explore economies and diseconomies of size and how they help explain changes in farm size and profitability

A good understanding of the costs of production is important in economics and very useful for making management decisions. Costs can be classified in different ways, depending on whether they are fixed or variable and cash or noncash. *Opportunity cost* is another type of cost not included in the accounting expenses for a business but an important economic cost nevertheless. It will be widely used in later chapters and will be the first cost discussed in this chapter.

OPPORTUNITY COST

Opportunity cost is an economic concept, not a cost that can be found in an accountant's ledger or on an income tax return. However, it is an important and basic concept that needs to be considered when making managerial decisions. Opportunity cost is based on the fact that once an input has been acquired, it may have one or more alternative uses. Once an input is committed to a particular use, it is no longer available for any other alternative use, and the income from the alternative must be foregone.

Opportunity cost can be defined in one of two ways:

1. The income that could have been earned by selling or renting the input to someone else or
2. The additional income that would have been received if the input had been used in its most profitable alternative use

The latter definition is perhaps the more common, but both should be kept in mind as a manager makes decisions on input use. The real cost of an input may not be its purchase price. Its real cost, or its opportunity cost, in any specific use is the income it would have earned in its next best alternative use. If this is greater than the income expected from the planned use of the input, the manager should reconsider the decision. The alternative appears to be a more profitable use of the input.

Opportunity costs are used widely in economic analysis. For example, the opportunity costs of a farm operator's labor, management, and capital are used in several types of budgets used for analyzing farm profitability.

Labor

The opportunity cost of a farm operator's labor (and perhaps that of other unpaid family labor) would be what that labor would earn in its next best alternative use. That alternative use could be nonfarm employment, but depending on skills, training, and experience, it might also be employment in another farm or ranch enterprise. Some operators state that their own time is *free*, but it should be given a value at least as high as the value that they put on leisure time.

Management

Management is the process of making and executing decisions—it is different from physical labor used in agriculture. The opportunity cost of management is difficult to estimate. For example, what is the value of management per crop acre? Some percentage of all other costs or the gross income per acre is often used, but what is the proper percent? In other cases, an annual opportunity cost of management is needed. Here and elsewhere, the analyst must be careful to exclude labor from the estimate. For example, if the opportunity cost of labor is estimated at $30,000 per year, and the individual could get a job in *management* paying $40,000 per year, the opportunity cost of management is estimated as the difference, or $10,000 per year. The opportunity cost of labor plus that of management cannot be greater than the total salary in the best alternative job. It is difficult to estimate the opportunity cost of management, so the opportunity costs of labor and management are often combined into one value.

Capital

Capital presents a different set of problems for estimating opportunity costs. There are many uses of capital and generally a wide range of possible returns. However, alternative uses of capital with higher expected returns may carry a higher degree of risk as well. To avoid the problem of identifying a use with a comparable level of risk, the opportunity cost of farm capital is often set equal to the interest rate on savings accounts or the current cost of borrowed capital.

This assumes that the capital invested in a farm or ranch enterprise could have been deposited in savings or used to pay down debt. This represents a minimum opportunity cost and is a somewhat conservative approach. If the expected rate of return on an investment with a comparable level of risk can be determined, it would be appropriate to use that rate when analyzing farm investments.

A special problem is how to determine the opportunity cost on the annual service provided by long-lived inputs such as land, buildings, breeding stock, and machinery. It may be possible to use rental rates for land and for machinery services. However, this does not work well for all such items, and the opportunity cost of these inputs often should be determined by the most profitable alternative use of the capital invested in them outside the farm or ranch business. This is the real or true cost of using inputs to produce agricultural products.

Some farm assets such as machinery, buildings, fences, and livestock equipment decrease in value over time; they depreciate. Their opportunity cost should be adjusted each year, by multiplying the opportunity cost rate of return by the value for which the asset could be sold, also called its salvage value. Sometimes a long-term investment analysis uses the average salvage value of an asset over its lifetime to calculate the opportunity cost of investing capital in it. This will be illustrated later in the chapter.

CASH AND NONCASH EXPENSES

Fixed costs can be either *cash* or *noncash* expenses. They can be easily overlooked or underestimated, because a large part of total fixed cost can be noncash expenses, as shown in Table 9-1.

Depreciation and opportunity costs are always noncash expenses, because there is no annual cash outlay for them. Repairs and property taxes are always cash expenses, and interest and insurance may be either. If money is borrowed to purchase the asset, there will be some cash interest expense.

TABLE 9-1	Cash and Noncash Expense Items		
Expense item		Cash expense	Noncash expense
Depreciation			X
Interest (own capital)			X
Interest (borrowed capital)		X	
Value of operator labor			X
Wages for hired labor		X	
Farm raised feed			X
Purchased feed		X	
Owned land			X
Cash rented land		X	
Seed, fertilizer, fuel, repairs		X	
Property taxes, insurance		X	

Box 9-1   Cash and Noncash Expenses

Any distinction between cash and noncash expenses does not imply that noncash expenses are any less important than cash expenses. In the short run, noncash expenses do mean that less cash is needed to meet current expenses. However, income must be sufficient to cover all expenses in the long run if the business is to survive, prosper, replace capital assets, and make an economic profit.

When the item is purchased entirely with the buyer's own capital, the interest charge would be the opportunity cost on this capital, and there is no cash payment to a lender. Insurance would be a cash expense if it is carried with an insurance company or noncash if the risk of loss is assumed by the owner. In the latter case, there would be no annual cash outlay, but the insurance charge should still be included in fixed costs to cover the possibility of damage to or loss of the item from fire, windstorm, theft, and so forth.

Some variable costs can be noncash costs as well. When crops are grown to feed livestock produced on the same farm, there is no cash outlay as there would be for purchased feed. However, there is an opportunity cost equal to the revenue that could have been received from selling the feed crop on the market.

FIXED, VARIABLE, AND TOTAL COSTS

Several important cost concepts were introduced in Chapter 7. Seven useful cost concepts and their abbreviations are

1. Total fixed cost (TFC)
2. Average fixed cost (AFC)
3. Total variable cost (TVC)
4. Average variable cost (AVC)
5. Total cost (TC)
6. Average total cost (ATC)
7. Marginal cost (MC)

These costs are *output* related. Marginal cost, also studied in Chapter 7, is the additional cost of producing an additional unit of output. The others are either the total cost or the cost per unit for producing a given amount of output.

Short Run and Long Run

Before these costs are discussed in some detail, it is necessary to distinguish between what economists call the short run and the long run. These are time concepts, but they are not defined as fixed periods of calendar time. The *short run*

is that period during which the available quantity of one or more production inputs is fixed and cannot be changed. For example, at the beginning of the planting season, it may be too late to increase or decrease the amount of cropland owned or rented. The current crop production cycle would be a short-run period, as the amount of available land is fixed.

Over a longer period, land may be purchased, sold, or leased, or leases may expire causing the amount of land available to increase or decrease. The *long run* is defined as that period during which the quantity of all necessary productive inputs can be changed. In the long run, a business can expand by acquiring additional inputs or go out of existence by selling all its inputs. Farm managers and employees may come or go. The actual calendar length of the long run as well as the short run will vary with the situation and circumstances. Depending on which inputs are fixed, the short run may be anywhere from several days to several years. One year and one crop or livestock production cycle are common short-run periods in agriculture.

Fixed Costs

The costs associated with owning a fixed input are called *fixed costs*. These are the costs that are incurred even if the input is not used. Depreciation, insurance, taxes (property taxes, not income taxes), and interest are the usual costs considered to be fixed. Repairs and maintenance may also be included as a fixed cost (see Box 9-2). Fixed costs do not change as the level of production changes in the short run but can change in the long run as the quantity of the fixed input changes. By definition there need not be any fixed resources owned in the long run, so fixed costs exist only in the short run.

Another characteristic of fixed costs is that they are not under the control of the manager in the short run. They exist, and stay at the same level regardless of how much or how little the resource is used. The only way they can be avoided is to sell the item, which can be done in the long run.

Total fixed cost (TFC) is the summation of the several types of fixed costs. Computing the average annual TFC for a fixed input requires finding the average annual depreciation and interest costs, among others. The straight-line depreciation method discussed in Chapter 4 also provides the average annual depreciation from the equation

$$\text{Depreciation} = \frac{\text{purchase price} - \text{salvage value}}{\text{useful life}}$$

where purchase price is the initial cost of the asset, useful life is the number of years the item is expected to be owned, and salvage value is its expected value at the end of that useful life. Although other methods can be used to estimate depreciation for each year of the asset's useful life, as discussed in Chapter 5, this equation can always be used to find the *average* annual depreciation.

Capital invested in a fixed input has an opportunity cost, so interest on that investment also is included as part of the fixed cost. However, it is not correct to charge interest on the purchase price or original cost of a depreciable asset every year, because its value is decreasing over time. Therefore, the interest component of TFC is commonly computed from the formulas

$$\text{Average asset value} = \frac{\text{purchase price} + \text{salvage value}}{2}$$

$$\text{Interest} = \text{average asset value} \times \text{interest rate}$$

where the interest rate is the cost of capital to the ranch or farm.[1] This equation gives the interest charged for the average value of the item over its useful life and reflects that it is decreasing in value over time. Depreciation is being charged to account for this decline in value. To find the

interest cost for the current year instead of the average cost over the entire useful life of an asset, use the following formula:

$$\text{Interest} = \text{current asset value} \times \text{interest rate}$$

Other ownership costs for fixed resources, such as property taxes and insurance, can be estimated as a percent of the average value of the asset over its useful life (long-term cost) or its current value (current year cost). The actual dollar amount of cost paid or to be paid for taxes and insurance can also be used.

As an example, assume the purchase of a harvesting machine for $120,000, with a salvage value of $50,000 and a useful life of five years. Annual property taxes are estimated to be $400, annual insurance is $500, and the cost of capital is 8 percent. Using these values and the two preceding equations results in the following annual TFC:

Average value	$= \dfrac{\$120,000 + 50,000}{2}$	$= \$85,000$
Interest	$= \$85,000 \times 8\%$	$= \$\ 6,800$
Depreciation	$= \dfrac{\$120,000 - 50,000}{5 \text{ Years}}$	$= \$14,000$
Taxes		400
Insurance		500
Annual total fixed cost		$\$ 21,700$

The annual total fixed cost is nearly 20 percent of the purchase price. It is often 15 to 25 percent of the purchase price for a depreciable asset.

Fixed cost can be expressed as an average cost per unit of output. Average fixed cost (AFC) is calculated using the equation

$$\text{AFC} = \frac{\text{TFC}}{\text{output}}$$

where output is measured in physical units such as bushels, bales, or hundred-weights.

[1]This equation is widely used but is only a close approximation of the true opportunity cost. The capital recovery methods discussed in Chapter 17 can be used to find the true charge, which combines depreciation and interest into one value. This value recovers the investment in the asset plus compound interest over its useful life.

> ### Box 9-2   **Repairs as a Fixed Cost?**
>
> Repairs can be found in some lists of fixed costs. However, repairs typically increase as use of the asset increases, which does not fit the definition of a fixed cost. The argument for including repairs in fixed costs is that some minimum level of maintenance is needed to keep an asset in working order, even if it is not being used. Therefore, this maintenance expense is more like a fixed cost. The practical problem with this is calculating what part of an asset's total repair and maintenance expense should be a fixed cost.
>
> In practice, all machinery repairs are generally considered variable costs, because most repairs are necessitated by use. Building repairs are more likely to be included as a fixed cost. These repairs are often more from routine maintenance caused by age and weather and less from use. This is particularly true for storage facilities and many general-purpose buildings. For convenience, all repair costs will be classified as variable costs when developing budgets in later chapters of this text.

Acres or hours are often used as the measure of output for machinery even though they are not units of production. By definition, TFC is a fixed or constant value, so AFC will decline continuously as output increases. One way to lower the per-unit cost of producing a given commodity is to get more output from the fixed resource. This will always lower the AFC per unit of output.

Variable Costs

Variable costs are those over which the manager has control at a given time. They can be increased or decreased at the manager's discretion and will increase as production is increased. Items such as feed, fertilizer, seed, pesticides, fuel, and livestock health expenses are examples of variable costs. A manager has control over these expenses in the short run, and they will not be incurred unless production takes place.

Total variable cost (TVC) can be found by summing the individual variable costs, each of which is equal to the quantity of the input used times its price. Average variable cost (AVC) is calculated from the equation

$$AVC = \frac{TVC}{output}$$

where output again is measured in physical units. AVC may be increasing, constant, or decreasing, depending on the underlying production function and the output level. For the production function illustrated in Figure 7-2, AVC will initially decrease as output is increased and then will increase beginning at the point where average physical product begins to decline.

Variable costs exist in both the short run and the long run. All costs can be considered to be variable costs in the long run, because there are no fixed inputs. The distinction between fixed and variable costs also depends on the exact time where the next decision is to be made. Fertilizer is generally a variable cost. Yet once it has been purchased and applied, the manager no longer has any control over the size of this expenditure. It must be considered a fixed cost for the remainder of the crop season, which can affect future decisions during that crop season. Labor cost and cash rent for land are similar examples. After a labor or lease contract is signed, the manager cannot change the amount of money obligated, and the salary or rent must be considered a fixed cost for the duration of the contract. Costs that are fixed because they have already been incurred or paid are sometimes called *sunk costs*.

Total Costs

Total cost (TC) is the sum of TFC and TVC (TC = TFC + TVC). In the short run, it will increase only as TVC increases, because TFC is a constant value. Average total cost (ATC) can be calculated by one of two methods. For a given output level, it is equal to AFC + AVC. It can also be calculated from the equation

$$ATC = \frac{TC}{output}$$

which will give the same result. ATC will typically be decreasing at low output levels, because AFC is decreasing rapidly and AVC may be decreasing also. At higher output levels, AFC will be decreasing less rapidly, and AVC will eventually increase and be increasing at a rate faster than the rate of decrease in AFC. This combination causes ATC to increase.

Marginal Costs

Marginal cost (MC) is defined as the change in total cost divided by the change in output:

$$MC = \frac{\Delta\ TC}{\Delta\ output} \ or\ MC = \frac{\Delta\ TVC}{\Delta\ output}$$

It is also equal to the change in TVC divided by the change in output. TC = TFC + TVC, and TFC is constant, so the only way TC can change is from a change in TVC. Therefore, MC can be calculated either way with the same result.

APPLICATION OF COST CONCEPTS

Table 9-2 is an example of some cost figures for the common problem of determining the profit-maximizing stocking rate of steer caves for a fixed amount of pasture land. It illustrates many similar problems where an understanding of the different cost concepts and relations will help a manager in planning and decision making.

The size of the pasture and the amount of forage available are both limited, so adding more and more steers will eventually cause the average weight gain per steer to decline over a fixed period. This is reflected in diminishing returns and a declining marginal physical product (MPP) when more than 30 steers are placed in the pasture. Total hundred-weights of beef sold off the pasture still increases, but at a decreasing rate as more steers compete for the limited forage.

TFCs are assumed to be $10,000 per year in the example. This would cover the annual opportunity cost on the land and any improvements, depreciation on fences and water facilities, and insurance. Variable costs are assumed to be $990 per steer (steers are the only variable input in this example). This includes the cost of the steer, transportation, health expenses, feed, interest on the investment in the steer, and any other expenses that increase directly along with the number of steers purchased.

The total and average cost figures in Table 9-2 have the usual or expected pattern as production or output is increased. TFC remains constant by definition, while both TVC and TC are increasing. AFC declines rapidly at first and then continues to decline but at a slower rate. AVC and MC are constant as the number of steers increases from 0 to 30, and then begin to increase.

Profit-Maximizing Output

The profit-maximizing output level was defined in Chapter 7 as one where MR = MC, but this exact point does not exist in Table 9-2. However, MC is less than MR when moving from 50 to 60 steers, but more than MR when moving from 60 to 70 steers. This makes 60 steers and 420 hundred-weights of beef the profit-maximizing levels of input and output for this example. When MC is greater than MR, the additional cost per hundred-weight of beef produced from the additional 10 steers is greater than the additional income per hundred-weight. Therefore, placing 70 steers on the pasture would result in less profit than placing 60 steers.

> ### Box 9-3   Is Labor a Variable or Fixed Cost?
>
> The cost of labor used in ranching and farming is not always easy to classify as a variable or fixed cost. Labor is not used unless a certain enterprise is carried out, and the amount of labor used depends on the size of the enterprise. This fits the definition of a variable cost. On the other hand, labor resources may have to be paid regardless of how much work is performed, and some farm work such as record keeping and maintenance may have to be done regardless of what enterprises are carried out.
>
> Several situations can occur:
>
> 1. Labor is hired only as needed and is paid by the hour, by the day, or by the amount of work accomplished. In this case it is a variable cost.
> 2. Labor is paid a fixed wage regardless of how much it is used, in which case it can be considered a fixed cost, at least for the period of the employment contract.
>
> 3. If labor is valued at its opportunity cost, as is often the case with unpaid operator labor, and using it in one enterprise reduces the labor available for another enterprise, then it can be considered a variable cost.
> 4. If permanent labor resources are in excess supply, that is, they have no significant opportunity cost, they can be treated as fixed resources and their cost can be ignored in short-run analysis.
>
> Some producers use family living costs as an estimate of unpaid labor value. While withdrawals made for family living should be included in a cash flow budget (see Chapter 13), the amount withdrawn will depend on family size, consumption patterns, location, and other sources of income available. There is no reason to believe that living costs accurately reflect the economic cost of labor used in the farming operation, though.

The profit-maximizing point will depend on the selling price, or MR, and MC. Selling prices often change, and MC can change with changes in variable cost (primarily from a change in the cost of the steer in this example). The values in the MC column indicate that the most profitable number of steers would be 70 if the selling price is over $180.00 but less than $198.00. That number would drop to 50 steers if the selling price dropped to anywhere between $152.31 and $165.00. As shown in Chapter 7, the profit-maximizing input and output levels will always depend on prices of both the input and the output.

Total Profit

For the selling price of $175.00 per hundred-weight and a stocking rate of 60 steers, the total revenue (TR) is $73,500 (420 hundred-weight × $175.00) and TC is $69,400, leaving a profit of $4,100.00. For a price of $156.00, however, the MR = MC rule indicates a profit-maximizing point of 50 steers and 360 hundred-weights. The ATC per hundred-weight at this point is $165.28, which would mean a loss of $9.28 per hundred-weight ($156.00 – $165.28 = –9.28) or a total loss of $3,340 ($9.28 × 360 hundred-weights). What should a manager do in this situation? Should any steers be purchased if the expected selling price is less than ATC and a loss will result?

The answer is *yes* for some situations and *no* for others. Data in Table 9-2 indicate that there would be a loss equal to the TFC of $10,000 if no steers were purchased. This loss would exist in the short run as long as the land is owned. It could be avoided in the long run by selling the land, which would eliminate the fixed costs. However, the fixed costs cannot be avoided in the short run, and the relevant question is, Can a profit be made or the loss reduced to less than $10,000 in the

TABLE 9-2	Illustration of Cost Concepts Applied to a Stocking Rate Problem*

Number of steers	Output (hundred-weight of beef)	MPP	Total costs			Average costs			Marginal costs		Total profit ($)
			TFC ($)	TVC ($)	TC ($)	AFC ($)	AVC ($)	ATC ($)	MC ($)	MR ($)	
0	0		10,000	0	10,000	—	—	—			(10,000)
		7.5							132.00 < 175.00		
10	75		10,000	9,900	19,900	133.33	132.00	265.33			(6,775)
		7.5							132.00 < 175.00		
20	150		10,000	19,800	29,800	66.67	132.00	198.67			(3,550)
		7.5							132.00 < 175.00		
30	225		10,000	29,700	39,700	44.44	132.00	176.44			(325)
		7							141.43 < 175.00		
40	295		10,000	39,600	49,600	33.90	134.24	168.14			2,025
		6.5							152.31 < 175.00		
50	360		10,000	49,500	59,500	27.78	137.50	165.28			3,500
		6							165.00 < 175.00		
60	**420**		**10,000**	**59,400**	**69,400**	**23.81**	**141.43**	**165.24**			**4,100**
		5.5							180.00 > 175.00		
70	475		10,000	69,300	79,300	21.05	145.89	166.95			3,825
		5							198.00 > 175.00		
80	525		10,000	79,200	89,200	19.05	150.86	169.90			2,675
		4.5							220.00 > 175.00		
90	570		10,000	89,100	99,100	17.54	156.32	173.86			650
		4							247.50 > 175.00		
100	610		10,000	99,000	109,000	16.39	162.30	178.69			(2,250)

*Total fixed cost is $10,000, and variable cost is $990 per steer. Selling price of the steers is assumed to be $175.00 per hundred-weight.

short run by purchasing some steers? Steers should not be purchased if it would result in a loss greater than $10,000, because the loss can be minimized at $10,000 by not purchasing any.

Variable costs are under the control of the manager and can be reduced to zero by not purchasing any steers. Therefore, no variable costs should be incurred unless the expected selling price is at least equal to or greater than the minimum AVC. This will provide enough total revenue to cover TVCs. If the selling price is greater than the minimum AVC but less than the minimum ATC, the income will cover all variable costs, with some left over to pay part of the fixed costs. There would be a loss, but it would be less than $10,000 in this example. To answer the previous question, yes, steers should be purchased when the expected selling price is less than the minimum ATC, but only if it is above the minimum AVC. This action will result in a loss, but it will be smaller than the loss that would result if no steers were purchased.

If the expected selling price is less than the minimum AVC, total revenue will be less than TVC: There will be a loss, and it will be greater than $10,000. Under these conditions, no steers should be purchased, which will minimize the loss at $10,000. In Table 9-2, the lowest AVC is $132.00 and the lowest ATC is $165.24. The loss would be minimized by not purchasing steers when the expected selling price is less than $132.00 and by purchasing steers when the expected selling price is between $132.00 and $165.24. In the last situation, the loss-minimizing output level is where MR = MC.

Production Rules for the Short Run

The preceding discussion leads to three rules for making production decisions in the short run. They are as follows:

1. Expected selling price is greater than minimum ATC (selling price of $165.24 in the following table). A profit can be made and is maximized by producing at the level where MR = MC.
2. Expected selling price is less than minimum ATC but greater than minimum AVC (selling price of $156.00 in the following table). A loss cannot be avoided but will be minimized by producing at the output level where MR = MC. The loss will be somewhere between zero and TFC.

3. Expected selling price is less than minimum AVC (selling price of $126.00 in the following table). A loss cannot be avoided but is minimized by not producing. The loss will be equal to TFC.

Selling price ($/cwt.)	Number of steers to buy	Total revenue ($)	Total costs ($)	Profit ($)
$175.00	60	$73,500	$69,400	$4,100
$156.00	50	$56,160	$59,500	($3,340)
$126.00	0	$ 0	$10,000	($10,000)

The application of these rules is graphically illustrated in Figure 9-1. With a selling price equal to MR_1, the intersection of MR and MC is well above ATC, and a profit is being made. When the selling price is equal to MR_2, the income will not be sufficient to cover total costs but will cover all variable costs, with some left over to pay part of fixed costs. In this situation, the loss is minimized by producing where MR = MC, because the loss will be less than TFC. Should the selling price be as low as MR_3, income would not even cover variable costs, and the loss would be minimized by stopping

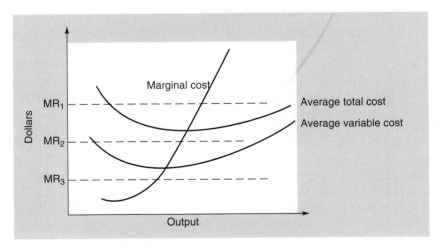

Figure 9-1 Illustration of short-run production decisions.

production altogether. This would minimize the loss at an amount equal to TFC.

Production Rules for the Long Run

The three production decision rules apply only to the short run where fixed costs exist. What about the long run where there are no fixed costs? Continual losses incurred by producing in the long run will eventually force the firm out of business.

There are only two rules for making production decisions in the long run:

1. Selling price is greater than ATC (or TR greater than TC). Continue to produce, because a profit is being made. This profit is maximized by producing at the point where MR = MC.
2. Selling price is less than ATC (or TR less than TC). There will be a continual loss. Stop production and sell the fixed asset(s), which eliminates the fixed costs. The money received should be invested in a more profitable alternative.

This does not mean that assets should be sold the first time a loss is incurred. Short-run losses will occur when there is a temporary drop in the selling price. Long-run rule number two should be invoked only when the drop in price is expected to be long lasting or permanent.

In some cases the selling price is known at the time the decision to produce or not is made, such as when a forward contract is available. Most of the time, though, the manager must decide based on a prediction or an estimate of the final selling price. Chapter 15 will discuss some techniques that a manager can use to deal with the effect of uncertain prices.

ECONOMIES OF SIZE

Economists and managers are interested in farm size and the relation between costs and size, for a number of reasons. The following are examples of questions being asked that relate to farm size and costs. What is the most profitable farm size? Can larger farms produce food and fiber more cheaply? Are large farms more efficient? Will family farms disappear and be replaced by large corporate farms? Will the number of farms and farmers continue to decline? The answers to these questions depend at least in part on what happens to costs and the cost per unit of output as farm size increases.

First, how is farm size measured? Number of livestock, number of acres, number of full-time workers, equity, total assets, profit, and other factors are used to measure size, and all have some advantages and disadvantages. For example, number of acres is a common and convenient measure of farm size but should be used only to compare farm sizes in a limited geographical area where farm type, soil type, and climate are similar. One hundred acres of irrigated vegetables in California is not the same size operation as 100 acres of arid range land in neighboring Arizona or Nevada.

Gross farm income, or total revenue, is a common measure of farm size. It has the advantage of converting everything into the common denominator of the dollar. This and other measures that are in dollar terms are better than any physical measure for determining and comparing farm size across different farming regions.

Size in the Short Run

In the short run, the quantity of one or more inputs is fixed, with land often being the fixed input. Given this fixed input, there will be a short-run average total cost curve, as shown in Figure 9-2. Short-run average cost curves typically will be U-shaped, with the average cost increasing at higher levels of production, because the limited fixed input makes additional production more and more difficult and therefore increases the average cost per unit of output.

For simplicity, size is measured as the output of a single product in Figure 9-2. The product can be produced at the lowest average cost per unit by producing the quantity 0a. However,

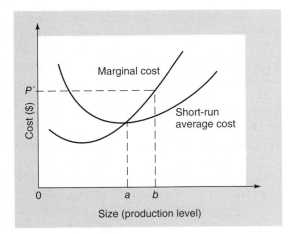

Figure 9-2 Farm size in the short run.

this may not be the profit-maximizing quantity, as profit is maximized at the output level where marginal revenue is equal to marginal cost. Marginal revenue is equal to output price, so a price of P' would cause profit to be maximized by producing the quantity $0b$. A higher or lower price would cause output to increase or decrease to correspond with the point where the new price is equal to marginal cost.

Because of a fixed input such as land, output can be increased in the short run only by intensifying production. This means more variable inputs such as fertilizer, chemicals, irrigation water, labor, and machinery time must be used. However, the limited fixed input tends to increase average and marginal costs as production is increased past some point and an absolute production limit is eventually reached. Additional production is possible only by acquiring more of the fixed input, a long-run problem.

Size in the Long Run

The economics of farm size is more interesting when analyzed in a long-run context. This gives the manager time to adjust all inputs to the level that will result in the desired farm size. One measure of the relation between output and costs

as farm size increases is expressed in the following ratio:

$$\frac{\text{Percent change in costs}}{\text{Percent change in output value}}$$

Both changes are calculated in monetary terms to allow combining the cost of the many inputs and the value of several outputs into one figure. This ratio can have three possible results called decreasing costs, constant costs, or increasing costs.

Ratio value	Type of costs
<1	Decreasing
=1	Constant
>1	Increasing

These three possible results are also called, respectively, increasing returns to size, constant returns to size, and decreasing returns to size. Decreasing costs means increasing returns to size, and vice versa. These relations are shown in Figure 9-3 using the long-run average cost curve per unit of output. When decreasing costs exist, the average cost per unit of output is decreasing, so that the average profit per unit of output is increasing. Therefore, increasing returns to size are said to exist. The same line of reasoning explains the relation between constant

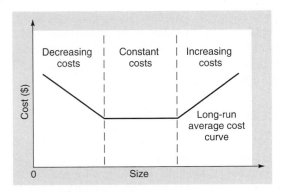

Figure 9-3 Possible size–cost relations.

costs and constant returns and between increasing costs and decreasing returns.

Economies of size exist whenever the long-run average cost curve is decreasing over some range of output (decreasing costs and increasing returns to size). *Diseconomies of size* exist when long-run average costs are increasing (increasing costs and decreasing returns to size). The existence or nonexistence of each and the range of farm sizes over which each occurs help explain and predict farm size. Farms of a size where long-run average costs are declining have an incentive to become larger. Any farms that may be so large that their long-run average costs are increasing will have less incentive to grow and will soon reach their long-run profit-maximizing size if they haven't already done so.

Causes of Economies of Size

Many studies of U.S. farms and ranches have documented the existence of economies of size over at least some initial size range. What is not always clear are the exact causes of these lower average costs as size increases. The following have been identified as possible reasons for economies of size.

Full Use of Existing Resources

One of the basic causes of economies of size is the more complete use of existing labor, machinery, capital, and management. These resources have fixed costs whether they are used or not. Their full use does not increase TFCs but does lower AFC per unit of output. Consider the farmer who rents some additional land and is able to farm it with existing labor and machinery. AFC per unit of output is now lower not only for the additional production but also for all of the original production.

Technology

New technology is often expensive but it can reduce the average cost per unit of output by the combination of replacing some current inputs and increasing production per acre or per head. However, because of the high initial investment, new technology must often be used over a larger number of acres or head to obtain these lower costs. This produces economies of size for larger farms and ranches but leaves smaller ones with older and less efficient technology.

Engineering Economies

As larger units of farm machinery are made available, the purchase price usually does not increase as fast as the capacity. This is because some components such as cabs, steering mechanisms, and electronic controls cost about the same regardless of the size of the machine. The same is true of many buildings. Only two end walls are needed regardless of the length of the building, and manure handling and feeding systems may not double even when floor space is doubled. This leads to lower fixed costs per unit of capacity.

Use of Specialized Resources

Labor and equipment on smaller farms must often be used for many different tasks and for several different enterprises, perhaps both crops and livestock. No single enterprise is large enough to justify specialized equipment that could do the job more efficiently and at less cost. Labor must perform so many different tasks that individuals do not have time to get sufficient experience to become skilled at any task. So little time may be spent on any single task during the year that it is difficult to justify the job training necessary to become more proficient. Larger farms can make full-time use of specialized equipment and individual workers can work full-time in one enterprise and perhaps at only one task. A large dairy where some workers can work full-time in the milking parlor and others in the feeding area is one example. Specialization often increases efficiency and reduces costs per unit of output.

Input Prices

Price discounts when purchasing large amounts of inputs and for buying in bulk are common in most industries including agriculture. Large farms and ranches can obtain substantial price discounts, for example, by buying feed by the truckload rather than a few bags at a time. Pesticides, fertilizer, seed, fuel, and supplies such as livestock vaccines and spare parts may also have price discounts for volume buying. These price discounts often are a result of lower transportation and handling costs per unit by the supplier or a vendor's desire to increase market share. Even if there is no price discount for bulk buying, the labor saved by the convenience, ease, and speed of handling the material can easily make up for additional storage and handling equipment needed.

Output Prices

Large-volume producers may also have a price advantage over smaller ones when selling their production. Grain producers may earn a premium price if they can deliver a large quantity at one time or guarantee the delivery of a fixed amount each month to a feedlot, feed mill, or ethanol plant. Ranchers who can deliver a full truckload of uniform weight cattle directly to a feedlot will usually receive a higher net price than those who sell a few at a time through a local auction barn.

In recent years many new, special-use grain varieties have been developed. Farmers receive a price premium for raising these specialized grains but they require special handling and separate storage away from all other varieties. Combines and trucks often need to be thoroughly cleaned before and after harvesting and hauling these varieties. This is another case where only larger-volume producers can justify the expense of separate and adequate storage and other related expenses to receive the premium for growing the crop.

Management

Many of the functions and tasks of management may contribute to economies of size. Purchasing inputs, marketing products, accounting, gathering information, planning, and supervising labor are examples. Doubling the size of the business may increase the time spent on each of these activities but not by 100 percent. Learning new management skills may take the same amount of time regardless of the number of units of production to which they are applied.

Causes of Diseconomies of Size

It is clear that economies of size exist over some initial range of farm sizes for nearly all types of farms and ranches. The existence of diseconomies of size is less clear as is the size at which diseconomies may begin. Diseconomies may result from any or all of the following.

Management

Limited management capacity has always been considered a prime cause of diseconomies of size. As a farm or ranch business becomes larger, it becomes more difficult for a manager to be knowledgeable about all aspects of all enterprises and to properly organize and supervise all activities. Multiple operators may have a harder time agreeing on management decisions or responding to problems quickly. Timeliness of operations and attention to detail begin to decline. The business becomes less efficient and average costs increase through some combination of increasing expenses and declining production.

Labor Supervision

Related to management is the increased difficulty of managing a larger labor force as size increases. In many types of agriculture, this is often further complicated by individuals working alone or in small groups spread over an entire field or many different fields on several different farms. This may require additional supervisors or considerable unproductive travel time by one supervisor.

Geographical Dispersion

Agricultural operations such as greenhouse production, cattle feedlots, confinement swine facilities, and poultry production can have a large operation concentrated in a relatively small area. However, increasing the size of a field crop operation requires additional land that may not always be available nearby. This increases the time and expense needed to move labor and equipment from farm to farm and the products from field to centralized storage or processing facilities.

Biohazards

Odor control, manure disposal, and increased risk of disease in large concentrations of livestock are potential sources of diseconomies in large livestock operations. State and federal regulations often require operations above a certain size to implement strict odor control and manure disposal procedures, which can increase costs compared to a smaller operation not subject to the regulations. Control of additional surrounding land may be required to keep odors from reaching neighbors and for manure disposal.

Long-Run Average Cost Curve

As farms and ranges increase in size, many of the economies and diseconomies of size occur simultaneously, and offset each other to some extent. The result is that efficiency as measured by long-run average cost (LRAC) per unit of output may stay fairly constant over a wide range of output levels.

Two possible LRAC curves are shown in Figure 9-4. Figure 9-4a assumes economies of size will dominate as a small business becomes larger. At some size all economies have been realized and LRAC is at a minimum. It is possible that this minimum cost may be constant or nearly

so over a range of sizes (see Figure 9-3). However, diseconomies will eventually dominate as size grows beyond some point and LRAC will begin to increase. Management as the limiting input is often cited as the reason for the increasing costs.

The LRAC curve in Figure 9-4b is often referred to as an L-shaped curve. It describes the results found in a number of cost studies on crop farms of different sizes. Average costs usually fall rapidly and reach a minimum at a size often associated with a full-time family farmer who hires at least some additional labor, either full- or part-time, and fully uses a set of machinery. As size increases beyond this point, average cost remains constant or nearly so over a wide range of sizes as managers replicate efficient sets of machinery, buildings, and workers. These studies show little or no increase in average costs over the range of sizes studied.

The data in Figure 9-5 show three examples of how costs per unit of production are related to farm size, based on recent farm record data. In Figures 9-5a and 9-5c spring wheat and dairy show decreasing average costs per unit of output across all farm sizes. Soybeans, on the other hand, shows some slight increase in cost of production across the three mid-sized groups, but there is a substantial reduction in costs when the smallest category is compared to the largest.

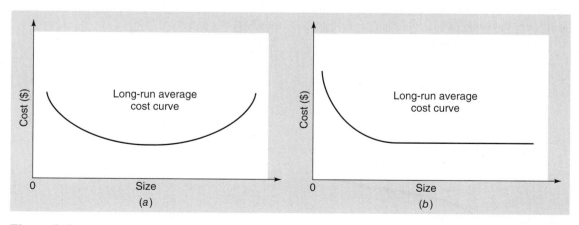

Figure 9-4 Two possible LRAC curves.

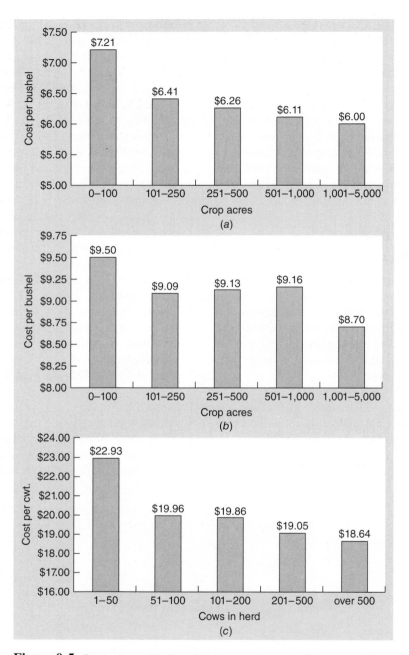

Figure 9-5 Costs per unit of production by farm size for (a) spring wheat; (b) soybeans; (c) milk.

Source: Center for Farm Financial Management, University of Minnesota, 2012.

SUMMARY

This chapter discussed the different economic costs and their use in managerial decision making. Opportunity costs are often used in budgeting and farm financial analysis. This noncash cost stems from inputs that have more than one use. Using an input one way means it cannot be put to any other alternative use, and the income from that alternative must be foregone. The income given up is the input's opportunity cost.

An analysis of costs is important for understanding and improving the profitability of a business. The distinction between fixed and variable costs is important and useful when making short-run production decisions. In the short run, production should take place only if the expected income will exceed the variable costs. Otherwise, losses will be minimized by not producing. Production should take place in the long run only if income is high enough to pay all costs. If all costs are not covered in the long run, the business will eventually fail or will be receiving less than the opportunity cost on one or more inputs.

An understanding of costs is also necessary for analyzing economies of size. The relation between cost per unit of output and size of the business determines whether there are increasing, decreasing, or constant returns to size. If unit costs decrease as size increases, there are increasing returns to size, and the business would have an incentive to grow, and vice versa. The type of returns that exist for an individual farm will determine in large part the success or failure of expanding farm size. Future trends in farm size, number of farms, and form of business ownership and control will be influenced by economies and diseconomies in farm and ranch businesses.

QUESTIONS FOR REVIEW AND FURTHER THOUGHT

1. How would you estimate the opportunity cost for each of the following items? What do you think the actual opportunity cost would be?

   a. Capital invested in land

   b. Your labor used in a farm business

   c. Your management used in a farm business

   d. One hour of tractor time

   e. The hour you wasted instead of studying for your next exam

   f. Your time earning a college degree

2. For each of the following, indicate whether it is a fixed or variable cost and a cash or noncash expense (assume short run).

	Fixed or variable?		Cash or noncash?	
a. Gas and oil	___	___	___	___
b. Depreciation	___	___	___	___
c. Property taxes	___	___	___	___
d. Salt and minerals	___	___	___	___
e. Labor hired on an hourly basis	___	___	___	___
f. Labor contracted for one year in advance	___	___	___	___
g. Insurance premiums	___	___	___	___
h. Electricity	___	___	___	___

3. Assume that Freda Farmer has just purchased a new combine. She has calculated total fixed cost to be $22,500 per year and estimates a total variable cost of $9.50 per acre.

   a. What will her average fixed cost per acre be if she combines 1,200 acres per year? 900 acres per year?

   b. What is the additional cost of combining an additional acre?

   c. Assume that Freda plans to use the combine only for custom work on 1,000 acres per year. How much should she charge per acre to be sure all costs are covered? If she would custom harvest 1,500 acres per year?

4. Assume the purchase price of a combine is $152,500. It is estimated to have a salvage value of $42,000 and a useful life of eight years. The opportunity cost of capital is 10 percent. Compute annual depreciation and interest.

5. Using the data in the table below, a price of $6 for the output, a cost of $10 per unit of variable input, and a TFC of $200, compute the three total costs (TVC, TFC, TC) and the three average costs (AVC, AFC, ATC).

Variable input (units)	Output (bushels)
0	0
10	35
20	75
30	105
40	130
50	140

   a. What is the maximum profit that can be made with the given prices?

   b. To continue production in the long run, the output price must remain equal to or above $_____.

   c. In the short run, production should stop whenever the output price falls below $_____.

6. Why is interest included as a fixed cost even when no money is borrowed to purchase the item?

7. What will profit be if production takes place at a level where MR = MC just at the point where ATC is minimum? At the point where MR = MC, and AVC is at its minimum?

8. Explain why and under what conditions it is rational for a farmer to produce a product at a loss.

9. Imagine a typical farm or ranch in your local area. Assume it doubles in size to where it is producing twice as much of each product as before.

   a. If total cost also doubles, is the result increasing, decreasing, or constant returns to size? What if total cost increases by only 90 percent?

   b. Which individual costs would you expect to exactly double? Which might increase by more than 100 percent? By less than 100 percent?

   c. Would you expect this farm or ranch to have increasing, decreasing, or constant returns to size? Economies or diseconomies of size? Why?

APPENDIX. COST CURVES

Relations among the seven output-related cost concepts can be graphically illustrated by a series of curves. The shape of these cost curves depends on the characteristics of the underlying production function. Figure 9-6 contains cost curves that represent the general production function shown in Figure 7-2. Other types of production functions would have cost curves with different shapes.

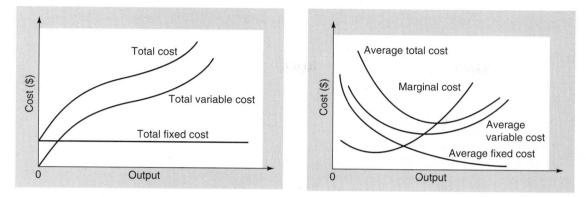

Figure 9-6 Typical total cost curves.

Figure 9-7 Average and marginal cost curves.

The relations among the three total costs are shown in Figure 9-6. Total fixed cost is constant and unaffected by output level. Total variable cost (TVC) is always increasing, first at a decreasing rate and then at an increasing rate. Total cost is the sum of total fixed cost and total variable cost, so its curve has the same shape as the total variable cost curve. However, it is always higher by a vertical distance exactly equal to total fixed cost.

The general shape and relation of the average and marginal cost curves are shown in Figure 9-7. Average fixed cost is always declining but at a decreasing rate. The average variable cost (AVC) curve is U-shaped, declining at first, reaching a minimum, and then increasing at higher levels of output. The average total cost (ATC) curve has a shape similar to that of the AVC curve. They are not an equal distance apart. The vertical distance between them is equal to average fixed cost (AFC), which changes with output level. This accounts for their slightly different shape and for the fact that their minimum points are at two different output levels.

The marginal cost curve will generally be increasing. However, for this particular production function, it decreases over a short range before starting to increase. The marginal cost curve crosses both average curves at their minimum points. As long as the marginal cost value is below the average cost value, the average cost will be decreasing, and vice versa. For this reason, the marginal cost curve will always cross the average variable and ATC curves at their minimum points.

Other Possible Cost Curves

As stated earlier, the shape of the cost curves is directly related to the nature of the underlying production function. The cost curves in Figures 9-6 and 9-7 are all derived from the shape of the generalized production function in Figure 7-2. Other types of production functions exist in agriculture, in particular, those which increase at a *decreasing* rate from the first unit of input. They do not have a Stage I and therefore begin with diminishing marginal returns. The data in Table 7-1 illustrate such a function.

Figure 9-8 shows the total, average, and marginal cost curves for this type of production function. The production function effectively begins in Stage II with diminishing marginal returns, so the TVC curve increases at an increasing rate from the beginning. This in turn causes the AVC curve to increase at an increasing rate throughout. However, because the ATC curve is the sum of AVC and AFC, it begins high due to the high AFC. Initially, AFC will be decreasing at a rapid rate and faster

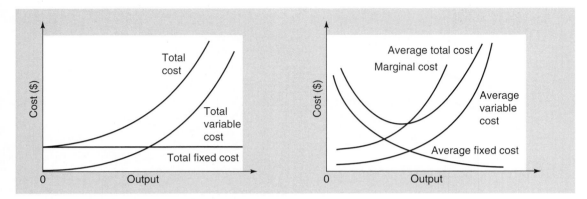

Figure 9-8 Cost curves for a diminishing marginal returns production function.

than AVC is increasing. This combination results in an ATC that decreases at first but eventually increases as the AVC curve begins increasing at a more rapid rate than the AFC curve is declining.

Cost curves with a different shape can result when output is measured in something other than the usual agricultural commodities. The output from machinery services is one example. It is difficult, if not impossible, to measure machinery output in bushels, pounds, or tons, particularly if the machine is used in the production of several outputs. Therefore, the output from tractors in particular, and often other machinery, is measured in hours of use or acres covered during a year. There is no declining marginal product in this case as another hour is another hour of the same length and the same amount of work can be performed in that hour. A constant marginal physical product results from each additional hour of use, that is, another hour of work performed.

Figure 9-9 illustrates the cost curves for this example. With a constant marginal physical product, TVC increases at a constant rate, which in turn causes AVC to be constant *per hour* of use. However, AFC is decreasing as hours of use increases. ATC is the sum of AVC and AFC, so it will also be continually decreasing as annual hours of use increases. (See Chapter 22 for more on machinery costs.)

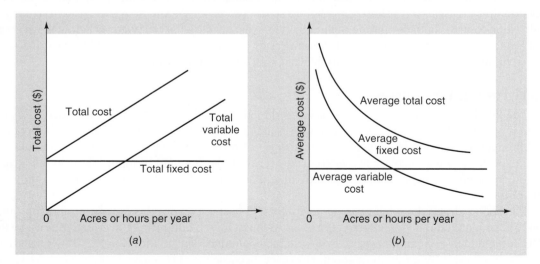

Figure 9-9 Cost curves for a production function with constant marginal returns.

© Digital Vision/Punchstock

BUDGETING FOR GREATER PROFIT

The previous section discussed some basic economic principles and cost concepts. One practical application of these concepts is their use in various types of budgets. Budgeting is a powerful forward-planning tool, used for comparing alternatives on paper before committing resources to a particular plan or course of action. Budgeting can be applied to a single input, an entire enterprise, or the whole-farm business.

Budgets reflect the manager's best estimate as to what will happen in the future if a certain plan is followed. Paper, pencils, calculators, and computers are the tools of budgeting and, as such, may be the most important tools a manager uses. Budgets can be used to compare the profitability of different enterprises and to find the best combination of enterprises for a farm. Budgets can also play a role in the control function of management. Actual outcomes can be compared to projections in the budgets and discrepancies examined.

Four types of budgeting are explored in Chapters 10, 11, 12, and 13. In Chapter 10, budgets for a single enterprise are explained and examined. In Chapter 11, whole-farm planning and budgeting are discussed. Partial budgets, the subject of Chapter 12, provide a framework for analyzing changes in the farm plan involving interactions between several enterprises. Cash flow budgeting is the topic of Chapter 13.

Economic principles, combined with enterprise, partial, whole-farm, and cash flow budgeting, provide the farm manager with a powerful set of tools for analyzing and selecting alternatives to be included in management plans. Careful budgeting can improve profits and prevent costly mistakes.

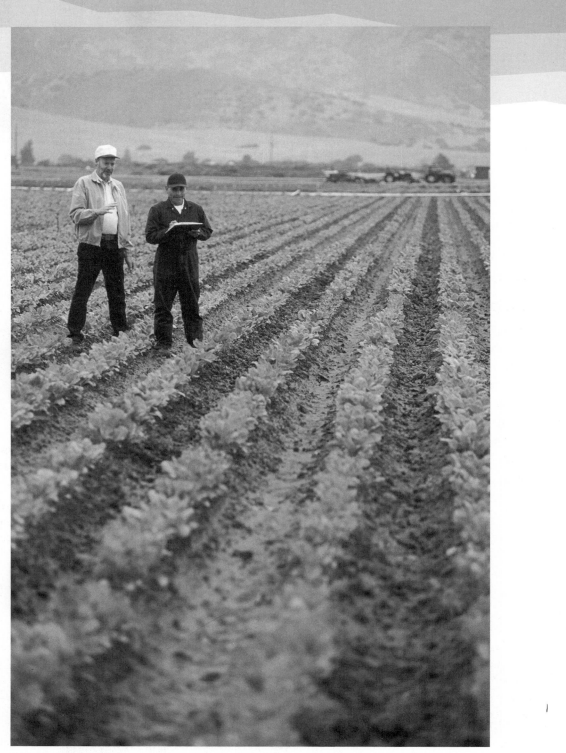

© Photo by Scott Bauer/USDA

ENTERPRISE BUDGETING

CHAPTER OUTLINE

Purpose, Use, and Format of Enterprise Budgets

Constructing a Crop Enterprise Budget

Constructing a Livestock Enterprise Budget

General Comments on Enterprise Budgets

Interpreting and Analyzing Enterprise Budgets

Summary

Questions for Review and Further Thought

CHAPTER OBJECTIVES

1. Define an enterprise budget and discuss its purpose and use

2. Illustrate the different sections of an enterprise budget

3. Learn how to construct a crop enterprise budget

4. Outline additional problems and steps to consider when constructing a livestock enterprise budget

5. Show how data from an enterprise budget can be analyzed and used for computing cost of production and breakeven prices and yields

An enterprise budget provides an estimate of the potential revenue, expenses, and profit for a single enterprise. Each crop or type of livestock that can be grown is an enterprise. Therefore, there can be enterprise budgets for cotton, corn, wheat, beef cows, dairy cows, farrow-finish hogs, watermelons, soybeans, peanuts, and so forth. They can be created for different levels of production

or types of technology, so there can be more than one budget for a given enterprise. The base unit for enterprise budgets is typically one acre for crops. For livestock, some managers develop enterprise budgets on a per-head basis, while others pick a *typical* size operation (e.g., a 30-head cow-calf operation) as the basis for a budget. Using common units permits an easy and fair comparison across different enterprises. Enterprise budgets for many enterprises are usually available from the county or state Cooperative Extension Service. These budgets, developed to represent a *typical* situation, may or may not be accurate for a particular operation and often need to be adjusted.

PURPOSE, USE, AND FORMAT OF ENTERPRISE BUDGETS

The primary purpose of enterprise budgets is to estimate the projected costs, returns, and profit per unit for the enterprises. Once this is done, the budgets have many uses. They help identify the more profitable enterprises to be included in the whole-farm plan. A whole-farm plan often consists of several enterprises, so enterprise budgets are often called the *building blocks* of a whole-farm plan and budget. (See Chapter 11 for more on whole-farm budgeting.) Although constructing an enterprise budget requires a large amount of data, once completed, it is a source of data for other types of budgeting. A manager will refer often to enterprise budgets for information and data when making many types of decisions. An illustration of other uses for enterprise budgets will be postponed until after discussing their organization and construction.

Example of an Enterprise Budget

Table 10-1 is an example that will be used to discuss the content, organization, and structure of a crop enterprise budget. Although no single organization or structure is used by everyone, most budgets contain the sections or parts included in Table 10-1. This example does not include all the detail on physical quantities and prices usually found on an enterprise budget, but it will serve to illustrate the basic organization and content.

The name of the enterprise being budgeted and the budgeting unit are shown first. Most enterprise budgets cover a year or less, but for

TABLE 10-1	Example Enterprise Budget for Watermelon Production (One Acre)

Item		Value per acre
Revenue		
400 cwt. @ $16 per cwt.		$6,400.00
Variable costs		
Seed and plants	$ 600.00	
Fertilizer and lime	300.00	
Pesticides	350.00	
Plastic mulch	130.00	
Machinery fuel, lube, and repairs	100.00	
Harvesting, hauling, and marketing	3,200.00	
Irrigation variable costs	130.00	
Other variable costs	390.00	
Labor	200.00	
Interest @ 5% for six months	135.00	
Total variable cost		$5,535.00
Income above variable cost		865.00
Fixed costs		
Machinery depreciation, interest, taxes and insurance	$ 155.00	
Land charge	120.00	
Overhead	350.00	
Total fixed costs		$ 625.00
Total costs		$6,160.00
Estimated profit (return to management)		$ 240.00

some enterprises, with a long production process, a multiyear budget is more useful. If the period is longer than a year, it is useful to state the period being covered. Income or revenue from the enterprise is typically shown next. Quantity, unit, and price should all be included to provide full information to the user. The cost section comes next and is generally divided into two parts: variable (or operating) costs and fixed (or ownership) costs. Some budgets further divide variable costs into preharvest variable costs, or those that occur before harvest, and harvest costs, or those that are a direct result of harvesting. Income or revenue above variable costs is an intermediate calculation and shows the revenue remaining to be applied to fixed costs. Income above variable costs is sometimes called the *gross margin* of an enterprise.

Fixed costs in a crop enterprise budget typically include the fixed costs for the machinery needed to produce the crop and a charge for land use. These costs are probably the most difficult to estimate, and discussion of the procedures to be used will be postponed until the next section.

The estimated profit per unit is the final value and is found by subtracting total costs from gross revenue. Everyone is interested in this value, and it is important that it be interpreted correctly.

Economic Budgeting

Most enterprise budgets, such as the one in Table 10-1, are economic budgets. This means that, in addition to cash expenses and depreciation, some opportunity costs are also included. Typically, there would be opportunity costs for operator labor, capital used for variable costs, capital invested in machinery, and possibly capital invested in land. Therefore, the profit or return shown on an enterprise budget is an estimated *economic* profit. This profit is different from accounting profit, where opportunity costs are not recognized. When working with enterprise budgets developed by others, managers should be careful to check whether opportunity costs have been included in the cost figures.

Enterprise and other types of budgets often use a slightly different terminology to describe the different types of costs. Economic variable costs may be called *operating costs* or *direct costs*, emphasizing that they arise from the actual operation of the enterprise. These costs would not exist except for the production from this enterprise.

Fixed costs may be called *ownership costs* or *indirect costs*. The term ownership costs refers to the fixed costs that arise from owning machinery, buildings, or land. They result from *owning* assets and would exist even if they were not used for this enterprise. They include such things as the usual economic fixed costs and other farm expenses such as property and liability insurance, legal and accounting fees, pickup truck expenses, and subscriptions to agricultural publications. These expenses are necessary and appropriate, but they are not directly tied to any single enterprise. It is therefore difficult to assign these expenses correctly when making enterprise budgets. They are often prorated on some basis to all enterprises to make sure all farm expenses, direct and indirect, are charged to the farm enterprises. Machinery expenses, for example, can be prorated based on the number of hours a machine is used for a particular enterprise. Other general operating expenses are sometimes assigned based on that enterprise's share of either the farm's total variable costs or its total gross revenue.

It is important to remember that an enterprise budget is a projection of what the manager believes gross revenue, costs, and profits will be in

---

Box 10-1   Opportunity Cost of Management

The opportunity cost of management is often omitted from an enterprise budget. While it would be appropriate to include it, it is a difficult opportunity cost to estimate. This may explain its frequent omission. If no opportunity cost for management is shown in the budget, the estimated profit should be interpreted as *estimated return to management and profit*. This terminology is used on some enterprise budgets.

a future time period. In contrast, a cost and return estimation (CARE) deals with the actual costs incurred by producers in a given time period.

CONSTRUCTING A CROP ENTERPRISE BUDGET

The first step in constructing a crop enterprise budget is to determine tillage and agronomic practices, input levels, type of inputs, and seeding

rate, fertilizer levels, type and amount of herbicide, number and type of tillage operations, and so forth. All of the agronomic, production, and technical decisions must be made before beginning work on the enterprise budget.

Table 10-2 is an enterprise budget for corn that will be used to discuss the steps in constructing a crop enterprise budget. This budget assumes an owner/operator paying all the expenses and receiving the entire crop.

TABLE 10-2	Enterprise Budget for Corn (One Acre)			
Item	**Unit**	**Quantity**	**Price**	**Amount**
Revenue				
Corn grain	bu	105	$5.00	$525.00
Gross revenue				$525.00
Operating expenses				
Seed	thousands	28	$2.95	$ 82.60
Seed treatment	acre	1	9.60	9.60
Fertilizer: Nitrogen	lb	140	0.50	70.00
Phosphorous	lb	60	0.50	30.00
Potash	lb	60	0.45	27.00
Micronutrients	acre	1	10.00	10.00
Lime (prorated)	tons	0.33	35.00	11.55
Pesticides	acre	1	28.50	28.50
Machinery variable costs	acre	1	33.43	33.43
Labor	hr	2.1	12.00	25.20
Crop insurance	acre	1	26.00	26.00
Interest (operating expenses for 6 months)	$	$176.94	6.0%	10.62
Drying and hauling	bu	105	0.50	52.50
Total operating expense				417.00
Income above variable costs				108.00
Ownership expenses				
Machinery depreciation	acre	1	11.00	11.00
Machinery interest	acre	1	10.50	10.50
Machinery taxes & insurance	acre	1	2.50	2.50
Land charge	acre	1	65.00	65.00
Miscellaneous overhead	acre	1	4.00	4.00
Total ownership expenses				93.00
Total expense				510.00
Profit (return to management)				$ 15.00

Revenue

The revenue section should include all cash and noncash revenue from the crop. Some crops have two sources of revenue, such as cotton lint and cottonseed, oat grain and oat straw. In some years for some crops, government program payments may be included as a source of revenue. All cash revenue should be included, and the concept of opportunity cost should be used to value any noncash sources of revenue such as crop residue used by livestock.

The accuracy of the projected profit for the enterprise may depend more on the estimates made in this section than in any other. It is important that both yield and price estimates be as accurate as possible. Projected yield should be based on historical yields, yield trends, and the type and amount of inputs to be used. The appropriate selling price will depend somewhat on whether the budget is for only the next year or for long-run planning purposes. For a budget to be used for planning for the next crop season, futures prices or forward contract prices may provide a good estimate of selling price. For a budget for long-term planning, a review of historical prices should be conducted.

Operating or Variable Expenses

This section includes those costs that will be incurred only if this crop is produced. The amount to be spent in each case is under the control of the decision maker and can be reduced to $0 by not producing this crop.

Seed, Fertilizer, Lime, and Pesticides

Costs for these items are relatively easy to determine once the levels of these inputs have been selected. Prices can be found by contacting input suppliers, and the total per-acre cost for each item is found by multiplying the quantity by the price.

Machinery Variable Costs

Machinery variable costs include costs for fuel, oil, and lubricants, as well as machinery repairs. Fuel and lubricants expense is related to the type and size of machinery used and to the number and type of machinery operations performed for this crop. A quick and simple way to obtain this value is to divide the total farm expense for fuel and lubricants by the number of crop acres. However, this method is not accurate if some machinery is also used for livestock production and if some crops take more machinery time than others.

A more accurate method is to determine fuel consumption per acre for each machine operation and then sum the fuel usage for all the operations scheduled for this crop. The result can be multiplied by the price of fuel to find the per-acre cost. Another method is to compute fuel consumption per hour of tractor use and then determine how many hours will be needed to perform the machine operations. This method is used by many computer programs written to calculate enterprise budget costs.

Estimating per-acre machinery repairs has many of the same problems as estimating fuel use. A method must be devised that allocates repair expense relative to the type of machinery used and the amount of use. Any of the methods discussed for estimating fuel expense can also be used to estimate machinery repair expense. Chapter 22 contains more detailed methods for estimating repair costs for all types of machinery.

Labor

Some enterprise budgets go into enough detail to divide labor requirements into that provided by the farm operator and that provided by hired labor. However, most use one estimate for labor with no indication of the source. Total labor needed for crop production is heavily influenced by the size of the machinery used and the number of machine operations. Besides the labor needed to operate machinery in the field, care should be taken to include time needed to get to and from fields, adjust and repair machinery, and perform other tasks related directly to the crop in the enterprise budget.

The opportunity cost of farm operator labor is often used to value labor. If some hired labor is used, the opportunity cost should certainly be at least equal to the cost of hired labor, including any fringe benefits. When estimating the opportunity cost of labor, it is important to include only the cost of labor, not management. A management charge can be shown as a separate item in the budget or, more commonly, included in the estimated net return. Therefore, the *bottom line* is labeled as a return to management.

Interest

This interest expense is for the capital tied up in operating expenses. For annual crops, it is generally less than a year from the time of the expenditure until harvest when income is or can be received. Therefore, interest is charged for a period of less than a year. This time should be the average length of time between when the operating costs are incurred and the harvest. Interest is charged on operating expenses without regard to how much is borrowed or even if any is borrowed. Even if no capital is borrowed, there is an opportunity cost on the farm operator's capital. If the amount of borrowed capital and equity capital that will be used is known, a weighted average of the interest rate on borrowed money and the opportunity cost of equity capital can be used. In our sample corn budget, interest on operating expenses is calculated for a six-month period.

Drying and Hauling, and Crop Insurance

There may be expenses related to drying the crop and hauling the crop to market. These expenses are normally related to the volume of production. In the example, interest is not charged on drying and hauling because these expenses would be incurred at the very end of the production cycle, so capital would not be tied up for any significant period of time. The cost of fuel and other factors will affect these charges.

Crop insurance is available for most major crops in the United States through private insurance providers. Purchasing crop insurance is a means of reducing the risk associated with agricultural production. Chapter 15 provides more information on risk management.

Income Above Variable Costs

This value, which may also be called *gross margin*, shows how much an acre of this enterprise will contribute toward payment of fixed or ownership expenses. It also shows how much revenue could decrease before this enterprise could no longer cover its variable or operating expenses.

Ownership or Fixed Expenses

This section includes those costs that would exist even if the specific crop were not grown. These are the costs incurred due to ownership of machinery, equipment, and land used in crop production.

Machinery Depreciation

The amount of machinery depreciation to charge to a crop enterprise will depend on the size and type of machinery used and the number and type of machine operations. As with machinery operating expenses, the problem is the need to properly allocate the total machinery depreciation to a specific enterprise. The first step is to compute the average annual depreciation on each machine. This can be quickly and easily done with the straight-line depreciation method. Annual depreciation on each machine can then be converted to a per-acre or per-hour value based on acres or hours used per year. Next, it can be prorated to a specific crop enterprise based on use.

Machinery Interest

Interest on machinery is based on the average investment in the machine over its life and is computed the same way no matter how much, if any, money was borrowed to purchase it. The equation used in Chapter 9 to compute the interest component of total fixed costs should be used to find the average annual interest charge. Next, this interest charge should be prorated to each enterprise using the same method as was used for depreciation.

Machinery Taxes and Insurance
Machinery is subject to personal property taxes in some cases, and most farmers carry some type of insurance on their machinery. The annual expense for these items should be computed and then allocated, using the same method as for other machinery ownership costs.

Land Charge
There are several ways to calculate a land charge: (1) what it would cost to cash rent similar land; (2) the net cost of a share-rent lease for this crop on similar land; and (3) for owned land, the opportunity cost of the capital invested, that is, the value of an acre multiplied by the opportunity cost of the owner's capital. The three methods can give greatly different values. This is particularly the case for the third method if it is used during periods of rapidly increasing land values. During inflationary periods, land values reflect both the appreciation potential of land and its value for crop production.

Most enterprise budgets use one of the rental charges even if the land is owned. Assuming a short-run enterprise budget, the land owner/operator could not sell the acre and invest the resulting capital. As long as the land is owned, if it is not farmed by the owner, the alternative is to rent it to another farm operator. The rental amount then becomes the short-run opportunity cost for the land charge.

Miscellaneous Overhead
Many enterprise budgets contain an entry for miscellaneous overhead expense. This entry can be used to cover many expenses such as a share of pickup truck expenses, farm liability insurance, farm shop expenses, and so on. These expenses cannot be directly associated with a single enterprise but are necessary and important farm expenses. They are often allocated based on the enterprise's share of either gross revenue or total variable costs.

Profit and Return to Management
The estimated profit is found by subtracting total expenses from gross revenue. If a charge for management has not been included in the budget, this value should be considered the return to management *and* profit. Management is an economic cost and should be recognized in an economic budget either as a specific expense or as part of the residual net return or loss. In the case of our example corn budget, the return to management is positive, meaning that the revenue generated by selling corn was sufficient to cover all the costs, including the opportunity costs, of producing the crop.

Other Considerations for Crop Enterprise Budgets
The corn example used to describe the process of constructing a crop enterprise budget is a fairly simple example. Other crops may have many more variable inputs or revenue and expenses specific to that crop. Special problems that may be encountered when constructing crop enterprise budgets include the following:

1. Double cropping, where two crops are grown on the same land in the same year. In this case, budgets should be developed for each crop and the annual ownership costs for land divided between the two crops.
2. Storage, transportation, and marketing expenses may be important for some crops. Most enterprise budgets assume sale at harvest and do not include storage costs. They are assumed to be part of a marketing decision and not a production decision. However, there may still be transportation and marketing expenses even with a harvest sale. If storage costs are included in the budget, then the selling price should be that expected at the end of the storage period, not the harvest price.

Box 10-2   Computers and Enterprise Budgets

Computers and spreadsheets are well adapted and widely used to develop enterprise budgets. Many agricultural universities and extension services use computers to develop annual enterprise budgets for the major enterprises in their state. Their computer programs often calculate machinery and building costs based on typical costs, sizes, use, and other agricultural engineering factors. Many individual farmers and ranchers have developed their own enterprise budgeting spreadsheet templates.

Enterprise budgets are the starting point in some large farm financial planning software programs. Once these budgets are prepared, a farm plan is produced by selecting the number of acres or head for each enterprise. The program then calculates income and expenses from the budgets and transfers the results to a cash flow budget, projected income statement, and other projected financial statements. It then becomes easy to make small changes to the farm plan and observe the results on cash flows and profit before selecting a final farm plan.

---

3. Establishment costs for perennial crops, orchards, and vineyards present another problem. These and other crops may take a year or more to begin production, but their enterprise budgets are typically for a year of full production. It is often useful to develop separate budgets for the establishment phase and the production phase. The latter budget must include an expense for a prorated share of the establishment costs and any other costs incurred before receiving any income. This is done by accumulating costs for all years before the onset of production and then determining the present value of these costs. This value can be used to determine the annual equivalent, included as an annual expense on the enterprise budget. (See Chapter 17 for a discussion of present value and annual equivalent.)

4. The methods for computing machinery depreciation and interest discussed earlier are easy to apply and widely used. However, they do not result in the exact amounts needed to cover depreciation and provide interest on the machine's remaining value each year. The capital recovery method, while more complex, does provide the correct amount. Capital recovery combines depreciation and interest into one value.

The general equation for the annual capital recovery amount is

$$(\text{TD} \times \text{amortization factor}) + (\text{salvage value} \times \text{interest rate})$$

where TD is the total depreciation over the life of the machine or the capital that must be recovered. An amortization factor corresponding to the interest rate used and the life of the machine can be found in Appendix Table 1 or from a financial calculator. The capital recovery amount obtained from this equation is money required at the end of each year to pay interest on the remaining value of the machine *and* recover the capital lost through depreciation. Because the salvage value will be recovered at the end of the machine's useful life, only interest is charged on this amount. (See Chapters 17 and 22 for discussions of capital recovery.)

CONSTRUCTING A LIVESTOCK ENTERPRISE BUDGET

Livestock enterprise budgets include many of the same entries and problems as crop enterprise budgets. However, livestock budgets can also have some unique and particular problems, which will be discussed in this section. The cow/calf budget in Table 10-3 will be used as the basis for the discussion.

Unit

The budgeting unit for livestock is usually one head, but different units such as one litter for swine, one cow unit, or 100 birds for poultry

TABLE 10-3	Example of a Cow/Calf Budget for One Cow Unit*			
Item	Unit	Quantity	Price	Amount
Revenue				
Cull cow (0.10 head)	cwt.	11.00	88.00	$96.80
Heifer calves (0.33 head)	cwt.	5.20	176.00	302.01
Steer calves (0.45 head)	cwt.	5.50	195.00	482.63
Gross revenue				881.44
Operating expenses				
Hay	ton	1.75	100.00	175.00
Supplement	cwt.	1.50	15.00	22.50
Salt, minerals	cwt.	0.80	32.00	25.60
Pasture maintenance	acre	2.00	115.00	230.00
Veterinary and health expense	head	1.00	26.00	26.00
Livestock facilities repair	head	1.00	12.00	12.00
Machinery and equipment	head	1.00	17.00	17.00
Breeding expenses	head	1.00	19.55	19.55
Labor	hours	6.00	12.00	72.00
Auction and hauling	head	1.00	35.00	35.00
Miscellaneous	head	1.00	25.00	25.00
Interest (on half of operating expenses)	$	329.83	6.0%	19.79
Total operating expense				679.44
Income above operating expenses				202.00
Ownership expenses				
Interest on breeding herd	$	1,250.00	4.5%	56.25
Livestock facilities				
Depreciation and interest	head	1.00	12.60	12.60
Machinery and equipment				
Depreciation and interest	head	1.00	28.90	28.90
Land charge	acre	2.00	35.00	70.00
Total ownership expenses				167.75
Total expenses				847.19
Profit (return to management)				34.25

handwritten annotations: "Represents Short-Run" pointing to 679.44; "Represents Long-Run" pointing to 847.19

*1 cow unit = 1 cow, 0.04 bull, 0.9 calf, 0.12 replacement heifer.

can also be used. In some instances, livestock budgets may be developed for several different typical sizes of the enterprise (e.g., 30 head, 50 head, and so on), to reflect economies of size.

Period

Although many livestock enterprises are budgeted for one year, some feeding and finishing enterprises require less than a year. Some types of breeding livestock, such as swine, produce offspring more than once a year. Whatever the period chosen, it is important that all costs and revenue in the budget be calculated for the same period.

Multiple Products

Many livestock enterprises will have more than one product producing revenue. For example, a dairy would have revenue from cull cows, calves, and milk, while a sheep flock would have revenue from cull breeding stock, lambs, and wool. Some operations may have revenue from manure sales, which should also be included in the enterprise budget. All sources of revenue must be identified and then prorated correctly to an average individual animal in the enterprise.

The cow/calf budget in Table 10-3 shows average revenue from cull cows, heifer calves, and steer calves *per cow unit* in the producing herd. A cow unit includes the cow and a portion of the calves, bulls, and replacement heifers. There is an implied 10 percent replacement rate each year based on the 0.10 cull cows sold per producing cow. Less than one calf is sold per cow unit due to several assumptions: (1) calving percentage is less than 100 percent, (2) some heifer calves are retained for replacing cull cows, and (3) some death loss is incurred.

Breeding Herd Replacement

An important consideration in enterprise budgets for breeding herds is properly accounting for replacements for the producing animals. The example in Table 10-3 assumes that female replacements are raised rather than purchased. With a 90 percent calving rate, 0.45 steer calves are available for sale per cow, and there should also be 0.45 heifer calves. However, at least 0.10 heifer calves per cow must be retained as replacements. This example shows that 0.12 (0.45 available less 0.33 sold) are retained. The additional heifer calves retained would cover a 2 percent death loss among the replacements and the producing herd. Whenever replacement animals are raised, the number of female offspring sold must be adjusted to reflect the percentage retained for replacing animals culled from the breeding herd and death loss.

If replacements are purchased, the enterprise budget would show revenue from selling all female offspring as calves. The net cost of the replacements can be included in one of two ways. First, an annual depreciation charge, computed using the straight-line method, can be included as an expense. This will account for the decline in value from purchase price to market value at the time of culling for the typical cow in the herd. No revenue from cull animals nor expense for purchasing replacements would be shown. As an alternative, the purchase cost of a replacement divided by the useful life can be shown as an expense, and the revenue from culled animals divided by useful life can be shown as revenue. The difference between these two values should be the same value as the annual depreciation in the first method.

Feed and Pasture

Many livestock enterprises will consume both purchased feed and farm-raised feed. Purchased feed is easily valued at cost. Farm-raised feed should be valued at its opportunity cost or what it would sell for if marketed off the farm. Care should be taken to include expenses for salt and minerals, any annual charges for establishment costs, fertilizer, spraying or mowing of pastures, and the feed needed to maintain the replacement herd if replacements are being raised. Pasture costs would also include a charge for the land used, based either on the cost of renting the pasture or, if owned, on its opportunity cost.

Livestock Facilities

Livestock facilities include buildings, fences, pens, working chutes, feeders, waterers, wells, windmills, feed storage, milking equipment, and other specialized items used for livestock production. Operating expenses for these items include repairs and any fuel or electricity required to operate them.

These items also incur fixed costs and present some of the same computational and allocation problems outlined when discussing crop machinery. Annual depreciation, interest, taxes, and insurance should be computed for each livestock facility. For specialized items used only by one livestock enterprise, the total annual fixed costs can simply be divided by the number of head that use it each year to get the per-head charge. When more than one livestock enterprise uses the item, some method must be used to allocate the fixed costs among the enterprises.

Machinery and Equipment

Tractors, pickups, and other machinery and equipment may be used for both crop and livestock production. Both operating and ownership expenses should be divided between crop and livestock enterprises according to the proportional use of the item.

GENERAL COMMENTS ON ENTERPRISE BUDGETS

Several factors need to be considered when constructing and using enterprise budgets. The economic principles of marginal value product (MVP) equals marginal input cost (MIC) and least-cost input combinations should be considered when selecting input levels for a budget. However, it should not be assumed that all third-party published budgets were constructed using these principles. Even if they were, these levels may not be correct for a specific, individual situation. The typical or average input levels used in many published budgets may not be the profit-maximizing levels for any individual farm.

It is possible to use many different input levels and input combinations, even on a single farm, so there is not a single budget for each enterprise. There are as many potential budgets as there are possible input levels and combinations. This again emphasizes the importance of selecting the profit-maximizing input levels for the individual situation or, if capital is limited, selecting the input levels that will satisfy the equal marginal principle.

The fixed-cost estimates for enterprise budgets are usually based on an assumed farm size or level of use. Different estimates may be needed if fixed costs are spread over significantly more or fewer output units than the level assumed in the budget.

Enterprise budgets require a large amount of data. Past farm or ranch records are an excellent source if they are available in sufficient detail and for the enterprise being budgeted. Many states publish an annual summary of the average income and expenses for farms participating in a statewide record-keeping service.

These summaries may contain sufficient detail as a useful source of data for enterprise budgeting. Research studies conducted by agricultural universities, the U.S. Department of Agriculture, and agribusiness firms are reported in bulletins, pamphlets, special reports, and farm magazines. This information often includes average yields and input requirements for individual enterprises.

The appropriate price and yield data may depend on the purpose of the budget. A budget to be used only for making adjustments in next year's crop and livestock plan should contain estimates of next year's price and expected yield. Estimates of long-run prices and yields should be used in budgets constructed to assist in developing a long-run plan for the business. The appropriate yield for a particular enterprise

---

---

budget will also depend on the types of inputs included in the budget and on the input levels. Higher input levels should be reflected in higher yields, and vice versa.

INTERPRETING AND ANALYZING ENTERPRISE BUDGETS

Any economic enterprise budget must be interpreted correctly. An economic budget includes opportunity costs on labor, capital, land, and perhaps management as expenses. The resulting profit (or loss) is the revenue remaining after covering all expenses, including opportunity costs. This figure can be thought of as an *economic* profit, which will not be the same as *accounting* profit. The latter would not include any opportunity costs as operating expenses. Rather, accounting profit is the revenue remaining to pay management, unpaid labor, and equity capital for their use. A projected economic profit of zero is not as bad as it might seem. This result means all labor, capital, and land are just earning their opportunity costs—no more, no less. A positive projected profit means one or more of these factors are earning more than their opportunity cost.

The data in an enterprise budget can be used to perform several types of analyses. These

analyses include calculating cost of production and computing break-even prices and yields.

Cost of Production

Cost of production is a term used to describe the average cost of producing one unit of the commodity. It is the same as average total cost discussed in Chapter 9, provided the same costs and production level are used to compute each. The cost of production equation for crops is

$$\text{Cost of production} = \frac{\text{total cost}}{\text{yield}}$$

which is the same as for average total cost, with *output* and *yield* being interchangeable terms. For the example in Table 10-2, the cost of production for corn is $510 divided by 105 bushels, or $4.86 per bushel. Cost of production will change if either costs or yields change.

Cost of production is a useful concept, particularly when marketing the product. Any time the product can be sold for more than its cost of production, a profit is being made. If opportunity costs are included in the expenses, the profit is an economic profit. The resulting accounting profit will be even higher.

Break-Even Analysis

The data contained in an enterprise budget can be used to do a break-even analysis for prices and yields. The formula for computing the *break-even yield* is

$$\text{Break-even yield} = \frac{\text{total cost}}{\text{output price}}$$

This is the yield necessary to cover all costs at a given output price. For the example in Table 10-2, it would be $510 divided by $5.00, or 102 bushels per acre. The output price is only an estimate, so it is often useful to compute the break-even yield for a range of possible prices, as shown here:

Price per bushel ($)	Break-even yield (bu)
4.00	127.5
4.50	113.3
5.00	102.0
5.50	92.7
6.00	85.0

This approach often provides some insight into how sensitive the break-even yield is to changes in the output price. As shown in this table, the break-even yield can be quite sensitive to changes in the output price.

The *break-even price* is the output price needed to just cover all costs at a given output level, and it can be found from the equation

$$\text{Break-even price} = \frac{\text{total cost}}{\text{expected yield}}$$

Again using the example in Table 10-2, the break-even price would be $510.00 divided by 105 bushels, or $4.86. The break-even price is the same as the cost of production. They are only two different ways of looking at the same value.

The break-even price can also be computed for a range of possible yields, as in the following table. Different yields cause different break-even prices (and cost of production), and these prices can vary widely depending on the yield level.

Yield (bu)	Break-even price ($)
80	6.38
100	5.10
120	4.25
140	3.64
160	3.19

The yield and output price in an enterprise budget are estimated rather than actual values, so the calculation of break-even yields and prices can aid managerial decision making. By studying the various combinations of break-even yields and prices, managers can form their own expectations about the probability of obtaining a price and yield combination that would just cover total costs. Break-even prices and yields can also be calculated from total variable costs rather than total costs. These results can help managers make the decisions discussed in Chapter 9 concerning continuing or stopping production to minimize losses in the short run. If a break-even analysis calculated on variable costs shows that producers cannot cover variable costs, their best economic move is usually not to produce that enterprise that year.

When there are multiple sources of revenue for an enterprise, break-even analysis can be conducted for one product by holding yields and prices of the other products constant. For example, if a cotton producer sells both cotton lint and cottonseed, there will be two sources of revenue. If the expected cottonseed yield is 1,250 pounds per acre and the expected cotton seed price is $0.12 per pound, then cottonseed will contribute an expected $150 per acre to gross revenue. If the total cost per acre is $700, then the following equation would set gross revenue equal to total cost:

Price of lint × yield of lint + 150 = $700

Price of lint × yield of lint = $550

The break-even yield of lint can be calculated by dividing $550 by the expected price. If the expected price is $0.75, for example, the break-even lint yield is 733 pounds per acre. The break-even price of lint can be found in a similar fashion. If the expected yield is 800 pounds per acre, the break-even price is 68.75 cents per pound.

The break-even calculations are relevant only for the input levels assumed in the enterprise budgets. If input levels are increased or decreased, not only costs but yields would also be expected to change. Also, when calculating break-even yields and selling prices for enterprises with more than one source of revenue, subtract the estimated value of the minor revenue sources from total costs before dividing by the expected selling price for the major product (to get break-even yield) or dividing by the expected yield of the major product (to get break-even selling price for the major product). An example, as already shown, is cotton, where cottonseed is the minor revenue source and lint is the major revenue source. The same would apply to USDA commodity payments.

Box 10-4   Break-Even Yield and Price When Some Costs Are Output Related

When some costs of production depend on the level of production achieved, the break-even approach presented in this chapter only approximates the true break-even yield and price. For example, in the corn budget in Table 10-2 the drying and hauling costs depend directly on the yield of corn.

Total costs of production, excluding drying and hauling, are estimated to be $457.50 per acre. Drying and hauling costs are $0.50 per bushel. If gross revenue is equal to price (P) times yield (Y), the true break-even relationship would be

$$P \times Y = \$457.50 + \$0.50Y$$

The break-even yield would be found by solving the following equation:

$$Y = (\$457.50 + 0.50Y)/P$$

For an assumed price of $5.00 per bushel, for example, the break-even yield would be

$$Y = (\$457.50 + 0.50Y)/\$5.00$$

Solving this equation gives a break-even yield of 101.7, which is very close to the break-even yield of 102 that was calculated when the yield-related costs were ignored.

Likewise, the equation for calculating the break-even price needed to cover total costs would be

$$P = (\$457.50 + \$0.50Y)/Y$$

For an assumed yield of 100 bushels per acre, for example, the break-even price would be

$$P = (\$457.50/100) + \$0.50 = \$5.08$$

This value is very close to the break-even price of $5.10 calculated for a 100 bushel expected yield, when the yield-related costs were ignored. (Note that break-even price for the 105 bushel yield would not change when yield-related costs are considered because $457.50 + $0.50 × 105 = $510.00, the total expenses in the budget.)

When yield-related costs are a small fraction of total costs, the approximate break-even values will usually be close enough for planning purposes.

SUMMARY

Enterprise budgets are an organization of projected income and expenses for a single enterprise. They are constructed for a single unit of the enterprise, such as one acre for a crop and one head for a livestock enterprise. Most enterprise budgets are economic budgets and, as such, include all variable or operating expenses, all fixed or ownership expenses, and opportunity costs on factors such as operator labor, capital, and management.

Enterprise budgets can be used to compare the profitability of alternative enterprises and are particularly useful when developing a whole-farm plan. They can also be used to make minor year-to-year adjustments in the farm plan in response to short-run price and yield changes. Once completed, an enterprise budget contains the data needed to compute cost of production, break-even yield, and break-even price.

QUESTIONS FOR REVIEW AND FURTHER THOUGHT

1. Review Chapter 9 and suggest how the *return above variable costs* value found on most enterprise budgets can be used to make some short-run production decisions.

2. Should the economic principles for determining profit-maximizing input levels be applied before or after completing an enterprise budget? Why?

3. An enterprise budget for soybeans shows a yield of 46 bushels, a selling price of $11.80 per bushel, and total cost of $460.20 per acre. What is the cost of production? The break-even yield? The break-even price?

4. Why should the opportunity costs of a farmer's labor, capital, and management be included on an enterprise budget?

5. If the land is owned, should a land charge be included on an enterprise budget? Why?

6. Would you expect two farms of widely different size to have the same fixed costs on their enterprise budgets for the same enterprise? Might economies or diseconomies of size explain any differences?

7. There are potentially many different enterprise budgets for a single enterprise. Defend or refute that statement.

8. How might an agricultural loan officer use enterprise budgets? A farm real estate appraiser? A farmer when ordering input supplies for the coming year?

9. How would an enterprise budget for a perennial or long-term crop differ from one for an annual crop?

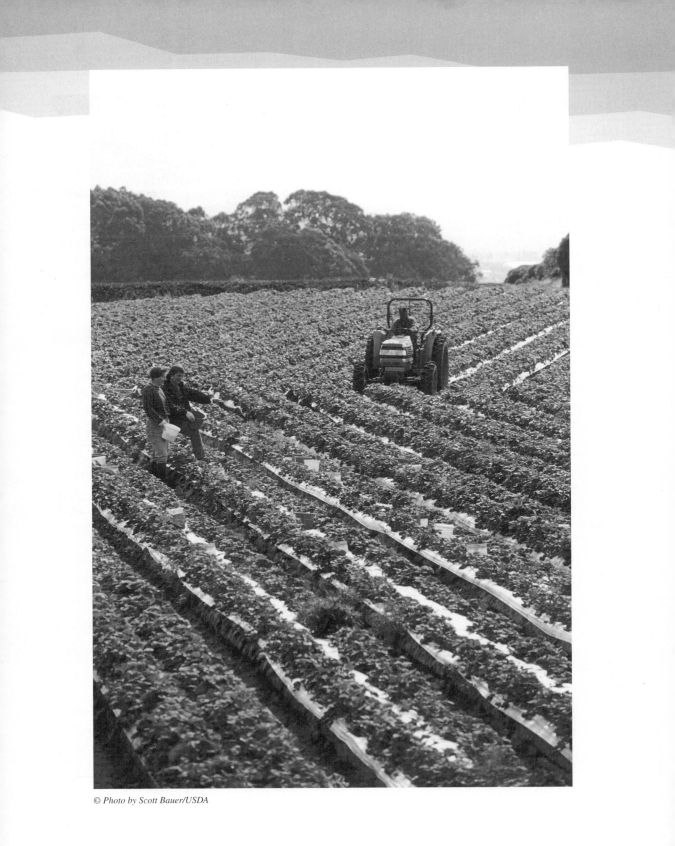

© Photo by Scott Bauer/USDA

WHOLE-FARM PLANNING

CHAPTER OUTLINE

What Is a Whole-Farm Plan?
The Planning Procedure
Example of Whole-Farm Planning
Other Issues
Summary
Questions for Review and Further Thought
Appendix. Graphical Example of Linear
    Programming

CHAPTER OBJECTIVES

1. Show how whole-farm planning differs from the planning of individual enterprises

2. Learn the steps and procedures to follow in developing a whole-farm plan

3. Understand the various uses for a whole-farm plan and budget

4. Introduce linear programming as a tool for choosing the most profitable combination of enterprises

5. Compare the assumptions used for short-run and long-run budgeting

Once management has developed a strategic plan for a farm or ranch business, the next logical step is to develop a tactical plan to carry it out. Every manager has a plan of some type—about what to produce, how to produce, and how much to produce—even if it has not been fully developed on paper. However, a systematic procedure for developing a whole-farm plan may well result in one that will increase farm profits or come closer to attaining other goals.

WHAT IS A WHOLE-FARM PLAN?

As the name implies, a *whole-farm plan* is an outline or summary of the production to be carried out on the entire farm and the resources needed to do it. It may contain sufficient detail to include fertilizer, seed, and pesticide application rates and actual feed rations for livestock, or it may simply list the enterprises to be carried out and their desired levels of production.

When the expected costs and returns for each part of the plan are organized into a detailed projection, the result is a *whole-farm budget*.

Chapter 10 discussed enterprise budgeting. Enterprise budgets can be used as building blocks for the development of a whole-farm plan and its associated budget. The whole-farm plan can be designed for the current or upcoming year, or it may reflect a typical year over a longer period. In some instances, a transitional plan may be needed for the period it takes to fully implement a major change in the operation. The effects of alternative plans on the financial position and risk exposure of the business also need to be considered.

THE PLANNING PROCEDURE

The development of a whole-farm plan can be divided into six steps, as shown in Figure 11-1: (1) determine objectives and specify goals; (2) take an inventory of the physical, financial, and human resources available; (3) identify possible enterprises and their technical coefficients; (4) estimate gross margins for each enterprise; (5) choose a plan—the feasible enterprise combination that best meets the specified goals; and (6) develop a whole-farm budget that projects the profit potential and resource needs of the plan. Each step will be discussed and illustrated by an example.

Often the farm plan is developed for one year, and resources such as land, full-time labor, breeding livestock, machinery, and buildings are considered to be fixed in quantity. The usual objective for a one-year plan is to maximize the total gross margin for the farm. However, alternatives that assume an increase in the supply of one or more of the key resources can also be considered as can a complicated multiyear plan. Some more advanced techniques for analyzing the profitability of new capital investments are discussed in Chapter 17.

Determine Objectives and Specify Goals

Most planning techniques assume that managers primarily seek profit maximization, but this objective is often subject to a number of

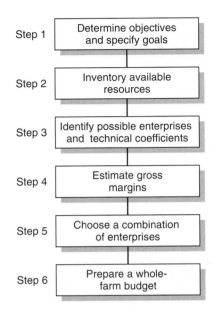

Figure 11-1 **Procedure for developing a whole-farm plan.**

personal and societal restrictions. Maintaining the long-term productivity of the land, protecting the environment, guarding the health of the operator and workers, maintaining financial independence, and allowing time for leisure activities are concerns that affect the total farm plan. Certain enterprises may be included in the plan because of the satisfaction the operator receives from them regardless of their economic results, while others may be excluded for personal reasons even though they could be profitable.

After identifying the overall objectives, both business and personal, a manager should be able to specify a set of performance goals. These can be defined in terms of crop yields, livestock production rates, costs of production, net income, or other measures.

Inventory Resources

The second step in the development of a whole-farm plan is to complete an accurate inventory of available resources. The type, quality, and quantity of resources available determine which enterprises can be considered in the whole-farm plan and which are not feasible.

Land

The land resource is generally the most valuable resource and one of the most difficult to alter. Land is also a complex resource with many characteristics that influence the type and number of enterprises to be considered. The following are some of the important items to be included in the land inventory:

1. Total number of acres available in cropland, pasture, orchards and vineyards, timber, and wasteland
2. Climatic factors, including temperature, annual rainfall, and length of the growing season
3. Soil types and factors such as slope, texture, and depth
4. Soil fertility levels and needs. A soil testing program may be needed as part of the inventory
5. The current water supply and irrigation system or the potential for developing it
6. Drainage canals and tile lines in existence, and any current or potential surface and subsurface drainage problems
7. Soil conservation practices and structures, including any current and future needs for improvement
8. The current soil conservation plan and any limitations it may place on land use or technology
9. Crop bases, established yields, long-term contracts, or other characteristics related to government programs or legal obligations
10. Existing and potential pest and weed problems that might affect enterprise selection and crop yields
11. Tenure arrangements and lease terms that may affect production decisions

This is also a good time to draw up a map of the farm showing field sizes, field layouts, fences, drainage ways and ditches, tile lines, and other physical features. A map can assist in planning changes or documenting past practices. If available, the cropping history of each field, including crops grown, yields obtained, fertilizer and lime applied, and pesticides used, can be recorded on a copy of the field map or in a computer database. This information is useful for developing a crop program where a crop rotation is desirable or herbicide carryover may be a problem. Many counties now have extensive computerized databases of soil and land characteristics that can be accessed through state extension specialists or U.S. Department of Agriculture offices.

Buildings

The inventory of buildings should include a list of all structures, along with their condition, capacity, and potential uses. Livestock enterprises and crop storage may be severely limited by the facilities available. Feedhandling equipment, forage and grain storage, water supply, manure handling, and arrangement and capacity of livestock facilities should be noted in the inventory.

The potential for livestock production may also be affected by the location of the farmstead. Close proximity to streams, lakes, or nearby residences may restrict the type and volume of animals that can be raised or finished. In addition, adequate land area should be available for environmentally sound manure disposal.

Machinery

Machinery can be a fixed resource in the short run, and the number, size, and capacity of the available machinery should be included in the inventory. Particular attention should be given to any specialized, single-purpose machines. The capacity limit of a specialized machine, such as a cotton picker, will often determine the maximum size of the enterprise in which it is used.

Capital

Capital for both short-run and long-run purposes can be another limiting resource. The lack of ready cash or limited access to operating credit can affect the size and mix of enterprises chosen. Reluctance to tie up funds in fixed assets or to leverage the business through long-term borrowing may also limit expansion of the farming operation or the purchase of labor-saving technology.

Labor

The labor resources should be analyzed for quantity and quality. Quantity can be measured in months of labor currently available from the operator, family members, and hired labor, including its seasonal distribution. The availability and cost of additional full-time or part-time labor should also be noted, as the final farm plan might profitably use additional labor. Labor quality is more difficult to measure, but any special skills, training, and experience that would affect the possible success of certain enterprises should be noted.

Management

The last part of a resource inventory is an assessment of the management skills available for the business. What are the age and experience of the manager? What are the past performance of the manager and his or her capacity for making management decisions? What special skills or critical weaknesses are present? If a manager has no training, experience, or interest in a certain enterprise, that enterprise is likely to be inefficient and unprofitable. The quality of the management resource should be reflected in the technical coefficients incorporated into the farm budgets. Past success and records are the best indicators of future performance.

Other Resources

The availability of local markets, transportation, consultants, marketing quotas, or specialized inputs are also important resources to consider when developing the whole-farm plan.

Identify Enterprises and Technical Coefficients

For some producers, the decision about which enterprises to include in the farm plan has already been determined by personal experience and preferences, fixed investments in specialized equipment and facilities, or the regional comparative advantages for certain products. For these producers, the whole-farm planning process focuses on preparing the whole-farm budget for their plan. Other managers, though, will want to experiment with different enterprise combinations by developing a series of budgets and comparing them.

The resource inventory will show which new crop and livestock enterprises are feasible. Those requiring an unavailable resource can be eliminated from consideration, unless purchasing or renting this resource is possible. Custom and tradition should not be allowed to restrict the list of potential enterprises, only the resource limitations. Many farms have incorporated alternative or nontraditional enterprises into their farm plans, and some have been quite profitable.

The *technical coefficients* for an enterprise indicate how much of a resource is required to produce one unit of an enterprise. These technical coefficients, or resource requirements, are important in determining the maximum possible size of enterprises and the final enterprise combination.

The technical coefficients are developed to correspond to the budgeting unit for each enterprise, which would typically be 1 acre for crops and 1 head for livestock. For almost all enterprises, technical coefficients will be needed for land, labor, and capital. Technical coefficients for 1 beef cow, for example, might be 2 acres of pasture, 6 hours of labor, and $679 of operating capital. For some enterprises, machinery time or building use may also be critical resources that will need to be explicitly considered in the farm planning process. Thus, the technical coefficients for an acre of corn might be 1 acre of land, 2.1 hours of labor, $417 of operating capital, and 1.1 hours of tractor time.

Accurate figures for technical coefficients, such as the amount of labor needed per crop acre or the amount of purchased feed needed per livestock unit, can often be obtained from detailed farm records kept by the manager. For new enterprises, information about technical coefficients can sometimes be obtained from third-party enterprise budgets, the Extension Service, or commodity organizations. Inaccurate figures for technical coefficients can lead to distorted or misleading farm plans; hence, accurate technical coefficients are essential for a sound planning process.

Estimate the Gross Margin per Unit

Enterprise budgets, discussed in detail in Chapter 10, are important tools for farm planning. An enterprise budget is required for every enterprise that might be chosen for the farm plan, those already being produced and any new alternatives being considered. Enterprise budgets provide the estimates of gross margin needed in the farm-planning process, and they also give information that can be used to develop the technical coefficients previously discussed.

As shown in Chapter 10, gross margin is the difference between gross income and variable costs. It represents how much each unit of an enterprise contributes toward fixed costs and profit, after the variable costs of production have been paid. If maximization of profit is the goal, using gross margin, instead of net profit, may seem strange. However, because the plan is for the short run, fixed costs are constant regardless of the farm plan selected. Any positive gross margin represents a contribution toward paying those fixed costs. Therefore, in the short run, maximizing gross margin is equivalent to maximizing profit (or minimizing losses), because the fixed costs will not change.

Accurate enterprise budgets are of great importance in whole-farm planning. A good estimate of the gross margin for an enterprise requires the manager's best estimates of gross revenue and variable costs. It is important that estimates of selling prices and yields be accurate and that the yield estimates reflect the production practices employed. The variable cost estimates also need to reflect production practices, and these estimates require identifying the amount of each variable input used as well as the purchase price per unit. Using inaccurate enterprise budgets in the farm planning process will result in a less-than-optimal farm plan and a reduction in profit.

Choose the Enterprise Combination

Given the gross margins of the enterprises, the amount of each resource available on the farm, and the amount of each resource required per unit of each enterprise being considered, managers will attempt to find the combination of enterprises that provides the highest total profit for the farm. If there are more than a few enterprises to consider, or many resource restrictions, using paper and pencil to find the best combination would be difficult, if not impossible. Fortunately, computer software programs exist to aid farm managers in their decision making. *Linear programming* (LP) is a mathematical technique that can be used to find the optimal combination of enterprises within the resource limits of the farm. Linear programming software is now widely available as an *add-on* to electronic spreadsheet programs. Linear programming will be discussed in detail later in this chapter.

Prepare the Whole-Farm Budget

The last step in the planning process is to prepare the whole-farm budget. A whole-farm budget can be used to

1. Estimate the expected income, expenses, and profit for a given farm plan
2. Estimate the cash inflows, cash outflows, and liquidity of a given farm plan
3. Compare the effects of alternative farm plans on profitability, liquidity, and other considerations
4. Evaluate the effects of expanding or otherwise changing the present farm plan
5. Estimate the need for, and availability of, resources such as land, capital, labor, livestock feed, or irrigation water
6. Communicate the farm plan to a lender, landowner, partner, or stockholder

The procedure for developing a whole-farm budget from a whole-farm plan will be shown later in this chapter.

Forms for organizing and recording whole-farm budgets are often available in farm record books. The Extension Service in each state may also have publications, forms, or computer software available. Use of such forms or computer programs will save time and improve the accuracy of the budget estimates.

EXAMPLE OF WHOLE-FARM PLANNING

The procedure used in whole-farm planning can be illustrated through the following example.

Objectives

The manager of the example farm wishes to choose a combination of crop and livestock activities that will maximize total gross margin for the farm for the coming year. For agronomic reasons, the operator wishes to carry out a crop mix that does not include more than 50 percent of the Class A cropland acres planted to cotton in any year.

Resource Inventory

Table 11-1 contains the resource inventory for the example farm. The land resource has been divided into three types: 520 acres of Class A cropland, 200 acres of Class B cropland, and 380 acres of land in permanent pasture. Labor is the only other limited resource, with 3,700 hours available per year. Capital and machinery are available in adequate amounts, and the few buildings available are sufficient for the livestock enterprises to be considered.

Enterprises and Technical Coefficients

Potential crop and livestock enterprises are identified and listed in Table 11-2. Any enterprise that is obviously unprofitable (e.g., has

TABLE 11-1	Resource Inventory for Example Farm
Resource	**Amount and comments**
Class A cropland	520 acres (not over 50% in cotton production)
Class B cropland	200 acres
Pasture	380 acres
Buildings	Only hay shed and cattle shed are available
Labor	3,700 hours available annually from operator and family members
Capital	Adequate for any farm plan
Machinery	Adequate for any potential crop plan
Management	Manager appears capable and has experience with crops and beef cattle
Other limitations	Any hay produced must be fed on farm, not sold

a negative gross margin) or that requires an unavailable resource should be eliminated. However, all feasible enterprises with a positive gross margin should be considered even if, at first glance, an enterprise appears less profitable than some others. The availability of surplus resources will sometimes cause one of the apparently less profitable enterprises to be included in the final plan.

As shown in Table 11-2, the manager of the example farm has identified three potential crop enterprises on Class A cropland (cotton, soybeans, and corn) and two on Class B cropland (soybeans and corn). Livestock enterprises are limited to beef cows or stocker steers because of the buildings available. The beef cow enterprise also requires some Class B land for hay production. Resource requirements per unit of each enterprise, or technical coefficients, are also shown in Table 11-2. For example, an acre of cotton requires 1 acre of Class A cropland, 4 hours of labor, and $490 of operating capital.

TABLE 11-2	Potential Enterprises and Resource Requirements

| Resource | Quantity available | Class A cropland | | | Class B cropland | | Livestock (per head) | |
		Cotton*	Soybeans	Corn	Soybeans	Corn	Beef cows	Stocker steers
Class A cropland (acres)	520	1	1	1	—	—	—	—
Class B cropland (acres)	200	—	—	—	1	1	0.2	—
Pasture (acres)	380	—	—	—	—	—	2	2
Labor (hours)	3,700	4	1.8	2.1	1.8	2.1	6	4
Operating capital ($)		490	364	417	274	370	679	992

*Limited to one-half the class A cropland for crop rotation needs.

TABLE 11-3	Estimating Gross Margins per Unit

| | Class A cropland | | | Class B cropland | | Livestock | |
	Cotton (acre)	Soybeans (acre)	Corn (acre)	Soybeans (acre)	Corn (acre)	Beef cows (head)	Stocker steers (head)
Yield	780 lb	40 bu	105 bu	28 bu	80 bu	—	—
Price ($)	0.80	12.00	5.00	12.00	5.00	—	—
Gross income ($)	624	480	525	336	400	881	1,050
Total variable costs ($)	490	364	417	274	370	679	992
Gross margin ($)	134	116	108	62	30	202	58

Even though operating capital is not limited in this example, it is included for estimating cash flow needs.

Gross Margins

Table 11-3 contains the estimated income, variable costs, and gross margins for the seven potential enterprises to be considered in the whole-farm plan. Detailed breakdowns of the variable costs have been omitted to save space but would be included in the enterprise budgets, which are done first. Budgets for corn on Class A cropland and for the beef cow (cow calf) enterprise are shown in Chapter 10.

Careful attention should always be paid to estimating yields, output prices, input levels, and input prices. The accuracy of the whole-farm plan depends heavily on the estimated gross margins.

The Enterprise Combination

In Chapter 8, a method for finding the most profitable combination of two enterprises was presented. In reality, however, producers often choose between many different enterprises, and they usually do not have perfect knowledge of the production possibility curves for their farms. Linear programming (LP) is a mathematical procedure that uses a systematic technique to find the *best* (e.g., most profitable) possible combination of enterprises. Linear programming models have a linear *objective function*, which is maximized or minimized subject to the resource restrictions.

Linear Programming Basics

In farm planning, normally the objective will be to maximize the total gross margin. In mathematical terms, we can write the objective function (OBJ) this way:

$$OBJ = GM(1) \times UNITS(1) +$$
$$GM(2) \times UNITS(2) +$$
$$GM(3) \times UNITS(3)$$

where $GM(1)$ is the gross margin per unit (e.g., acre of crop or head of livestock) of the first enterprise and $UNITS(1)$ is the number of units of this enterprise (acres or head) produced and so on for all the enterprises being considered in the plan. Each enterprise (corn, soybeans, stocker steers, etc.) must be represented by a term in the equation. The order of the enterprises does not matter, but it must be consistent throughout the linear programming model. Here, three enterprises are being considered. For more enterprises, more terms must be added.

Developing the resource restrictions for the linear programming model requires the use of the technical coefficients, discussed previously in this chapter, and the overall resource limits. The general form of the restrictions is

$$X1 \times UNITS(1) +$$
$$X2 \times UNITS(2) +$$
$$X3 \times UNITS(3) \leq RESOURCE$$

where $X1$ represents the amount of a particular resource needed to produce one unit of the first enterprise, and so on. The total amount of the resource available for all uses, represented by RESOURCE in the equation, limits the number of units of the three possible enterprises that the

farm can produce. Each enterprise being considered needs to be included in the equations. A restriction needs to be developed for every limited resource. The manager may also impose subjective constraints by arbitrarily setting a minimum or maximum level on some enterprises. A restriction to limit an enterprise, say enterprise 3, to no more than a certain amount would look like

$$UNITS(3) \leq LIMIT$$

where LIMIT is the maximum amount of enterprise 3 the manager will allow.

A linear programming model with two enterprises can be solved graphically, as shown in the Appendix to this chapter. Small linear programs can also be solved by hand using matrix algebra, but the process is tedious and, if there are more than a few enterprises or many restrictions, it is easy to make mistakes. A computer program can quickly and accurately solve large linear programming problems with many enterprises and complex restrictions.

The Linear Programming Tableau

Linear programming software is often available as an *add-on* to an electronic spreadsheet program. To solve an LP problem using these programs requires that the information be put in an organized form, often called the LP *tableau*. An example tableau is shown in Table 11-4. The first step in developing the tableau is to set up a column in the spreadsheet for each possible activity or enterprise. Two other columns are also needed: one for the type of restriction (less than or equal to, greater than or equal to, or equal to) and one for the limit on each resource. The limit column is sometimes called the *RHS* because, in the mathematical relation described previously, this value sits on the right-hand side of the inequality sign.

The first row in the LP tableau will hold the gross margin coefficients from the objective function (OBJ). The type is *MAX* for

TABLE 11-4	Linear Programming Tableau for the Farm Planning Example

	Units	Class A cotton (acre)	Class A soybeans (acre)	Class A corn (acre)	Class B soybeans (acre)	Class B corn (acre)	Beef cows (head)	Stocker steers (head)	Type	Limit
Gross margin	$/unit	$134	$116	$108	$62	$30	$202	$58	MAX	
Class A land	acre	1	1	1	0	0	0	0	LE	520
Class B land	acre	0	0	0	1	1	0.2	0	LE	200
Pasture	acre	0	0	0	0	0	2	2	LE	380
Labor	hour	4	1.8	2.1	1.8	2.1	6	4	LE	3,700
Rotation limit	acre	1	0	0	0	0	0	0	LE	260

maximization. There will be no entry in the *Limit* columns in the OBJ row. The subsequent rows will represent the resource restrictions. The column for each enterprise contains the technical coefficients, the amount of that resource required to produce one unit of enterprise. The *Type* column shows the type of restriction (e.g., *less than or equal to*), and the Limit column shows the total amount of that resource available.

The example tableau in Table 11-4 uses the information provided in Tables 11-1, 11-2, and 11-3. The first row is the objective function, so the values are the gross margins for each enterprise, taken from Table 11-3. The next row is the first restriction, Class A cropland. As shown in Table 11-2, each Class A crop requires one acre of this cropland. A "1" is therefore entered in the columns for Class A cotton, Class A soybeans, and Class A corn. The amount of Class A land used must be less than or equal to the total amount available, so an LE for *less than or equal to* is entered in the Type column. Finally, this row is completed by putting the total amount of Class A land available, 520 acres, in the Limit column.

The next row, for the restriction on Class B cropland, is completed in the same manner. Each acre of soybeans Class B and corn Class B will require one acre of Class B land. In addition, each beef cow unit will require 0.2 acres of Class B land, as shown in Table 11-2. The total

amount of Crop B land available is 200 acres, so 200 is the limit on this resource restriction.

For pasture, as shown in Table 11-2, each beef cow requires 2 acres and each stocker steer requires 2 acres. The limit on pasture is 380 acres. These technical coefficients and the limit appear in the corresponding columns of the pasture-restriction row. Labor requirements for each enterprise are found in Table 11-2, and the total amount available, 3,700 hours, is found in Table 11-1. These values are used to fill in the labor-restriction row.

The final row of the tableau is a limit on cotton acreage, imposed by rotational considerations. Capital, machinery, and management available are adequate for any farm plan, so no restrictions are developed for these resources in this case. In other situations, restrictions for these resources might be included.

Table 11-5 shows the solution to the example LP problem. The exact format of the computer output for the solution will depend on the computer package used to solve the problem; however, most packages will give information similar to that shown here. For the example, the optimal enterprise combination is 260 acres of cotton on Class A land, 260 acres of soybeans on Class A land, 162 aces of soybeans on Class B land, and 190 beef cows. Total gross margin (the objective value) for the farm is $113,424. The slack column indicates how

TABLE 11-5	Linear Programming Solution to the Farm Planning Example

Activity	Optimum level	Reduced cost ($)	Rows	Level of use	Slack (unused)	Shadow price ($)
Cotton—Class A	260	0.00	Total gross margin	$113,424	—	—
Soybeans—Class A	260	0.00	Class A crop land	520	0	116.00
Corn—Class A	0	−8.00	Class B crop land	200	0	62.00
Soybeans—Class B	162	0.00	Pasture	380	0	94.80
Corn—Class B	0	−32.00	Labor	2,940	760	0.00
Beef cows	190	0.00	Rotation limit	260	0	18.00
Stocker steers	0	−131.60				

much of a resource would be left over if this farm plan were followed. In this example, all resources except labor are completely used. For labor, 2,940 of the available 3,700 hours are used, leaving a slack value of 760 hours. The other entries in this table will be explained later in this chapter.

Additional Features of Linear Programming

Because computerized linear programming routines can quickly solve large problems, many different enterprises can be considered and detailed resource restrictions can be included. For example, the labor resources could be defined on a monthly basis instead of annually, resulting in 12 labor restrictions instead of one. This refinement ensures that potential labor bottlenecks during peak seasons are not overlooked. Limits on operating capital can be defined in a similar manner, and the land resource can be divided into several classes. Additional restraints that reflect desirable land use practices, environmental considerations, farm program provisions, or subjective preferences of the manager can also be incorporated. Several enterprises can be defined for each crop or type of livestock, representing different types of technology, levels of production, or stages.

In addition to limitations imposed by fixed resources such as owned land and permanent labor, linear programming can incorporate activities that increase the supplies of certain resources. These activities could include renting land, hiring additional labor, or borrowing money. The optimal solution to the planning problem will then include the acquisition of additional resources, to the extent that total gross margin can be increased by doing so. For example, if we wanted to allow labor to be hired, we would include a column in our tableau labeled *Hire labor*. If labor costs $12.00 per hour to hire, our objective function value in this column would be −$12.00. The objective function value for hiring labor would be negative because hiring labor is a cost. In the labor restriction row, we would put a technical coefficient with the value −1 in the Hire labor column. This coefficient would be negative because hiring labor increases the amount of labor available rather than using up an hour. The negative number on the left side of the inequality would thus be equivalent to adding the hours on the right-hand side.

It is also possible to transfer production from one enterprise to the resource supply needed by another enterprise and hence increase the amount of that resource available. For example, corn could be transferred from a corn enterprise to a feed resource to increase the supply of feed available for livestock production. A transfer of this type requires some careful thinking to be certain that the model does not also allow the sale of the same corn that is transferred to the feed resource.

Shadow Prices and Reduced Costs

Besides selecting the enterprises for the optimal farm plan, linear programming routines provide other useful information. Of particular use to many managers are values called *dual values*. There are two types of dual values, *shadow prices* and *reduced costs*. Shadow prices are the dual values on the resource restrictions, and reduced costs are dual values on the activities. Dual values—shadow prices and reduced costs—for our LP example can be found in Table 11-5.

For every resource restriction in the LP tableau, the LP software will calculate a shadow price, which represents the amount by which total gross margin would be increased if one more unit of that resource were available. For resources not completely used, this value would be zero. If a resource is completely used up by the enterprises selected in the plan, the shadow price tells the manager whether it may or may not be a good idea to acquire more of that resource. If the shadow price on a resource is higher than the cost of acquiring one more unit of that resource, the manager might consider obtaining more of that resource. Conversely, if the shadow price is less than the cost, it would not be profitable to acquire more of that resource.

The shadow prices shown in the last column of Table 11-5 indicate that having one more acre of Class B land, for example, would increase gross margin by $62, exactly the same value as the gross margin for 1 acre of soybeans on Class B land, the most profitable crop on Class B land. On the other hand, the value of one more hour of labor is zero, because not all of the original labor supply is used. Similar interpretations can be placed on the other resource shadow prices. This information is useful for deciding how much the manager can afford to pay for additional resources. For example, it would be profitable to rent more Class B land only if the cost were less than $62 per acre. It would not be profitable to hire extra labor given the current unused labor hours.

The reduced costs, found next to the enterprise levels in Table 11-5, tell the manager how much the total gross margin (e.g., the objective value) would be reduced by forcing into the plan one unit of an enterprise that was not included in the optimal plan. Only those enterprises not chosen in the optimal plan will have a nonzero reduced cost. The reduced cost of any enterprise that appears in the solution (at any level) is always zero. Hence, Class A cotton, Class A soybeans, Class B soybeans, and beef cows have values of $0.00 for their reduced costs.

In our example, Class A corn, Class B corn, and stockers all have nonzero reduced costs because these enterprises were not part of the optimal solution. Some software packages report all nonzero reduced costs as negative numbers (as seen in Table 11-5), while others may report them as positive numbers. The interpretation is the same, whether the number is reported with a negative sign or a positive sign.

The reduced cost for corn on Class A land indicates that forcing one acre of corn on Class A land into the solution would reduce total gross margin by $8.00. Similarly, the reduced costs for corn on Class B land and for stocker steers indicate that forcing one acre of Class B corn or one head of stocker steers into the solution would reduce total gross margin by $32.00 and $131.60, respectively. Before these enterprises could compete for a place in the farm plan, the gross margin per unit from these enterprises would have to increase by these amounts or more.

While linear programming is a good tool for finding the best enterprise combination, it does not substitute for human judgment. Most programming models do not consider goals other than maximization of gross margin, although some of the more sophisticated programming techniques can take into account such concerns as reduction of risk, conservation of resources, interactions among enterprises, and optimal growth strategies over time. Model results are sometimes highly sensitive to the estimated gross margins or the technical coefficients. Sometimes, a manager may accidentally omit an important restriction, such as management time or talent, leading to implausible results. The results of these computer models should therefore be interpreted cautiously. If the results do not make sense, the manager should check the gross margins, the technical coefficients, and the resource limits for accuracy and rerun the model as needed.

Developing the Whole-Farm Budget

Once the producer has a whole-farm plan, the final step is to develop a whole-farm budget. The income and variable costs used for computing enterprise gross margins provide important information for the whole-farm plan. As shown in Figure 11-2, these values are multiplied by the number of units of each enterprise

in the plan to get a first estimate of total gross income and total variable costs. When linear programming is used to develop the whole-farm plan, the objective function of the linear programming model should be very close in value to the difference between total gross income and total variable costs for the whole-farm plan, excluding any income and costs from activities not included in the linear programming process.

Other farm income that does not come directly from the budgeted enterprises, such as custom work income and fuel tax refunds, should also be added into the budget. Past records are a good source of information for estimating these additional sources of income. Any costs not already included in the enterprise variable costs should now be added. In practice, expenses such as building repairs, car and truck expenses, interest, utilities, insurance, and other overhead costs are difficult to allocate to specific enterprises and are affected very little by the final enterprise combination. These fixed costs are often called indirect costs. Other expenses, such as cash rent or property taxes, may apply to part of the acres in the plan but not

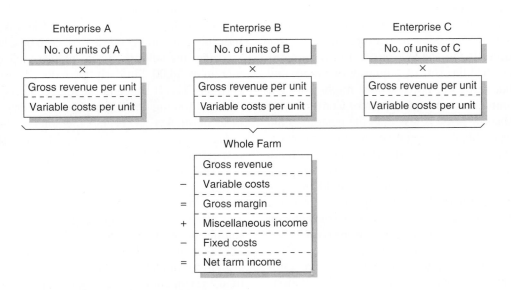

Figure 11-2 Constructing the whole-farm budget.

to others. Although these and other fixed costs do not affect the selection of a short-term profit-maximizing plan, they may be a major portion of total expenses and should be included in the whole-farm budget.

Budgets based on farm plans that include new investments in additional fixed assets should have their fixed costs revised accordingly. Opportunity costs may or may not be included in the whole-farm budget. If they are included, the budgeted profit is economic profit. If not, it is accounting profit or net farm income.

The whole-farm budget for our example farm is shown in Table 11-6 under the Plan 1 heading. Income and variable costs for individual enterprises were calculated by multiplying the income and expenses per unit by the number of units to be produced. Estimated total gross margin is $113,424, identical to the value of the objective function for the linear programming model.

An estimate of income from other sources, such as custom work done for neighbors, is included in the budget. Also shown are other expenses that do not vary directly with the number of crop acres produced or number of livestock raised. These items include property taxes, property and liability insurance, interest on fixed debt, depreciation, and miscellaneous expenses. Interest on borrowed operating capital may be shown here if a general operating loan or credit line for the entire farm is used. However, interest on an operating loan tied to an individual enterprise, such as for the purchase of feeder cattle, should be included in the enterprise budget data. Opportunity costs are not included in this example. The estimated net farm income from following this plan is $91,905.

Alternative Plans

Table 11-6 also contains whole-farm budgets for two other farm plans. Plan 2 involves renting an additional 300 acres of Class A cropland to be equally divided between cotton and soybeans. This plan has a projected net farm income of $108,590, an increase of $16,685 over

Plan 1. There are additional expenses for cash rent and for additional labor, as the surplus labor available from the original plan would not be sufficient to cover all the additional crop acres.

Plan 3 involves renting an additional 200 acres of pasture land and adding 100 more beef cows to the plan. Twenty acres of Class B land are shifted from soybeans to produce more raised feed needed by the additional cows. In this case, it is assumed that the labor available is sufficient for this plan and no additional labor needs to be hired. The projected net farm income from this plan was $98,240, $6,335 more than the original plan.

OTHER ISSUES

The whole-farm budget projects the most likely profit picture from following a particular farm plan. However, further analysis can provide some additional information about risk and liquidity.

Sensitivity Analysis

Although the whole-farm budget may project a positive net income, unexpected changes in prices or production levels can quickly turn that into a loss. Analyzing how changes in key budgeting assumptions affect income and cost projections is called *sensitivity analysis*.

At the end of the example in Table 11-6, a very simple sensitivity analysis was performed by reducing the anticipated gross farm income by 10 percent and recalculating the net farm income. This reduction could be caused by a decrease in production, or selling prices, or both. Additional sensitivity analysis could be carried out by constructing several entire budgets using different values for key prices and production rates. In Chapter 15, several more advanced approaches to evaluating the risk of a particular farm plan will be discussed.

Analyzing Liquidity

Liquidity refers to the ability of the business to meet cash flow obligations as they come due. A whole-farm budget can be used to analyze

TABLE 11-6	Example of a Whole-Farm Budget

		Plan 1		Plan 2		Plan 3	
	$/Unit	Units	Total	Units	Total	Units	Total
Gross income							
Cotton A	$624	260	$162,240	410	$255,840	260	$162,240
Soybeans A	480	260	124,800	410	196,800	260	124,800
Corn A	525	0	0	0	0	0	0
Soybeans B	336	162	54,432	162	54,432	142	47,712
Corn B	400	0	0	0	0	0	0
Beef cows	881	190	167,390	190	167,390	290	255,490
Stocker steers	1,050	0	0	0	0	0	0
Total gross income			$508,862		$674,462		$590,242
Variable costs							
Cotton A	$490	260	$127,400	410	$200,900	260	$127,400
Soybeans A	364	260	94,640	410	149,240	260	94,640
Corn A	417	0	0	0	0	0	0
Soybeans B	274	162	44,388	162	44,388	142	38,908
Corn B	370	0	0	0	0	0	0
Beef cows	679	190	129,010	190	129,010	290	196,910
Stocker steers	992	0	0	0	0	0	0
Total variable costs			$395,438		$523,538		$457,858
Total gross margin			$113,424		$150,924		$132,384
Other income			$ 22,500		$ 22,500		$ 22,500
Other expenses							
Property taxes			$ 2,700		$ 2,700		$ 2,700
Insurance			3,240		3,240		3,240
Interest			9,799		9,799		15,424
Hired labor			0		1,315		0
Depreciation			16,000		16,000		16,000
Cash rent			8,500		28,000		15,500
Miscellaneous			3,780		3,780		3,780
Total other expenses			$ 44,019		$ 64,834		$ 56,644
Net farm income			$ 91,905		$108,590		$ 98,240
10% reduction in gross income			−50,886		−67,446		−59,024
Revised net farm income			$ 41,019		$ 41,144		$ 39,216

liquidity as well as profitability. This is especially important when major investments in fixed assets or major changes in noncurrent debt are being considered. Table 11-7 shows how net cash flow can be estimated from the whole-farm budget. Besides cash farm income, income from nonfarm work and investments can be added to total cash inflows. Cash outflows include cash farm expenses (but not noncash expenses such as depreciation), cash outlays to replace capital assets, principal payments on term debts (interest has already been included in cash farm expenses), and nonfarm cash expenditures for family living and income taxes.

Profitable farm plans will not always have a positive cash flow, particularly when they include a heavy debt load. Interest payments will be especially large during the first few years of the loan period, so it is wise to analyze liquidity both for the first year or two of a plan and for an average year. A negative cash flow projection indicates that some adjustments to the plan are needed if the business is going to be able to meet all of its obligations on time. All three plans in the example have a projected positive net cash flow, but all three would have negative cash flows if prices or yields fell 10 percent below expectations.

Long-Run Versus Short-Run Budgeting

Short-run budgets that assume some resources are fixed should generally incorporate assumptions about prices, costs, and other factors that are expected to hold true over the next production period. However, when major changes in the supply of land, labor, or other assets, or in the way they are financed, are being contemplated, a longer-run perspective is needed. Few farms or ranches are profitable every year, but a plan that involves a long-term investment or financing decision should project a positive net income in an average or typical year.

The following procedures should be used for developing a typical year budget:

1. Use average or long-term planning prices for products and inputs, not prices expected for the next production cycle.

TABLE 11-7	Example of Liquidity Analysis for a Whole-Farm Budget		
	Plan 1	Plan 2	Plan 3
Cash inflows			
Cash farm income	$531,362	$696,962	$612,742
Nonfarm income	9,500	9,500	9,500
	$540,862	$706,462	$622,242
Cash outflows			
Cash farm expenses	$423,457	$572,372	$498,502
Term debt principal	22,250	22,250	22,250
Equipment replacement	15,000	15,000	15,000
Nonfarm expenses	36,000	36,000	36,000
	$496,707	$645,622	$571,752
Net cash flow	$44,155	$60,840	$50,490
10% reduction in cash income from enterprises	–50,886	–67,446	–59,024
Revised net cash flow	–$6,731	–$6,606	–$8,534

In particular, use prices that accurately reflect the long-term price relations among various products and inputs.

2. Use average or long-term crop yields and livestock production levels. Use past records as a guide. For new enterprises, use conservative performance rates.

3. Ignore carryover inventories of crops and livestock, accounts payable and receivable, or cash balances when estimating income, expenses, and cash flows. In the long run, these will cancel out from year to year. Assume that units sold are equal to production within a typical year.

4. Assume that the borrowing and repayment of operating loans can be ignored when projecting cash flows in a typical year, because they will offset each other. If significant short-term borrowing is anticipated, however, the interest cost

Box 11-1   Whole-Farm Resource Budgets

Although most whole-farm budgets are used to project income and expenses for a particular farm plan, a similar procedure can be used to estimate the quantities of key resources needed. Examples include the use of

- Labor
- Feedstuffs
- Irrigation water
- Machinery time

A detailed whole-farm resource budget can be used to estimate needs not only for the entire year but also for critical periods within the year. These can be compared to the fixed quantities of resources available to see if potential bottlenecks exist. If shortages are projected, they can be resolved by

- Changing the mix of enterprises to use less of the limited resource

- Changing the technical coefficients (e.g., feeding a different ration or using herbicides to replace cultivation)
- Temporarily increasing the supply of the resource by actions such as hiring part-time labor, leasing or custom hiring machinery, or buying feed from outside sources.

As more enterprises are included in the whole-farm plan that compete for the same resources, it becomes more important to create a budget that projects their requirements. This is especially the case when the resource needs are quite variable throughout the year, such as labor for harvesting or irrigation water in mid-summer. By planning ahead, provisions can be made well in advance to have enough resources available at the right time. Table 21-4 shows how a detailed resource budget for labor can be developed for a whole farm.

Box 11-2   Whole-Farm Planning and Farming Systems

Most whole-farm planning and budgeting procedures treat the various crop and livestock enterprises under consideration as independent activities. At most, they may be considered to compete for some of the same fixed resources. However, there may be positive or negative interactions between enterprises that also need to be taken into account. Some examples include

- Fixation of nitrogen by legume crops, making nitrogen available for other crops that follow or are planted in conjunction with the legumes
- Interruption of pest cycles by rotating dissimilar crops
- Shading of low-growing crops by higher-growing crops in an intercropping system

- Providing feed for livestock from crops on the same farm and nutrients to crops through use of livestock manure
- Passing young livestock from a breeding enterprise into a feeding enterprise on the same farm

The term *farming systems analysis* is given to the study of how various agricultural activities interact to achieve the overall goals of the farm or ranch. Although such interactions may at times be rather complex, ignoring them may lead to plans that fall considerably short of the farm's potential. Techniques such as input/output analysis and linear programming allow planners to model such interactions and create more realistic whole-farm plans and budgets.

that results should be incorporated into the estimate of cash expenses.

5. Assume that enough capital investment is made each year to at least maintain depreciable assets at their current level, to replace those that wear out.

6. Assume that the operation is neither increasing nor decreasing in size. This is especially critical when projecting liquidity for a typical year.

In some cases, the farm business may require several years to move from the current plan to a future plan. Profits and cash flows may be reduced temporarily because of inventory buildup, start-up costs, low production levels while learning new technology, and rapid debt repayment. Several transitional budgets may be needed to analyze conditions that will exist until the new farm plan is fully implemented.

The actual profitability of the operation may fall below levels projected by the whole-farm budget in some years. If this is due to unfavorable weather or low price cycles, however, the whole-farm plan chosen may still be the most profitable one for the long run.

SUMMARY

Whole-farm planning and the resulting whole-farm budget analyze the combined profitability of all enterprises in the farming operation. Planning starts with determining objectives, setting goals, and taking an inventory of the resources available. Feasible enterprises must be identified and their gross income per unit, variable costs, and gross margins computed.

Linear programming can be used to select the combination of enterprises that maximizes gross margin without exceeding the supply of resources available. It can handle large, complex planning problems quickly and accurately and also provides information such as the value of obtaining additional resources or the penalties for including certain enterprises.

The combination of enterprises chosen can then be used to prepare a whole-farm budget. Income and variable costs per unit are multiplied by the number of units to be produced and then combined with other farm income, fixed costs, and any additional variable costs to estimate net farm income. The completed whole-farm budget is an organized presentation of the sources and amounts of income and expenses. Whole-farm budgets can be based on either short-run or long-run planning assumptions and can also be used to evaluate liquidity.

QUESTIONS FOR REVIEW AND FURTHER THOUGHT

1. Why do goals have to be set before a whole-farm plan can be developed?

2. What should be included in a resource inventory needed for whole-farm planning?

3. How does a whole-farm budget differ from an enterprise budget?

4. Why can fixed costs be ignored when developing the whole-farm plan, but are included in the whole-farm budget?

5. Give some examples in which *fixed* costs would change when comparing whole-farm budgets for alternative farm plans.

6. Use linear programming software to find the profit-maximizing whole-farm plan using the following information:

		Resource requirements per acre	
Resource limit	Resource limit	Corn	Soybeans
Land (acre)	800	1	1
Capital ($)	90,000	150	100
Labor (hour)	3,500	5	2.5
Gross margin per acre ($)		100	80

7. What values could you change in a whole-farm budget to perform a sensitivity analysis?

8. What is the difference between analyzing profitability and analyzing liquidity?

9. What prices and yields would you use when developing a long-run whole-farm plan? A short-run plan? What other assumptions would change?

APPENDIX. GRAPHICAL EXAMPLE OF LINEAR PROGRAMMING

The basic logic for solving a linear programming problem can be illustrated in graphical form for a small problem involving two enterprises, corn and soybeans, and three limited resources. The necessary information is shown in Table 11-8, with land, labor, and operating capital as the limiting resources. Gross margins and the technical coefficients are also shown in the table.

Resource limits and the technical coefficients are used to graph the possible enterprise combinations shown in Figure 11-3. The supply of land limits corn and soybeans to a maximum of 120 acres each. These points, A and A', are found on the axes and connected with a straight line. Any point on line AA' is a possible combination of corn and soybeans, given only the land restriction. Labor, however, restricts corn to a maximum of 100 acres (500 hours divided by 5 hours per acre) and soybeans to 166.7 acres (500 hours divided by 3 hours per acre). These points on the axes are connected by line BB'. Any point on line BB' is a possible combination of corn and soybeans permitted by the labor restriction. In a similar manner, line CC' connects the maximum corn acres permitted by the operating capital restriction ($30,000 ÷ $200 per acre = 150 acres) with the maximum soybean acres ($30,000 ÷ $160 per acre = 187.5 acres). Line CC' identifies all the possible combinations based only on the operating capital restriction.

The sections of lines AA', BB', and CC' closest to the origin of the graph, or line BDA', represent the maximum possible combinations of corn and soybeans when all limited resources are considered together. Line BDA' is a segmented production possibility curve similar to those for competitive enterprises discussed in Chapter 8 (see Figure 8-2). The graph reveals that operating capital is not a limiting resource. It is fixed in amount, but $30,000 is more than sufficient for any combination of corn and soybeans permitted by the land and labor resources.

The next step is to find which of the possible combinations of corn and soybeans will maximize total gross margin. Total gross margin from planting 100 acres of corn, the maximum possible amount if no soybeans are grown (represented by point B), is $12,000. Adding 1 acre of soybeans increases the gross margin by $96, but requires 3 hours of labor, which in turn causes 0.60 less acre of corn to be grown (remember, no extra labor is available). This subtracts ($0.60 × $120 = $72) from total gross margin, for a net increase of $24.

This substitution can be continued until no more unused land is available (point D in Figure 11-3). At this point, increasing soybeans by 1 more acre requires 1 less acre of corn to be grown, resulting in a net decrease in gross margin of $120 – $96 = $24. Thus, point D represents the combination of the two crops that maximizes gross margin. This combination is 70 acres of corn and 50 acres of soybeans, with a total gross margin of $13,200. Producing either more acres of corn or more acres of soybeans would only reduce the total gross margin.

TABLE 11-8	Information for Linear Programming Example

Resource	Resources limit	Resource requirements (per acre) Corn	Soybeans
Land (acres)	120	1	1
Labor (hours)	500	5	3
Operating capital ($)	30,000	200	160
Gross margin ($)		120	96

Figure 11-4 shows the graphical solution to this example. Only the relevant production possibility curve, or line segment BDA′, is shown. The graphical solution to a profit-maximizing linear programming problem is the point where a line containing points of equal total gross margin just touches or is tangent to the production possibility curve on its upper side. This is point D in Figure 11-4, just tangent to a line representing all possible combinations of corn and soybeans producing a total gross margin of $13,200. Higher gross margins are not possible, because they require combinations of enterprises or enterprise levels not permitted by the limited resources. Combinations other than 70 acres of corn and 50 acres of soybeans are possible, but they would have a total gross margin of less than $13,200.

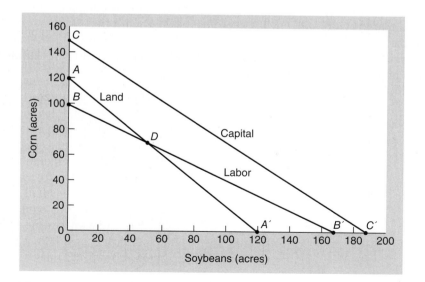

Figure 11-3 Graphical illustration of resource restrictions in a linear programming problem.

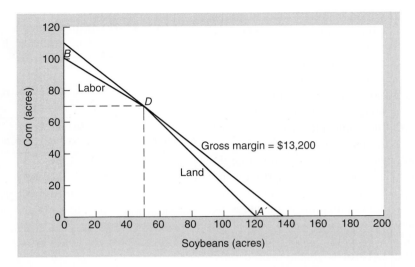

Figure 11-4 Graphical solution for finding the profit-maximizing plan using linear programming.

The solution at point D was found in a manner similar to that used to find the profit-maximizing enterprise combination in Chapter 8, where the substitution ratio was equated with the profit ratio. One basic difference is that linear programming generates a production possibility curve with linear segments, rather than a smooth, continuous curve, as in Figure 8-2. The solution will generally be at one of the corners or points on the production possibility curve, so the substitution ratio will usually not exactly equal the profit ratio. Only the variable costs change as the number of acres of each crop changes, so the ratio of the gross margins per acre for each crop is compared instead of the profit ratio. The gross margin ratio will fall between the substitution ratios for the two most limiting resources. For the example, the substitution ratio of soybeans for corn is 0.60 along segment BD and 1.0 along segment DA′. The gross margin ratio is $96 ÷ $120 = 0.80, which falls between the two substitution ratios.

© Photo by Ken Hammond/USDA

PARTIAL BUDGETING

CHAPTER OUTLINE

Uses of a Partial Budget

Partial Budgeting Procedure

The Partial Budget Format

Partial Budgeting Examples

Factors to Consider When Computing
   Changes in Revenue and Costs

Sensitivity Analysis

Limitations of Partial Budgeting

Final Considerations

Summary

Questions for Review and Further Thought

CHAPTER OBJECTIVES

1. Discuss the purpose of a partial budget

2. Emphasize the many possible uses
   of a partial budget

3. Illustrate the format of a partial budget

4. Show what types of entries are made
   on a partial budget

5. Note the importance of including only
   changes in revenue and expenses on a
   partial budget

6. Demonstrate the use of partial budgeting
   with several examples

Enterprise budgets are useful, but they do have limitations because they are restricted to one enterprise. A *partial budget* is often the appropriate way to analyze changes involving interactions between several enterprises.

Many of the day-to-day management decisions made by farmers and ranchers are really adjustments to, or fine-tuning of, an existing farm plan. Even the best farm plan will need some occasional fine-tuning as changes occur and new information becomes available. These adjustment decisions often affect revenue and expenses. A convenient and practical method for analyzing the profit potential of these partial changes in the overall whole-farm plan is the use of a partial budget.

USES OF A PARTIAL BUDGET

Examples of decisions that can be analyzed with a partial budget are whether to increase the size of, or to eliminate, a small herd of beef cows; to own harvesting equipment or custom hire harvesting; or to plant more barley and less wheat. Most of these decisions could be evaluated by comparing two whole-farm budgets, but time and effort would be wasted collecting and organizing information that will not change and therefore would not affect the decision.

A partial budget provides a formal and consistent method for calculating the expected change in profit from a proposed change in the farm business. It compares the profitability of one alternative, typically what is being done now, with a proposed change or new alternative. Throughout the discussion of partial budgeting, the emphasis will be on *change* in revenue and expenses. The final result is the expected change in profit.

Designed to analyze relatively small changes in the farm business, partial budgeting is really a form of marginal analysis. Figure 12-1 illustrates this point by showing how typical changes analyzed by partial budgeting relate to a production function, an isoquant, and a production possibility curve. Assuming that the current input/output combination is point A, the production function in the first panel of Figure 12-1 shows possible increases or decreases in that combination. Examples would be using more or less fertilizer, irrigation water, labor, or capital and analyzing the effects on output, revenues, expenses, and profit.

The second panel shows possible movements up or down an isoquant or different combinations of two inputs to produce a given amount of output. Possible changes in input combinations can be analyzed easily with a partial budget. Substituting larger machinery for less labor would be an example. Another typical use of a partial budget is to analyze the change in profit from substituting more of one enterprise for another. This adjustment is shown in the third panel by possible movements up or down the production possibility curve from the current combination at point A. A fourth general type of alternative adapted to partial budget analysis is expanding or contracting one or more enterprises. This would be illustrated by moving to a higher or lower isoquant or a higher or lower production possibility curve.

PARTIAL BUDGETING PROCEDURE

Steps in the tactical decision-making process discussed in Chapter 2 included identifying and defining the problem, identifying alternatives,

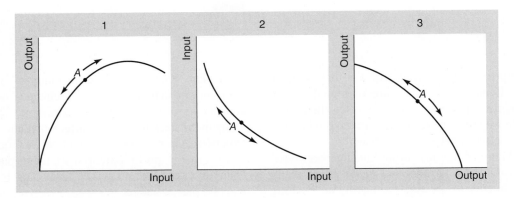

Figure 12-1 Partial budgeting and marginal analysis.

collecting data and information, and analyzing alternatives. Partial budgeting fits this process, with one modification. It is capable of analyzing only two alternatives at a time: the current situation and a single proposed alternative. Several partial budgets can be used to evaluate a number of alternatives. Identifying the alternative to be analyzed before gathering any information reduces the amount of information needed. The only information required is changes in costs and revenues *if* the proposed alternative is implemented. There is no need for information on any other alternative or for information about costs and revenue that will not be affected by the proposed change.

The changes in costs and revenues needed for a partial budget can be identified by considering the following four questions. They should be answered on the basis of what would happen *if* the proposed alternative was implemented.

1. What new or additional costs will be incurred?
2. What current costs will be reduced or eliminated?
3. What new or additional revenue will be received?
4. What current revenue will be lost or reduced?

For many problems, it will be easier to first identify all physical changes that would result if the alternative was adopted. These can then be assigned a dollar value to use in the partial budget.

THE PARTIAL BUDGET FORMAT

The answers to the preceding questions are organized within one of the four categories shown on the partial budget form in Table 12-1. There are different partial budgeting forms, but all have these four categories arranged in some manner. For each category, only the *changes* are included, not all costs or revenues.

Additional Costs

These are costs that do not exist at the current time with the current plan. A proposed change may cause additional costs because of a new or expanded enterprise that requires the purchase of additional inputs. Other causes would be increasing the current level of input use or substituting more of one input for another for an existing enterprise. Additional costs may be either variable or fixed, because there will be additional fixed costs whenever the proposed alternative requires additional capital investment. These additional fixed costs would include depreciation, interest (opportunity cost), taxes, and insurance for a new depreciable asset.

Reduced Revenue

This is revenue currently being received but that will be lost or reduced should the alternative be adopted. Revenue may be reduced if an enterprise is eliminated or reduced in size, if the change causes a reduction in yields or production levels, or if the selling price will decrease. Estimating reduced revenue requires careful attention to information about yields, livestock birth and growth rates, and output selling prices.

Additional Revenue

This is revenue to be received only if the alternative is adopted. It is not being received under the current plan. Additional revenue can be received if a new enterprise is added; if there is an increase in the size of a current enterprise; or if the change will cause yields, production levels, or selling price to increase. As with reduced revenue, accurate estimates of yields and prices are important.

Reduced Costs

Reduced costs are those now being incurred that would no longer exist under the alternative being considered. Cost reduction can result from eliminating an enterprise, reducing the size of

TABLE 12-1	Partial Budget Form

PARTIAL BUDGET

Alternative:

Additional Costs:	**Additional Revenue:**
Reduced Revenue:	**Reduced Costs:**
A. Total additional costs and reduced revenue $____	B. Total additional revenue and reduced costs $____
	$____
Net change in profit (B minus A) $____	

Box 12-1	Can Fixed Costs Really *Change*?

It may seem strange to talk about computing changes in fixed costs when Chapter 9 emphasized that fixed costs do not change. The explanation is in the difference between short run and long run. Fixed costs do not change in the short run. However, analyzing the purchase or sale of a capital asset is a long-run decision with respect to that asset, and fixed costs can change from one time to another. Computing the fixed costs that would exist at one time and those that would exist at another after a purchase or sale would indicate a difference. It is that difference, or change, that should be included on a partial budget.

an enterprise, reducing input use, substituting more of one input for another, or being able to purchase inputs at a lower price. Reduced costs may be either fixed or variable. A reduction in fixed costs will occur if the proposed alternative will reduce or eliminate the current investment in machinery, equipment, breeding livestock, land, or buildings.

The categories on the left-hand side of the partial budget in Table 12-1 are the two that reduce profit—additional costs and reduced revenue. On the right-hand side of the budget are the two categories that increase profit—additional revenue and reduced costs. Entries on the two sides of the form are summed and then compared to find the net change in profit. If the total of additional revenue and reduced costs is greater than the total of additional costs and reduced revenue, then the net change in profit will be positive and profit will increase if the change were made. In the opposite case, net change in profit will be negative and profits would fall if the change were made. Whenever opportunity costs are included on a partial budget, the result is the estimated change in *economic profit*. This will not be the same as the change in *accounting profit*.

PARTIAL BUDGETING EXAMPLES

Two examples will illustrate the procedure and possible uses of partial budgeting. The first is a relatively simple partial budget, which examines the alternative already considered in the third column of Table 11-6, adding 100 cows to the example farm by renting 200 acres of pasture and converting 20 acres of cropland to feed production. It is assumed that the additional pasture can be rented for $35.00 per acre. Gross returns per head of beef cattle are $881 per head, as shown in Table 10-3, the corresponding enterprise budget, and also in Table 11-3. Variable costs per head would equal $679, as shown in Table 11-3.

The additional costs of adding 100 beef cows can be divided into fixed and variable costs. Fixed costs would include an additional annual interest charge of $5,625 on the additional cows. Variable costs include the increased variable costs of $67,900 (100 × $679) plus the pasture rent of $7,000 ($35 × 200). No new labor needs to be hired because the original farm plan had surplus labor of 760 hours, enough to support the additional beef cows.

The additional revenue from adding 100 beef cows would total $88,100 (100 × $881). In Table 12-2, this revenue is divided into that received from cull cows, heifer calves, and steer calves, following the format in the beef cattle enterprise budget, Table 10-3.

The reduced revenue and reduced costs are associated with the 20-acre reduction in plantings of soybeans on Class B land. Gross revenue and variable costs for this enterprise are shown in Table 11-3, $336 and $274 per acre, respectively. If the 20 acres are converted to feed production, then the reduced revenue would equal $6,720 (20 × $336). Reduced costs would total $5,480 (20 × $274).

To complete the partial budget, the first step is to sum additional costs and reduced revenue. Together they total $87,245. Next, additional revenue and reduced costs are totaled. Together these sum to $93,623. Finally, the sum of total additional costs and reduced revenue is subtracted from the total additional revenue and reduced costs. In this example, the net change in profit is $6,335, the same figure we derived in the whole-farm budget format. Hence, the change is slightly more profitable than the original farm plan.

The example shown in Table 12-2 is very simple and includes only broad categories of costs and returns. A more detailed breakdown may be useful for planning purposes. Also, to be accurate, a partial budget should include an interest charge on the variable costs. The interest charge for the production costs is already included in the $679 per-head figure used to calculate the production costs, as shown in Table 10-3. However an interest charge for the money tied up in the pasture rent has not been included. Assuming

TABLE 12-2	**Partial Budget for Adding 100 Cows**

PARTIAL BUDGET

Alternative: Add 100 beef cows on 200 acres rented land and convert 20 acres of cropland to feed production

Additional Costs:		Additional Revenue:	
Fixed costs		10 cull cows	$ 9,700
Interest on cows/bulls	$ 5,625	33 heifer calves	30,200
		45 steer calves	48,200
Variable costs			
Pasture rent	7,000		
Production costs	67,900		
Reduced Revenue:		**Reduced Costs:**	
Soybean sales, 20 acres	6,720	Soybean production costs, 20 acres	5,480
A. Total additional costs and reduced revenue	$87,245	B. Total additional revenue and reduced costs	$93,580
			$87,245
		Net change in profit (B – A)	$ 6,335

an interest rate of 5 percent and that the rent is paid at the beginning of the year, an additional annual interest charge of $350 should probably be included for greater accuracy.

A More Detailed Example

The second and more detailed example of a partial budget is shown in Table 12-3. This partial budget does not apply to our sample farm from Chapter 11. Here it is assumed that a producer is considering switching from dryland cotton to irrigated cotton. This producer currently farms 320 acres of dryland cotton and would convert the whole 320 acres. Irrigation equipment necessary for the change would have an original cost of $360,000, a salvage value of $50,000 after 20 years, and an expected economic useful life of 20 years. Additional insurance for the machinery would cost $3,075 per year. Additional variable costs of irrigation are estimated at $95 per acre.

The dryland cotton has a yield of 750 pounds per acre. The irrigated cotton is expected to have a yield of 1,100 pounds per acre. Costs of fertilizer and fuel and chemicals are expected to increase with irrigation, as shown below.

	Dryland	**Irrigated**
Fertilizer	$142/acre	$182/acre
Fuel and chemicals	$75/acre	$89/acre

Cotton is expected to sell for $0.80 per pound, regardless of whether it is dryland or irrigated.

The proposed change will cause additional fixed costs as well as additional variable costs. Depreciation on the equipment is calculated using straight-line depreciation at $15,500 per year. Average annual interest is estimated using the formula presented in Chapter 9:

$$\frac{(\text{Original Cost} + \text{Salvage Value})}{2} \times R$$

TABLE 12-3	Partial Budget for Converting Dryland Cotton to Irrigated Cotton

PARTIAL BUDGET

Alternative: Convert 500 acres of dryland cotton to irrigated cotton

Additional Costs:		Reduced Costs:	
Fixed costs		Variable costs	
Depreciation	$ 15,500	Fertilizer	$ 45,440
Interest	9,225	Fuel and chemicals	24,000
Insurance	3,075	Interest on variable costs	2,083
Variable costs			
Fertilizer	58,240		
Fuel and chemicals	28,480		
Irrigation variable costs	30,400		
Interest on variable costs	3,514		
Reduced Revenue:		**Additional Revenue:**	
Dryland cotton production 320 acres × 750 lbs. × $0.80	$192,000	Irrigated cotton production 320 acres × 1,100 lbs. × $0.80	$281,600
A. Total additional costs and reduced revenue	**$340,434**	**B. Total additional revenue and reduced costs**	**$353,123**
			$340,434
		Net change in profit (B – A)	**$ 12,689**

Using an assumed 4.5 percent interest rate for the purchase of the irrigation equipment, fixed interest comes out to $9,225 per year. The insurance costs are also fixed costs and are included in that section of the partial budget.

Additional variable costs include the cost of fertilizer, fuel and chemicals, and labor, as well as direct variable costs associated with the irrigation. Interest on variable costs is calculated on half their total value, times the assumed short-term interest rate of 6 percent. Interest is calculated only on half of the production costs because they are assumed to be incurred for six months. Other variable costs of production are assumed not to change and hence are not included in the partial budget.

The reduced revenue would result from the loss of the sale of dryland cotton. There are 320 acres multiplied by 750 pounds per acre

multiplied by $0.80 per pound for a total of $192,000 in reduced revenue. The sum of additional costs and reduced revenue equals $340,434.

Additional revenue will be received from the sale of the irrigated cotton. A total of $281,600 in additional revenue results from multiplying the 320 acres by the yield of 1,100 pounds per acre and by the price of $0.80 per pound. Reduced costs are calculated for fertilizer and fuel and chemicals on the dryland acres. As before, a variable interest charge is calculated on half the total of these costs multiplied by an assumed 6 percent interest rate.

The sum of additional revenue and reduced costs is $353,123. From this total, the sum of additional costs and reduced revenue, $340,434, is subtracted to find the expected net change in profit of $12,689. Because the difference is positive, the proposed change would increase profit.

FACTORS TO CONSIDER WHEN COMPUTING CHANGES IN REVENUE AND COSTS

In addition to the usual problem of acquiring good information and data, there are several other potential problems when doing a partial budget. The first is nonproportional change in costs and revenue. This problem will occur more often with costs, but it is possible for revenues. Assume the proposed change is a 20 percent increase (decrease) in the size of an enterprise. It would be easy to take the totals for each existing expense and revenue and assume that each will be 20 percent higher (lower). This could be wrong for two reasons. Fixed costs would not change unless the 20 percent change caused an increase or decrease in capital investment. Many relatively small changes will not. Even variable costs may not change proportionally. For example, adding 20 cows to an existing beef or dairy herd of 100 cows will increase labor requirements but probably by something less than 20 percent. Also, as we saw in Table 12-2, if there is an unused resource, such as labor, an enterprise can be expanded up to a point with no increase in the associated cash cost of that resource. Economies and diseconomies of size must be considered when estimating cost and revenue changes.

Opportunity costs are other easily overlooked items. They should be included on a partial budget to permit a fair comparison of the alternatives. This is particularly important if the difference in capital or labor requirements is large. Additional variable costs represent capital that could be invested elsewhere, so opportunity cost on them should be included as another additional cost. The reverse of this argument holds true for reduced variable costs, so an opportunity cost on these should be included as a reduced cost. Opportunity cost on any additional capital investment becomes part of additional fixed costs and likewise should be a part of reduced fixed costs if the capital investment will be reduced.

Opportunity cost on the farm operator's labor may also be needed on a partial budget. However, several things should be considered when estimating this opportunity cost. Is there really an opportunity cost for using additional labor if it is currently unused? Free or leisure time would be given up, and there may be an opportunity cost on this time. Alternatively, the farm operator may desire some minimum return before using any excess labor in a new alternative. The same question exists in reverse if the alternative will reduce labor requirements. Is there a productive use for an additional 50 or 100 hours of labor, or will it just be additional leisure time? What will it earn in the alternative use, or what is the value of an additional hour of leisure time? Answers to these questions help determine the appropriate opportunity cost of operator labor on a partial budget.

Another consideration is the unit of change used in the partial budget. Is the budget based on changes in total farm revenue and expenses, or is it for one acre of crop or one head of livestock? In other words, is the unit the whole farm or some smaller unit? Some alternatives can be analyzed either way if they have a common physical unit such as acres. Others where the alternative involves both changes in acres of crops and head of livestock have no common unit of measurement. They must be budgeted on a whole-farm basis. Budgeting on a whole-farm basis is always the safer method to prevent any confusion about the budgeting unit.

SENSITIVITY ANALYSIS

It is often difficult to estimate the average prices and yields needed in a partial budget. Estimation is particularly difficult if the budget projects well into the future. Yet the accuracy of the analysis and of the resulting decision directly depends on these values. A sensitivity analysis of the budget can provide some additional information on just how dependent the results are on the prices and yields used.

Sensitivity analysis consists of doing the budget computations several times, each with a different set of prices or yields. The results show how sensitive the estimated change in profit is to changes in these values. One way to perform a basic sensitivity analysis is to use low, average, and high prices each in a different partial budget. The same could be done for low, average, and high yields, if appropriate. A comparison of the results will show how sensitive the expected change in profit is to price or yield variation. This will provide the manager with some idea of the risk involved in the proposed change.

Another way to do sensitivity analysis would be to look at prices that, for example, are 10, 20, and 30 percent higher and lower than the expected average price. Using this method, one of the prices may result in an expected change in profit of somewhere near zero, meaning it is close to the break-even price. For some types of partial budgeting problems, it is possible to compute the break-even price or yield directly, which simplifies the calculations. Once a break-even value is calculated, the manager can decide if the future price or yield is more likely to be above or below that value. This information can help make the final decision.

The example in Table 12-3 can be used to illustrate sensitivity analysis. The key figure in this problem is the expected yield of the irrigated cotton. Current dryland yields are known, and information on the cost of the irrigation equipment can be obtained fairly easily. We would expect higher irrigated cotton yields to favor purchasing the equipment and vice versa. For example, an expected irrigated cotton yield of 1,000 pounds per acre rather than 1,100 pounds per acre would lower additional revenue by $25,600, making the net change in profit a negative $12,911.

The break-even irrigated cotton yield is the value that makes the net change in profit equal to zero. This requires a reduction of $12,689 in additional revenue. Therefore, an irrigated yield of around 1,050 pounds per acre is the break-even value. Any irrigated yield higher than this value would make it profitable to convert to irrigated production.

In a similar fashion, sensitivity analysis for the price of cotton could be developed. Higher prices would favor conversion to irrigated production, and vice versa. If the cotton price falls below $0.764, irrigated cotton production would be less profitable than dryland, given the assumptions about yields and equipment costs.

Sensitivity analysis and break-even calculations can also be done on the budget in Table 12-2, but it is more difficult because of the greater number of prices and yields. However, it can be done by holding all price and yield values constant but the one of most interest.

The additional profit from adding 100 beef cows was $6,335, which is about 7.2% of the expected revenue. If cattle prices were to fall by more than this percentage, adding 100 cows would not be a profitable alternative.

Performing a sensitivity analysis and computing break-even values can require numerous calculations. However, a partial budget is relatively easy to set up on a computer spreadsheet. Once this is done, it is quick and easy to change a value and observe the result.

LIMITATIONS OF PARTIAL BUDGETING

Partial budgets are easy to use, require minimal data, and are readily adaptable to many types of management decisions. However, partial budgeting does have some limitations. It can only compare the present management plan with one alternative at a time. This requires many budgets when there are many alternatives to consider. Partial budgeting can still be used in this situation but it can be time consuming.

The data in partial budgets are expected average annual changes in economic revenue and expenses. While an alternative may increase profit based on average changes, there are other factors to consider when the changes are not constant from year to year. An example would be the planting of an orchard or other crop where no revenue is expected for several years and then expected to increase annually for several years before reaching a maximum level. Though it may be a profitable alternative based on *average* annual values, it may be difficult to meet the crop expenses in the early years and to make any loan payments. In other words, the cash shortage in the early years is not reflected in the partial budget. Any type of change requiring a large capital investment and revenues that vary over time should be analyzed by more detailed procedures that should include a cash flow projection. (See Chapters 13 and 17.)

Partial budgets should include appropriate opportunity costs to account for all economic costs. However, they are not included as accounting costs, so the expected change in net profit shown on a partial budget should not be interpreted as the expected change in accounting profit. The expected change in net profit must be adjusted for any opportunity costs included in its calculation to find the expected change in accounting profit.

FINAL CONSIDERATIONS

The partial budget in Table 12-3 can be used to illustrate two additional factors to be considered in the decision. Before adopting a proposed change that appears to increase profit, any additional risk and capital requirements should be carefully evaluated. If risk is measured in terms of annual variability in profit, is the profit from the irrigated acreage more or less variable than the profit from the nonirrigated acres? Irrigation should reduce the yield variability, but the increased costs make the producer more vulnerable to price downturns. The decision maker must evaluate the potential effects of additional risk on the financial stability of the business. Is the additional average profit worth the additional risk or variability of profit? Chapter 15 discusses risk in more detail.

Purchasing irrigation equipment will require an additional capital investment. Is the capital for acquiring the equipment available or can it be borrowed? If it is borrowed, how will this affect the financial structure of the business, risk, cash flow requirements, and repayment ability? Will this additional investment cause a capital shortage in some other part of the business? These questions need to be evaluated carefully before making the final decision to adopt the change. A profitable change may not be adopted if the increase in profit is relatively small, it increases risk, or an additional capital investment is required. Potential changes requiring additional capital investment can also be analyzed in other ways. See Chapter 17 for capital budgeting and other more comprehensive investment analysis methods.

SUMMARY

A partial budget is an extremely useful type of budget. It can be used to analyze many of the common, everyday problems and opportunities that confront the farm and ranch manager. Partial budgets are intended to analyze the profitability of proposed changes in the operation of the business where the change affects only part of the farm plan or organization. The current situation is compared to the expected situation after implementing a proposed change.

Data requirements are rather small, because only changes in costs and revenues are included on a partial budget. The sum of additional costs and reduced revenue is subtracted from the sum of additional revenue and reduced costs to find the estimated change in profit. A positive result indicates that the proposed change would increase profit. However, any additional risk and capital requirements should be considered before making the final decision.

QUESTIONS FOR REVIEW AND FURTHER THOUGHT

1. Will partial budgets ever contain some fixed ownership costs? If yes, give an example of a partial budget that might contain some fixed ownership costs.

2. List the types of changes that would appear in a partial budget for determining the profitability of participating in a government farm program. The program requires that 10 percent of your crop land be left idle in exchange for a lump-sum payment.

3. Why are changes in opportunity costs included in partial budgets?

4. Assume that a proposed change would reduce labor requirements by 200 hours. If this was the farm operator's labor rather than hired hourly labor, would you include a reduced cost for labor? What factors would determine the value to use?

5. Partial budgeting can be used to develop a whole-farm plan. True or false? Explain?

6. Besides additional profit, what other factors should a farm operator take into account when evaluating a proposed change?

CASH FLOW BUDGETING

CHAPTER OUTLINE

Features of a Cash Flow Budget
Constructing a Cash Flow Budget
Uses for a Cash Flow Budget
Monitoring Actual Cash Flows
Investment Analysis Using a Cash Flow Budget
Summary
Questions for Review and Further Thought

CHAPTER OBJECTIVES

1. Identify cash flow budgeting as a tool for financial decision making and business analysis

2. Understand the structure and components of a cash flow budget

3. Illustrate the procedure for completing a cash flow budget

4. Describe the similarities and differences between a cash flow budget and an income statement

5. Discuss the advantages and potential uses of a cash flow budget

6. Show how to use a cash flow budget when analyzing a possible new investment

Even the most profitable farms and ranches occasionally find themselves short of cash. Anticipating these shortages and having a plan to deal with them is an important management activity. A cash flow budget is a financial analysis tool with applications in both forward planning and the ongoing analysis of a farm or ranch business. Preparing one is the next logical step after finishing the whole-farm plan and budget. The cash flow budget will provide answers to some remaining questions. Is the plan financially feasible? Will there be sufficient capital

227

available at the specific times it will be needed? If not, how much will need to be borrowed? Will the plan generate the cash needed to repay any new loans? These types of questions can be answered by preparing and analyzing a cash flow budget.

FEATURES OF A CASH FLOW BUDGET

A cash flow budget is a summary of the projected cash inflows and cash outflows for a business over a given period. This period is typically a future accounting period and is divided into quarters or months. As a forward planning tool, its primary purpose is to estimate the amount and timing of future borrowing needs and the ability of the business to repay these and other loans on time. Given the large amount of capital today's commercial farms and ranches require and often must borrow, cash flow budgeting is an important budgeting and financial management tool.

A discussion of cash flow budgeting must continually emphasize the word *cash*. All cash flows must be identified and included on the budget. Cash flows into a farm business from many sources throughout the year, and cash is used to pay business expenses and to meet other needs for cash. Identifying and measuring these sources and uses of cash is the important first step in constructing a cash flow budget. The concept of cash flows is shown graphically in Figure 13-1. It assumes that all cash moves through the business checking account, making it the central point for identifying and measuring the cash flows.

Two things make a cash flow budget substantially different from a whole-farm budget. First, a cash flow budget contains all cash flows, not just revenue and expenses, and it does not include any noncash items. For example, cash inflows would include cash from the sale of capital items and proceeds from new loans, but not inventory changes. Principal payments on debt and the full cost of new capital assets would be included as cash outflows, but depreciation would not. The

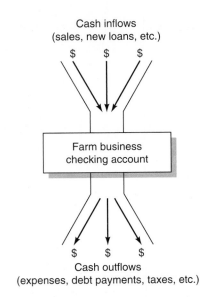

Figure 13-1 Illustration of cash flows.

emphasis is on cash flows, regardless of source or use and whether or not they are business revenue or expenses. For this reason, personal and non-farm cash revenue and expenses are also included in a cash flow budget, because they affect the amount of cash available for farm business use.

The second major difference between a whole-farm budget and a cash flow budget is the latter's concern with the timing of revenue and expenses. A cash flow budget also includes *when* cash will be received and paid out as well as *for what* and *how much*. This timing is shown by preparing a cash flow budget on a quarterly or monthly basis. Farming and ranching are seasonal businesses, so most agricultural cash flow budgets are done on a monthly basis to permit a detailed analysis of the relation between time and cash flows.

The unique characteristics of a cash flow budget do not allow it to replace any other type of budget or any of the records discussed in Part II. It fills other needs and is used for different purposes. However, it will be shown shortly that much of the information needed for a cash flow budget can be found in the whole-farm budget and farm records.

Actual Versus Estimated Cash Flows

By definition, a cash flow budget contains estimates of cash flows for a future period. However, it is possible to record and organize the actual cash flows for some past period. The Farm Financial Standards Council recommends that a *Statement of Cash Flows* be developed as part of the standard end-of-year set of financial statements for the farm business. Table 5-4 provided an example of a Statement of Cash Flows. Recall that the Statement of Cash Flows is a financial record, recording what has happened, not a budget, which projects what will happen in the future. In addition, compared to a cash flow budget, the Statement of Cash Flows is generally greatly condensed, typically recording only the annual totals of cash received and cash expended in various broad categories.

Keeping good records of actual cash flows is important for several reasons. First, if cash flow records are recorded and summarized monthly, the monthly cash flow can be compared with monthly budgeted values at the end of each month. This comparison can provide an early warning of any substantial deviations while there is still time to determine causes and make corrections. Second, detailed records of actual cash flows can provide useful insight into the financial structure of the business and show how the operating, financing, and investing activities combine and interact as sources and uses of cash. Third, good records of actual cash flows provide a good starting point for developing the next annual cash flow budget. With the totals and timing of past cash flows, it is relatively easy to make the adjustments needed to project cash flows into the future. Reviewing past cash flows also prevents some important items from being overlooked on the new budget.

Structure of a Cash Flow Budget

The structure and format of a cash flow budget are shown in Table 13-1 in condensed form. This condensed budget illustrates the sources and uses of cash, which need to be included on

TABLE 13-1	Simplified Cash Flow Budget	
	Period 1	**Period 2**
1. Beginning cash balance	$ 1,000	$ 500
Cash inflows:		
2. Farm product sales	$ 2,000	$12,000
3. Capital sales	0	5,000
4. Miscellaneous cash income	0	500
5. Total cash inflow	$ 3,000	$18,000
Cash outflows:		
6. Farm operating expenses	$ 3,500	$ 1,800
7. Capital purchases	10,000	0
8. Miscellaneous expenses	500	200
9. Total cash outflow	$14,000	$ 2,000
10. Cash balance	−11,000	16,000
(line 5 – line 9)		
11. Borrowed funds needed	$11,500	0
12. Loan repayments	0	
(principal and interest)		11,700
13. Ending cash balance	500	payback
(line 10 + line 11 – line 12)		4,300
14. Debt outstanding	$11,500	$ 0

any cash flow form. There are five potential sources of cash:

1. The beginning cash balance, or cash on hand at the beginning of the period
2. Farm product sales, or cash revenue from the operation of the farm business
3. Capital sales, or the cash received from the sale of capital assets such as land, machinery, breeding livestock, and dairy cattle
4. Nonbusiness cash receipts, which would include nonfarm cash income, cash gifts, and other sources of cash
5. New borrowed capital or loans received

The last source cannot be included in the cash inflow section, because borrowing requirements are not known until the cash outflows are matched against the cash inflows.

Table 13-1 also shows the four general uses of cash. They are

1. Farm operating expenses, the normal and usual cash expenses incurred in producing the farm revenue
2. Capital purchases, the full purchase price of any new capital assets such as land, machinery, and dairy or breeding livestock
3. Nonbusiness and other expenses, which would include cash used for living expenses, income and Social Security taxes, and any uses of cash not covered elsewhere
4. Principal payments on debt; interest payments should also be included here, unless they were included as part of the operating expenses

The difference between total cash inflows and total cash outflows for any period is shown as the ending cash balance for that period.

Only two periods are shown in Table 13-1, but once the basic procedure is understood, the budgeting process can be extended to any number of periods. In the first period, the total cash inflow of $3,000 includes the beginning cash balance. The total cash outflow is $14,000, leaving a projected cash balance of –$11,000. This deficit will require borrowing $11,500 to provide a $500 minimum ending cash balance.

Total cash inflow for the second period is estimated to be $18,000, resulting in a cash balance of $16,000 after subtracting a total cash outflow of $2,000. This large cash balance permits paying off the debt incurred in the first period, estimated at $11,700 when interest of $200 is included. The result is an estimated $4,300 cash balance at the end of the second period. Following this same procedure for all subsequent periods traces out the projected level and timing of borrowing and debt repayment potential.

CONSTRUCTING A CASH FLOW BUDGET

Considerable information is needed to construct a cash flow budget. For simplicity and to ensure accuracy, a logical, systematic approach should be followed. The following steps summarize the process and information needs:

1. Develop a whole-farm plan. It is impossible to estimate cash revenues and expenses without knowing what crops and livestock will be produced.
2. Take inventory. This should include existing crops and livestock available for sale during the budget period.
3. Estimate crop production and, for combination crop and livestock farms, estimate livestock feed requirements. This step projects crops available for sale after meeting livestock feed requirements. It may show a need to purchase feed if production plus beginning inventory is less than what is needed for livestock.
4. Estimate cash receipts from livestock enterprises. Include both livestock included in the beginning inventory and those to be produced and sold within the year. Sales of livestock products such as milk and wool should also be included.
5. Estimate cash crop sales. First, determine a desired ending inventory for feed use and sale next year. Next, compute the amount available for sale as beginning inventory *plus* production *less* livestock feed requirements *less* desired ending inventory.
6. Estimate other cash income such as revenue from custom work or government farm program payments. If the budget includes both business and personal cash flows, include rent, interest, or dividends on nonfarm investments plus any other nonfarm sources of cash revenue.
7. Estimate cash farm operating expenses. Reviewing actual cash flows for the past year will prevent overlooking items such as property taxes, insurance, repairs, and other cash expenses not directly related to crop and livestock production.
8. Estimate personal and nonfarm cash expenses. Included here would be cash needed for living expenses, income and

Social Security taxes, and any other nonfarm cash expenses. If the budget is for the farm business only, simply estimate the cash withdrawals that will be needed.

9. Estimate purchases and sales of capital assets. Include the full purchase price of any planned purchases of machinery, buildings, breeding livestock, and land, as well as the total cash to be received from the sale of any capital assets.

10. Find and record the scheduled principal and interest payments on existing debt. This will be primarily noncurrent debt, where the amounts and dates of each payment are shown on a repayment schedule. Any carryover current debt from the previous year should also be included here.

The data estimated and organized in these steps can now be entered on a cash flow form. With a few more calculations, the result will be an estimate of borrowing needs for the year, the ability of the business to repay these loans, and the timing of each. Key assumptions about selling prices, input costs, and production levels should be well documented before the plan is presented to a lender.

A Cash Flow Budgeting Form

Printed forms for completing a cash flow budget are available from many sources, including lending agencies and the agricultural extension service in most states. These sources, as well as commercial software firms, may have computer programs that can be used. A cash flow budget can also be constructed by anyone familiar with any of the spreadsheet software programs. Table 13-2 illustrates one type of form for a cash flow budget. To save space, only the headings for the first three columns are shown. Other forms may differ in organization, headings, and details, but all provide the same basic information.

The first 13 lines of the form are for recording cash inflows projected from both farm and nonfarm sources. Both nonfarm cash income and expenses will affect the cash available for farm business use, so these items are included in the budget, even though they are not directly related to the farm business. The total annual amount expected for each cash inflow is recorded in the *Total* column. This amount is then allocated to the month or months when it will be received.

Cash farm operating expenses are listed on lines 14 through 34, with the total on line 35. As with all entries on a cash flow budget, the projected total for each expense item is put in the *Total* column, and this amount is then allocated to the month or months when the cash will be needed. The sum of total expenses for the individual months should always be compared with the sum in the *Total* column on line 35. This cross-check will show whether any errors were made when allocating individual cash expenses to specific months.

Several other possible cash outflows are shown on lines 36 through 47. Capital expenditures for replacement or expansion of machinery and equipment, breeding livestock, land, and buildings require cash. The full purchase price for any capital expenditures should be recorded, even when borrowing will be used to make the purchase. One purpose of the budget is to estimate the necessary amount of such borrowing. Family living expenses, income and Social Security taxes, and other nonfarm cash expenses should be entered on lines 39 through 43. All personal expenses, such as automobile expenses and medical and life insurance premiums, should be included in family living expenses.

Lines 44 through 47 are used to enter the scheduled principal and interest payments on debt incurred in past years. Entries here would be payments on noncurrent debt, as well as on any current debt carried over from the past year. Only scheduled payments on old debt are entered in this section, because payments on any new debt incurred during the coming year will be computed and entered in a later section.

Lines 35 through 47 are summed to get total annual cash outflows and totals for each month,

TABLE 13-2	Form for a Cash Flow Budget

Cash Flow Budget

Name: I. M. Farmer	Total	Jan	Feb	March
1 Beginning cash balance				
Operating receipts:				
2 Grain and feed				
3 Feeder livestock				
4 Livestock products				
5 Other				
6				
Capital receipts:				
7 Breeding livestock				
8 Machinery and equipment				
9				
Nonfarm income:				
10 Wages and salary				
11 Investments				
12				
13 Total cash inflow				
(add lines 1–12)				
Operating expenses:				
14 Seed				
15 Fertilizer and lime				
16 Pesticides				
17 Other crop expenses				
18 Fuel, oil, lubricants				
19 Hired labor				
20 Machine hire				
21 Feed and grain				
22 Feeder livestock				
23 Livestock expenses				
24 Repairs—machinery				
25 Repairs—buildings				
26 Cash rent				
27 Supplies				
28 Property taxes				
29 Insurance				
30 Utilities				
31 Auto and pickup (farm share)				
32 Other farm expenses				
33				
34				
35 Total cash operating expenses				
(add lines 14–34)				

TABLE 13-2	*(CONTINUED)*

Cash Flow Budget

Name: I. M. Farmer		Total	Jan	Feb	March
	Capital expenditures:				
36	Machinery and equipment				
37	Breeding livestock				
38					
	Other expenditures:				
39	Family living expenses				
40	Income tax and Social Security				
41	Other nonfarm expenses				
42					
43					
	Scheduled debt payments:				
44	Current debt—principal				
45	Current debt—interest				
46	Noncurrent debt—principal				
47	Noncurrent debt—interest				
48	Total cash outflow				
	(add lines 35–47)				
49	Cash available (line 13 – line 48)				
	New borrowing:				
50	Current				
51	Noncurrent				
52	Total new borrowing				
	Payments on new current debt				
53	Principal				
54	Interest				
55	Total debt payments				
	(line 53 + line 54)				
56	Ending cash balance				
	(lines 49 + 52 – 55)				
	Summary of debt outstanding				
57	Current				
58	Noncurrent				
59	Total debt outstanding				

[handwritten annotation:] 49: cash in – cash out = cash available (if negative, need to borrow)

with the amount entered on line 48. Next, the total estimated cash available at the end of the month is calculated by subtracting line 48 from line 13, with the result entered on line 49. If the total cash outflow is greater than the total cash inflow, the cash available will be negative, and new borrowing or other adjustments will be needed to obtain a positive ending cash balance.

Any new borrowing is entered on line 50 or 51, depending on the type of loan. The total new borrowing is added to the cash available on line 49 to find the ending cash balance for the month. No cash would be available to pay down new current debt, so line 55 will be zero in this case.

If the cash available on line 49 is greater than zero, the total cash inflow for the month was greater than the total cash outflow. This amount can be used to pay off part or all of any new current debt incurred earlier in the year or carried over to a future period. The principal and interest are entered on lines 53 and 54, and these amounts are adjusted to permit a positive ending cash balance. After finding the ending cash balance by subtracting line 55 from line 49 (line 52 is zero in this case), the result is entered on line 56. The same amount is transferred to line 1 of the next month as its beginning cash balance.

Lines 57 through 59, summary of debt outstanding, are not a necessary part of a cash flow budget. However, they summarize the debt situation for the business and provide some useful information. For each type of debt, the amount outstanding at the end of any month will equal the amount at the end of the previous month, plus any new debt, minus any principal payments made during the month. Lines 57 and 58 should be summed for each month, and the total debt outstanding should be entered on line 59. This figure shows the pattern of total debt and its changes throughout the annual production cycle. This pattern will often repeat itself each year because of the seasonal nature of agricultural production, income, and expenses. Some lenders specify a maximum amount of debt the operator can have at any given time during the year. This part of the budget helps project whether this limit will allow sufficient capital to be borrowed.

Example of a Cash Flow Budget

A completed cash flow budget is shown in Table 13-3. It will be used to review the budgeting steps and to point out several special calculations. The first estimates needed are the beginning cash balance on January 1 and all sources and amounts of cash inflows for the year. These amounts are entered in the Total column and then allocated to the month or months when the cash will be received. This example shows estimated total annual cash inflow of $541,180, including the beginning cash balance of $7,000. The total cash inflow for any month other than January cannot be determined until its beginning cash balance is known. The ending cash balance for each month becomes the beginning cash balance for the next month. This requires completing the budget for one month before beginning the next month.

The next step is to estimate total cash operating expenses by type, placing each amount in the Total column. Each expense estimate is then allocated to the appropriate month(s), and the total expenses for each month are entered on line 35. In the example, the total annual cash operating expenses are projected to be $350,000, with the largest expenditures in April and September. The same procedure is followed for capital expenditures, family living expenses, income taxes, and Social Security payments. Another important requirement for cash is the scheduled principal and interest payments on debt outstanding at the beginning of each year. These amounts are shown on lines 44 through 47. For some term debt, the total payment remains the same each month, meaning that as the amount owed falls, interest owed each month also falls, so that more principal is paid each time. In other cases, equal principal payments may be due each time, meaning the total payment will decrease over time as interest falls. In this example, machinery debt, paid monthly, is assumed to have equal total payments while the farm mortgage, paid twice a year, is assumed to have equal principal payments.

Total cash outflow, calculated by summing lines 35 through 47, is $514,927 in the example. These steps should be completed before any of the calculations on lines 50 to 59 are attempted.

Beginning with the month of January, total cash inflow is $48,140 and total cash outflow is

$36,730, leaving $11,410 of cash available at the end of January (line 49). The total cash outflow includes scheduled debt payments of $9,500 for current debt with $475 in interest also due on this debt, plus $9,625 of noncurrent debt with associated interest of $12,040. Because cash available is a positive figure, there is no need to take out any new loans in January and lines 50 through 55 are all zero. $11,410 becomes the ending cash balance, as shown on line 56. The first column of lines 57 and 58 indicate there was a current debt balance of $9,500 and a noncurrent debt of $217,750 at the beginning of the year. In January, $9,500 in principal was paid on the current debt and $9,625 on the noncurrent debt, bringing the balances to $0 for current debt and $208,125 for noncurrent debt.

The next step is to transfer the January ending cash balance of $11,410 to the beginning cash balance for February. Calculations for February are similar to those for January. Cash inflow is sufficient to cover cash outflow and no new debt need be incurred. The ending balance of $6,360 becomes the beginning balance of March.

In March, an old piece of machinery is sold for $12,000, and a new one costing $52,000 purchased, partly with a cash payment of $12,000, obtained from the sale of the old machinery and partly by an additional noncurrent loan of $40,000. The new noncurrent loan brings the ending cash balance to a positive figure, so no new current borrowing is needed. The ending cash balance of $4,440, after the noncurrent loan is included, becomes the beginning cash balance in April. The noncurrent debt increases in March to $247,484 because the additional loan secured of $40,000 is considerably larger than the scheduled noncurrent debt principal payment of $321.

In April, cash outflow exceeds cash inflow by $84,994. To cover the shortfall, a new current loan of $85,500 will be needed. The loan will leave a small projected ending balance of $506. In case of unexpected cash needs, the farm operator in this example always wants at least $500

in cash available in each month. After the appropriate adjustments in the loan balances on lines 57 and 58, April will end with an estimated $332,110 of debt outstanding.

This example shows that additional new borrowing will also be needed in May, June, and July. Payments can be made in August when the feeder livestock and cull cows are sold. A total of $77,920 in cash (line 49) is available. All of this cannot be used for principal payments, however, as there will be some interest due, and a positive ending balance of at least $500 is needed.

In this example, $85,500 is borrowed in April, $46,700 in May, $24,300 in June, and $8,200 in July. An interest rate of 6 percent per year is assumed to be in place for these loans. Interest due is calculated using the equation

$$\text{Interest} = \text{principal} \times \text{rate} \times \text{time}$$

	Principal	Interest
April		
Borrowing	$ 85,500 × 6% × 4/12 of a year =	$1,710
May		
Borrowing	$ 46,700 × 6% × 3/12 of a year =	$ 701
June		
Borrowing	$ 24,300 × 6% × 2/12 of a year =	$ 243
July		
Borrowing	$ 8,200 × 6% × 1/12 of a year =	$ 41
Total	$164,700	$2,695

Sufficient cash is not available in August to repay all new current debt to date, including the interest due of $2,695. However, sufficient cash is available to pay $74,700 of the current debt principal, reducing the balance to $90,000.

In September, sufficient cash is on hand to repay the remaining current debt balance, including one month's interest on the total. In October, November, and December, cash inflow exceeds cash outflow, so no new borrowing is needed. The end-of-the-year cash balance is estimated to be $63,108.

The large cash balances toward the end of the year suggest several alternatives that could

TABLE 13-3 | Cash Flow Budget

Name: I. M. Farmer

Cash Flow Budget

#		Total	Jan	Feb	March	April	May	June	July	Aug	Sept	Oct	Nov	Dec
1	Beginning cash balance	7,000	7,000	11,410	6,360	4,440	506	522	508	514	525	30,351	10,566	121,722
	Operating receipts:													
2	Crop sales	362,680	40,640	0	0	0	0	0	0	0	181,640	0	140,400	0
3	Feeder livestock	117,500	0	0	42,000	0	0	0	0	75,500	0	0	0	0
4	Livestock products	0	0	0	0	0	0	0	0	0	0	0	0	0
5	Other	22,500	0	0	0	0	0	0	0	0	0	22,500	0	0
6														
	Capital receipts:													
7	Breeding livestock	10,000			0	0	0	0	0	10,000	0	0	0	0
8	Machinery and equipment	12,000			12,000	0	0	0	0	0	0	0	0	0
9														
	Nonfarm income:													
10	Wages and salary	6,000	500	500	500	500	500	500	500	500	500	500	500	500
11	Investments	3,500	0	0	875	0	0	875	0	0	875	0	0	875
12														
13	Total cash inflow (add lines 1–12)	541,180	48,140	11,910	61,735	4,940	1,006	1,897	1,008	86,514	183,540	53,351	151,466	123,097
	Operating expenses:													
14	Seed	59,000	0	0	16,000	34,000	0	0	0	0	0	0	0	9,000
15	Fertilizer and lime	64,000	0	0	19,000	15,000	0	0	0	0	0	0	0	30,000
16	Chemicals	65,000	0	0	0	25,000	25,000	0	0	0	0	0	15,000	0
17	Other crop expenses	57,240	0	0	0	6,200	7,800	5,000	1,500	1,500	12,000	8,240	7,500	7,500
18	Fuel, oil, lubricants	15,700	0	0	0	3,000	2,000	2,200	750	750	4,000	1,500	1,000	500
19	Hired labor	2,250	0	0	0	0	1,000	0	0	0	1,250	0	0	0
20	Machine hire	5,000	0	0	0	0	0	0	0	0	2,000	3,000	0	0
21	Feed and grain	5,000	0	0	0	0	0	0	0	0	0	2,500	0	2,500
22	Feeder livestock	0	0	0	0	0	0	0	0	0	0	0	0	0
23	Livestock expenses	18,000	1,500	1,500	1,500	1,500	1,500	1,500	1,500	1,500	1,500	1,500	1,500	1,500
24	Repairs—machinery	3,450	50	100	600	500	300	800	200	100	500	300	0	0
25	Repairs—buildings	0	0	0	0	0	0	0	0	0	0	0	0	0
26	Cash rent	32,500	0	0	0	0	0	0	0	0	32,500	0	0	0
27	Supplies	1,080	90	90	90	90	90	90	90	90	90	90	90	90
28	Property taxes	2,700	0	0	0	0	1,350	0	0	0	0	1,350	0	0
29	Insurance	3,280	0	0	820	0	0	820	0	0	820	0	0	820
30	Utilities	4,200	350	350	350	350	350	350	350	350	350	350	350	350
31	Auto and pickup (farm share)	1,200	100	100	100	100	100	100	100	100	100	100	100	100
32	Other farm expenses	10,400	0	0	800	0	0	8,000	0	0	800	0	0	800
33														
34														

Name: I. M. Farmer

Cash Flow Budget

#		Total	Jan	Feb	March	April	May	June	July	Aug	Sept	Oct	Nov	Dec
35	Total cash operating expenses	350,000	2,090	2,140	39,260	85,740	39,490	18,860	4,490	4,390	55,910	18,930	25,540	53,160
	(add lines 14–34)													
	Capital expenditures:													
36	Machinery and equipment	52,000	0	0	52,000	0	0	0	0	0	0	0	0	0
37	Breeding livestock	3,500	0	0	0	0	3,500	0	0	0	0	0	0	0
38														
	Other expenditures:													
39	Family living expenses	36,000	3,000	3,000	3,000	3,000	3,000	3,000	3,000	3,000	3,000	3,000	3,000	3,000
40	Income tax and Social Security	10,500	0	0	2,625	0	0	2,625	0	0	2,625	0	0	2,625
41	Other nonfarm expenses	0	0	0	0	0	0	0	0	0	0	0	0	0
42														
43														
	Scheduled debt payments:													
44	Current debt — principal	9,500	9,500	0	0	0	0	0	0	0	0	0	0	0
45	Current debt — interest	475	475	0	0	0	0	0	0	0	0	0	0	0
46	Noncurrent debt — principal	27,804	9,625	320	321	874	897	910	914	917	921	10,250	925	930
47	Noncurrent debt — interest	25,148	12,040	90	89	320	297	294	290	287	283	10,605	279	274
	Total cash outflow													
48	(add lines 35–47)	514,927	36,730	5,550	97,295	89,934	47,184	25,689	8,694	8,594	62,739	42,785	29,744	59,989
49	Cash available (line 13–line 48)	26,253	11,410	6,360	−35,560	−84,994	−46,178	−23,792	−7,686	77,920	120,801	10,566	121,722	63,108
	New borrowing:													
50	Current	164,700	0	0	0	85,500	46,700	24,300	8,200	0	0	0	0	0
51	Noncurrent	40,000	0	0	40,000	0	0	0	0	0	0	0	0	0
52	Total new borrowing	204,700	0	0	40,000	85,500	46,700	24,300	8,200	0	0	0	0	0
	Payments on new current debt													
53	Principal	164,700	0	0	0	0	0	0	0	74,700	90,000	0	0	0
54	Interest	3,145	0	0	0	0	0	0	0	2,695	450	0	0	0
	Total new debt payments													
55	(line 53 1 line 54)	167,845	0	0	0	0	0	0	0	77,395	90,450	0	0	0
	Ending cash balance													
56	(lines 49 1 52 – 55)	63,108	11,410	6,360	4,440	506	522	508	514	525	30,351	10,566	121,722	63,108
	Summary of debt outstanding													
57	Current (beginning of $9,500)	9,500	0	0	0	85,500	132,200	156,500	164,700	90,000	0	0	0	0
58	Noncurrent (beginning of $217,750)	217,750	208,125	207,805	247,484	246,610	245,713	244,803	243,889	242,972	242,051	231,801	230,876	229,946
59	Total debt outstanding	227,250	208,125	207,805	247,484	332,110	377,913	401,303	408,589	332,972	242,051	231,801	230,876	229,946

be considered. First, arrangements might be made to invest this money for several months rather than have it sit idle in a checking account. Any interest earned would be additional income. Second, the excess cash could be used to make early payments on the noncurrent loans to reduce the interest charges. A third alternative relates to the $40,000 of new noncurrent debt used to purchase machinery in March. This budget projects that the additional debt could be paid off by November. Hence, the operator could have used current borrowing, rather than a new noncurrent loan, to acquire the machinery.

This last alternative would require paying for the entire cost of the new machinery in the year of purchase and would add considerable risk to the operation. If prices and yields were below expectations, sufficient cash might not be available to repay all the new current debt. For this reason, using current debt to finance the purchase of long-lived items such as machinery and buildings is not recommended. A better practice would be to obtain a noncurrent loan and then make additional loan payments in years when extra cash is available.

This example assumed that all monthly cash shortages would be met by new current borrowing. However, other potential solutions to monthly cash shortages should be investigated first. For example, could some products be sold earlier than projected to cover negative cash balances? Could some expenditures, particularly large capital expenditures, be delayed until later in the year? Would it be possible to change the due date of some expenditures such as loan payments, insurance premiums, and family expenditures to a later date? These adjustments would reduce or eliminate the amount of new current borrowing needed for some months and reduce interest expense. Using a software program or electronic spreadsheet makes it relatively easy to estimate how such changes would affect the total interest paid over the year. For example, interest savings from fewer new loans might be more than enough to offset a reduced price from selling a little earlier than planned.

Uses for a Cash Flow Budget

The primary use of a cash flow budget is to project the timing and amount of new borrowing that the business will need during the year and the timing and amount of loan repayments. Other uses and advantages are as follows:

1. A borrowing and debt repayment plan can be developed to fit an individual farm business. The budget can prevent excessive borrowing, and it shows how repaying debts as quickly as possible will save interest.

2. A cash flow budget may suggest ways to rearrange purchases and scheduled debt repayments to minimize borrowing. For example, capital expenditures and insurance premiums might be moved to months where a large cash inflow is expected.

3. A cash flow budget can combine business and personal financial affairs into one complete plan.

4. A bank or other lending agency is better able to offer financial advice and spot potential weaknesses or strengths in the business based on a completed cash flow budget. A realistic line of credit can be established.

5. By planning ahead and knowing when cash will be available, managers can obtain discounts on input purchases by making a prompt cash payment.

6. A cash flow budget can also have a payoff in tax planning by pointing out the income tax effects of the timing of purchases, sales, and capital expenditures.

7. A cash flow budget can help spot an imbalance between current and noncurrent debt and suggest ways to improve the situation. For example, too much current debt relative to noncurrent debt can create cash flow problems.

Box 13-1   Does Positive Cash Flow Mean a Profit?

Not necessarily! While a positive net cash flow is certainly preferred to a negative one, it doesn't always mean a positive profit. There are too many differences between a cash flow budget and an income statement. A cash flow budget does not include noncash items such as inventory changes and depreciation, which do affect profit. It does include cash items such as loan payments, new loans, and withdrawals for personal use, which do not affect profit.

A large positive net cash flow and a low profit can occur whenever there are receipts from sales of capital assets such as land and breeding livestock, cash gifts or inheritances, large nonfarm revenue, new loans, large depreciation, or inventory decreases. Negative cash flow and a good profit could result from substantial inventory increases, cash purchases of capital assets, and larger-than-usual debt payments and personal withdrawals

---

Other uses and advantages of a cash flow budget could be listed, depending on the individual financial situation. The alert manager will find many ways to improve the planning and financial management of a business through the use of a cash flow budget.

MONITORING ACTUAL CASH FLOWS

A cash flow budget can also be used as part of a system for monitoring and controlling the cash flows during the year. The control function of management was discussed earlier, and one control method using a cash flow budget will be illustrated here. A form such as the one shown in Table 13-4 provides an organized way to compare budgeted cash flows with actual amounts.

Budgeted total annual cash flow for each item can be entered in the first column. The total budgeted cash flow to date is entered in the second column, and the actual cash flow to date in the third column. If the figures are updated monthly, or at least quarterly, it is easy to make a quick comparison of budgeted and actual cash flows.

This comparison is a means of monitoring and controlling cash flows, particularly cash outflows, throughout the year. Outflows exceeding budgeted amounts are quickly identified, and

action can be taken to find and correct the causes. Estimates for the rest of the year can also be revised. Actual results at the end of the year can be used to make adjustments in budgeted values for next year, which will improve the accuracy of future budgeting efforts. In addition, monitoring and updating actual cash flows throughout the year makes it easy to put together the annual Statement of Cash Flows for the business, such as the one seen in Table 5-4.

INVESTMENT ANALYSIS USING A CASH FLOW BUDGET

The discussion to this point has been on using a cash flow budget that covers the entire business for one year. There is another important use for a cash flow budget. Any major new capital investment, such as the purchase of land, machinery, or buildings, can have a large effect on cash flows, particularly if additional capital is borrowed to finance the purchase.

Borrowed capital requires principal and interest payments, which represent new cash outflows. The question to answer before making the new investment is, Will the new investment generate enough additional cash income to meet its additional cash requirements? In other words, is the investment financially feasible, as opposed to

TABLE 13-4	A Form for Monitoring Cash Flows

Name _____   Year _____

		Annual budget	Budget to date	Actual to date	Deviation from budget
1	**Beginning cash balance**				
	Operating receipts:				
2	Crop sales				
3	Market livestock sales				
4	Livestock product sales				
5	Other sales and receipts				
	Capital receipts:				
6	Breeding livestock sales				
7	Machinery and equipment sales				
8	Other capital asset sales				
	Nonfarm income:				
9	Wages and salaries				
10	Investments and rents				
11	Other nonfarm income				
12	**Total cash inflows (add lines 1–11)**				
	Operating expenses:				
13	Seed				
14	Fertilizer and lime				
15	Pesticides				
16	Other crop expenses				
17	Fuel and lubricants				
18	Wages and salaries, benefits				
19	Machine hire				
20	Feed purchases				
21	Feeder livestock purchases				
22	Livestock expenses				
23	Repairs to machinery				
24	Repairs to buildings, improvements				
25	Cash rent paid				
26	Supplies				
27	Property taxes				
28	Insurance				
29	Utilities				
30	Legal and business fees				
31	Other farm expenses				
	Capital investments:				
32	Machinery and equipment				
33	Buildings and improvements				

TABLE 13-4	*(CONTINUED)*

Name _____   Year _____

		Annual budget	Budget to date	Actual to date	Deviation from budget
34	Breeding livestock				
	Nonfarm expenses and withdrawals:				
35	Family living and income taxes				
36	Other withdrawals				
37	**Total cash outflows (add lines 13–36)**				
38	**Ending cash balance (line 12 – 37)**				

economically profitable? An example illustrates this use of a cash flow budget. Suppose a farmer is considering the purchase of an irrigation system to irrigate 120 acres. Information has been gathered to use in completing the cash flow analysis:

Cost of irrigation system	$120,000
Down payment in cash (30%)	36,000
Capital borrowed (@ 4.5% for 3 years)	84,000
Additional crop income from yield increase ($204 per acre)	24,480
Additional crop expenses from higher input levels ($62 per acre)	7,440
Irrigation expenses ($45 per acre)	5,400

Table 13-5 summarizes this information in a cash flow format. This is a long-lived capital investment, so it is important to look at the cash flow over a number of years, rather than month by month for one year as was done for the whole-farm cash flow budget.

The new loan requires a principal payment of $28,000 each year of the three-year loan, plus interest on the unpaid balance. This obligation generates a large cash outflow requirement during the first three years, causing a negative net cash flow for these years. Once the loan is paid off in the third year, there is a positive net cash flow in the following years. This result is common when a large part of the purchase price is

TABLE 13-5	**Cash Flow Analysis for an Irrigation Investment**

	Year				
	1	**2**	**3**	**4**	**5**
Cash inflow:					
Increase in crop income	$24,480	$24,480	$24,480	$24,480	$24,480
Cash outflow:					
Additional crop expenses	7,440	7,440	7,440	7,440	7,440
Irrigation expenses	5,400	5,400	5,400	5,400	5,400
Principal payments	28,000	28,000	28,000	0	0
Interest payments	3,780	2,520	1,260	0	0
Total cash outflow	44,620	43,360	42,100	12,840	12,840
Net cash flow	–20,140	–18,880	–17,620	11,640	11,640

Box 13-2   **Cash Flow by Enterprise**

Cash flow budgets can be created for each enterprise. These can be done on a unit basis (e.g., for one acre of a crop or one head of livestock) and then multiplied by the number of units. Then, these enterprise-by-enterprise cash flow budgets can be used as the basis for completing the overall whole-farm cash flow budget. One advantage of this method is that it is harder to overlook items when dealing with a single enterprise. As a result, the overall cash flow budget would be likely to be more accurate. Another advantage is that the effect on cash flows of eliminating, adding, or changing the level of an enterprise, or changing key commodity prices, can easily be estimated.

borrowed and the loan must be paid off in a relatively short time.

This investment is obviously going to cause a cash flow problem the first three years. Does this mean the investment is a bad one? Not necessarily. The irrigation system should last for more than the five years shown in the table and will continue to generate a positive cash flow in later years. Over the total life of the irrigation system, there would be a positive net cash flow, perhaps a substantial one. The problem is how to get by the first three years.

At this point, the purchase of the irrigation system should be incorporated into a cash flow budget for the entire farm. This budget may show that other parts of the farm business are generating enough excess cash to meet the negative cash flow that would result from the purchase of the irrigation system. If not, one possibility would be to negotiate with the lender for a longer loan with smaller annual payments. This solution would help reduce the cash flow problem but would extend principal and interest payments over a longer period and increase the total amount of interest paid.

This cash flow analysis did not include any effects the investment would have on income taxes. However, a new investment can have a large impact on income taxes to be paid, which in turn affects the net cash flows. Accuracy is improved if *after-tax* cash flows are used in this type of analysis. (See Chapter 16 for a discussion of income taxes.)

Land purchases often generate negative cash flows for a number of years, unless a substantial down payment is made. New livestock facilities may require significant cash for construction, breeding stock, and feed before any additional cash income is generated. Orchards and vineyards may take a number of years before they become productive and generate cash revenue. These examples illustrate the importance of analyzing a new investment with a cash flow budget. It is always better to know about and solve a cash flow problem ahead of time than to be faced with an unpleasant surprise. In some cases, there may be no way to solve a projected negative cash flow associated with a new investment. In this situation, the investment would be financially infeasible, because the cash flow requirements cannot be met.

SUMMARY

A cash flow budget is a summary of all cash inflows and outflows for a given future period. The emphasis is on cash and all cash flows, regardless of type, source, or use. Both farm and personal cash needs are included on a cash flow budget. No noncash entries are included. A cash flow budget also includes the timing of cash inflows and outflows to show how well they match up. The result provides an estimate of the borrowing needs of the business, its debt repayment capacity, and the timing of both.

A cash flow budget can also be used to do a financial feasibility analysis of a proposed investment. The investment may cause major changes in cash income and cash expenses, particularly if new borrowing is used to finance the purchase. A cash flow analysis concentrating on just the cash flows resulting from the purchase will indicate whether the investment will generate the cash needed to meet the cash outflows it causes. If not, the purchase needs to be reconsidered, or further analysis done, to be sure the cash needed is available from other parts of the business.

QUESTIONS FOR REVIEW AND FURTHER THOUGHT

1. Why is machinery depreciation not included on a cash flow budget?

2. Identify four sources of cash inflows that would not be included on an income statement but that would be on a cash flow budget. Why are they on the cash flow budget?

3. Identify four types of cash outflows that would not be included on an income statement but that would be on a cash flow budget. Why are they on the cash flow budget?

4. Identify four noncash entries found on an income statement but not on a cash flow budget.

5. Discuss the truth or falsity of the following statement: A cash flow budget is used primarily to show the profit from the business.

6. Discuss how you would use a cash flow budget when applying for a farm business loan.

7. Assume you would like to make an investment in agricultural land in your area. Determine local land prices, cash rental rates, and the cash expenses you would have as the land owner. Construct a five-year cash flow budget for the investment, assuming that 60 percent of the purchase price is borrowed with a 20-year loan at 10 percent interest. Would the investment be financially feasible without some additional source of cash inflow?

8. A feasible cash flow budget should project a positive cash balance for each month of the year, as well as for the entire year. When making adjustments to the budget to achieve this, should you begin with the annual cash flow or the monthly values? Why?

IMPROVING MANAGEMENT
SKILLS

T he basic management skills presented in Parts I, II, III, and IV will now be extended and applied to specific areas of the farm or ranch business. In addition, some of the simplifying assumptions, such as knowing prices and production in advance, will be relaxed, and decision making in a risky environment will be discussed.

A beginning farmer or rancher must decide on what type of business organization to use, and this decision should be reviewed throughout the life of the business. Business organization refers to the legal and operational framework in which management decisions are made and carried out. The basic choices include sole proprietorship, partnership, corporation, and others. They are discussed in Chapter 14. Over time, changes in family size and involvement, goals, tax rules, and financial conditions will require adjustments to the organizational structure of the business.

Agriculture is a risky business. Few decisions can be made with complete information about the future. Usually, there is uncertainty about prices, yields, and other production and financial conditions. In Chapter 15, some methods are presented to improve decision making under uncertainty, and some techniques are discussed for reducing the risk inherent in agricultural production, marketing, and finance.

A profitable farm or ranch business eventually will have to use part of its profits to meet its income tax obligations. This affects the cash flow of the business and slows the growth in equity. Chapter 16 discusses how management decisions affect income taxes and why a manager should consider the tax effects of all decisions made throughout the year. It also discusses some basic tax management strategies that may be useful to meet a goal of maximizing after-tax profit.

Many resources used in agriculture are long-term investments that tie up large amounts of capital. These investment decisions can affect the financial condition of the business for many years. Chapter 17 explains various investment analysis techniques, such as capital budgeting, net present value, and internal rate of return. Using these tools will help farm managers make better decisions about long-term capital investments and financing methods.

Chapter 18 presents information on enterprise analysis, or how to analyze individual enterprises as profit and/or cost centers. Procedures for analyzing the profitability and efficiency of the entire business were discussed in Chapter 6. However, a problem may be caused by only one or two of the enterprises. Enterprise analysis helps pinpoint problems in the business and as such is an important tool for the control function of management.

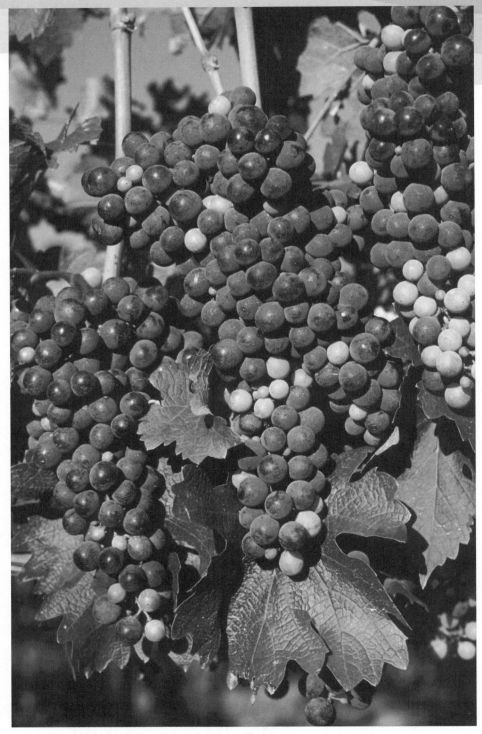

© *Flat Earth Images*

FARM BUSINESS ORGANIZATION AND TRANSFER

CHAPTER OUTLINE

Life Cycle
Sole Proprietorship
Joint Ventures
Operating Agreements
Partnerships
Corporations
Limited Liability Companies
Cooperatives
Transferring the Farm Business
Summary
Questions for Review and Further Thought

CHAPTER OBJECTIVES

1. Describe the sole proprietorship, partnership, corporation, limited liability company, and cooperative as primary forms of business organization available to farmers

2. Discuss the organization and characteristics of each form of business organization

3. Compare the advantages and disadvantages of each form of business organization

4. Show how income taxes are affected by the form of business organization

5. Summarize the important factors to be considered when selecting a form of business organization

6. Compare alternative arrangements for transferring the income, ownership, and management of a farm business from one generation to the next

Any business, including a farm or ranch, can be organized in a number of different legal and business formats. Many managers change the type of organization during the life of their business to better meet changing goals and objectives.

The five most common forms of business organization for farms and ranches are: (1) *sole (or individual) proprietorship*, (2) *partnership*, (3) *corporation*, (4) *limited liability company* (LLC), and (5) *cooperative*. Each one has different legal and organizational characteristics and is subject to different income tax regulations. According to the 2012 U.S. Census of Agriculture, nearly 87 percent of U.S. farms are organized as sole proprietorships, approximately 7 percent are partnerships, and just over 5 percent are corporations. Of the farms organized as corporations, the majority are family held with 10 or fewer shareholders. In addition, a few farms and ranches are organized as limited liability companies or cooperatives or have operating agreements with other businesses.

The proper choice of organization depends on factors such as the size of the business, the number of people involved in it, the career stage and age of the primary operators, and the owners' desires for passing on their assets to their heirs. A final choice of business organization should be made only after analyzing all the possible long-run effects on the business and on the individuals involved.

LIFE CYCLE

Each farm business has a life cycle, with four stages: entry, growth, consolidation, and exit. These stages are depicted in Figure 14-1. The *entry* stage includes choosing farming as a career, selecting enterprises, acquiring and organizing the necessary resources, and establishing a financial base.

The *growth* stage involves expansion in the size of the business, typically through purchasing or leasing additional land or by increasing the scale of the livestock enterprises. Growth can also come about by intensifying production on a fixed land base. This stage often uses debt capital to finance the expansion and requires good financial planning and risk management. It may even include merging with another operator. Following the growth stage, the emphasis usually turns toward *consolidation* of the operation. Debt reduction becomes a priority, and increased efficiency is preferred to increased size. Early planning and merging of the next generation into the business, however, may allow it to continue in the growth or consolidation stage for some time without showing a decline in size.

As the farm operator nears retirement, attention turns toward reducing risk, decreasing labor needs, liquidating the business, or transferring property to the next generation. In this *exit* stage,

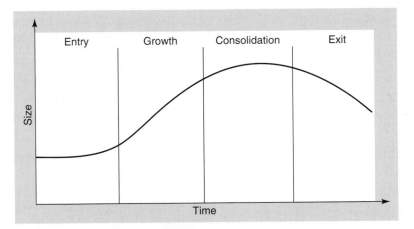

Figure 14-1 Illustration of the life cycle of a farm business.

Box 14-1 | The Life Cycle of a Farm Business

Arthur and Beatrice Campbell entered farming by renting 240 acres from Arthur's uncle when they were in their early twenties. They also milked several cows that Beatrice had acquired during her 4-H years. Ten years later, they bought the land from Arthur's uncle and rented another 160 acres. They kept adding cows until the herd numbered 60 head.

When they reached their early fifties, they decided that their farming operation had grown large enough. They concentrated on paying off all the debts they had incurred, making their milking system more efficient, and improving the genetics of their herd.

None of their children had an interest in farming as a career, so they sold off the dairy herd when they reached the age of 60. A few years later, they sold their machinery and rented their land to a younger farmer who lives nearby. However, they continue to live on the farm and take an active role in their community.

---

business size may decline. The income tax consequences of liquidation or transfer must be considered, along with the need for adequate retirement income. Another concern is to provide equitable treatment to all the children who may choose farming as a career, as well as to those who pursue other interests.

The life-cycle stage in which the farm is currently operating is an important consideration in selecting the form of business organization. Total capital invested, size of debt, income tax considerations, the goals of the owners and operators, and other factors are likely to be different in each stage. Farmers and ranchers may find that a different form of business organization is appropriate for different stages in the life cycle. However, careful and early thought must be given to how the change from one type of organization to another will be handled.

SOLE PROPRIETORSHIP

A sole proprietorship is easy to form and easy to operate, which accounts for much of its popularity.

Organization and Characteristics

In a sole proprietorship, the owner owns and manages the business, assumes all the risks, and receives all profits and losses. Its distinguishing characteristic is the single owner, who acquires and organizes the necessary resources, provides the management, and is solely responsible for the success or failure of the business, as well as for servicing all business debts. Often the sole proprietorship could be more accurately called a family proprietorship, because on many farms and ranches, the husband, wife, and children are all involved in ownership, labor, and management.

A sole proprietorship is established by starting to operate the business. No special legal procedures, permits, or licenses are required. A sole proprietorship is not limited in size by either the amount of inputs that can be used or the amount of commodities produced. The business can be as large or small as the owner desires. There can be any number of employees. Additional management may be hired, and property may even be co-owned with others. A sole proprietorship does not necessarily need to own any assets; it can exist even when all land and machinery are leased.

Income Taxes

The owner of a business organized as a sole proprietorship pays income taxes on any business profit at the tax rates in effect for individual or joint returns. Business profits and capital gains are added to any other taxable income earned to determine the individual's total taxable income. Earned income is subject to self-employment and Medicare taxes.

Advantages

The advantages of a sole proprietorship are its simplicity and the freedom the owner has in operating the business. No other operator or owner needs to be consulted when a management decision is made. The owner is free to organize and operate the business in any legal manner, and all business profits and losses belong to the sole proprietor.

A sole proprietorship is also flexible. The manager can quickly make decisions regarding investments, purchases, sales, enterprise combinations, and input levels, based solely on his or her best judgment. Assets can be quickly purchased or sold, money borrowed, or the business even liquidated if necessary, although the latter may require the concurrence of a landlord or lender.

Disadvantages

The management freedom inherent in a sole proprietorship also implies several responsibilities. Owners of sole proprietorships are personally liable for any legal difficulties and debts related to the business. Creditors have the legal right to attach not only the assets of the business but also the personal assets of the owner to fulfill any unpaid financial obligations. This feature of a sole proprietorship can be an important disadvantage for a large, heavily financed business where the owner has substantial personal and nonfarm assets. Business failure can result in these assets being acquired by creditors to pay the debts of the sole proprietorship.

The size of a sole proprietorship is limited by the capital available to the single owner. If only a limited amount is available, the business may be too small to realize many economies of size, making it difficult to compete with larger and more efficient farms. At the other extreme,

the management abilities and time of the single owner may be insufficient for a large business, making it difficult to become an expert in any one area. Thus, a large sole proprietorship may need to hire additional management expertise or consulting time.

Another disadvantage of a sole proprietorship is a lack of business continuity. It is difficult to bring children into the business on any basis other than as employees or tenants. Death of the owner also means the business may have to be liquidated or reorganized under new ownership. This can be time consuming and costly, resulting in a smaller inheritance and less income for the heirs during the transition period.

JOINT VENTURES

While a sole proprietorship offers the manager maximum flexibility and independence, other forms of business organization are possible. They allow two or more operators to combine their respective abilities and assets to achieve levels of efficiency or other goals that might be unattainable on their own.

Such businesses are called *joint ventures*. Of the several types of joint ventures, those most often used in agriculture include

1. Operating agreements
2. Partnerships
3. Corporations
4. Limited liability companies (LLC)
5. Cooperatives

Each type of joint venture has unique characteristics related to property ownership, distribution of income, taxation, continuity, liability, and formal organization. In many cases, only part of the total farm business is included in the joint venture. Sometimes the joint venture affects only one enterprise. In other cases, some assets, such as land, are excluded to protect them from the unlimited liability of the joint venture or to make ownership shares more equal.

Particular care should be taken to distribute the income earned by a joint venture in an equitable manner. Businesses with an operating agreement may divide gross income, as will be illustrated later. In more complex businesses, owners of assets that are used but not owned by the joint venture may be paid a fair market rental rate. People who lend money to the business receive interest payments. People who contribute labor should be paid a fair wage, even if they are also owners of the business. And, finally, profits earned by the business are distributed on the basis of ownership share, patronage rate, or some other mutually agreeable criterion.

It is important to distinguish among payments made as returns to rented assets, borrowed capital, labor, and equity capital, even though the recipients may be some or all of the same people. Figure 14-2 illustrates this relation.

The primary advantages of a joint venture over a sole proprietorship involve combining capital and management. Pooling the capital of the members of the joint venture allows a larger business to be formed, which can be more efficient than two or more smaller businesses. It can also increase the amount of credit available, allowing further increases in business size. The total supply of management and labor is also increased by pooling the capabilities of all the members. Management efforts can be divided, with each person specializing in one area of the business, such as crops, livestock, marketing, or

accounting. It is also easier for one operator to be absent from the business when another operator is available to fill in.

OPERATING AGREEMENTS

Sometimes two or more sole proprietors carry on some farming activities jointly while maintaining individual ownership of their own resources. Such an activity is often called an *operating agreement*. They tend to be informal, limited arrangements, and often involve only a single farming enterprise.

In most operating agreements, all parties pay the costs related to ownership of their own assets, such as property taxes, insurance, maintenance, and interest on loans. Operating expenses such as the cost of seed, fertilizer, veterinarian fees, or utilities may be shared among the parties in a fixed proportion, often in the same proportion as the fixed costs. In other cases, one party or another may pay for all of certain operating costs, such as fuel or feed, as a matter of convenience. In either case, the general principle for an operating agreement is to share income in the same proportion as total resources are contributed, including both fixed assets and operating costs.

An enterprise budget can be a useful tool for comparing the resource contributions by each party in an operating agreement. The example in

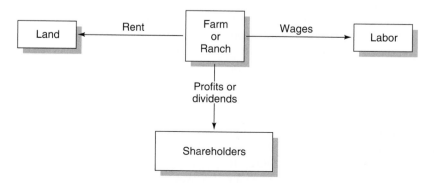

Figure 14-2 Distribution of income from a joint venture.

TABLE 14-1	Example Budget for a Cow-Calf Joint Enterprise (One Head)		

Item	Value ($)	Party A ($)	Party B ($)
Operating expenses			
Hay	175.00		175.00
Supplement	22.50		22.50
Salt, minerals	25.60		25.60
Pasture maintenance	230.00	230.00	
Veterinary & health expense	26.00		26.00
Livestock facilities repair	12.00	12.00	
Machinery & equipment	17.00		17.00
Breeding expenses	19.55	19.55	
Labor	72.00		72.00
Auction and hauling	35.00	17.50	17.50
Miscellaneous	25.00		25.00
Interest (on half of operating expenses)	19.79	8.37	11.42
Ownership expenses			
Interest on breeding herd	56.25	56.25	
Livestock facilities			
Depreciation & interest	12.60	12.60	
Machinery & equipment			
Depreciation & interest	28.90		28.90
Land charge	70.00	70.00	
Total expenses	847.19	426.27	420.92
Percent contribution	100%	50%	50%

Table 14-1 is based on the cow/calf enterprise budget found in Table 10-3. In this situation, Party A owns all the pasture, buildings, and handling facilities for the cow herd, as well as all the breeding livestock. Party A will pay the costs for pasture maintenance, repair of fences and facilities, and breeding expenses.

Party B will provide all the feed and labor for the cow/calf enterprise and pay all other variable costs except auction and hauling fees, which they agree to divide equally. How should the income from the cattle sold be divided? Table 14-1 shows that Party A is contributing approximately 50 percent of the total costs, and Party B is contributing the other 50 percent. Income can be divided in the same proportion as total costs

or, in this case, equally to each person, because the results are close to a 50/50 division of costs.

An alternative would be to sum the value of the unpaid resources contributed by each party (land, facilities, livestock, machinery, and labor, in this example) and share all cash operating expenses and income in the same proportion. In the example, Party A's annual contributions of livestock, facilities, and land are valued at a total of $138.85 per cow unit, while the labor and machinery contribution of Party B is valued at $100.90. Party A is providing 58 percent of the fixed resources, and Party B, 42 percent. They could agree to divide all the remaining costs, as well as gross income, in the same proportion. This may not be convenient, however, unless a joint operating bank account is established. Moreover, the arrangement then begins to look like a partnership, which may be neither intended nor desirable.

Two cautions should be observed when valuing resource contributions. The opportunity costs for all unpaid resources, such as labor and capital assets, must be included along with cash costs, and all costs must be calculated for the same period, usually one year. It would not be correct, for example, to value land at its full, current market value when labor and operating costs are included for only one year. An annual interest charge or rental value for the land should be used instead.

PARTNERSHIPS

A partnership is an association of two or more persons who share the ownership of a business to be conducted for profit. Partnerships can be organized to last for a brief time or for a long duration.

Two types of partnerships are recognized in most states. The *general partnership* is the most common, but the *limited partnership* is used for some situations. Both types have the same characteristics, with one exception. The limited partnership must have at least one general partner, but can have any number of limited partners. Limited partners cannot participate in the management of the partnership business, and in exchange, their financial liability for partnership debts and obligations is limited to their actual investment in the partnership. The liability of general partners, however, can extend even to their personal assets.

Limited partnerships are most often used for businesses such as real estate development and cattle feeding, where investors want to limit their financial liability and do not wish to be involved in management. Most farm- and ranch-operating partnerships are general partnerships. For this reason, the discussion in the remainder of this section will apply only to general partnerships.

Organization and Characteristics

There are many possible patterns and variations in partnership arrangements. However, the partnership form of business organization has three basic characteristics:

1. A sharing of business profits and losses
2. Shared control of property, with possible shared ownership of some property
3. Shared management of the business

The exact sharing arrangement for each of these characteristics is flexible and should be outlined in the partnership agreement.

Oral partnership agreements are legal in most states but are not recommended. Important points can be overlooked, and fading memories over time on the details of an oral agreement can create friction between the partners. Problems can also be encountered when filing the partnership income tax return if arrangements are not well documented.

A written partnership agreement is of particular importance if the profits will not be divided equally. The courts and the Uniform Partnership Act, a set of regulations enacted by most states to govern the operation of partnerships, assume an equal share in everything, including management decisions, unless there is a written agreement specifying a different arrangement. For any situation not covered in the written agreement, the regulations in the state partnership act apply.

The written agreement should cover at least the following points:

1. *Management:* State who is responsible for which management decisions and how they will be made.
2. *Property ownership and contribution:* List the property that each partner will contribute to the partnership, describe how it will be owned, and state the beginning value of each asset.
3. *Share of profit and losses:* Carefully describe the method for calculating profits and losses and the share going to each partner, particularly if there is an unequal division.

Box 14-2	An Example of a Partnership

McDonald and Garcia formed a partnership for the purpose of finishing feeder pigs. They agreed to put in $30,000 of operating capital each and the feed they had on hand. McDonald will put up the finishing building and pay for the feeders, waterers, fans, and other equipment. Garcia will contribute a tractor and machinery to handle manure. The table summarizes the value of their contributions.

Assets contributed	Total	McDonald	Garcia
Operating capital	$60,000	$30,000	$30,000
Current feed inventory	25,600	9,500	16,100
Livestock equipment	21,500	20,200	1,300
Tractor and machinery	27,000	0	27,000
Building and manure storage	85,000	85,000	0
Total	$219,100	$144,700	$74,400
Ownership share	100%	66%	34%

After looking at the values, they agreed to divide net income and losses from the partnership in the ratio of two-thirds to McDonald and one-third to Garcia. However, Garcia will spend about 25 hours per week operating the finishing facility, and McDonald will work only 5 hours per week, so they agreed that the partnership would pay each of them a fair wage for their labor before net income is calculated. They also agreed to leave one-half of the net income in the partnership the first year to increase their operating capital.

4. *Records:* Records are important for the division of profits and for maintaining an inventory of assets and their ownership. Designate who will keep what records as part of the agreement.
5. *Taxation:* Include a detailed account of the tax basis of property owned and controlled by the partnership and copies of the partnership information tax returns.
6. *Termination:* State the date that the partnership will be terminated, if one is known or can be determined.
7. *Dissolution:* The termination of the partnership on either a voluntary or an involuntary basis requires a division of partnership assets. The method for making this division should be described to prevent disagreements and an unfair division.

Partners may contribute land, capital, labor, management, and other assets to the partnership capital account. Compensation for labor contributed or assets rented to the partnership should be paid first, because these may not be in the same proportions as the ownership. The remaining profits generally are divided in proportion to the ownership of the partnership, usually based on the initial value of the assets contributed to the business. Some partners prefer to share partnership profits equally, however, regardless of contributions.

Property may be owned by the partnership, or the partners may retain ownership of their individual property and rent it to the partnership. When the partnership owns property, any partner may sell or dispose of any asset without the consent and permission of the other partners. This aspect of a partnership suggests that retaining individual ownership of some assets and renting them to the partnership may be desirable.

A partnership can be terminated in a number of ways. The partnership agreement may specify a termination date. If not, a partnership will terminate upon the incapacitation or death of a partner, bankruptcy, or by mutual agreement between the partners. Termination upon the death of a partner can be prevented by placing provisions in the written agreement that allow the deceased partner's share to pass to the estate and hence to the legal heirs.

In addition to the three basic characteristics mentioned earlier, factors that indicate a particular business arrangement is legally a partnership including the following:

1. Joint ownership of assets in the partnership capital account
2. Operation under a firm name
3. A joint bank account
4. A single set of business records
5. Management participation by all parties
6. Sharing of profits and losses

Joint ownership of property does not by itself imply a partnership. However, a business arrangement with all or most of these characteristics may be a legal partnership, even though a partnership was not intended. These characteristics should be kept in mind for landlord-tenant arrangements, for example. A partnership can be created by a leasing arrangement based on shares, unless steps are taken to avoid most of the conditions mentioned.

Income Taxes

A partnership does not pay any income taxes directly. Instead it files an information income tax return reporting the income and expenses for the partnership. Income from the partnership is then reported by the individual partners on their personal tax returns, based on their respective shares of the partnership income as stated in the partnership agreement.

In a 50/50 partnership, for example, each partner reports one-half of the partnership income, capital gains, expenses, depreciation, losses, and so forth. These items are included with any other income the partner has to determine the total income tax liability. Partnership income, therefore, is taxed at individual tax rates, with the exact rate depending on the amount of partnership income and other income earned by each partner/taxpayer.

Advantages

A partnership is easier and cheaper to form than a corporation. It may require more records than a sole proprietorship, but not as many as a corporation. While each partner may lose some individual freedom in making management decisions, a carefully written agreement can maintain most of this freedom.

A partnership is a flexible form of business organization, in which many types of arrangements can be accommodated and included in a written agreement. It fits situations such as when parents desire to bring children and their spouses into the business. The children may contribute only labor to the partnership initially, but the partnership agreement can be modified over time to allow for their increasing contributions of management and capital.

Disadvantages

The unlimited liability of each general partner is an important disadvantage of a partnership. A partner cannot be held personally responsible for the *personal* debts of the other partners. However, each partner can be held personally and individually responsible for any lawsuits and financial obligations arising from the operation of the partnership. If the partnership does not have sufficient assets to cover its legal and financial obligations, creditors can bring suit against all partners individually to collect any money due them. In other words, a partner's personal assets can be claimed by a creditor to pay partnership debts.

This disadvantage takes on a special significance, considering that any partner, acting individually, can act for the partnership in legal and financial transactions. For this reason, if for no other, it is important to know and trust your partners and to have the procedures for making management decisions included in the partnership agreement. Too many partners or an unstructured management system can easily create problems.

Like a sole proprietorship, a partnership has the disadvantage of poor business continuity. It can be unexpectedly terminated by the death of one partner or a disagreement among the partners. Dissolution of any business is generally time consuming and costly, particularly when it is caused by the death of a close friend and partner or by a disagreement that results in bad feelings among the partners. The required sharing of management decision making and the loss of some personal freedom in a partnership are always potential sources of conflict between the partners.

Laws governing the formation and taxation of partnerships are less detailed than for corporations. Unfortunately, this allows many farm partnerships to operate with minimal records and little documentation of how resources were contributed and how income has been divided. This can make a fair dissolution of the partnership difficult.

Partnership Operation

A partnership is usually formed for a relatively long period, so contributions to the partnership can be valued on the basis of current market values for the assets contributed by each party at the time the partnership was started. All operating expenses should be paid by the partnership, usually from a separate partnership account. If either or both partners contribute labor, they should be paid a representative and fair wage by the partnership.

In some partnerships, major assets such as land and buildings are not included in partnership property. Instead, the partnership pays rent to the owners of these assets (who may also be the partners) for their use. In such cases, the rented assets are not included when calculating the relative ownership shares of the partnership.

Net income may either be paid out or left in the partnership account to increase business equity. As long as all withdrawals (not including wages or rent) are made in the same proportion as the ownership of the partnership, the ownership shares will not change. However, if one partner leaves more profits in the business, such as by retaining more breeding stock, or contributes additional assets, the ownership percentages will change. Sometimes this is done as a way to move toward more equal ownership of the partnership.

When a partnership is liquidated, the proceeds should be distributed in the same proportion as the ownership. For this reason, it is important to keep detailed and accurate records of the property used by a partnership, and of how wages, rent, and net income are distributed to the partners.

CORPORATIONS

A corporation is a separate legal entity that must be formed and operated in accordance with the laws of the state in which it is organized. It is a legal *person*, separate and apart from its owners, managers, and employees. This separation of the business entity from its owners distinguishes a corporation from the other forms of business organization. As a separate business entity, a corporation can own property, borrow money, enter into contracts, sue, and be sued. It has most of the basic legal rights and duties of an individual.

The number of farm and ranch corporations, though still small, has been increasing. The majority are classified as family farm corporations with a relatively small number of related shareholders.

Organization and Characteristics

The laws affecting the formation and operation of a corporation vary somewhat from state to state. In addition, several states have laws that prevent a corporation from engaging in farming or ranching, or that place special restrictions on farm and ranch corporations. For this and other reasons, competent legal advice should be sought before attempting to form a corporation.

While state laws do vary, some basic steps will generally apply to forming a farm corporation:

1. The incorporators file a preliminary application with the appropriate state official. This step may include reserving a name for the corporation.
2. The incorporators draft a pre-incorporation agreement outlining the major rights and duties of the parties after the corporation is formed.
3. The articles of incorporation are prepared and filed with the proper state office.
4. The incorporators turn property and/or cash over to the corporation in exchange for shares of stock representing their ownership share of the corporation.
5. The shareholders meet to organize the business and elect directors.
6. The directors meet to elect officers, adopt bylaws, and begin business in the name of the corporation.

Three groups of individuals are involved in a farming corporation: shareholders, directors, and officers. The shareholders own the corporation. Stock certificates are issued to them in exchange for property or cash transferred to the corporation. The number of shareholders may be as few as one in some states, while three is the minimum number in several other states. As owners, the shareholders have the right to direct the affairs of the corporation, done through the elected directors and at annual meetings. Each shareholder has one vote for each share of voting stock owned. Therefore, any shareholder with 51 percent or more of the outstanding voting stock has majority control over the business affairs of the corporation. In some corporations, contributions of capital are exchanged for bonds instead of stock. Bonds carry a fixed return but no voting privileges.

The directors are elected by the shareholders at each annual meeting, and they hold office for the following year. They are responsible to the shareholders for the management of the business. The number of directors is normally fixed by the articles of incorporation. Meetings of the directors are held to conduct the business affairs of the corporation and to set broad management policy to be carried out by the officers.

The officers of a corporation are elected by the board of directors and may be removed by them. They are responsible for the day-to-day operation of the business within the guidelines established by the board. The officers' authority flows from the board of directors, to whom they are ultimately responsible. A corporation president may sign certain contracts, borrow money, and perform other duties without board approval but will normally need board approval before committing the corporation to large financial transactions or performing certain other acts.

In many small family farm corporations, the shareholders, directors, and officers are all the same individuals. To an outsider, the business may appear to be operated like a sole proprietorship or partnership. Even the directors' meetings may be held informally around a kitchen table, but a set of minutes must be kept for each official meeting.

Two types of corporations are recognized by the Internal Revenue Service (IRS) for federal income tax purposes. The first is the regular corporation, also called a C corporation. A second type of corporation is the S corporation, sometimes referred to as a tax-option corporation. Both types have many of the same characteristics. However, there are certain restrictions on the formation of S corporations. Some of the more important restrictions include the following:

1. There is a limit on the number of shareholders. The federal limit is 100, but some states set lower limits, such as 75. A husband and wife may be considered one shareholder, even if both own stock.
2. Shareholders are limited to individuals (foreign citizens are excluded), estates, and certain types of trusts. Other corporations cannot own stock in an S corporation.
3. There may be only one class of stock, but the voting rights may be different for some shareholders.

4. All shareholders must initially consent to operating as an S corporation.

There are also certain limitations on the types and amounts of income that a corporation can receive from specified sources and still maintain the S corporation status.

Income Taxes

A disadvantage of a C corporation is the potential double taxation of income. After the corporation pays tax on its taxable income, any after-tax income distributed to the shareholders as dividends is considered taxable income to the shareholders. It is taxed at their applicable individual rates. Many small-farm corporations avoid some of this double taxation by distributing most of the corporation income to the shareholders as rent, wages, salaries, and bonuses. These items are tax-deductible expenses for the corporation but taxable income for the shareholders/employees. However, any wages and salaries paid must be reasonable for bona fide work performed for the corporation and not a means to avoid double taxation.

The S corporations do not pay income tax themselves but are taxed like a partnership; hence, the name *tax-option* corporation. The corporation files an information tax return, but the shareholders report their share of the corporation income, expenses, and capital gains according to the proportion of the total outstanding stock they own. This income (or loss) is included with the shareholders' other income, and tax is paid on the total based on the applicable individual rates. The income tax treatment of an S corporation avoids the dividend double-taxation problem of a C corporation.

Only a partial treatment of the income tax regulations applicable to both types of corporations is included here. There are many rules pertaining to special situations and special types of income and expenses. All applicable tax regulations should be reviewed with a qualified tax consultant before the corporate form of business organization is selected.

Advantages

Corporations provide limited liability for all the shareholder/owners. They are legally responsible only to the extent of the capital they have invested in the corporation. Personal assets of the shareholders cannot be attached by creditors to meet the financial obligations of the corporation. This advantage may be negated if a corporation officer is required to personally co-sign a note for corporation borrowing. In this case, the officer can be held personally responsible for the debt if the corporation cannot meet its responsibilities.

A corporation, like a partnership, provides a means for several individuals to pool their resources and management. The resulting business, with a larger size and the possibility of specialized management, can provide greater efficiency than two or more smaller businesses. Credit may also be easier to obtain because of the business continuity advantage of a corporation. The business is not terminated by the death of a shareholder, because the shares simply pass to the heirs and the business continues. However, a plan for management continuity should exist by having more than one person involved in management and capable of taking over responsibility.

A corporation provides a convenient way to divide and transfer business ownership. Shares of stock can be easily purchased, sold, or given as gifts without transferring title to specific parcels of land or other assets. Transferring shares does not disrupt or reduce the size of the business and is a convenient way for a retiring farmer to gradually transfer part of an ongoing business to the next generation.

There can be income tax advantages to incorporation, depending on the size of the business, how it is organized, and the income level of the shareholders. Any C corporation is a separate tax-paying entity and, as such, is subject to different tax rates than individuals. Current personal and corporate tax rates are shown in Table 14-2.

One advantage of a C corporation is that the owners are usually employees as well. This allows the tax deductibility of certain fringe benefits provided to the shareholder/employees, such as

TABLE 14-2	**Personal and Corporate Income Tax Rates (2015)**

Personal			Corporate		
Taxable income ($)*		Marginal tax rate (%)	Taxable income ($)		Marginal tax rate (%)
Single	Joint				
0 to 9,225	0 to 18,450	10	0 to 50,000		15
9,225 to 37,450	18,450 to 74,900	15	50,000 to 75,000		25
37,450 to 90,750	74,900 to 151,200	25	75,000 to 100,000		34
90,750 to 189,300	151,200 to 230,450	28	100,000 to 335,000		39
189,300 to 411,500	230,450 to 411,500	33	Over 335,000		34 to 38
411,500 to 413,200	411,500 to 464,850	35			
Over 413,200	Over 464,850	39.6			

* The personal tax brackets will be increased in future years to adjust for the effects of inflation.

premiums for health, accident, and life insurance. It also is easier to allocate income among individuals by setting salaries, rents, and dividends.

Disadvantages

Corporations are more costly to form and maintain than sole proprietorships and partnerships. Certain legal fees are necessary when organizing a corporation, and legal advice will be needed on a continuing basis to ensure compliance with state regulations. An accountant may also be needed during the formation period and throughout the life of the corporation to handle financial records and tax-related matters. Most states require various fees when filing the articles of incorporation and some type of annual operating fee or tax on corporations, which are not assessed on the other forms of business organization.

Doing business as a corporation requires that shareholders and directors meetings are held, minutes are kept of directors meetings, and annual reports are filed with the state. However, if forming a corporation results in better business and financial records being kept, this might be viewed as an advantage rather than a disadvantage, because better information for making management decisions will be available.

Corporation Operation

A corporation can be formed and operated in a manner similar to a partnership. The shares of the corporation are divided in the same proportion as the original contributions of equity capital. Partnerships and corporations can receive contributions of debt from the partners, as well as assets.

The number of shares to be issued is an arbitrary decision by the stockholders. The initial value of each share is found by dividing the beginning equity of the corporation by the number of shares to be issued. As with a partnership, the corporation can pay salaries for labor contributed by shareholders, as well as rent for the use of assets not owned by the corporation. Distributions of net income are in the form of dividends.

LIMITED LIABILITY COMPANIES

A limited liability company (LLC) is a type of business organization that closely resembles a partnership, but offers its members the advantage of limited liability. This means that creditors or other claimants can pursue the assets of the LLC to satisfy debts or other obligations, but cannot pursue personal or business assets owned individually by the members of the LLC, unless personal

Box 14-3   An Example of a Corporation

Suppose McDonald and Garcia had decided to form a corporation for finishing feeder pigs instead of a partnership. The assets they contributed and their value would be exactly the same as in the partnership example. These are summarized in the following table:

Assets contributed	Total	McDonald	Garcia
Operating capital	$ 60,000	$ 30,000	$30,000
Current feed inventory	25,600	9,500	16,100
Livestock equipment	21,500	20,200	1,300
Tractor and machinery	27,000	0	27,000
Building and manure storage	85,000	85,000	0
Total	$219,100	$144,700	$74,400
Liabilities contributed			
Term loan on buildings	56,000	56,000	0
Equity contributed	$163,100	$88,700	$74,400
Number of shares @ $100	1,631	887	744

McDonald had to borrow $56,000 to construct the finishing building. In this case, the shareholders agreed to have the corporation take over this debt and make the principal and interest payments as they come due.

Including the $56,000 loan in the capitalization of the corporation lowered the initial equity of the business to $163,100. McDonald and Garcia agreed to issue 1,631 shares of stock with a book value of $100 each. These shares are divided in proportion to the initial equity contributions: 887 shares to McDonald and 744 shares to Garcia.

The corporation will pay each of them wages based on their contribution of labor. Any dividends paid out will be divided in proportion to the number of shares owned.

guarantees have been pledged. This is a significant advantage to potential investors.

LLCs can include any number of members, all of whom can participate in management. Ownership is distributed according to the fair market value of assets contributed, as in a partnership. Likewise, net farm income and other income tax attributes from an LLC are passed to the individual tax returns of the members in proportion to their shares of ownership. An LLC is chartered by state law, so organizational requirements vary. However, it is always a good idea to keep a careful record of contributions and distributions of assets and income and to write down the agreed-on rules of operation.

Unlike a corporation, an LLC cannot deduct the cost of fringe benefits, such as insurance plans or use of vehicles, provided to employees who are also members of the LLC. Neither does it automatically continue in the event of the death of a member.

Some states allow formation of a limited liability partnership, similar to an LLC. Limited liability companies and partnerships offer an attractive alternative for farm families who desire the simplicity and flexibility of a partnership

TABLE 14-3	Comparison of Forms of Farm Business Organization

Category	Sole proprietorship	Partnership	Corporation	Limited liability company
Ownership	Single individual	Two or more individuals	Separate legal entity owned by shareholders	Separate legal entity owned by shareholders
Life of business	Terminates on death	Agreed-on term, or terminates at death of a partner	Forever unless fixed by agreement	Fixed by agreement
Liability	Proprietor is liable	Each partner is liable for all partnership obligations, even to personal assets (except limited partners)	Shareholders are not personally liable for corporation obligations unless they co-sign notes	Shareholders are not personally liable for LLC obligations unless they co-sign notes
Source of capital	Personal investments, loans	Partnership contributions, loans	Shareholder's contributions, sales of bonds, and loans	Shareholder's contributions and loans
Management decisions	Proprietor	Agreement of partners	Shareholders elect directors, who manage the business or hire a manager	Agreement of shareholders or by designated manager
Income taxes	Business income is combined with other income on individual tax returns	Partners' shares of partnership income and losses are added to their individual tax returns	Regular C: files a tax return and pays income tax; shareholders pay taxes on dividends received Tax-option S: shareholders report their shares of income and losses on their individual returns	Shareholders report their shares of income, operating losses, and capital gains on their individual returns

combined with the limited financial liability offered by a corporation.

Table 14-3 summarizes the important features of each of the four main forms of business organization. The advantages and disadvantages of each feature should be evaluated carefully before one is selected.

COOPERATIVES

Farmers have used cooperatives for many years as a means of obtaining inputs and services or for marketing products jointly. In some countries, cooperatives for the purpose of farm production have been formed by groups of small-scale landholders or farm workers to gain efficiencies in production. Many of the state and collective farms that existed in eastern European countries were reorganized into large production cooperatives, with memberships made up of their former farm workers.

In the United States, farm cooperatives that produce grain or livestock have been rare, but some states have passed laws that encourage the formation of small-scale cooperatives for this purpose. Often they are made up of independent farmers who wish to carry out one particular operation jointly. Examples include sow cooperatives that produce feeder pigs for their members to finish or a cooperative that grows high-quality seed for its members.

Cooperatives are a special type of corporation. They require articles, bylaws, and detailed records. Members who contribute capital enjoy limited liability on those contributions. Net income is passed on to members before it is subject to income tax. Cooperatives can also provide tax deductible benefits to owner/members.

Unlike other corporations, cooperatives can pay a maximum return of 8 percent annually to shareholders. Remaining net income is distributed to members as *patronage refunds*, based on how much business each member does with the cooperative, rather than on shares of ownership. Control of a cooperative also differs from other forms of business organizations, in that all members have one vote each when it comes to making decisions, regardless of how much of the cooperative they own. Most agricultural cooperatives limit membership to active farmers.

In recent years a new form of farmer cooperative, known as a *closed* or *new generation* cooperative, has become popular. Members must contribute a significant amount of equity capital to join and agree to sell a certain volume of production to the cooperative on a fixed schedule. Memberships can be bought and sold. These new cooperatives have been formed mostly for value-added processing, but also include livestock finishing and egg-laying operations.

Perhaps the key factor in making a farming cooperative successful is a true desire to cooperate. Each member must perceive that working together will help obtain benefits that outweigh the necessity of giving up some degree of managerial independence.

TRANSFERRING THE FARM BUSINESS

At the beginning of this chapter, the life cycle of a farm business was illustrated (Figure 14-1). If the exit stage of one operator coincides with the entry stage of the next operator, the transfer process may be relatively simple. Livestock, equipment, and machinery may be sold or leased to the new operator or dispersed by an auction.

Land may be sold outright, sold on an installment contract, or rented to the new operator. The outward size and structure of the business may change very little in the process.

In many family farming situations, however, the next generation is ready to enter the business while the current generation is still in the growth or consolidation stage. This brings up several important questions.

1. Is the business large enough to productively employ another person or family? If the labor supply will be decreased through retirement of the present operators or a reduction in the amount of labor hired, then a new operator can be used efficiently. If not, the business may have to expand to provide more employment.
2. Is the business profitable enough to support another operator? Adding more labor does not necessarily produce more net income. If an additional source of income such as Social Security payments will become available to the older generation, then income from the farm can be more easily diverted to the new operator.

   If adding another operator will require expanding the business, a detailed whole-farm budget should be completed. Liquidity, as well as profitability, should be analyzed, especially if additional debt will be used to finance the expansion. The net cash flow that will be available in a typical year should be projected for each person, as well as for the business as a whole, to avoid unexpected financial problems.
3. Can management responsibilities be shared? If the people involved do not have compatible personalities and mutual goals, then even a profitable business may not provide a satisfying career. Farmers and ranchers accustomed to working alone and making decisions independently may find it difficult to accommodate a partner. Parents and children may find it especially difficult to work as equal partners in management.

Key Areas to Transfer

Three key areas of a farm business must be transferred: income, ownership, and management. Typically they are not all transferred at the same time, however.

Income can be transferred initially by paying the new operator a wage, which could also include some type of bonus or profit sharing. As the younger individual acquires more assets, part of the income can be based on a return on investment as well.

Ownership can be transferred by allowing the younger generation to gradually acquire property, such as by saving breeding stock or investing in machinery. Longer-life assets can be sold on contract or by outright sale. Property can also be gifted to the eventual heirs, but if annual giving exceeds certain limits, federal estate taxes due at the death of the giver may be increased. Gifting of assets also involves finding a way to treat nonfarm heirs fairly. If the farm business is organized as a corporation, individual shares of stock can be sold or given away over time. This is much easier than transferring specific assets or ownership shares of assets. Finally, careful estate planning will assure an orderly transfer of property upon the death of the owner.

Management may be the most difficult part of a farming operation to transfer. The new operator may be given responsibility for one particular enterprise or for a certain management area, such as feeding, breeding, or record keeping. Allowing younger operators to rent additional land or produce a group of livestock on their own while using the farm's machinery or facilities helps them learn management skills without putting the entire operation at risk.

The older generation must understand that there are many different ways to carry out farming enterprises. Knowing when to offer advice and when to keep silent will help make the management transfer occur smoothly.

Stages in Transferring the Business

The specific arrangements chosen for transferring income, ownership, and management depend on the type of organization that is ultimately desired and how quickly the family wants to complete the transfer. A *testing stage* of one to five years is recommended, during which the entering operator is employed and receives a salary and bonus or incentive and begins to own or rent some additional assets. Forms of business organization and acquisition of assets that would be difficult to liquidate should be avoided until both parties have determined their ultimate goals and know if their work and management styles are compatible. Following the testing stage, at least three alternative types of arrangements are possible, as illustrated in Figure 14-3.

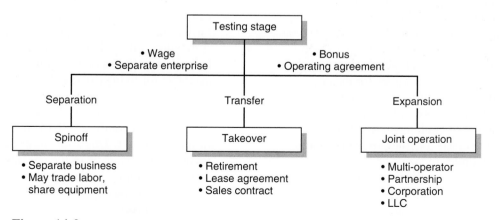

Figure 14-3 **Alternatives for farm business transfer.**

Box 14-4 **Death and Taxes**

When a person dies, ownership of real property and other assets can be passed to heirs by a will, deeds, insurance, trusts, gifts, and contracts. However, assets transferred to heirs may be subject to federal, and, in some cases, state taxes. Estate taxes are excise taxes on the transfer of real estate and other assets to heirs. Inheritance taxes are those placed on the privilege of receiving property.

CURRENT LAWS

Graduated federal estate tax must be paid on estates exceeding a legislated threshold value. Legislation passed in 2013 increased the value of estates exempt from the tax. For persons dying in 2014, the first $5,340,000 of assets can be passed on to their heirs with no federal estate tax. This threshold is increased annually based on the current rate of inflation. The Tax Policy Center estimates that fewer than 1 percent of farm estates are affected by federal estate tax obligations.

Gifting of assets may reduce the amount of federal estate tax exemption to which the giver is entitled. In 2013, $14,000 of property could be given in any year to any individual without affecting the giver's estate tax exemption. Spousal gifts are separate, which essentially doubles the amount that can be given by a husband and wife to an individual without affecting their estate tax exemption. In addition, up to $5,000,000 in gifts over a lifetime is exempt from the gift deduction. Unlimited gifts are allowed between spouses.

SPECIAL USE VALUATION FOR FARMS AND SMALL BUSINESS

Assets are generally valued at their fair market value for estate tax purposes. To lower the assessed value of the property and hence the potential taxes owed, a *special use valuation* may be available to farms and small businesses. Agricultural assets and land may be valued at their use in agricultural production instead of at fair market value, which could be influenced by non-agricultural uses. Special IRS requirements must be met to qualify for this tax treatment.

Publication 950 from the IRS, "Introduction to Estate and Gift Taxes," provides an overview of current laws. However, estate tax law is periodically revised. Readers should always check for the most up-to-date laws and limits.

---

1. The *spin-off* option involves the separation of the operators into their own farming operations. A spin-off works best when both operators wish to be financially and operationally independent, or when expansion possibilities for the original business are limited. The separate operations may still trade labor and the use of equipment, or even own some assets together, to realize some economies of size.

2. The *takeover* option occurs when the older generation phases out of active labor and management, usually to retire or enter another occupation. Expansion of the business may not be necessary. Transfer can take place by renting or selling the farm or ranch to the younger generation. A detailed budget should be developed to make sure that enough income will be available for living expenses after lease or mortgage payments are made.

3. A *joint operation* may be developed when both generations wish to continue farming together. This arrangement often involves an expansion of the business to adequately employ and support everyone. The farm may be organized with a joint operating agreement, or as a partnership, LLC, or corporation. These multi-operator

family farming operations include some of the most profitable and efficient farm businesses to be found. However, it is crucial that effective personal working and management relationships are developed.

The testing stage may also end in a decision by the younger generation to not enter farming as a career. This should not be considered a failure but a realistic assessment of a person's values and goals in life.

SUMMARY

This chapter relates to Part VI, which discusses acquiring and managing the resources necessary to implement a farm or ranch production plan. Before the resources to implement the plan are acquired, a form of business organization must be selected. The selection should be reviewed from time to time as the business grows and moves through the stages of the life cycle.

A farm or ranch business can be organized as a sole proprietorship, a partnership, a corporation, a limited liability company, or a cooperative. There are advantages and disadvantages to each form of business organization, depending on the size of the business, stage of its life cycle, and the desires of the owners. The right business organization can make transferring farm income, ownership, and management to the next generation easier. A testing stage in which members of the younger generation work as employees can lead to a separate spin-off, a takeover of the existing business, or a joint operation.

QUESTIONS FOR REVIEW AND FURTHER THOUGHT

1. What are the differences among the four stages in the life cycle of a farm business? Think about a farm or ranch with which you are familiar. In which stage is it?
2. Why do you think the sole proprietorship is the most common form of farm business organization?
3. What general advantages does a joint venture have over a sole proprietorship? Disadvantages?
4. How does an operating agreement differ from a partnership?
5. For the joint operating agreement example shown in Table 14-1, how would you divide gross income if Party A and Party B each owned one-half of the livestock?
6. Explain the importance of putting a partnership agreement in writing. What should be included in a partnership agreement?
7. Explain the difference between a general partnership and a limited partnership.
8. Does a two-person partnership have to be 50/50? Can it be a 30/70 or a 70/30 partnership? How should the division of income be determined?
9. Explain the differences between C and S corporations. When would each be advantageous?
10. Why might a partnership or corporation keep more and better records than a sole proprietorship?
11. Compare the personal and corporate tax rates shown in Table 14-2. Would a sole proprietor (joint return) or a C corporation pay more income taxes on a taxable income of $25,000? On $50,000? On $150,000?
12. What special characteristics do limited liability companies and cooperatives have?
13. What form of business organization would you choose if you were beginning a small farming operation on your own? What advantages would this form have for you? What disadvantages?
14. What form of business organization might be preferable if you had just graduated from college and were joining your parents or another established operator in an existing farm? What advantages and disadvantages would there be to you? To your parents?

Design Pics/Carson Ganci

MANAGING RISK AND UNCERTAINTY

CHAPTER OUTLINE

Sources of Risk and Uncertainty
Risk-Bearing Ability and Attitude
Expectations and Variability
Decision Making Under Risk
Tools for Managing Risk
Summary
Questions for Review and Further Thought

CHAPTER OBJECTIVES

1. Identify the sources of risk and uncertainty that affect farmers and ranchers

2. Show how risk and uncertainty affect decision making

3. Discuss how personal values and financial stability affect risk bearing

4. Illustrate several ways to measure the degree of risk attached to alternative management actions

5. Demonstrate several methods that can be used to help make decisions under risky conditions

6. Discuss tools that can be used to reduce risk or control its effects

Decision making was discussed in Chapter 2 as a principal activity of management, and Chapters 7 through 13 introduced some useful principles and techniques for making good management decisions. These earlier chapters implicitly assumed that all the necessary information about input prices, output prices, yields, and other technical data was available, accurate, and known with certainty. This assumption of perfect knowledge simplifies the understanding of a new principle or concept, but it seldom holds true in the real world of agriculture.

We live in a world of uncertainty. There is an old saying, "Nothing is certain except death and taxes." Many agricultural decisions have outcomes that take place months or years after the initial decision is made. Managers find that their best decisions often turn out to be less than perfect because of changes that take place between the time the decision is made and the time the outcome of that decision is finalized.

Crop farmers must make decisions about what crops to plant and what seeding rates, fertilizer levels, and other input levels to use early in the cropping season. The ultimate yields and prices obtained will not be known with certainty until several months later or even several years later in the case of perennial crops. A rancher who has decided to expand a beef cow herd by raising replacement heifers must wait several years before the first income is received from the calves of the heifers kept for the herd expansion. Unfortunately, farmers and ranchers can do little to speed up the biological processes in crop and livestock production or to make them more predictable.

When an outcome is more favorable than expected, a manager may regret not having implemented the decision more aggressively or on a larger scale. However, in this case, the financial health of the operation has been enhanced, not threatened. The real risk comes from unexpected outcomes with adverse results, such as low prices, drought, or disease. Risk management is mostly concerned with reducing the possibility of unfavorable outcomes or at least softening their effects.

SOURCES OF RISK AND UNCERTAINTY

Risk is a term used to describe situations in which the possible outcomes and the chances of each one occurring are known. *Uncertainty*, on the other hand, characterizes a situation where both the possible outcomes and their probabilities of occurring are unknown.

The sources of risk and uncertainty in agriculture are many. What are the risks associated with selecting enterprises, determining the proper levels of feed and fertilizer to use, hiring a new employee, or borrowing money? What makes the outcomes of these decisions difficult to predict? The sources of risk in agriculture are diverse, but they can be summarized into five broad management areas: production and technical, price and market, financial, legal, and personal.

Production and Technical Risk

Manufacturing firms know that the use of a certain collection of inputs will almost always result in the same quantity and quality of output, with only minor deviations. This is not the case with most agricultural production processes. Crop and livestock performances depend on biological processes affected by weather, diseases, insects, weeds, metabolism, and genetics. These factors cannot be predicted with certainty.

The relative importance of various causes of insured crop losses is shown in Figure 15-1. Nearly all of them are related to weather. Cropping programs must be chosen, fertilizer applied, and money borrowed, all before the weather and its effects on production are known. Horticultural crops in particular are susceptible to unexpected freezes and frost and a whole array of pests and diseases. Livestock producers also face important production risks. Cold, wet weather in the spring or dry weather in the summer can be devastating to range production of sheep and cattle. Disease outbreaks may force a producer to liquidate an entire flock or herd.

Another source of production risk is new technology. There is always some risk involved when changing from tested and reliable production methods to something new. Will the new technology perform as expected? Has it been thoroughly tested? What if it costs more? These and other questions must be considered before a

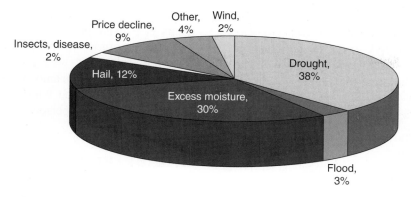

Figure 15-1 **Causes of insured crop losses (2001–2010).**
Source: Risk Management Agency, U.S. Department of Agriculture.

new technology is adopted. However, not adopting a successful new technology means the operator may miss out on additional profits and become less competitive. The risk associated with adopting new technology is captured in an old saying, "Be not the first upon whom the new is tried nor the last to lay the old aside."

New crops may be touted as having high profit potential. But their reaction to dry weather or insects may be untested, and markets may be unreliable.

Price and Market Risk

Price variability is another major uncertainty in agriculture. Commodity prices vary from year to year, as well as day to day, for reasons beyond the control of an individual producer. The supply of a commodity is affected by farmers' production decisions, the weather, and government policies. Demand for a commodity is a result of consumer preferences and incomes, exchange rates, export policies, the strength of the general economy, and the price of competing products. Some price movements follow seasonal or cyclical trends, which can be predicted, but even these movements exhibit a great deal of volatility.

Sometimes market access is a source of risk. A processor or packer may go out of business and leave no viable marketing channel. Or buyers may impose quality standards or quantity restrictions that a producer cannot easily meet.

Costs represent another source of price risk. Input prices tend to be less variable than output prices but still add to uncertainty. Several times in recent decades, shortages of petroleum have caused sudden increases in energy costs, which carried through to fuel, fertilizer, and pesticide prices. Likewise, livestock producers who purchase feeder animals and/or feed are especially susceptible to volatile input prices.

Financial Risk

Financial risk is incurred when money is borrowed to finance the operation of the business. This risk is caused by uncertainty about future interest rates, a lender's willingness to continue lending at the levels needed now and in the future, changes in market values of loan collateral, and the ability of the business to generate the cash flows necessary for debt payments. In Chapter 19, the financial risk created by using borrowed capital is discussed in detail.

Production, marketing, and financial risk exist on most farms and ranches and are interrelated. The ability to repay debt depends on production levels and the prices received for the production. Financing the production and storage of commodities depends on the ability to borrow the necessary capital. Therefore, all three types of risk need to be considered together, particularly when developing a whole-farm risk management plan.

Legal Risk

As farmers and nonfarmers come into closer contact in rural areas, more regulation of agricultural production can be expected. Moreover, increased awareness of food safety is affecting how products are being grown and processed. Livestock producers need to be aware of withdrawal periods for antibiotics, as well as rules about locating production sites and handling manure. Violations can bring expensive fines and lawsuits. Losses also occur when milk must be thrown out due to high levels of residue, or when livestock carcasses are condemned.

Farmers can also be subject to legal actions or liability suits for accidents caused by machinery or livestock or for violating laws regarding health and safety or treatment of hired workers. Ignorance of the law is not an acceptable excuse for not following it, so good managers need to be informed about current rules and regulations.

Personal Risk

No matter how much capital is invested in land, livestock, or machinery, the most irreplaceable assets on a ranch or farm are the manager and key employees. The risk that one of them could suffer a sudden injury or illness is real—farming is traditionally a hazardous occupation. Some health problems appear only after prolonged exposure to dust, odors, or sun.

Key employees also can be lost due to retirement, a career change, or relocation. If no one else is informed about or skilled in the employee's area of responsibility, significant production losses can occur before a replacement is hired.

Finally, family disputes or divorce settlements can divert property, financial assets, or cash from a farm business and bring about economic losses, as well as personal stress. Mental health problems among farmers often go untreated and unreported, but can eventually lead to severe financial and personal losses.

RISK-BEARING ABILITY AND ATTITUDE

Ranchers and farmers vary greatly in their willingness to take risks and in their abilities to survive any unfavorable outcomes of risky actions. Therefore, the level of risk that a farm business should accept is an individual decision. Certainly, good risk management does not mean eliminating all risk. Rather, it means limiting risk to a level that the operators are willing and able to bear.

Ability to Bear Risk

Financial reserves play a big part in determining an operation's risk-bearing ability. Farms with a large amount of equity capital can stand larger losses before they become insolvent. Highly leveraged farms, with a high value of debt relative to assets, can quickly lose equity, because their volume of production is high relative to their capital. These farms are also more vulnerable to financial risks such as rising interest rates.

Cash flow commitments also affect risk-bearing ability. Families with high fixed living expenses, educational expenses, or health care costs are less able to withstand a low-income year and should not expose themselves to as much risk as other operations. Farmers who have more of their assets in a liquid form, such as a savings account or marketable grain and livestock; have secure off-farm employment; or can depend on relatives or friends to assist them in a financial emergency also have greater risk-bearing ability.

Willingness to Bear Risk

Some farmers and ranchers refuse to take risks even though they have no debt and a strong cash flow. They may have experienced financial setbacks in the past or be concerned about having enough income for retirement. Most farm and ranch operators are risk avoiders. They are willing to take some risks, but only if they can reasonably expect to increase their long-run returns by doing so. Age, equity, financial commitments, past financial experiences, the size of the gains or losses involved, family responsibilities, familiarity with the risky proposition, emotional health, cultural values, and community attitudes are all factors that influence the amount of risk producers are willing to take.

EXPECTATIONS AND VARIABILITY

The existence of risk adds complexity to many decisions. When managers are uncertain about the future, they often use some type of average or *expected* values for yields, costs, or prices. There is no assurance that this value will be the actual outcome each time, but decisions must be made based on the best information available. To analyze risky decisions, a manager needs to understand how to form expectations, how to use probabilities, and how to analyze the whole distribution of potential outcomes.

Probabilities are useful when forming expectations. The true probabilities for various outcomes are seldom known, but subjective probabilities can be derived from whatever information is available plus the experience and judgment of the individual. The probability of rain in a weather forecast or odds of a futures contract exceeding a certain price are examples of subjective probabilities. Each individual has had different experiences and may interpret the available information differently, so subjective probabilities will vary from person to person. This is one reason why different individuals make different decisions when they are faced with the same risky alternatives.

Forming Expectations

Several methods can be used to form expectations about future events. Once a *best estimate* is chosen, it can be used for planning and decision making until more information allows a better estimate to be made.

Most Likely

One way to form an expectation is to choose the value most likely to occur. This procedure requires a knowledge of the probabilities associated with each possible outcome, either actual or subjective. This can be based on past occurrences or on analysis of current conditions. The outcome with the highest probability is selected as the most likely to occur. An example is contained in Table 15-1, where six possible ranges of wheat yields are shown along with the estimated probabilities of the actual yield falling into each one. Using the *most likely* method to form an expectation, a yield of 29 to 35 bushels per acre would be selected. For budgeting purposes, we could use the midpoint of the range, or 32 bushels. There is no assurance the actual yield will be between 29 and 35 bushels per acre in any given year, but if the probabilities are correct, it will occur about 35 percent of the time over the long run. The *most likely* method is

TABLE 15-1	Using Probabilities to Form Expectations

Possible wheat yields (bushels/acre)	Number of years actual yield was in this range	Probability (%)
0–14	1	5
15–21	2	10
22–28	5	25
29–35	7	35
36–42	4	20
43–51	1	5
Total	20	100

especially useful when there is only a small number of possible outcomes to consider.

Averages

Two types of averages can be used to form an expectation. A *simple average*, or *mean*, can be calculated from a series of past results. The primary problem is selecting the length of the data series to use in calculating the simple average. Should the average be for the past 3, 5, or 10 years? As long as the fundamental conditions that affect the observed outcomes have not changed, as many observations should be used as are available.

In some cases, the fundamental conditions have changed. New technology may have increased potential crop yields, and long-term changes in supply and demand may have affected market prices. In these cases, a method that gives recent values more importance than the older ones can be used to calculate a *weighted average*. A weighted average can also be used when true or subjective probabilities of the expected outcomes are available, but are not all equal. A weighted average that uses probabilities as weights is also called an *expected value*. The expected value is an estimate of what the average outcome would be if the event were repeated many times. The accuracy of the expected value depends on the accuracy of the probabilities used, though.

Table 15-2 shows an example of using both simple and weighted average methods. Price information from five previous years is used to predict the average selling price for beef cattle during the coming year. One projection is the simple average of the past five years, or $99.32. Alternatively, the most recent values can be assigned greater weights than the observations that occurred longer ago, as shown in the last column of Table 15-2. The assigned weights should always add to 1.00. Each price is multiplied by its assigned weight, and the results are summed. The expected price is $104.25 using the weighted average method. This method assumes that recent prices (which have been higher) more accurately reflect current supply and demand conditions, while the simple average treats each year's result with equal importance.

Expert Opinions

Many types of newsletters, weather reports, and electronically delivered predictions are available to help producers project supply and demand conditions. Some even offer probabilities that certain outcomes will occur. Professional forecasters usually have access to more information and more sophisticated analysis tools than individual producers.

Table 15-2	**Using Averages to Form an Expected Value for the Price of Beef Cattle**		

| | | Weighted average | |
Year	Average annual price	Weight	Price × weight
5 years ago	$ 89.10	0.10	$ 8.91
4 years ago	80.30	0.15	12.05
3 years ago	92.20	0.20	18.44
2 years ago	113.00	0.25	28.25
Last year	122.00	0.30	36.60
Summation	496.60	1.00	$104.25

Simple average: $\dfrac{\$496.60}{5} = \$99.32$ Weighted average = $104.25

However, their recommendations may not fit a particular farm's production situation or risk-bearing ability.

Futures Markets and Forward Contracts

Many agricultural commodities are bought and sold for future delivery at a central location called a *futures market*. Several contracts for each product are traded, each with a different delivery date. The futures prices represent the approximate prices that the people buying and selling contracts collectively think will exist at a future date. Later, the role that futures markets play in helping producers reduce price risk will be explained.

Local buyers may also offer prices for future delivery of crops and livestock, through *forward contracts*. Forward contract prices also represent the buyers' best estimates of what prices will be in the future, but are adjusted for local demand and transportation costs.

Variability

A manager who must select from two or more alternatives should consider another factor in addition to the expected values. The *variability* of the possible outcomes around the expected value is also important. For example, if two alternatives have the same expected value, most managers will choose the one whose potential outcomes have the least variability, because there will be smaller deviations with which to deal.

Range

Variability can be measured in several ways. One simple measure of variability is the difference between the lowest and highest possible outcomes, or the *range*. Alternatives with a smaller range are usually preferred over those with a wider range if their expected values are the same. Range is not the best measure of variability, however, because it does not consider the probabilities associated with the highest and lowest values, nor the other outcomes within the range and their probabilities.

Standard Deviation

A common statistical measure of variability is the *standard deviation*.[1] It can be estimated from a sample of past actual outcomes for a particular event, such as historical price data for a certain week of the year. A larger standard deviation indicates a greater variability of possible outcomes and, therefore, a greater likelihood that the actual outcome will differ from the expected value.

Coefficient of Variation

The standard deviation is difficult to interpret, however, when comparing two types of occurrences that have different means. The occurrence with the higher mean value often has a larger standard deviation, but is not necessarily more risky. In this situation, it is more useful to look at the relative variability. The *coefficient of variation* measures variability relative to the mean and is found by dividing the standard deviation by the mean. Smaller coefficients of variation indicate that the distribution has less variability compared to its mean than other distributions.

$$\text{Coefficient of variation} = \frac{\text{standard deviation}}{\text{mean}}$$

Table 15-3 shows historical data for corn and soybean yields on an individual farm. If we assume that production potential has not changed over time, we can use the simple means as estimates of the expected yields for next year and the variations from the mean to calculate the standard deviations. Corn had a greater standard deviation than soybeans. However, calculating the coefficients of variation shows that soybean yields were more variable than corn yields when compared to the mean. Thus, an operator who

---

[1]Standard deviation is equal to the square root of the variance. The equation for variance is

$$\text{Variance} = \frac{\sum_{i=1}^{n}\left(X_i - \bar{X}\right)^2}{n-1}$$

where X_i is each of the observed values, $\bar{X}$ is the mean of the observed values, and n is the number of observations.

TABLE 15-3	**Historical Corn and Soybean Yields for an Individual Farm**

Year	Corn (bushels/acre)	Soybeans (bushels/acre)
1	165	45
2	185	55
3	181	48
4	128	38
5	145	43
6	169	54
7	158	50
8	115	31
9	172	47
10	167	58
Mean (expected value)	158.5	46.9
Standard deviation	22.7	8.2
Coefficient of variation	0.14	0.17

wished to reduce yield risk might prefer corn to soybeans.

Cumulative Distribution Function

Many risky events in agriculture have an almost unlimited number of possible outcomes, and the probability of any one of them occurring becomes very small. A useful format for portraying a large number of possible outcomes is a *cumulative distribution function (CDF)*. The CDF is a graph of the values for all possible outcomes for an event against the probability that the actual outcome will be equal to or less than each value. The outcome with the smallest possible value has a cumulative probability of nearly zero, while the largest possible value has a cumulative probability of 100 percent.

The steps in creating a CDF are as follows:

1. List a set of possible values for the outcome of an event or strategy, and estimate their probabilities. For example, the data from Table 15-3 can be used as a

set of possible values for corn and soybean yields. If it is assumed that each of the 10 historical observations has an equal chance of occurring again, each one represents 10 percent of the total possible outcomes or distribution.
2. List the possible values in order from lowest to highest, as shown in Table 15-4.
3. Assign a *cumulative probability* to the lowest value equal to one-half of the range it represents. Each observation represents one segment or range out of the total distribution, so it can be assumed that the observation falls in the middle of the range. For the example, the lowest yield observed represents the first 10 percent of the distribution, so it can be assigned a cumulative probability of 5 percent.
4. Calculate the cumulative probabilities (probability of obtaining that value or a smaller one) for each of the other values by adding the probabilities represented by all the smaller values to that value's own probability. In the example, the remaining

TABLE 15-4	**Cumulative Probability Distributions for Corn and Soybean Yields**

Corn (bushels/acre)	Soybeans (bushels/acre)	Cumulative probability (%)
115	31	5
128	38	15
145	43	25
158	45	35
165	47	45
167	48	55
169	50	65
172	54	75
181	55	85
185	58	95

observed yields would have cumulative probabilities of 15 percent, 25 percent, and so on.

5. Graph each pair of values and connect the points, as shown in Figure 15-2.

The cumulative distribution function permits a view of all possible results for a certain event. The more vertical the graph, the less variability among the possible outcomes. The upper portions of the graphs in Figure 15-2 are steeper than the lower portions, indicating that the positive yield responses to good weather are not as significant as the negative responses to poor growing conditions.

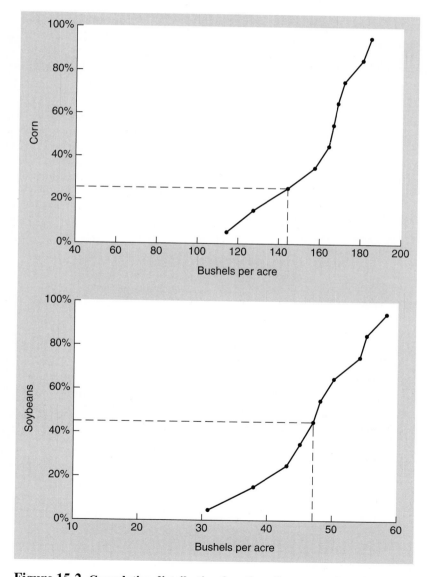

Figure 15-2 Cumulative distribution functions for corn and soybean yields.

DECISION MAKING UNDER RISK

Making risky decisions requires careful consideration of the various strategies available and the possible outcomes of each. The process can be broken down into several steps:

1. Identify an *event* that could be a possible source of risk.
2. Identify the possible *outcomes* that can occur from the event, such as various weather conditions or prices, and their probabilities.
3. List the alternative *strategies* available.
4. Quantify the consequences or *results* of each possible outcome for each strategy.
5. Estimate the risk and expected returns for each strategy, and evaluate the trade-offs among them.

Box 15-1   Yield Risk and Trends in Yields

When yield of a crop is increasing over time, variability of the yield can be overestimated if the general upward trend is not recognized. Statistical techniques can allow producers to distinguish between a predictable yield trend and random variability.

Consider the case of a peanut producer, who has 19 years of yield data for this crop. Using the actual yield data, the producer finds a standard deviation of 591 pounds.

The following graph shows actual yields as dots.

Using statistical techniques, the solid line was fit to the data to distinguish the upward trend in yields from random year-to-year changes. The difference between the actual yield and the expected yield, the corresponding point on the trend line, is a better estimate of variability for this enterprise. Once the predicted portion of the yield change is accounted for, the average deviation from the trend line is found to be only 395 pounds.

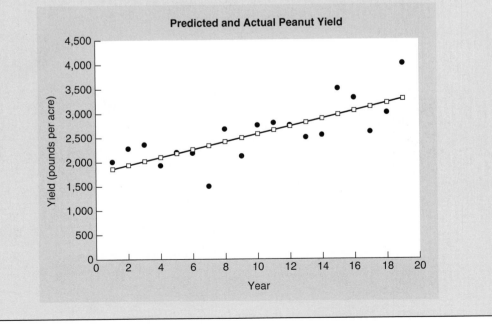

Predicted and Actual Peanut Yield

An example can be used to illustrate these steps. Assume that a wheat farmer plants a given number of acres of wheat in the fall. Stocker steers can be purchased in the fall and grazed on the wheat over the winter and sold at a known contract price in the spring. The farmer's main source of risk (*step 1*) is the weather event, because it affects how much grazing will be available. Assume that there are three possible outcomes for this event— good, average, or poor weather (*step 2*)—and their probabilities are estimated to be 20, 50, and 30 percent, respectively. The selection of probabilities is important. They can be estimated by studying past weather events as well as near-term forecasts.

If too few steers are purchased and the weather is *good*, excess grazing will be available, and an opportunity for additional profit will be lost. If too many steers are purchased and the weather is *poor*, there will not be enough

grazing, extra feed will have to be purchased, and profit will be reduced or a loss incurred. The third possibility is for *average* weather to occur with a typical supply of feed available.

The farmer is considering three alternative actions: purchase 300, 400, or 500 steers (*step 3*). These choices are the *decision strategies*. The same three weather outcomes are possible for each strategy, which creates nine potential combinations of results to consider. Once the elements of the problem are defined, it is helpful to organize the information in some way. Two ways of doing this are with a decision tree or with a payoff matrix.

Decision Tree

A *decision tree* is a diagram that traces several possible management strategies, the potential outcomes from an event, and their results. Figure 15-3 is a decision tree for the previous example. It

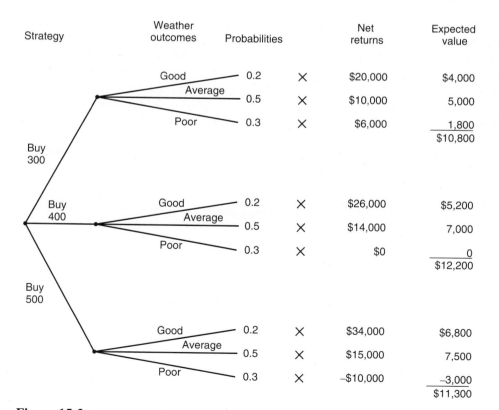

Figure 15-3 Decision tree for stocker steer example.

shows three potential weather outcomes for each of three strategies, the probability for each outcome (which is the same regardless of the strategy chosen), and the estimated net returns for each of the nine potential consequences. For example, if 300 steers are purchased, the net return is $20,000 with good weather, $10,000 with average weather, and only $6,000 with poor weather (*step 4*).

The expected value for each strategy is the weighted average of the possible outcomes, calculated by multiplying each outcome by its probability and summing the results. Based on these values alone, it might be expected that the farmer would select the *Buy 400* strategy, as it has the highest expected value, $12,200. However, this overlooks the possibility of experiencing poor weather and only breaking even. Ways of making a decision taking this risk into account will be shown later (*step 5*).

Payoff Matrix

A *payoff matrix* contains the same information as a decision tree but is organized in the form of a contingency table. The top part of Table 15-5 shows the consequences of each strategy for each of the potential weather outcomes for the stocker steer example. The expected values, as well as the minimum and maximum values and range of outcomes, are shown in the bottom part of the payoff

matrix. Both decision trees and payoff matrices can be used to organize the consequences of one or more related events. However, if each event has many possible outcomes, the diagram or table can become quite large.

Decision Rules

Several decision rules have been developed to help choose the appropriate strategy when faced with a decision involving risk. Using different rules may result in the selection of different strategies. The appropriate rule to use will depend on the decision maker's attitude toward risk, the financial condition of the business, cash flow requirements, and other individual factors.

These factors vary widely among decision makers, so it is impossible to say that a certain rule is best for everyone.

Most Likely Outcome

This decision rule identifies the outcome most likely to occur (has the highest probability) and chooses the strategy with the best consequence for that outcome. Table 15-5 indicates that average weather has the highest probability, at 0.5, and the *Buy 500* strategy has the greatest net return for that outcome, $15,000. This decision rule is easy to apply, but does not consider the variability of the consequences nor the probabilities of other possible outcomes.

TABLE 15-5	**Payoff Matrix for Stocker Steer Problem**			
		Net return for each purchase strategy		
Weather outcomes	**Probability**	**Buy 300**	**Buy 400**	**Buy 500**
Good	0.2	$20,000	$26,000	$34,000
Average	0.5	10,000	14,000	15,000
Poor	0.3	6,000	0	−10,000
Expected value		10,800	12,200	11,300
Minimum value		6,000	0	−10,000
Maximum value		20,000	26,000	34,000
Range		14,000	26,000	44,000

Maximum Expected Value

This decision rule says to select the strategy with the highest expected value. The expected value represents the weighted average result for a particular strategy based on the estimated probabilities of each possible result occurring.

Both Figure 15-3 and Table 15-5 show that the *Buy 400* strategy has the highest expected value, $12,200, so it would be selected using this rule. This rule will result in the highest average net return over time, but it ignores the variability of the outcomes. In the example, poor weather has only a 30 percent chance of occurring. However, there is no guarantee that it will not occur two or three years in a row, with the resulting $0 net return each year. The rule of choosing the maximum expected value disregards variability and should be used only by managers who have good risk-bearing ability and are not strong risk avoiders.

Risk and Returns Comparison

Managers who do not have high risk-bearing ability must consider the risk associated with various strategies, as well as the expected returns. Any strategy that has both a lower expected return and higher risk than another strategy should be rejected. Such is the case with the Buy 500 strategy in Table 15-5. It has a lower expected value than the Buy 400 strategy, $11,300, and also more risk, because it is the only alternative that could incur a loss. The *Buy 300* strategy also has a lower expected return than the Buy 400 strategy but is less risky. Therefore, managers who are risk conscious might prefer this strategy.

Safety First

The safety first rule concentrates on the worst possible outcome for each strategy and ignores the other possible outcomes. The decision maker assumes that better-than-expected outcomes pose no serious problems, whereas the unfavorable outcomes are of real concern. Therefore, the strategy with the best possible result among the worst-case outcomes is selected, the one with the *least bad* value. Referring to Table 15-5, application of the safety first rule would result in selection of the Buy 300 strategy, because its minimum consequence of a $6,000 profit is higher than the other minimums, a $0 profit for buying 400 steers and a loss of $10,000 for buying 500 steers. The safety first rule is appropriate for a business that is in such poor financial condition that it could not survive the consequences of even one bad year.

Break-Even Probability

Knowing the probability that each strategy will result in a financial loss can also help the decision maker choose among them. Suppose that the farmer with the corn and soybean yield histories shown in Table 15-3 calculated that a yield of 145 bushels per acre for corn or 47 bushels per acre for soybeans was needed to just pay all production costs. From the CDF graphs in Figure 15-2, the probability of realizing less than the break-even level of production can be estimated by drawing a vertical line from the break-even yield (x-axis) to the CDF line, then a horizontal line to the cumulative probability scale (y-axis). In this example, the probability of suffering an economic loss (producing a below break-even yield) is approximately 25 percent for corn and 45 percent for soybeans. Thus, soybeans carry more financial risk. However, the risk of a loss also must be weighed against the expected average return from each crop.

TOOLS FOR MANAGING RISK

Fortunately, farmers and ranchers have a variety of management tools available to them for softening the consequences of taking risky actions. Some of the tools are used to reduce the amount of risk the manager faces, while others help soften the impact of an undesirable result. They all follow one of four general approaches, however.

1. Reduce the variability of possible outcomes. The probability of a bad result is decreased, but the probability of a good result is often reduced, as well.

2. Set a minimum income or price level, usually for a fixed charge. Most insurance programs operate this way. The cost of the risk reduction is known, and the probability of achieving a better-than-average result is not affected.
3. Maintain flexibility of decision making. Managers do not *lock in* decisions for long periods, in case price or production prospects change.
4. Improve the risk-bearing ability of the business, so that an adverse result is less likely to affect the survival of the farming operation.

Farmers and ranchers use many examples of all four types of risk management tools.

Production Risk Tools

Variable crop yields, uncertain livestock production rates, and uneven product quality are the evidence of production risk. Several strategies can be used to reduce production risk.

Stable Enterprises

Some agricultural enterprises have a history of producing more stable income than others. Modern technology may be able to control the effects of weather on production, or government programs and marketing orders may control prices or the quantity of a commodity that can be sold. For example, irrigation will produce more stable crop yields than dryland farming in areas where rainfall is marginal or highly variable during the growing season. At the other extreme, enterprises such as livestock finishing, where both buying and selling prices can vary, and growing perishable products, such as flowers, fruits, and vegetables, tend to have highly variable incomes.

Diversification

Many farms produce more than one product to avoid having their income depend totally on the production and price of a single commodity. If profit from one product is poor, profit from producing and selling other products may prevent total profit from falling below acceptable levels. In agricultural production, diversifying by producing two or more commodities will reduce income variability if all prices and yields are not low or high at the same time.

Table 15-6 shows an example of how diversification can work to reduce income variability. Based on the average net farm income for a 21-year period for a sample of Kansas farms, beef cow farms had the most variable net income per operator, as measured by the coefficient of variation. The general farms had the lowest variability. Lack of a strong income correlation among enterprises *smooths out* annual income under diversification.

How much will diversification reduce income variability in an actual farm situation?

TABLE 15-6	Comparison of Specialized and Diversified Farms

	Farm type				
	Beef cow	Irrigated crops	Dryland crops	Beef and crop	General farm
Net farm income per operator (average, 1992–2012)	$21,338	$125,481	$70,518	$39,184	$51,426
Standard deviation	$24,557	$123,895	$51,537	$29,944	$32,962
Coefficient of variation	1.15	0.99	0.73	0.76	0.64

Source: Kansas Farm Management Association

The answer depends on the price and yield correlations among the enterprises selected. If prices or yields for all of the enterprises tend to move up and down together, little is gained by diversifying. The less these values tend to move together, or the more they move in opposite directions, the more income variability will be reduced by diversification. Likewise, adding a highly variable enterprise to a stable one may increase whole-farm risk.

Weather is the primary factor influencing crop yields, so crops with the same growing season tend to have a strong positive yield correlation. The yield correlations among crops that have different growing seasons and are susceptible to different insects and diseases will be somewhat lower. Production rates among different types of livestock are less closely correlated, and there is little correlation between crop yields and livestock performance.

Most studies on the price correlations for major agricultural commodities show that pairs of commodities with a strong yield correlation often have a positive price correlation as well, because year-to-year production changes have a major impact on prices. Some specialty crops such as fruits and vegetables, however, may show a weak or even negative price correlation with some of the major field crops, and crop and livestock prices are mostly independent of each other.

Diversification plans can include nonfarm activities as well. Investing in stocks and bonds, carrying out a part-time business unrelated to agriculture, or holding a nonfarm job can improve the stability of family income. Diversification may also require giving up the benefits of specializing in one enterprise to gain the benefits from less variability in net income.

Insurance

There are several types of insurance that will help reduce both production and financial risk. Formal insurance can be obtained from an insurance company to cover events that could threaten business equity and survival. An alternative is for the business to be self-insured, to maintain some type of readily available or liquid financial reserves in case a loss does occur. Without these financial reserves, a crop failure, major windstorm, or fire might cause such a large financial setback that the business could not continue.

Financial reserves held in a form that can be easily liquidated, such as a savings account, often earn a lower rate of return than the same capital would if it were invested in the farm business or some other long-term investment. This sacrifice in earnings is the opportunity cost of being self-insured and should be compared to the premium cost for an insurance policy that would provide the same protection.

Farmers and ranchers use several common types of insurance against production risks:

1. *Property insurance*: It protects against the loss of buildings, machinery, livestock, and stored grain from fire, lightning, windstorm, theft, and other perils. Property insurance is relatively inexpensive, while the loss from a serious fire or windstorm can be devastating. Therefore, most farmers and ranchers choose to carry at least a minimum level of property insurance on their more valuable assets.

2. *Multiple peril crop insurance (MPCI)*: This is an insurance program backed by the U.S. Department of Agriculture, with policies sold through private insurers. Production guarantees can be purchased for a portion of the *proven yield* for the insured farm, up to 85 percent in some cases. When actual yields fall short of the guarantee, the producer is paid a fixed price for each bushel of loss. Examples of covered losses would be those due to drought, floods, hail, an early freeze, and insect damage. Added coverage for specific perils such as hail and fire is also available. A number of private companies provide this type of coverage, and the cost will depend on the amount of coverage desired and the past frequency and intensity of hailstorms in the area.

3. *Revenue insurance*: This is another type of MPCI insurance policy that allows crop producers to purchase a guarantee for a certain level of gross income per acre. If the producer's actual yield multiplied by the actual market price at harvest is below the guarantee, the insurance company pays the farmer the difference. Thus, the producer can count on at least a minimum amount of gross income per acre. Some crop revenue insurance policies increase the level of the revenue guarantee if market prices rise from the time the insurance is purchased until harvest. This feature is especially useful for producers who forward price much of their crop before harvest or who need a guaranteed supply of feed for their livestock.

The concept of revenue insurance has been extended to beef cattle, sheep, dairy, and hogs as well. However, only protection against lower-than-expected prices is included—production risk is not addressed. Livestock Risk Protection (LRP) allows a livestock feeder to purchase insurance for a minimum selling price, for any quantity of sales up to a maximum level. Livestock Gross Margin (LGM) insurance is similar to LRP, but allows producers to guarantee a minimum margin between their selling price and the cost of major feedstuffs.

Another program called Adjusted Gross Revenue insurance allows a farmer to purchase a guaranteed minimum gross income for the entire farm. It is especially useful for crop and livestock commodities not covered by other types of revenue insurance.

Extra Production Capacity

When poor weather delays planting or harvesting of crops, many farmers depend on having excess machinery or labor capacity to help them catch up. They incur higher-than-necessary machinery ownership costs or wages in some years as insurance against crop losses that could occur because of late planting or harvesting in other years. Some operators also prefer to own

newer machinery to reduce the risk of breakdowns at crucial times or unexpected repair bills.

Share Leases

In many states, *crop share leases* and *livestock share leases* are common. The landowner usually pays part of the operating expenses and receives a portion of the crops or livestock produced instead of a cash rental payment. In this way, the risk of poor production, low selling prices, or high input costs is shared between the tenant and the owner. The tenant also needs less operating capital under a share lease than a cash lease. Some tenants use a *variable cash lease* to achieve similar risk reduction. Both types of leases are described in Chapter 20.

Custom Farming and Feeding

Rather than risk uncertain prices and yields, some operators prefer to custom farm cropland. They perform all machinery field operations for a landowner in exchange for a fixed payment. The landowner takes all the price and yield risk.

Custom feeding is a similar arrangement. Livestock producers feed animals or poultry owned by investors in their own facilities for a fixed price per head or per space, or for a fixed rate per day. Contracts for raising or caring for breeding livestock are also available. Although custom livestock production contracts pass on most production risk to the livestock owner, some do contain penalties for excessive death loss. All custom agreements should be analyzed carefully to compare their potential risks and returns to those of being an independent producer and marketer.

Input Procurement

Some livestock feeders depend on a reliable source of feed or feeder livestock to keep their facilities full. A long-term contract with a supplier reduces the risk of having to operate below capacity. The price may be determined by a set formula based on quality factors and current

---

Box 15-2   **Managing Risk with Crop Insurance**

Paul Edmundson raises 1,200 acres of soybeans each year in the Delta. One year, to protect his investment, he purchases multiple peril crop insurance. His proven average yield based on past production records is 40 bushels per acre. He chooses to insure at the 75 percent production level, so he receives a yield guarantee of (40 × 75%) = 30 bushels. He also chooses the maximum price guarantee available that year, $11.50 per bushel. His premium cost for this coverage is $16.00 per acre. Later in the year, a late-summer flood cuts his average yield to only 21 bushels per acre. He receives a loss payment from his insurance company for (30 − 21) = 9 bushels, at $11.50 per bushel, or $103.50 per acre.

The next year, he decides to switch to crop revenue insurance. His insurance company offers him a policy for 75 percent of his projected gross income, based on his proven yield of 40 bushels per acre and that year's insurance price of $12.00 per bushel. His revenue guarantee is (40 × $12.00 × 75%) = $360 per acre.

In the fall, Edmundson's soybeans yield 40 bushels per acre, but the market price drops to $8.25, so his actual gross income is only (40 × $8.25) = $330 per acre. He receives an insurance payment for his revenue deficit equal to ($360 − 330) = $30 per acre.

---

market prices. Other key inputs may be secured by advance contracts, as well, such as labor crews for harvesting fruits and vegetables.

Market Risk Tools

Market risk exists because of the variability of commodity prices and because the manager does not know what future prices will be when making the decision to produce a commodity. Several methods can be used to reduce price variability or to set a satisfactory price in advance of when crops or livestock are ready for sale.

Spreading Sales

Instead of selling all of a crop at one time, many farmers prefer to sell part of it at several times during the year. For example, 25 percent of the crop may be sold every three months or one-sixth every two months. Spreading sales avoids selling all the crop at the lowest price of the year but also precludes selling all of it at the highest price. The result of spreading sales should be an

average price received near the average annual price for the commodity.

Livestock sales can also be spread throughout the year. This can be accomplished by feeding several groups during the year or by calving or farrowing several times per year. Spreading sales of dairy and egg products occurs naturally, due to the continuous nature of their production.

Contract Sales

Producers of crops such as seed, nursery stock, and fruits and vegetables often sign a contract with a buyer or processor before planting the crop. This contract will usually specify certain management practices to be followed, as well as the price to be received for the crop and possibly the amount to be delivered. A contract of this type removes the price risk at planting time and guarantees the producer will have a market. However, quality standards may be strict, creating added production risk.

It is also possible to obtain a *forward price contract* for many field crops and some types of livestock. Many grain and livestock buyers will contract to purchase a given amount of these commodities at a set price for delivery at a later date or at regular intervals. These contracts are often available during the growing season as well as after the crop has been harvested and stored. Contract sales remove price uncertainty but do not allow selling at a higher price if markets rise later in the year. An exception is a *minimum price contract*, which guarantees the seller a certain price, usually slightly below expected levels, but allows the commodity to be sold at the actual market price if it is above the minimum. The contract may impose a penalty if the producer cannot deliver the agreed-on quantity or quality of commodity due to production problems.

Hedging

A market price can be established in advance by *hedging* on a commodity futures market. Hedging is possible before the crop is planted as well as during the growing season or while grain is stored. Livestock can also be hedged at the time of purchase, or at any time during the feeding period.

Hedging involves selling a commodity futures contract instead of the actual commodity, usually because the manager is currently unable or unwilling to deliver the commodity at that time. The contract is purchased by a buyer at a futures market exchange somewhere, who may represent a processor who wants to lock in the price of the commodity for future use or a speculator hoping to sell the contract later for a higher price. Although futures contracts for some commodities allow for delivery when the contract expires, the contract usually is repurchased and the commodity is sold on the local cash market. Cash and futures prices usually move up or down together. Thus, any gain or loss that occurred because the cash market went up or down is offset by a corresponding loss or gain on the futures contract being held.

Before beginning a hedging program, a manager should carefully study the hedging process and the futures market and have a good understanding of *basis*, or the normal difference between the futures contract price and the local cash market price. The basis can become wider or narrower while the futures contract is being held, meaning that gains and losses in the cash and futures markets do not exactly offset each other. This variation is called *basis risk*. Basis risk should be taken into account, but basis is less variable and more predictable than cash prices.

Hedging can also be used to lock in the price of inputs to be purchased in the future, such as feedstuffs or feeder livestock. In this case, a futures contract is purchased (rather than sold) in advance, before it is feasible to take delivery of the input, then resold when the actual input is purchased.

Commodity Options

Many marketers dislike the fact that although forward contracting or hedging protects them against falling prices, it also prevents them from benefiting from rising prices (see the example in Box 15-3). They prefer to use commodity options that set a minimum price in exchange for paying a set fee or premium, but still allow them to sell at a higher price should it occur.

A manager who wants to protect against a price decline will generally buy the right to sell a futures contract at a specified price, called a *put option*. If the market goes down (both cash and futures), the value of the put option goes up and offsets the loss in value of the commodity the farmer is holding. If the market goes up, the value of the put option goes down and may eventually reach zero. If the market rises even further, the value of the commodity being held continues to rise, but there is no further decline in the value of the put (because it is already at zero), and the farmer gains. When it is time to sell the commodity, the put is also sold, if it still has some value. Thus, for the cost of buying the put (the premium), the farmer is protected against falling prices but can still benefit from rising prices. The commodity option provides a type of price insurance.

Call options are similar to put options, but give the purchaser the right to buy a futures contract at a specified price. They can be used to set a maximum purchase price for a commodity, but allow the buyer to benefit if the price goes down. Call options are useful for producers who want to establish a maximum price for inputs such as feeder livestock or livestock feed.

U.S. Department of Agriculture Programs

The U.S. Department of Agriculture (USDA) has a long history of helping farmers manage both price and production risk. Disaster programs have been approved by Congress many times over the years when widespread drought or disease occurred. However, ad hoc disaster programs have largely been replaced by the permanent multiple peril crop insurance programs described earlier. For some specific commodities, additional risk management programs are offered by the USDA. Some of these programs offer supplemental payments to producers when market prices are below certain preset reference levels. Payments are equal to the difference between the actual price and the reference price, times the farm's historical yields and planted acres for that crop. The actual market price is usually an average of prices received during the 12 months following the harvest period, called the *marketing year price*.

Other programs protect against low gross revenue rather than low prices. Revenue guarantees are based on historical national prices and farm-level or county-level yields for eligible crops. Actual revenue is based on the marketing year price and the farm or county yield for each year. The producer receives a payment if the actual revenue is below the guaranteed revenue, similar to that received under a crop revenue insurance policy. Payments are usually limited to a fixed percent of the guarantee, so do not entirely replace crop insurance. Some programs cover specific crops while others apply to the aggregate income received for all crops grown on the farm.

Price protection programs for certain livestock commodities, such as milk and wool, are also available. In addition, disaster assistance is available when livestock producers suffer losses from adverse weather, fire, or other hazards.

The specific details of USDA commodity programs are adjusted every few years, so producers should review information from the Farm Service Agency of the USDA for current provisions. Participation in such programs should be a part of the farm manager's overall risk strategy, but should not be expected to eliminate all chances of financial loss.

Flexibility

Some management strategies allow the operator to change a decision if price trends or weather conditions change. Planting annual crops instead of perennial or permanent crops is one example.

Investing in buildings and equipment that can be used for more than one enterprise is another. Many grain producers build storage bins so they can delay marketing until prices are more favorable. In the case of livestock, animals may be sold as feeder livestock or finished to slaughter weight. Renting certain assets such as land or machinery instead of buying them is another example of maintaining management flexibility.

Financial Risk Tools

Reducing financial risk requires strategies to maintain liquidity and solvency. Liquidity is needed to provide the cash to pay debt obligations and to meet unexpected financial needs in the short run. Solvency is related to long-run business survival or having enough assets to adequately secure the debts of the business.

Fixed Interest Rates

Many lenders offer loans at either fixed or variable interest rates. The fixed interest rate may be higher when the loan is made but prevents the cost of the loan from increasing if interest rates trend upward.

Self-Liquidating Loans

Self-liquidating loans are those that can be repaid from the sale of the loan collateral. Examples are loans for the purchase of feeder livestock and crop production inputs. Personal loans and loans for land or machinery are examples of loans that are not self-liquidating. The advantage of self-liquidating loans is that the source of the cash to be used for repayment is known and relatively dependable.

Liquid Reserves

Holding a reserve of cash or other assets easily converted to cash will help the farm weather the adverse results of a risky strategy. However, there may be an opportunity cost for keeping funds in reserve rather than investing them in the business or other long-term assets.

Credit Reserve

Many farmers do not borrow up to the limit imposed on them by their lender. This unused credit or credit reserve means additional loan funds can be obtained in the event of some unfavorable outcome. This technique does not directly reduce risk but does provide a safety margin. However, it also has an opportunity cost, equal to the additional profit this unused capital might have earned in the business.

Box 15-3	Reducing Price Risk by Hedging

Plainview Feeders, Inc. has just filled one of their feedlots with a new set of cattle that have an expected cost of production as follows:

Average purchase cost of cattle per head	$1,000
Expected feed cost (500 pounds of gain @ $.90)	$ 450
Nonfeed costs (200 days @ $.10)	$ 20
	$1,470

At an average selling weight of 1,150 pounds, their break-even selling price is $128 per hundredweight (cwt.). The finished cattle market is about $133 per cwt. now, and June cattle futures contracts are selling at $137 per cwt. They decide to *hedge* the cattle by selling a futures contract.

By June, when the cattle are ready to sell, the market has trended downward somewhat. Plainview sells the cattle for $128 per cwt. and buys back the futures contract for $132 per cwt. Their net price is

Sold futures contract	+$137 per cwt.
Bought back futures contract	−$132
Sold for cash price	+$128
	= $133 per cwt.

Although the market declined by $5, they still netted $133 due to the $5 gained on the futures contract.

What if the market had trended upward instead? Suppose by June the futures market price was $143 per cwt. and the cash price was $138. Their net price would have been

Sold futures contract	+$137 per cwt.
Bought back futures contract	−$143
Sold for cash price	+$138
	= $132 per cwt.

Their net selling price would have been nearly the same regardless of whether the cattle market went up or down. The only thing affecting it was the basis, or the difference between the futures price and the cash price at the time the cattle were sold, $4 in the first example and $5 in the second one.

Owner Equity

In the final analysis, it is the equity or net worth of the business that provides its solvency and much of its liquidity. Therefore, equity should be increased steadily, particularly during the early years of the business, by retaining profits in the business or attracting outside capital.

Legal Risk Tools

Business Organization

Farms and ranches can be organized under several different legal forms. Some of them, such as corporations, limited liability companies, and cooperatives, offer more protection from legal liabilities and damages than others. Chapter 14 contains more details.

Estate Planning

Having a will and an estate plan that provide for the orderly transition of a ranch or farm business to heirs will often save thousands of dollars in estate and income taxes or revenue lost due to interrupted management and division of an efficiently sized business. Having all farm leases and contracts in writing will also reduce legal problems down the road.

Liability Insurance

Liability insurance protects against lawsuits by third parties for personal injury or property damage for which the insured or employees may be liable. Liability claims on a farm may occur when livestock wander onto a road and cause an accident or when someone is injured on the farm property. The risk of a liability claim may be small, but some of the damages awarded have been very large. Most people find liability insurance an inexpensive method to provide some peace of mind and protection against unexpected events.

Personal Risk Tools

Health Insurance

Self-employed farmers and ranchers often have difficulty obtaining health insurance coverage at a reasonable cost. The after-tax cost has been reduced somewhat by making more of the premiums tax deductible. Raising the levels at which coverage begins and obtaining policies through farm organizations can also reduce costs. Given the high cost of medical treatments and unpredictability of health problems, no prudent manager should be without some type of health insurance. This also applies to workers' compensation insurance for employees.

Life Insurance

Life insurance is available to provide protection against losses that might result from the untimely death of the farm operator or a member of the family. The insurance proceeds can be used to meet family living expenses, pay off existing debts, pay inheritance taxes, and meet other expenses related to transferring management and ownership of the business. Care should be taken to select the most suitable type of life insurance for each individual's needs and interests.

Safety Precautions

Common sense, attention to the job at hand, and not getting in a hurry will go a long way toward avoiding accidents and injuries. Common safety measures include keeping machinery guards in place, shutting off machinery before making repairs or adjustments, not moving equipment on public roads after dark, following prescribed procedures for applying pesticides and fertilizers, and avoiding proximity to dangerous livestock.

Backup Management

When only one person is informed about key aspects of the farm business, a sudden accident or illness could seriously disrupt day-to-day and long-term operations. Key employees, spouses, and attorneys should know the location of tax, financial, and legal records and be able to step in when the main operator is unable to continue.

Which of these many risk management tools are employed by a particular farm or ranch will depend on the type of risks being faced, the financial stability of the business, and the risk-bearing attitude of the manager.

SUMMARY

W e live in a world of uncertainty. Rarely do we know the exact what, when, where, how, and how much of any decision and its possible outcomes. Decisions must still be made, however, using whatever information and techniques are available. No one will make a correct decision every time, but decision making under uncertainty can be improved by knowing how to identify possible events and strategies, estimate the value of possible outcomes, and analyze their variability.

Decision trees, payoff matrices, and cumulative distribution functions can be used to organize the outcomes of different strategies. Several decision rules can be used to choose among risky alternatives. Some consider only expected returns, some take into account the variability of outcomes both above and below the mean, and some look only at adverse results.

Production, marketing, financial, legal, and personal risk can be reduced or controlled using a number of techniques. Some reduce the range of possible outcomes, while others guarantee a minimum result in exchange for a fixed cost, provide more flexibility for making decisions, or increase the risk-bearing ability of the business.

QUESTIONS FOR REVIEW AND FURTHER THOUGHT

1. List at least five sources of risk and uncertainty for farmers in your area. Classify them as production, price, financial, legal, or personal risk. Which are the most important? Why?

2. How might a young farmer with heavy debt view risk, compared with an older, established farmer with little debt?

3. Might these same two farmers have different ideas about the amount of insurance they need? Why?

4. How do subjective probabilities differ from true probabilities? What sources of information are available to help form them?

5. Select five prices that you feel might be the average price for lean market hogs (or some other familiar commodity) next year, being sure to include both the lowest and the highest price you would expect. Ask a classmate to do the same.

   a. Who has the greatest range of expected prices?

   b. Calculate the simple average for each set of prices. How do they compare?

   c. Assign subjective probabilities to each price on your list, remembering that the sum of the probabilities must equal 1.0. Compare them.

   d. Calculate the expected value for each set of prices, and compare them.

   e. Would you expect everyone in the class to have the same set of prices and subjective probabilities? Why or why not?

6. Assume that average annual wheat prices in your region for the past 10 years have been as follows:

Year 1	$3.63	Year 6	$7.39
Year 2	3.51	Year 7	5.26
Year 3	3.70	Year 8	6.54
Year 4	4.48	Year 9	8.38
Year 5	7.16	Year 10	8.27

a. Calculate the simple average or mean. Calculate the coefficient of variation given that the standard deviation is 1.95.

b. Draw a cumulative distribution function (CDF) for the price of wheat using the data in the table.

c. From the CDF, estimate the probability that the annual average price of wheat will be $6.00 or less in a given year.

7. Identify the steps in making a risky decision, and give an example.

8. Suppose that a crop consultant estimates that there is a 20 percent chance of a major insect infestation, which could reduce your per-acre gross margin by $60. If no insect damage occurs, you expect to receive a gross margin of $120 per acre. Treatment for the insect is effective but costs $15 per acre. Show the possible results of treating or not treating in a decision tree and in a payoff matrix. What is the expected gross margin for each choice? What is the range between the high and low outcomes for each choice?

9. Would a manager who did not need to consider risk choose to treat or not treat for insects in question 8? What about a manager who could not afford to earn a gross margin of less than $90 per acre?

10. Give two examples of risk management strategies that fall into each of the following general categories:

a. Reduce the range of possible outcomes.

b. Guarantee a minimum result for a fixed cost.

c. Increase flexibility of decision making.

d. Improve risk-bearing ability.

11. Describe one important risk management tool or strategy that will help cope with each of the following types of risk on a ranch or farm:

a. Production

b. Market

c. Financial

d. Legal

e. Personal

MANAGING INCOME TAXES

CHAPTER OUTLINE

Types of Income Taxes

Objectives of Tax Management

The Tax Year

Tax Accounting Methods

The Tax System and Tax Rates

Some Tax Management Strategies

Depreciation

Capital Gains

Summary

Questions for Review and Further Thought

CHAPTER OBJECTIVES

1. Show the importance of income tax management on farms and ranches

2. Identify some objectives of income tax management

3. Point out the differences between the cash and accrual methods of computing taxable income and the advantages and disadvantages of each

4. Explain how the marginal tax rates and Social Security tax rates are applied to taxable income

5. Review some tax management strategies that can be used by farmers and ranchers

6. Illustrate how depreciation is computed for tax purposes and how it can be used in tax management

7. Show the difference between ordinary and capital gains income and how each is taxed

Federal income taxes have been collected in the United States since 1913. Many states also tax personal and business income. Taxes affect all types of businesses, including farms and ranches. Income taxes are an unavoidable result of operating a profitable business, and are necessary to keep public institutions and services operating. However, management decisions made over time can have a large impact on the timing and amount of income tax due. For this reason, a farm manager needs to have a basic understanding of the tax regulations and to be able to analyze possible tax consequences of management decisions. These decisions should be made throughout the year. Income management is a year-round process, not something to be done only when the tax return is completed.

Farm managers do not need and cannot be expected to have a complete knowledge of all the tax regulations. However, a basic understanding and awareness of the tax topics to be discussed in this chapter will enable a manager to recognize the possible tax consequences of some common business decisions. It will also help a manager identify decisions that may have large and complex tax consequences. This should be the signal to obtain the advice of a tax accountant or attorney experienced in farm tax matters. Expert advice *before* the management decision is implemented can be well worth the cost and can save time, trouble, and money at a later date.

TYPES OF INCOME TAXES

Three types of taxes are imposed on farm income (Figure 16-1). Ordinary income tax is paid on most net farm income after it is combined with other types of taxable income. Some income from the sale of certain assets, called capital gains, is reported separately and often taxed at a lower rate than ordinary income. Finally, income earned through self-employment activities, including farming and ranching, is also subject to a tax to support the Social Security and Medicare systems.

Income tax regulations are numerous, complex, and subject to change as new tax legislation is passed by Congress. Only some of the more basic principles and regulations can be covered in this chapter. They will be discussed from the standpoint of a farm business operated as a sole proprietorship. Farm partnerships and corporations are subject to different regulations in some cases, and anyone contemplating using these forms of business organization should obtain competent tax advice before making the decision. The reader is also advised to check the current tax regulations for the latest rules. Tax rules and regulations can and do change frequently.

Most states also impose a tax on individual income. Some states have progressive rates—that is, higher incomes are taxed at higher rates—while other states impose a flat rate on all levels of income. Due to the wide diversity of state

Figure 16-1   Three types of income taxes.

---

income tax rules, the discussion in the rest of this chapter will apply only to federal income taxes.

OBJECTIVES OF TAX MANAGEMENT

Any manager with a goal of profit maximization needs to modify that goal when considering income taxes. This goal should now become *the maximization of long-run, after-tax profit.* Income tax payments represent a cash outflow for the business, leaving less cash available for other purposes. The cash remaining after paying income taxes is the only cash available for family living expenses, debt payments, and new business investment. Therefore, the goal should be the maximization of after-tax profit, because it is this amount that is finally available to use at the manager's or owner's discretion. The goal is not to minimize income taxes. This goal could be met by having little or no net farm income year after year.

A short-run objective of tax management is to minimize the taxes due on a given year's income at the time the tax return is completed and filed. However, this income is the result of decisions made throughout the year and perhaps in previous years. It is too late to change these decisions. Effective tax management requires a continuous evaluation of how each decision will affect income taxes, not only in the current year but in future years. It should be a long-run rather than a short-run objective, and certainly not something to be thought of only once a year at tax-filing time. Therefore, tax management consists of managing income, expenses, and the purchases and sales of capital assets in a manner that will maximize long-run after-tax profits.

Tax management is not tax evasion but rather avoiding the payment of more taxes than are legally due, as well as postponing the payment of taxes when possible. Several tax management strategies are available that tend to postpone or delay tax payments but not necessarily reduce the amount paid over time. However, any tax payment that can be put off until later represents additional cash available for business use for one or more years, resulting in less need to borrow funds.

THE TAX YEAR

Most farmers and ranchers use a tax accounting year that is the same as the calendar year. However, they can ask permission from the Internal Revenue Service (IRS) to use a *fiscal year* for filing taxes, which starts on some date other than January 1 and ends one year later. Businesses that have a production cycle that regularly begins in one calendar year and ends in the next may prefer to have a fiscal year that ends after most of their production is sold (see the discussion of the accounting period in Chapter 3).

Farmers and ranchers have a slightly different filing schedule for federal income taxes than salaried individuals. If their tax year is the same as the calendar year, they must file their return and pay any tax due by March 1 of the following year, instead of April 15. An exception is granted if they file an estimate of the tax they will owe by January 15 and send a payment for that amount. In those cases the final return and payment are not due until April 15. Taxpayers who use a fiscal year face similar deadlines relative to the beginning of their tax years.

TAX ACCOUNTING METHODS

A unique feature of farm business income taxes is the choice of tax accounting methods allowed. Farmers and ranchers are permitted to report their taxable income using either the *cash* or the *accrual* method of accounting. All other taxpayers engaged in the production and sale of goods where inventories may exist are required to use the accrual method. Even though farmers and ranchers often have inventories, they may elect to use either method. The choice is made when the individual or business files the *first* farm tax return. Whichever method is used on this first tax return must be used in the following years.

It is possible to change tax accounting methods by obtaining permission from the IRS and paying a fee. The request must be filed on a special form and in a timely manner. It must also include the reason for the desired change. If the request is granted, it may add complexity to the accounting procedures during the change-over period. Therefore, it is advisable to study the advantages and disadvantages of each method carefully and make the better choice when the first farm tax return is filed.

The Cash Method

Under the cash accounting method, income is taxable in the year it is received as cash, or *constructively received*. Income is constructively received when it is credited to an account or is available for use before the end of the taxable year. An example of the latter situation would be a check for the sale of grain that the elevator was holding for the customer to pick up. If the check was available on or before December 31 but was not picked up until several days later, it would still be considered income constructively received in December. The check was available in December, and the fact it was not picked up does not defer the taxable income to the following year.

Under cash accounting, expenses are generally tax deductible in the year they are actually paid, regardless of when the item was purchased or used. An example would be a bill for veterinary services rendered in December but not paid until the following January. This would be a tax-deductible expense in the following year and not in the year the service was actually provided and charged. An exception to this rule is the cost of items purchased for resale, which includes feeder livestock. These expenses can be deducted only in the taxable year in which the items are sold. This means that the expense of purchasing feeder cattle or other items purchased for resale needs to be held over into the next year if the purchase and sale do not occur in the same tax year.

Inventories are not used to determine taxable income with the cash method of tax accounting. Income is taxed when received as cash and not as it accumulates in an inventory of crops and livestock.

Advantages

A large majority of farmers and ranchers use the cash method for calculating taxable income. A number of advantages make this the most popular method.

1. *Simplicity*: The use of cash accounting requires fewer records, primarily because there is no need to track inventories for tax purposes.
2. *Flexibility*: Cash accounting provides maximum flexibility for tax planning at

the end of the year. Taxable income for any year can be adjusted by the timing of sales and payment of expenses. A difference of a few days can put the income or expense into this year's or next year's taxable income.

3. *Capital gains from the sale of raised breeding stock*: More of the income from the sale of raised breeding stock will generally qualify as long-term capital gain income under cash accounting. This income is not subject to self-employment taxes and is taxed at a lower rate than ordinary income. Capital gains will be discussed in detail in a later section.

4. *Delaying tax on growing inventory*: A growing business with an increasing inventory can postpone paying tax on the inventory until it is sold and converted into cash.

The Accrual Method

Income under the accrual method of tax accounting is taxable in the year it is earned or produced. Accrual income includes any change, from the beginning to the end of the taxable year, in the value of crops and livestock in inventory as well as any cash income received. Therefore, any inventory increase during the tax year is included in taxable income, and inventory decreases reduce taxable income.

Another difference from cash accounting is that the cost of items purchased for resale, such as feeder livestock, can be deducted in the year of purchase even though the items have not been resold. Their cost is offset by the increase in ending inventory value for the same items.

Advantages

Accrual tax accounting has several advantages over cash accounting, which should be considered when choosing a tax accounting method.

1. *More accurate measure of income*: Taxable income under accrual accounting is calculated much like net farm income

was in Chapter 5. For this reason, it is a more accurate method of computing net farm income than cash accounting and keeps income taxes paid up on a current basis.

2. *Reduces income fluctuations*: The inclusion of inventory changes prevents wide fluctuations in taxable income even if the marketing pattern results in the production from two years being sold in one year.

3. *Declining inventory*: During years when the inventory value is declining, as might happen when a farmer is slowly retiring from the business, tax will be paid later under accrual accounting. The decrease in inventory value offsets some of the cash receipts in the short run.

Tax Record Requirements

Regardless of the accounting method chosen, complete and accurate records are essential for both good tax management and proper reporting of taxable income. Both agricultural production and tax regulations have become too complex for records to be a collection of receipts and canceled checks tossed in a shoe box. Poor records often result in two related and undesirable outcomes: the inability to verify receipts and expenses in case of a tax audit and paying more taxes than may be legally due. In either case, good records do not cost; they pay, in the form of lower income taxes.

Complete tax records include a list of cash receipts and expenses for the year. A tax depreciation schedule is also needed for all depreciable property to determine annual tax depreciation and to calculate any gain, loss, or depreciation recapture when the item is sold. Permanent records should be kept on real estate and other capital items, including purchase date and cost, depreciation taken, cost of any improvements made, and selling price. These items are important to determine any gain or loss from sale or for gift and inheritance tax purposes.

A computerized record system can be helpful in keeping tax records. Accuracy, speed, and

---

Box 16-2   **How Long Should Tax Records Be Kept?**

This is a common question, and fortunately, the answer is not *forever*. A taxpayer has three years from the date the return was due or filed to amend it, and the IRS has three years from the same date to assess any additional tax. The possibility of either event means records should be kept a minimum of three years from the date the related return was filed. For example, the return for calendar year 2015 is due April 15, 2016. Therefore, records for year 2015 should be kept at least until April 15, 2019. Some tax practitioners recommend that a copy of the tax return be kept for some longer period. If a fraudulent return or no return was filed, the IRS has a longer period to assess any tax.

All records related to the cost, depreciation, purchase/sale dates, and sales price of land and depreciable assets should be kept for three years *after* their sale or other disposition. Land is often owned for many years, so records for land purchases will often need to be kept for many years.

---

convenience are some of the advantages. In addition to keeping a record of income and expenses, some computer accounting programs also have the ability to maintain a depreciation schedule and calculate each year's tax depreciation. Some can print out the year's income and expenses in the same format as a farm tax return, and even print a copy of the completed farm tax return.

THE TAX SYSTEM AND TAX RATES[1]

The federal income tax system in the United States is based on a number of *marginal tax rates* (the additional tax on an additional dollar of income within a certain range). While the actual tax rates and the range of taxable income for each rate change from time to time, the rates have always been marginal rates. They are called *progressive tax rates* because as an individual's taxable income increases, the rate at which it is taxed also increases. The following table shows the marginal tax rates for a joint and a single tax return in 2015.

[1]New tax legislation may change many rules and rates including marginal tax rates, tax brackets, tax rates on capital gain income, Section 179 expensing, and other items discussed in this chapter. The reader should always check current IRS regulations for the latest tax information.

Taxable income brackets*		Marginal tax rates
Married filing jointly ($)	**Single ($)**	**2015 (%)**
0–18,450	0–9,225	10
18,450–74,900	9,225–37,450	15
74,900–151,200	37,450–90,750	25
151,200–230,450	90,750–189,300	28
230,450–411,500	189,300–411,500	33
411,500–464,850	411,500–413,200	35
Over 464,850	Over 413,200	39.6

*The dividing points between each marginal tax rate are adjusted upward each year by an amount equal to the general inflation rate.

Taxable income includes farm income as well as income earned from other sources, such as salaries and wages, interest, rents and royalties, *minus* personal exemptions and deductions.

The higher rate for each bracket applies only to the dollars of income in that bracket. For example, a single taxpayer with $80,000 of taxable income would pay the following tax:

Income tax due	
10% on the first $9,225 =	$ 922
15% on the next $28,225 = ($37,450 – $9,225)	4,234
25% on the next $42,550 = ($80,000 – $37,450)	10,638
Total tax due	$15,794

Even though the taxpayer is in the 25 percent marginal tax bracket, only the income above $37,450 is taxed at the rate of 25 percent.

Income received from the sale of depreciable assets and land may qualify as long-term *capital gains* income. It is reported separately and is often taxed at a lower rate. Capital gains will be discussed in more detail later in this chapter.

In addition to income taxes, self-employed individuals such as farmers and ranchers are subject to *self-employment tax*. Unlike income tax, all eligible income from a ranch or farm business activity is subject to this tax. It is applied to income before personal exemptions or deductions are subtracted.

Self-employment tax includes both Social Security and Medicare taxes. For 2015, these combined tax rates were

Income subject to self-employment tax*	Rate
$0–118,500	15.3%
Over 118,500	2.9%

*These values are set by legislation and have generally increased every year. An additional 0.9% rate is assessed on self-employment income over $125,000 (married filing separately), $250,000 (married filing jointly), or $200,000 (single).

Because capital gains are not included, and no personal deductions and exemptions are subtracted, the income subject to self-employment tax will differ from that subject to income taxes. However, the combination of the two taxes creates a number of income brackets and combined marginal tax rates, as shown in Fig. 16-2.

SOME TAX MANAGEMENT STRATEGIES

A number of tax management strategies exist because of the marginal nature of the income tax rates and other features of the tax system. Tax management strategies are basically one of two types: those that reduce taxes and those that only postpone the payment of taxes. However, any tax payment that can be delayed until next year makes that sum of money available for business

use during the coming year. That amount is available for investment or to reduce the amount of borrowing needed, which either earns interest income or reduces interest expense.

Form of Business Organization

As discussed in Chapter 14, the form of business organization will have an effect on the taxes paid over time. The different forms of business organization should be analyzed both when starting a farm or ranch business and whenever there is a major change in the business. For example, Table 14-2 compares federal tax brackets and rates for personal and corporate returns. While taxes may not be the only reason, nor always the most important reason, for changing the type of business organization, it is always a factor to consider in the decision.

Income Leveling

There are two reasons for trying to maintain a steady level of taxable income. First, years of higher-than-normal income may put the taxpayer in a higher marginal tax bracket. More tax will be paid over time than with a level income taxed at a lower rate. Second, in years of low income, some personal exemptions and deductions may not be fully used. These cannot be carried over to future tax years.

For the 2015 tax year, each taxpayer gets a personal exemption of $4,000, and the same amount for each dependent. There is also a standard deduction of $12,600 for a married couple filing a joint return.[2]

Taxpayers who *itemize* their personal deductions may have an even larger deduction. Therefore, a married couple with two dependent children has exemptions and a minimum deduction of ($4,000 × 4) + $12,600 = $28,600. Taxable income is only the income above this amount. If the total family income is less than $28,600, no income taxes are due, but some (or

[2]Both the personal exemption and standard deduction are adjusted upward each year to account for the general rate of inflation in the economy.

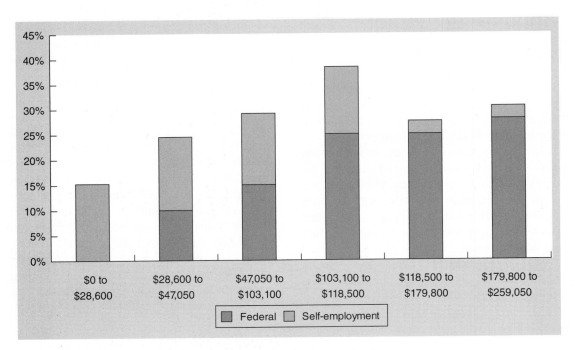

Figure 16-2 Example of marginal tax rates by level of gross income (after subtracting the standard deduction, four personal exemptions, and one-half of self-employment tax) for a married couple filing a joint return.

all if income is $0 or less) of the allowable exemptions and deductions are lost. They cannot be carried over to use next year.

Even though no income tax may be due, there may still be some self-employment tax to pay. One-half of the self-employment tax due can be deducted from ordinary taxable income, however.

Tax Credits

From time to time Congress makes available income *tax credits* as an incentive or reward for certain actions or types of taxpayers. Examples include credits for having dependent children, investing in energy-saving technology, and continuing formal education. Tax credits are much more powerful than tax deductions because each dollar of a credit reduces taxes by a full dollar instead of only a certain percentage. They are subtracted after the total amount of tax due has been calculated.

Income Averaging

If it is impossible to maintain a fairly level taxable income from year to year, income averaging is allowed. It is particularly useful in a year when taxable income is well above average and the past three years were average or below, as income can be moved from years with high marginal tax rates to years with lower rates.

To use income averaging, farmers and ranchers select the amount of the current year's taxable income they want to use in the averaging process. That amount is deducted from the current year's taxable income and one-third of it is added to each of the past three years' taxable income. Income taxes are then recalculated for the past three years as well as for the current year. There are some limitations on what can and cannot be averaged and, because of other provisions of the tax code, the tax savings may be less than expected.

Deferring or Postponing Taxes

Taxes can be deferred or postponed by delaying taxable income until a later year. This saves taxes in the current year, and the money saved can be invested or used to reduce borrowing during the next year. Taxable income can be deferred by increasing expenses in the current year and deferring sales until the next year. Tax regulations limit the amount of prepaid expenses that can be deducted but still allow considerable flexibility for many farmers and ranchers.

To level income or defer taxes requires flexibility in the timing of income and expenses. The taxpayer must be able to make the expense deductible and the income taxable in the year of choice. Flexibility is greatest with the cash method of computing taxable income. With the accrual method, many of the timing decisions on purchases and sales are offset by changes in inventory values.

The method of depreciation selected for newly acquired capital assets also can be used to affect the amount of taxable income that is deferred until future tax years. The use of Section 179 expensing is especially effective for postponing tax liabilities. It will be discussed in more detail later in the chapter.

Taxpayers must be cautious and consider reasons besides tax savings when adjusting the timing of purchases and sales. It is easy to make a poor economic or marketing decision in an effort to save or postpone a few dollars in taxes. Expected market trends are one factor to consider. For example, if commodity prices are expected to move lower into next year, a farmer may be ahead by selling now and paying the tax in the current year. Remember, the objective of tax management is to maximize long-run after-tax profit, not to minimize income taxes paid in any given year.

Net Operating Loss

Wide fluctuations in farm prices and yields can cause a *net operating loss (NOL)* in some years, despite farmers' best efforts to level out annual taxable income. Special provisions relating to net operating losses allow them to be used to reduce past and/or future taxable income. Any NOL from a farm business can first be used to offset taxable income from other sources. If the loss is greater than nonfarm income, any remaining qualifying farming loss can be carried back up to five years and then forward up to 20 years. Nonfarm losses can only be carried back up to two years. However, a taxpayer with a farming loss can elect to use the two-year rather than the five-year carryback.

When carried back, the NOL is used to reduce taxable income for these prior years, and a tax refund is requested. When carried forward, it is used to reduce taxable income, and therefore income taxes, in future years. A taxpayer can elect to forego the two- or five-year carry-back provision and apply the net operating loss only against future taxable income for up to 20 years. However, this election must be made at the time the tax return is filed, and it is irrevocable. In making this election, it is important to receive maximum advantage from the loss by applying it to years with above-average taxable income.

A net operating loss is basically an excess of allowable tax-deductible expenses over gross income. However, certain adjustments and special rules apply when calculating an NOL. Expert tax advice may be necessary to properly calculate and use an NOL to its best advantage.

Tax-Free Exchanges

The sale of an asset such as land will generally incur some taxes whenever the sale price is above the asset's cost or tax basis (its value for tax purposes). If the land being sold is to be replaced by the purchase of another farm, a tax-free exchange can eliminate, or at least reduce, the tax caused by the sale. There are some specific regulations regarding tax-free exchanges, so competent tax advice is necessary to be sure the exchange qualifies as tax free.

A tax-free exchange involves trading property. The farm you wish to own is purchased by the person who wants to purchase your farm.

You then exchange or trade farms, with both parties ending up with the farms they wish to own. Completion of a tax-free exchange transfers the basis on the original farm to the new farm, and no tax is due if the sale prices of both farms are equal. In practice, this rarely happens and some tax may be due. Again, good tax advice is necessary to compute any tax due on the exchange. Tax-free exchanges do not fit every situation because of the strict requirements. However, they are something to consider whenever an eligible asset is to be replaced by a new or different one.

DEPRECIATION

Depreciation plays an important role in tax management for two reasons. First, it is a noncash but tax-deductible expense, which means it reduces taxable income without requiring a matching cash outflow. Second, some flexibility is permitted in calculating tax depreciation, making it another tool that can be used for leveling income and postponing taxes. A working knowledge of tax depreciation, depreciation methods, and the rules for their use is necessary for good tax management.

The difference between tax depreciation and other calculations of depreciation should be kept in mind throughout the following discussion. Tax depreciation is depreciation expense computed for tax purposes. Depreciation for accounting and management purposes would normally use one of the depreciation methods discussed in Chapter 5. As will be shown, the usual tax depreciation method is different and may use different useful lives and salvage values than would be used for management accounting. Therefore, there may be a large difference in the annual depreciation for tax versus management uses.

Tax Basis

Every business asset has a tax basis, the asset's value for tax purposes at any time. At the time of purchase, it is called a *beginning tax basis*, and when this value changes, it becomes an *adjusted tax basis*. Any asset purchased directly, either new or used, has a beginning tax basis equal to its purchase price. When a new asset is purchased by trading in a used asset plus cash, the basis is determined in a different manner. Beginning basis on the new asset is equal to the cash paid to complete the trade, plus the remaining adjusted tax basis of the item traded, if any.

For example, assume an old tractor with an adjusted tax basis of $10,000 plus $80,000 cash is traded for a new tractor with a list price of $120,000. The $80,000 is the *boot* or the difference between the cost of the new tractor and the trade-in value of the old one. The beginning tax basis on the new tractor is $80,000, the cash boot paid, plus the $10,000 adjusted basis from the old tractor, or $90,000. This may seem low for a tractor with a list price of $120,000, but it is the result of one or both possibilities. First, the old tractor had a trade-in value higher than its adjusted tax basis, which is likely given the fast depreciation used for tax purposes. Second, the dealer was willing to sell the tractor for something less than full list price.

An asset's basis is adjusted downward each year by the amount of depreciation taken. The result is its adjusted tax basis. In the previous example, if $9,639 of tax depreciation is taken the first year, the tractor would have an adjusted tax basis of $90,000 − $9,639 = $80,361 at the end of the year. The basis would be adjusted downward each following year by the amount of tax depreciation taken that year. The adjusted basis is not the same as market value and may not be anywhere close to market value.

The basis on a nondepreciable asset such as land generally remains at its original cost for as long as it is owned. An exception would be if some capital improvement is made to the land, when the cost of the improvement is neither tax deductible in the current year nor depreciable. Terraces or earthen dams are possible examples. In this case, the cost of the improvement is *added* to the original basis to find the new adjusted basis.

An accurate record of the basis of each asset is important when it is sold or traded. The current adjusted basis is used to determine the gain or loss on the sale. Cost and basis information on assets such as land and buildings, which have a long life and are sold or traded infrequently, may need to be kept for many years. This is another reason for a complete, accurate, and permanent record-keeping system.

MACRS Depreciation

The current tax depreciation system is called the Modified Accelerated Cost Recovery System (MACRS). There are several alternatives under MACRS, but the discussion here will concentrate on the regular MACRS, used more frequently than the others. Alternative MACRS methods will be discussed briefly in a later section.

MACRS contains a number of property classes, which determine an asset's useful life. Each depreciable asset is assigned to a particular class, depending on the type of asset. Farm and ranch assets generally fall into the 3-, 5-, 7-, 10-, 15-, or 20-year classes. Some examples of assets in each class are

3-year: breeding swine

5-year: cars, pickups, breeding cattle, goats and sheep, dairy cattle, computers, trucks

7-year: most machinery and equipment, fences, grain bins, silos, furniture

10-year: single-purpose agricultural and horticultural structures, trees bearing fruit or nuts

15-year: paved lots, wells, drainage tile

20-year: general-purpose buildings such as machine sheds and hay barns

These are only some examples. The tax regulations should be checked for the proper classes for other specific depreciable assets.

Another characteristic of MACRS is an assumed salvage value of zero for all assets. Therefore, taxpayers do not have to select a useful life and salvage value when using MACRS. This method also includes an automatic one-half year of depreciation in the year of purchase, regardless of the purchase date. This is called the half-year convention, or half-year rule. It benefits a taxpayer who purchases an asset late in the year but is a disadvantage whenever an asset is purchased early in the year. This half-year rule applies unless more than 40 percent of the assets purchased in any year are placed in service in the last quarter of the year. In this case, the mid-quarter rule applies. It requires the depreciation on each newly purchased asset to begin in the middle of the quarter in which it was actually purchased.

MACRS depreciation is computed using fixed percentage recovery rates, shown in Table 16-1 for the three-, five-, and seven-year classes. For agricultural assets, these percentages are based on the use of a 150 percent declining balance method for each successive year, until the percentage using straight-line depreciation results in a higher depreciation value. At that point, straight-line rates are used.

Regular MACRS depreciation for each year of an asset's assigned life is found by multiplying the asset's beginning tax basis by the appropriate percentage. For example, if a new pickup is purchased (five-year class) and has a beginning

TABLE 16-1	Regular MACRS Recovery Rates for Farm and Ranch Assets in the 3-, 5-, and 7-Year Classes (Half-Year Convention)

	Recovery percentages		
Recovery year	3-year class	5-year class	7-year class
1	25.00	15.00	10.71
2	37.50	25.50	19.13
3	25.00	17.85	15.03
4	12.50	16.66	12.25
5		16.66	12.25
6		8.33	12.25
7			12.25
8			6.13

tax basis of $40,000, the depreciation would be computed as follows:

Year 1:	$40,000 \times 15.00\% = \$ 6,000$
Year 2:	$40,000 \times 25.50\% = \$10,200$
Year 3:	$40,000 \times 17.85\% = \$ 7,140$
Year 4:	$40,000 \times 16.66\% = \$ 6,664$
Year 5:	$40,000 \times 16.66\% = \$ 6,664$
Year 6:	$40,000 \times 8.33\% = \$ 3,332$
Total	$\$40,000$

Depreciation for assets in other classes would be computed in a similar manner using the appropriate percentages for the class.

Alternative Depreciation Methods

Some flexibility is allowed in computing tax depreciation, because there are alternatives to the regular MACRS method previously discussed. These alternatives all result in slower depreciation than regular MACRS, which is one reason they are not widely used. However, taxpayers who wish to delay some depreciation until later years can choose from one of the alternative methods:

- Regular MACRS class life or recovery period, with straight-line depreciation
- Alternative MACRS recovery periods (usually longer than regular MACRS) and 150 percent declining balance method
- Alternative MACRS recovery periods with straight-line depreciation

The latter would be the *slowest* of the four possible methods. Slower depreciation may be preferred by a young farmer who expects to have more taxable income and a higher marginal tax rate in future years than in the near term.

If an alternative method is used for one asset, all assets in that class purchased in the same year must be depreciated using the alternative method. Also, once an alternative method has been selected for an asset, it cannot be changed in a later year.

Bonus depreciation equal to a maximum percent of the initial basis is sometimes allowed in the year an asset is purchased and placed in service. The beginning basis is then reduced by the amount of bonus depreciation claimed before MACRS rates are applied.

Expensing

Section 179 of the tax regulations provides for a special deduction called *expensing*. While not specifically called depreciation, it has the same effect on basis and taxable income. Expensing is an *optional* deduction, which can be elected only in the year an asset is purchased. The maximum Section 179 expensing for one year allowed is set by Congress each year.

Property eligible for expensing is defined by the tax regulations somewhat differently than property eligible for depreciation. However, the general result is that most 3-, 5-, 7-, and 10-year class property acquired from an unrelated person is eligible. The amount eligible for expensing is equal to the cost or beginning basis on an outright purchase, but only the cash difference or *boot* is eligible when there is a trade.

Expensing is treated as depreciation, so it reduces an asset's tax basis. The beginning basis must be reduced by the amount of expensing taken before computing any remaining regular depreciation. From this procedure, it can be seen that an election to take expensing does not increase the total amount of depreciation taken over the asset's life. It only puts more depreciation in the first year (expensing plus bonus depreciation plus regular depreciation for Year 1), which leaves less annual depreciation for later years.

Assume a combine has been purchased for $158,000, which would be its beginning basis. If $108,000 of Section 179 expensing was taken (it is optional), the basis in the combine would be reduced to $158,000 − $108,000 = $50,000. This is all the depreciation remaining, so annual

MACRS depreciation for the first two years would be computed as follows:

Year 1:	$50,000 × 10.71% = $5,355
Year 2:	$50,000 × 19.13% = $9,565

Depreciation would continue for six more years using percentages for a seven-year class life asset. The combination of expensing and regular depreciation results in a tax expense of $108,000 + $5,355 = $113,355 in the first year. The choice to take expensing does not increase total depreciation over the life of the asset. It only moves more of the lifetime depreciation into the first year, which leaves less depreciation in later years than if no expensing had been taken. Assets with a purchase price equal to the Section 179 maximum or less can be completely expensed in the year of purchase, leaving no regular depreciation. Any remaining amount of the annual limit can be applied to other eligible assets.

There is a limit on the amount of eligible assets a business can purchase in any one year and still take the maximum expensing. Purchases over this limit reduce the maximum expensing that can be taken dollar for dollar. This limit may be adjusted from year to year by legislation.

Depreciation, Income Leveling, and Tax Deferral

The choices taxpayers are allowed in computing tax depreciation permit depreciation to be used as a means to level out taxable income or defer taxes.

Box 16-3 Income Leveling and Tax Postponement

It is early December and Jason and Kate Starko are meeting with their tax preparer. Her preliminary estimate shows that they will have taxable income of $50,000, which will produce a federal income tax liability of $6,577.

$18,450 taxed at 10% = $1,845
$31,550 taxed at 15% = $4,732
$50,000　　　　　　$6,577

However, she points out that because they use the cash accounting option, they have some alternatives. By postponing some grain sales, prepaying some expenses, and using the expensing option on some livestock equipment they bought, they can reduce their taxable income to zero for this year. Is this a good strategy?

First of all, they would have the $6,577 that they would have to pay in taxes to use for another year. This would help reduce their operating loan, accruing interest at 7.5 percent annually. Jason figures their interest savings will be $493.

$$\$6,577 \times 7.5\% = \$493$$

However, Kate points out that deferring that much income might put them in a higher tax bracket next year. If next year is similar to this year, they could have $100,000 of taxable income, which would produce a tax liability of $16,588.

$ 18,450 taxed at 10% = $ 1,845
$ 56,450 taxed at 15% = $ 8,468
$ 25,100 taxed at 25% = $ 6,275
$100,000　　　　　　$16,588

If instead they reported $50,000 of taxable income both years, their total taxes would be $6,577 each year, or $13,154. So, even with the $493 savings in interest, they would be worse off by $16,588 − $13,154 − $493 = $2,941 if they deferred all their income.

They decide to defer part of their taxable income, knowing that they can use some of the same tax deferral strategies again next year. If they have below-average income next year, they can elect not to defer income and keep their taxable income fairly level.

If eligible assets are purchased in a year when taxable income is high, expensing can be elected and the regular MACRS method can be used for depreciation to get maximum current tax deductions. In years when taxable income is below average, the optional expensing would not be taken and one of the slower alternative depreciation methods can be used. However, a taxpayer should always remember that due to the time value of money, a dollar of current tax savings is worth more than a dollar of tax savings in the future.

Depreciation Recapture

The combination of expensing, bonus depreciation, and fast depreciation under the regular MACRS method and the assumed zero salvage value means an asset's adjusted tax basis will often be less than its market value. Whenever a depreciable asset is sold for more than its adjusted basis, up to its original basis, the difference is called *depreciation recapture*. This amount represents excess depreciation taken, because the asset did not lose market value as fast as it was being depreciated for tax purposes. Depreciation recapture is included as taxable ordinary income for the year of sale, but it is not subject to self-employment taxes.

Suppose the pickup mentioned earlier was sold after three years for $25,000. The original basis was $40,000, but after three years a total of $23,340 of depreciation had been claimed, leaving an adjusted tax basis of $40,000 − $23,340 = $16,660. The sale price exceeded the adjusted basis, so recaptured depreciation of $25,000 − $16,660 = $8,340 is reported.

Depreciation recapture may be required on a direct sale of an asset but not if it is traded for a like asset. For a trade, the remaining basis will be added to the beginning basis of the new asset. For example, if the old pickup was traded for a new one, and $15,000 additional cash was paid, the beginning basis of the new pickup would be $16,660 + $15,000 = $31,660. Note that the stated purchase price and trade-in allowance do not affect the beginning basis of the new vehicle.

CAPITAL GAINS

The tax regulations recognize two types of income: ordinary income and capital gains. Ordinary income includes wages and salaries, interest, cash rents, and revenue from the sale of crops and feeder livestock. Capital gains can result from the sale or exchange of certain types of qualified assets. In simplified terms, capital gain is the gain or profit made by selling an asset at a price higher than the original purchase price. Technically, it is the difference between the selling price and the higher of the asset's cost or adjusted basis. For depreciable assets, that means depreciation recapture applies first, and capital gain is only the amount by which the sales price exceeds the original cost or beginning basis. It is also possible to have a *capital loss* if the selling price is less than the adjusted basis.

Two types of assets can qualify for capital gains treatment. Capital assets include primarily nonbusiness investments such as stocks and bonds. Of more importance to most farmers and ranchers are Section 1231 assets, defined as property used in a trade or business. In a farm or ranch business, Section 1231 assets would include buildings, fences, machinery, equipment, breeding livestock, and other depreciable assets used in the business, as well as land.

For income to qualify as long-term capital gain, the asset must have been owned for a minimum time, or else the gain will be short-term capital gain. The required holding period has changed from time to time but is currently 12 months for most assets. One exception is certain types of livestock, which will be discussed later.

Using the same three-year-old pickup with an adjusted basis of $16,660 as an example, suppose it was sold outright for $42,500. In this case, all $23,340 of depreciation that had been claimed over three years would have to be recaptured as ordinary income. In addition, the excess of the sale price over the original $40,000 basis, or $42,500 − $40,000 = $2,500, would be taxed as a capital gain.

If the pickup was sold for less than the adjusted basis, say $12,000, a capital loss of $16,660 – $12,000 = $4,660 would result.

Land is a special case, because it is not eligible for depreciation. For example, assume that farmland was purchased for $300,000 and sold 10 years later for $450,000. A $150,000 long-term capital gain would result from this transaction but no depreciation would be recaptured. However, improvements to the land such as fences and tile lines would have a portion of the selling price assigned to them, and the same treatment as described for other depreciable assets would be applied.

Taxation of Long-Term Capital Gains

The distinction between ordinary income and long-term capital gain income is important because of the different way the two types of income are taxed. The current tax advantages for long-term capital gain income are as follows:

1. Capital gain income (both short and long term) is not subject to self-employment tax.
2. With some exceptions not generally applicable to farms and ranches, long-term capital gain income is taxed at

Ordinary income marginal tax rate (%)	Capital gains tax rate (%)
10, 15	0
25, 28, 33, 35	15
39.6	20

These rates were in effect for 2014, and are subject to change by legislation.

Except for not being subject to Social Security tax, short-term capital gains income does not receive any special tax treatment—it is

---

Box 16-4   Example of Income and Self-Employment Taxes

Frank and Eileen Brown own a small dairy farm and also earn some wages away from the farm. They have three children at home. Part of their farming income came from selling their cull dairy cows, so $10,000 of it qualified to be reported as long-term capital gain income. The following example shows their taxable income and tax due for this year, based on the rates and limits shown in this chapter.

Ordinary farm income	$63,500
Off-farm wages	+16,300
Standard deduction for married couple filing jointly	–12,600
Personal exemptions ($4,000 × 5)	–20,000
One-half of self-employment tax payable	–4,858
Taxable income	$42,342
Tax due on first $18,450 (18,450 × 10%)	$ 1,845
Tax due on remaining taxable income (42,342 – 18,450 = $23,892 × 15%)	3,584
Ordinary income tax due	$ 5,429
Tax due on capital gains income ($10,000 × 0%)	0
Self-employment tax due ($63,500 × 15.3%)	$ 9,715
Total tax due	$15,144

Values have been rounded to the nearest dollar. The tax due may be reduced by any applicable tax credits.

taxed the same as ordinary income. This makes knowledge of, and a record of, the required holding period an important factor. The tax advantages for long-term capital gains income will be lost by selling even a day short of the required holding period. Also, if there are short-term as well as long-term gains and losses in a given tax year, special rules apply for offsetting the various gains with the losses. These rules should be studied carefully to obtain maximum advantage from the losses.

Capital Gains and Livestock

Farmers and ranchers often have more frequent opportunities to receive capital gains income from livestock sales than from selling other assets. Certain livestock are classified as Section 1231 assets, and their sale or exchange may result in a long-term capital gain or loss, provided they were held for draft, dairy, breeding, or sporting purposes. In addition, cattle and horses must have been owned for at least 24 months and hogs and sheep for at least 12 months. Eligibility depends on both requirements—the purpose and the holding period.

Raised Livestock

The Section 1231 tax provision is of special importance to cash-basis farmers and ranchers who raise their own replacements for a breeding or dairy herd and later sell them as cull animals. Raised replacements do not have a tax basis established by a purchase price, nor do they have a basis established by an inventory value as would occur with accrual tax accounting. Therefore, cash-basis taxpayers have a zero basis on raised replacement animals, and the entire income from their sale is treated as long-term capital gain (sale price minus zero basis). This provision of the tax law makes cash-basis accounting attractive for taxpayers with beef, swine, sheep, or dairy breeding herds who raise their own replacements. An accrual basis taxpayer loses most of this capital gain advantage on raised animals, because they have a tax basis derived from their last inventory value, and their increase in inventory value from year to year is taxed as ordinary income.

Purchased Livestock

It is also possible to receive capital gain income from the sale of purchased breeding livestock, but only if the selling price is above the original purchase price. For example, assume a beef heifer was purchased for $1,000 and depreciated to an adjusted basis of $0 over the five-year class life. If this animal is then sold for $1,400, the sale would result in $1,000 of depreciation recapture and $400 of long-term capital gain.

The capital gain that cash-basis taxpayers may receive from selling raised breeding and dairy animals does not necessarily mean that replacements should be raised rather than purchased. There are also some tax advantages from purchasing replacements. The initial cost of purchased replacements can be depreciated, including the use of the optional expensing. A replacement method should be selected after carefully considering all the costs and production factors, as well as the income tax effects. This is another example of the importance of income taxes and the various tax regulations in a manager's decision-making environment. To maximize long-run, after-tax profit, a manager is forced to consider income taxes along with the costs and technical production factors of the various alternatives being considered.

Box 16-4 contains an example of how ordinary farm income, self-employment income, and capital gains income are each taxed at a different rate to arrive at the total tax due. Off-farm income and certain deductions and exemptions affect ordinary taxable income only.

SUMMARY

A business can spend or invest only that portion of its profit that remains after income taxes are paid. Therefore, the usual goal of profit maximization should be maximization of long-run, after-tax profit. For a farm or ranch business, the tax management necessary to achieve this goal begins with selecting either the cash or the accrual tax accounting method. It should continue on a year-round basis because of the many production and investment decisions that have tax consequences.

A number of tax management strategies are possible. They include ways to take full advantage of existing tax regulations and to defer taxes. Some examples include income leveling, tax deferral, income averaging, proper use of a net operating loss, and tax-free exchanges. Appropriate and full use of depreciation and the related expensing option is another common tax management strategy.

Long-term capital gains income is not subject to self-employment taxes and is taxed at lower rates than ordinary income. This type of income may result from the sale or exchange of land or depreciable assets. Another common source of capital gains income is the sale of breeding or dairy animals. Careful planning may be required to qualify such income for the reduced tax rates applicable to long-term capital gain income.

QUESTIONS FOR REVIEW AND FURTHER THOUGHT

1. How does a farmer choose a tax accounting method? Can it be changed? How?

2. Which accounting method would you recommend for each of the following? Why?

   a. A crop farmer whose marketing policies cause wide variations in cash receipts and inventory values from year to year

   b. A rancher with a beef breeding herd who raises all the necessary replacement heifers

3. What advantage is there to deferring income taxes to next year? If they must be paid, why not pay them now?

4. Assume Fred Farmer purchases a new cotton harvester for $232,500 on March 1. Compute the maximum Section 179 expensing (assume a limit of $200,000 for the current year) and regular depreciation for the year of purchase using the regular MACRS method, and calculate the harvester's adjusted tax basis at the end of the year.

5. Repeat the calculations in question 4 without Section 179 expensing, and compare the annual depreciation.

6. In a year when farm prices and yields are very good, machinery and equipment dealers often experience large sales toward the end of the year. How would you explain this increase in sales?

7. Explain how a cash-basis farmer can raise or lower taxable income by purchase and selling decisions made in December. Can an accrual-basis farmer do the same thing? Why or why not?

8. Assume that a rancher purchased a young herd bull for $3,000. Three years later, when the bull had an adjusted tax basis of $1,785, he was sold for $5,000. How much depreciation recapture and/or capital gains income resulted from this sale?

9. Suppose a farmer made a mistake and counted $10,000 of long-term capital gains income as ordinary income on the farm tax return. How much additional income and self-employment tax would this mistake cost if the farmer had taxable income of $45,000? Taxable income of $125,000? (Use the single payer tax rates in "The Tax System and Tax Rates" section.)

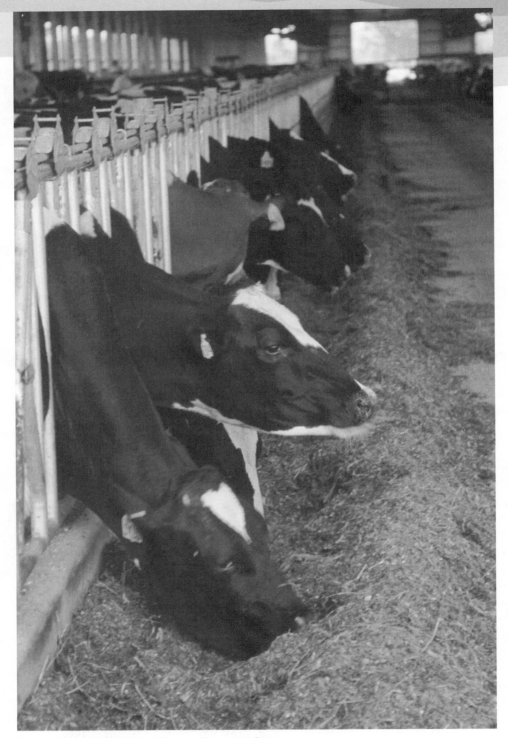

Photo by Bob Nichols, USDA Natural Resources Conservation Service

INVESTMENT ANALYSIS

CHAPTER OUTLINE

Time Value of Money
Investment Analysis
Financial Feasibility
Income Taxes, Inflation, and Risk
Summary
Questions for Review and Further Thought
Appendix. An Example of an Investment
   Analysis

CHAPTER OBJECTIVES

1. Explain the time value of money and its use in decision making and investment analysis

2. Illustrate the process of compounding or determining the future value of a present sum of money or series of payments

3. Demonstrate the process of discounting or determining the present value of a future sum of money or series of payments

4. Discuss payback period, simple rate of return, net present value, and internal rate of return, and compare their usefulness for investment analysis

5. Show how to apply these concepts to different types of investment alternatives

6. Explain how income taxes, inflation, and risk considerations affect investment analysis

On a farm or ranch, capital can be used to finance not only annual operating inputs such as feed, seed, fertilizer, pesticides, and fuel but capital assets such as land, machinery, buildings, and orchards as well. Different methods should be used to analyze each type of investment

311

because of differences in the timing of the expenses and their associated returns. Both expenses and income from investing in annual operating inputs occur within one production cycle, usually a year or less. In contrast, investing in a capital asset typically means a large initial expense, with the resulting returns spread over a number of future periods. Partial budgeting can be used to analyze these investments by looking at changes in costs and returns for an average year. Enterprise and partial budgets implicitly recognize the time value of money by including opportunity costs on funds invested in annual operating inputs, but the amounts are typically small due to the short time.

While time may be of minor importance when analyzing annual operating inputs, it becomes of major importance for capital assets. They usually require large sums of money, and the expenses and returns occur in different periods spread over many years. The amounts may be irregular as well. A proper analysis of these capital investments requires careful consideration of the size and timing of the related cash flows.

There are other reasons for carefully analyzing potential capital investments before they are made. Decisions about operating inputs can be changed annually. However, capital investments are, by definition, long-lasting assets, and investment decisions are made less frequently. It is more difficult to change a capital investment decision once the asset is purchased or constructed. Therefore, more time and more accurate analytical techniques should be used when making these decisions.

TIME VALUE OF MONEY

We all would prefer to have $100 today rather than $100 five years from now. People instinctively recognize that money received today is worth more than a dollar at some future date. Why? A dollar received today can be invested to earn interest and will therefore increase to a dollar *plus* interest by the future date. In other words, the interest represents the opportunity cost

of receiving the dollar in the future rather than now. This is the investment explanation of the time value of money. But there are other explanations of why interest is charged or paid on funds.

- Money can purchase goods and services for *consumption*. If the dollar were to be spent for consumer goods such as a new TV, automobile, or furniture, we would prefer to have the dollar now so we could enjoy the new item immediately.
- *Risk* is another reason for preferring the dollar now rather than later. Some unforeseen circumstance could prevent us from collecting it in the future.
- Finally, *inflation* in the general cost of goods and services may diminish what a dollar can buy in the future compared to today.

This chapter will focus on the investment explanation for the time value of money, although risk and inflation will be discussed as well. This concept is important for a manager who must make business investment decisions on a farm or ranch.

Terms, Definitions, and Abbreviations

A number of terms and abbreviations will be used in the discussion and equations throughout this section. They are as follows, with the abbreviations to be used later shown in parentheses:

- *Present value (PV):* The number of dollars available or invested at the current time, or the current value of some amount(s) to be received in the future.
- *Future value (FV):* The amount of money to be received at some future time, or the amount a present value will become at some future date when invested at a given interest rate.
- *Payment (PMT):* The number of dollars to be paid or received at the end of each of a number of time periods.
- *Interest rate (i):* Also called the discount rate in some applications, it is the interest rate used to find present and future values,

often equal to the opportunity cost of capital. In the mathematical formulas presented, *i* will be expressed as a decimal value rather than a percentage.

- *Time periods (n)*: The number of time periods to be used for computing present and future values. Time periods are often a year in length but can be shorter, such as a month. The annual interest rate, *i*, must be adjusted to correspond to the length of the time periods; that is, a monthly interest rate must be used if the time periods are months.
- *Annuity*: A term used to describe a series of equal periodic payments (PMT). The payments may be either receipts or expenditures. In the special case of a loan scheduled to be repaid with a series of equal payments (amortized), the periodic payments are an annuity (see Chapter 19 for a discussion).

Future Values

The future value of money refers to the value of an investment at a specific date in the future. This concept assumes that the investment earns interest during each time period, which is then reinvested at the end of each period so it will also earn interest in later time periods. Therefore, the future value will include the original investment and the interest it has earned, plus interest on the accumulated interest. The procedure for determining future values when accumulated interest also earns interest is called *compounding*. It can be applied to a one-time lump sum investment (PV) or to an investment that takes place through a series of payments (PMT) over time. Each case will be analyzed separately.

Future Value of a Present Value

Figure 17-1a graphically illustrates this situation. Starting with a given amount of money today, a PV, what will it be worth at some date in the future? The answer depends on three things: the PV, the interest rate it will earn, and the length of time it will be invested.

Assume you have just invested $1,000 in a savings account that earns 8 percent interest compounded annually. You would like to know what the future value of this investment will be after three years. The data in Table 17-1 illustrate how the balance in the account changes from year to year.

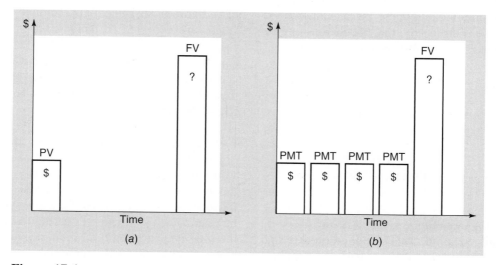

Figure 17-1 Illustration of the concept of future value for a present value (a) and for an annuity (b).

TABLE 17-1	Future Value of $1,000

Year	Value at beginning of year	Interest rate (%)	Interest earned ($)	Value at end of year ($)
1	1000.00	8	80.00	1,080.00
2	1080.00	8	86.40	1,166.40
3	1166.40	8	93.30	1,259.70

All interest earned is allowed to accumulate in the account, so it also earns interest for the remaining years. This is the principle of compounding and shows the results of using *compound interest*. In this example, a present value of $1,000 has a future value of $1,259.70 when invested at 8 percent interest for three years. Interest is compounded when the accumulated interest also earns interest in the following time periods.

If the interest had been withdrawn, only $1,000 would have earned interest each year. A total of $240.00 in interest would have been earned compared to the $259.70 of interest earned in this example.

This procedure for finding a future value can become tedious if the investment is for a long time. Fortunately, the calculations can be simplified by using the mathematical equation

$$FV = PV(1 + i)^n$$

where the abbreviations are as defined earlier. Applied to the example, the calculations would be

$$FV = \$1,000 \times (1 + 0.08)^3$$
$$= \$1,000 \times (1.2597)$$
$$= \$1,259.70$$

giving the same future value as before.

This equation can be applied quickly and easily using a calculator that will raise numbers to a power, a financial calculator with built-in functions, or a computer spreadsheet. Without such tools, the equation can be difficult to use, particularly when n is large. To simplify the calculation of future values, tables have been constructed giving values of $(1 + i)^n$ for different combinations of i and n. Appendix Table 2 is for finding the future value of a present value. Any future value can be found by multiplying the present value by the factor from the table that corresponds to the appropriate interest rate and length of the investment. The value for 8 percent and three years is 1.2597, which when multiplied by $1,000 in the example gives the same future value of $1,259.70 as before.

An interesting and useful rule of thumb is the *Rule of 72*. The *approximate* time it takes for an investment to double can be calculated by dividing 72 by the interest rate. For example, an investment at 8 percent interest would double in approximately nine years (the table value is 1.9990). At 12 percent interest, a present value would double in approximately six years (table value of 1.9738).

The concept of future value can be useful in a number of ways. For example, what is the future value of $1,000 deposited in a savings account at 8 percent interest for 10 years? The Appendix Table 2 value is 2.1589, which gives a future value of $2,158.90. Another use is to estimate future asset values. For example, if land values are expected to increase at an annual compound rate of 6 percent, what will land with a present value of $2,500 per acre be worth five years from now? The table value is 1.3382, so the estimated future value is $2,500 × 1.3382, or $3,345.50 per acre.

Future Value of an Annuity
Figure 17-1b illustrates the concept of a future value of a regular series of payments. What is the future value of a number of payments (PMT) made at the end of each year for a given number of years? Each payment will earn interest from the time it is invested until the last payment is made. Suppose $1,000 is deposited at the end of each year in a savings account that pays 8 percent annual interest. What is the value of

this investment at the end of three years? It can be calculated in the following manner:

First payment	$1,000	$1,000 (1 + 0.08)^2 = 1,166.40$
Second payment	$1,000	$1,000 (1 + 0.08)^1 = 1,080.00$
Third payment	$1,000	$1,000 (1 + 0.08)^0 = 1,000.00$
Future value		$3,246.40

The money is deposited at the end of each year, so the first $1,000 earns interest for only two years, the second $1,000 earns interest for one year, and the third $1,000 earns no interest. A total of $3,000.00 is invested, and the total interest earned is $246.40.

The future value of an annuity can be found using this procedure, but an annuity with many payments requires many computations. An easier method is to use the equation

$$FV = PMT \times \frac{(1 + i)^n - 1}{i}$$

where PMT is the amount invested at the end of each time period. It is even easier to use table values such as those in Appendix Table 3, which cover a range of interest rates and time periods. Continuing with the same example, the table value for 8 percent interest and three years is 3.2464. Multiplying this factor by the annual payment, or annuity, of $1,000 confirms the previous future value of $3,246.40.

Present Values

The concept of present value refers to the value today of a sum of money to be received or paid in the future. Present values are found using a process called *discounting*. The future value is discounted back to the present to find its current or present value. Discounting is done because a sum to be received in the future is worth less than the same amount available today. A present value can be interpreted as the sum of money that would have to be invested now at a certain interest rate to equal a given future value on the same date. When used to find present values, the interest rate is often referred to as the *discount rate*.

Compounding and discounting are opposite, or inverse, procedures, as shown in Figure 17-2. A present value is compounded to find its future value, and a future value is discounted to find its present value. These inverse relations will become more apparent in the following discussion.

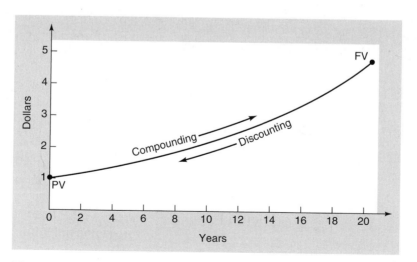

Figure 17-2 Relation between compounding and discounting ($1 at 8 percent interest).

Present Value of a Future Value

Figure 17-3a illustrates the concept of a present value. The future value is known, and the problem is to find the present value of that amount. Figure 17-3a is exactly like Figure 17-1a, except for the placement of the dollar sign and the question mark. This also illustrates the inverse relation between compounding and discounting.

The present value of a future value depends on the interest rate and the length of time before the payment will be received. Higher interest rates and longer time periods will reduce the present value, and vice versa. The equation to find the present value of a single payment to be received in the future is

$$PV = \frac{FV}{(1 + i)^n} \text{ or } FV \times \frac{1}{(1 + i)^n}$$

where the abbreviations are as defined earlier.

This equation can be used to find the present value of $1,000 to be received in five years using an interest rate of 8 percent. The calculations would be

$$PV = \$1,000 \times \frac{1}{(1 + 0.08)^5}$$
$$= \$1,000 \times (0.68058)$$
$$= \$680.58$$

A payment of $1,000 to be received in five years has a present value of $680.58 at 8 percent compound interest. Stated differently, $680.58 invested for five years at 8 percent compound interest would have a future value of $1,000. This again shows the inverse relation between compounding and discounting. A more practical explanation is that an investor should not pay more than $680.58 for an investment that will return $1,000 after five years, if an 8 percent return is desired. Investment analysis makes heavy use of present values, as will be shown later in this chapter.

Tables are also available to assist in calculating present values, such as Appendix Table 4. The factor for the appropriate interest rate and number of years is multiplied by the future value to obtain the present value. For example, the present value of $1,000 to be received in five years at 8 percent interest is equal to $1,000 multiplied by the table value of 0.68058, or $680.58.

Present Value of an Annuity

Figure 17-3b illustrates the common problem of determining the present value of an annuity or a number of payments to be received over time. Suppose a payment of $1,000 will be received at the end of each year for three years, and the interest rate is 8 percent. The present value of this

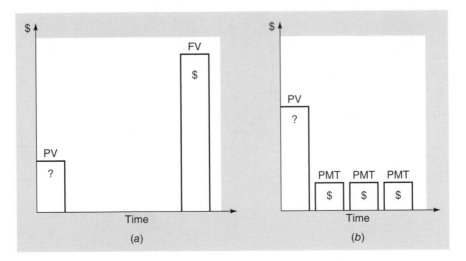

Figure 17-3 Illustration of the concept of future value for a present value (a) and for an annuity (b).

TABLE 17-2	Value of an Annuity	

Year	Amount ($)	Present value factor	Present value ($)
1	1,000	0.92593	925.93
2	1,000	0.85734	857.34
3	1,000	0.79383	793.83
Total			2,577.10

income stream or annuity can be found as shown in Table 17-2 using values from Appendix Table 4.

A present value of $2,577.10 represents the maximum an investor should pay for an investment that will return $1,000 at the end of each year for three years, if an 8 percent return is desired. Higher interest rates will reduce the present value, and vice versa.

The present value of an annuity can be found directly from the equation

$$PV = PMT \times \frac{1 - (1 + i)^{-n}}{i}$$

As before, it is much easier to use table values such as those in Appendix Table 5. The value corresponding to the proper interest rate and number of years is multiplied by the annual payment to find the present value of the annuity. The table value for 8 percent and three years is 2.5771, which is multiplied by $1,000 to find the present value of $2,577.10.

Present values are more useful than future values when analyzing investments. Not all investments have the same useful lives nor pattern of net cash flows. Future values that occur in different years and in different amounts are not directly comparable until they are discounted to a common time, such as the present.

INVESTMENT ANALYSIS

Investment, as the term will be used in this section, deals with other than short-term or annual expenditures. It refers to the addition of long-term

or noncurrent assets to the business. These assets last a number of years, or indefinitely in the case of land, so these investment decisions will have long-lasting consequences and often involve large sums of money. Such investments should be thoroughly analyzed before the investment decision is made.

Information Needed

Investment analysis, or *capital budgeting* as it is sometimes called, is the process of determining the profitability of an investment or comparing the profitability of two or more alternative investments. A thorough analysis of an investment requires four pieces of information: (1) the initial cost of the investment, (2) the annual net cash revenues realized, (3) the terminal or salvage value of the investment, and (4) the interest or discount rate to be used.

Initial Cost

The initial cost of the investment should be the actual total expenditure for its purchase, not the list price nor just the down payment if it is being financed. It should include sales tax paid, labor to install or set up equipment, and other start-up costs. Of the four types of information required, this will generally be the easiest to obtain.

Net Cash Revenues

Net cash revenues or cash flows must be estimated for each time period in the life of the investment. Cash receipts minus cash expenses equals the additional *net cash revenues* generated by the proposed investment. Depreciation is not included, because it is a noncash expense and is already accounted for by the difference between the initial cost and the terminal value of the investment. Any interest and principal payments on a loan needed to finance the investment are also omitted from the calculation of net cash revenue. Investment analysis methods are used to determine the profitability of an investment without considering the method or amount of financing needed to purchase it. However, investment analysis techniques can be used to compare several alternatives for financing an investment. An example of this for purchasing or leasing a new tractor is presented in Chapter 22.

Terminal Value

The *terminal value* will need to be estimated and is usually the same as the salvage value for a depreciable asset. For a nondepreciable asset such as land, the terminal value should be the estimated market value of the asset at the time the investment is terminated. Or, if the land will be held indefinitely, its terminal value can be ignored, because the net cash revenues are assumed to go on forever (see the income capitalization approach to land valuation in Chapter 20).

Discount Rate

The discount rate is often one of the more difficult values to estimate. It is the opportunity cost of capital, representing the minimum rate of return required to justify the investment. If the proposed investment will not earn this minimum rate, the capital should be invested elsewhere. If funds will be borrowed to finance the investment, the discount rate can be set equal to the cost of borrowed capital. If a combination of borrowed and equity capital will be used, a weighted average of the interest rate on the loan and the equity opportunity cost rate should be used.

For example, a certain ranch plans to spend $80,000 to upgrade its stock-handling facilities. The owners will invest $20,000 of their own funds, which are earning a 2 percent rate of interest in a savings account. The remaining $60,000 will be borrowed at a 6 percent annual interest rate. The weighted cost of capital (discount rate) is thus equal to (2% × 20,000/80,000) + (6% × 60,000/80,000) = 5.0%.

Risk must also be considered, so the discount rate should be equal to the rate of return expected from an alternative investment of equal risk. Adjustment of the discount rate for income taxes, risk, and inflation will be discussed later in this chapter.

An example of the information needed for investment analysis is contained in Table 17-3.

TABLE 17-3	Net Cash Revenues for Two $10,000 Investments (No Terminal Value)

	Net cash revenues ($)	
Year	Investment A	Investment B
1	3,000	1,000
2	3,000	2,000
3	3,000	3,000
4	3,000	4,000
5	3,000	6,000
Total cash revenues	15,000	16,000
Less initial investment	−10,000	−10,000
Net cash revenues	5,000	6,000
Average net revenue/year	1,000	1,200

Investment A yields a fixed return of $3,000 per year for five years. Investment B also lasts five years, but net cash revenues are lower at first. They increase each year, though, and actually exceed those of investment A over the entire five years.

For simplicity, the terminal values are assumed to be zero. Whenever terminal values exist, they should be added to the net cash revenue for the past year, because they represent an additional cash receipt. The information for the two potential investments in Table 17-3 will be applied to illustrate four methods that can be used to analyze and compare investments: (1) the payback period, (2) the simple rate of return, (3) the net present value, and (4) the internal rate of return.

Payback Period

The payback period is the number of years it would take an investment to return its original cost through the annual net cash revenues it generates. If net cash revenues are constant each year, the payback period can be calculated from the equation

$$PP = \frac{C}{ER}$$

where PP is the payback period in years, C is the initial cost of the investment, and ER is the expected annual cash revenue. For example, investment A in Table 17-3 would have a payback period of $10,000 divided by the net cash revenue of $3,000 per year, or 3.33 years.

When the annual net cash revenues are not equal, they should be summed year by year to find the year in which the total is equal to the amount of the investment. For investment B in Table 17-3, the payback period would be four years, because the accumulated net cash revenues reach $10,000 in the fourth year. In this case investment A is preferred over investment B, because it has the shorter payback period.

The payback period method can be used to rank investments, as shown. Limited capital can be invested first in the highest ranked investment and then on down the list until the investment capital is exhausted. Another application is to establish a maximum payback period and reject all investments with a longer payback. For example, a manager may select a four-year payback as a standard and invest only in alternatives with a payback of four years or less.

The payback method is easy to use and quickly identifies the investments with the most immediate cash returns. However, it also has several serious disadvantages. This method ignores any cash flows that occur after the end of the payback period, as well as the timing of cash flows during the payback period. Selecting investment A using this method ignores the higher returns from investment B in Years 4 and 5, as well as its greater total return. The payback method does not really measure profitability but is more a measure of how quickly the investment will contribute to the liquidity of the business. For these reasons, it should be used only to compare investments with similar lives and relatively constant net cash revenues.

Simple Rate of Return

The simple rate of return expresses the average annual net revenue as a percentage of the original investment. Net revenue is calculated by subtracting the initial investment from the total of the cash revenues, and dividing by the number of periods. In Table 17-3, the average annual net revenue for investment A is shown to be $1,000, and for investment B, $1,200. The simple rate of return is calculated from the equation

$$\text{Rate of return} = \frac{\text{average annual net revenue}}{\text{initial cost}}$$

Applying the equation to the example in Table 17-3 gives the following results:

$$\text{Investment A} = \frac{\$1,000}{\$10,000} = 10\%$$

$$\text{Investment B} = \frac{\$1,200}{\$10,000} = 12\%$$

This method would rank investment B higher than A, a different result than that obtained from the payback method. The simple rate of return method is a better indicator of profitability than the payback method, because it considers an investment's earnings over its entire life. Its biggest drawback is its use of average annual earnings, which fails to consider the size and timing of the earnings and can therefore cause errors in selecting investments. This is particularly true when there are increasing or decreasing net revenues. For example, investment A would have the same 10 percent rate of return if it had no net cash revenue the first four years and $15,000 in Year 5, because the average annual net revenue would still be $1,000. However, a consideration of the time value of money would show the present value of these investments to be greatly different. Because of this shortcoming, the simple rate of return method is not generally recommended for analyzing investments with variable annual net cash revenues.

Net Present Value

The net present value (NPV) method is a preferred method of evaluation, because it does consider the time value of money as well as the size of the stream of cash flows over the entire life of the investment. It is also called the *discounted cash flow* method.

An investment's NPV is the sum of the present values for each year's net cash flow (or net cash revenue) minus the initial cost of the investment. The equation for calculating the NPV of an investment is

$$NPV = \frac{P_1}{(1 + i)^1} + \frac{P_2}{(1 + i)^2} + \cdots + \frac{P_n}{(1 + i)^n} - C$$

where NPV is net present value, P_n is the net cash flow in year n, i is the discount rate, and C is the initial cost of the investment. An example of calculating net present values using a 10 percent discount rate is shown in Table 17-4. The present value factors are table values from Appendix Table 4. Summing the NPVs for each investment and subtracting the initial cost of $10,000 results in a total NPV of $1,370 for investment A and $1,272 for investment B. In this example, investment A would be preferred, because it has a higher NPV.

Investments with a positive NPV would be accepted using this procedure, if investment capital is unlimited. Those with a negative NPV would be rejected, and a zero value would make the investor indifferent. The rationale behind accepting investments with a positive NPV can be explained in two ways. First, it means that the actual rate of return on the investment is greater than the discount rate used in the calculations. In other words, the percent return is greater than the cost of capital. A second explanation is that the investor could afford to pay more for the investment and still achieve a rate of return equal to the discount rate used in calculating the NPV. In Table 17-4, an investor could afford an initial cost of up to $11,370 for investment A and $11,272 for investment B and still receive at least a 10 percent return on invested capital. This method assumes that the annual net cash flows can be reinvested each period to earn a rate of return equal to the discount rate being used.

Both investments in Table 17-4 show a positive NPV using a 10 percent discount rate. In any determination of a present value, the selection of the discount rate has a major influence on the results. The net present values in Table 17-4 would have been lower if a higher discount rate had been used, and vice versa. At some higher discount rate, the net present values would fall

TABLE 17-4	Net Present Value and Internal Rate of Return for Two Investments of $10,000 (10 Percent Discount Rate and No Terminal Values)

	Investment A			Investment B		
Year	Net cash flow ($)	Present value factor	Present value ($)	Net cash flow ($)	Present value factor	Present value ($)
1	3,000	0.909	2,727	1,000	0.909	909
2	3,000	0.826	2,478	2,000	0.826	1,652
3	3,000	0.751	2,253	3,000	0.751	2,253
4	3,000	0.683	2,049	4,000	0.683	2,732
5	3,000	0.621	1,863	6,000	0.621	3,726
		Total	11,370		Total	11,272
		Less initial cost	10,000		Less initial cost	10,000
		Net present value	1,370		Net present value	1,272
		Internal rate of return	15.2%		Internal rate of return	13.8%

to zero, and at an even higher rate, they would become negative. Therefore, care must be taken to select the appropriate discount rate.

Annual Equivalent and Capital Recovery

The NPV of an investment can be converted to an annual equivalent using the amortization factors found in Appendix Table 1. The annual equivalent is an annuity that has the same present value as the investment being analyzed.

For example, investment A in Table 17-4 had a NPV of $1,370, using a discount rate of 10 percent over five years. The amortization factor for 10 percent and five years is 0.2638 (Appendix Table 1). The annual equivalent is

$$\$1,370 \times 0.2638 = \$361.41$$

The annual equivalent for investment B is

$$\$1,272 \times 0.2638 = \$335.55$$

In other words, the net returns from investment A would be equivalent to receiving $361.41 per year for five years and reinvesting it at a 10 percent return. Investment B would be equivalent to receiving $335.55 per year and would still be the less desirable alternative.

Computing the annual equivalent value makes it possible to compare investments with different lives. A different amortization factor would have to be used to convert each NPV. Implicitly, it is assumed that each investment could be repeated with the same results over time.

Annual equivalent values can also be used to convert the initial investment cost of a capital asset to an annual cost. This is an alternative to calculating depreciation and interest using the methods explained in Chapter 9 and is called *capital recovery*. The capital recovery value is the annual amount that recovers the initial cost of the asset over its life, including interest on the unrecovered balance (salvage value), and takes the place of the normal depreciation and interest costs. An example of using the capital recovery

approach to estimate annual ownership costs for farm machinery is shown in Chapter 22.

The annual equivalent value factors can also be used to calculate the periodic payment needed to repay a loan amortized by the equal total payment method (see example in Chapter 19).

Internal Rate of Return

The time value of money is also used in another method of analyzing investments, the *internal rate of return (IRR)*. It provides some information not available directly from the NPV method. Both investments A and B have a positive NPV using the 10 percent discount rate. But what is the actual rate of return on these investments? It is the discount rate that makes the NPV just equal to zero. The actual rate of return on an investment, with proper accounting for the time value of money, is the IRR.

The equation for finding IRR is the same as for NPV, with two differences. First, NPV is set equal to zero, and second, the equation is solved for *i*. NPV is zero, so the initial cost of the investment, C, is often moved to the left of the equality sign. In this arrangement, the solution to the equation is the interest rate, *i*, that sets the NPV of the investment's net cash flows just equal to its initial cost. The IRR is usually calculated by using a computer program or a financial calculator.

$$C = \frac{P_1}{(1+i)^1} + \frac{P_2}{(1+i)^2} + \cdots + \frac{P_n}{(1+i)^n}$$

The two investments shown in Table 17-4 both have a positive NPV when discounted at a rate of 10 percent. That means their IRR must be greater than 10 percent. Investment A is found to have an IRR of 15.2 percent, making it preferred again over B, which has an IRR of only 13.8 percent.

The IRR method can be used several ways in investment analysis. Any investment with an IRR greater than the discount rate would be a profitable investment; that is, it has a positive

NPV. Some investors select an arbitrary cutoff value for IRR and invest only in those projects with a higher value.

The IRR method has several limitations as well. It implicitly assumes that the annual net returns or cash flows from the investment can be reinvested each period to earn a return equal to the IRR. If the IRR is fairly high, this may not be possible, causing the IRR method to overestimate the actual rate of return. Another drawback is that IRR says nothing about the size of the initial investment. A small investment may generate a high IRR but have only a small NPV in absolute dollars. The investment analysis approaches presented do not always give consistent rankings, as shown in the summary in Table 17-5. All of them should be taken into account before a decision is made. In the example, investment A appears more favorable than investment B for all the approaches except the simple rate of return, and generally would be the preferred investment to make.

FINANCIAL FEASIBILITY

The investment analysis methods discussed so far are methods to analyze *economic profitability*. They are meant to answer the question, Is the investment profitable? The question of how the investment was financed was ignored, except for calculating the discount rate. However, when the method and amount of financing used to make the investment are included in the analysis, investments identified as profitable may have years of negative cash flows. Thus, besides the profitability of an investment, an equally important question may be, Is the investment *financially feasible*? In other words, will the investment generate sufficient cash flows at the right time to meet the required cash outflows, including loan payments? This potential problem was discussed in Chapter 13 but is worthy of further discussion here. Determination of financial feasibility should be the final step in any investment analysis.

A potential problem is illustrated in Table 17-6 for the example investments A and B used throughout this chapter. Assume that each investment is

TABLE 17-5	Comparison of Two Investments	
Measure	**Investment A**	**Investment B**
Payback period	3.33 years	4.00 years
Simple rate of return	10%	12%
Net present value	$1,370.00	$1,272.00
Annual equivalent	$ 361.41	$ 335.55
Internal rate of return	15.2%	13.8%

TABLE 17-6	Cash Flow Analysis of Investments A and B

	Investment A ($)			Investment B ($)		
Year	**Net cash revenue**	**Debt payment***	**Net cash flow**	**Net cash revenue**	**Debt payment***	**Net cash flow**
1	3,000	2,800	200	1,000	2,800	−1,800
2	3,000	2,640	360	2,000	2,640	−640
3	3,000	2,480	520	3,000	2,480	520
4	3,000	2,320	680	4,000	2,320	1,680
5	3,000	2,160	840	6,000	2,160	3,840

*Assumes a $10,000 loan at 8 percent interest with equal principal payments over five years.

totally financed with a $10,000 loan at 8 percent interest, to be repaid over five years with equal annual principal payments, plus interest. Both interest and principal are included in the debt payment column and are subtracted from the net cash revenue to find the net cash flow.

Investment A shows a positive net cash flow for each year, because the net cash revenue is greater than the debt payment. However, investment B has lower net cash revenues the first two years, which causes negative cash flows in those years. Both investments had positive net present values using a 10 percent discount rate. However, it is not unusual to find profitable investments that have negative cash flows in the early years if the net cash revenues start slowly and the investment requires a large amount of borrowed capital. The problem can be further compounded if the loan must be paid off in a relatively short time.

If a project such as investment B is undertaken, something must be done to make up for the negative cash flows. Several possibilities exist, which can be used individually or in combination. First, some equity capital could be used for part or all of the initial cost of the investment to reduce the size of the loan and the annual debt payments. Second, the payment schedule for the loan could be lengthened to make the debt payments more nearly equal to the net cash revenues. Arranging for smaller payments with a balloon payment at the end, or interest-only payments for the first few years, would be other possibilities.

If none of these alternatives is feasible, the cash deficits will have to be made up by using excess cash from other parts of the business or from savings. This will require a cash flow analysis of the entire business to see whether cash will be available in sufficient amounts and at proper times to meet the temporary deficits from the investment.

Financing and Net Present Value

Should cash flows related to financing ever be included in the calculation of the NPV of an investment? Generally not, because the decision to make an investment and the question of how to finance it should be considered separately. Occasionally, though, an investment and the method to finance it may be closely linked. An example would be comparing the purchase of a farm that can be obtained only with a seller installment contract, versus another farm that can be purchased only with a conventional loan. In this case, the net cash flows used to calculate the NPVs should include the down payment, if any, as well as the principal and interest payments, because the investment and the financing decisions are tied together.

In some cases, it may be possible to finance the same investment several different ways. Once the investment is accepted, the financing alternatives can be compared by discounting the payment streams for each one and selecting the one with the *lowest* NPV. An example of using this procedure to compare a lease-or-buy choice for acquiring a tractor is shown in Chapter 22.

INCOME TAXES, INFLATION, AND RISK

So far, only the basic procedures and methods used in investment analysis have been discussed. Several additional factors must be included in a thorough analysis of any investment or when comparing alternative investments. These include income taxes, inflation, and risk.

Income Taxes

The examples used to illustrate the methods of investment analysis did not consider the effects of income taxes on the net cash revenues. Taxes were omitted to simplify the introduction and discussion of investment analysis. However, many investments generate taxable income, as well as tax-deductible expenses. Income taxes can substantially change the net cash revenues, depending on the investor's marginal tax bracket and the type of investment. Different investments may affect income taxes differently, so they should be compared on an *after-tax basis*.

Income taxes will reduce the net cash revenues when taxable income generated by the investment exceeds deductions, and vice versa. The actual value of the change depends on the marginal tax bracket of the taxpayer. For example, when taxable net cash flows are positive, an investor in the 28 percent marginal tax bracket will have 28 percent less net cash revenues on an after-tax basis.

If all cash inflows are taxable and all cash outflows are tax deductible, the NPV is simply reduced by the marginal tax rate. However, the net taxable income is often not the same as the net cash revenue. For example, depreciation is not included in calculating net cash revenue, because it is a noncash expense. However, depreciation is a tax-deductible expense that reduces taxable income and therefore income taxes. The tax savings resulting from any depreciation associated with a new investment should be added to the net cash revenues. Certain investments will further reduce income taxes if they are eligible for any special tax credits or deductions that may be in effect at the time of purchase.

Some cash inflows may be taxed differently than others. For example, capital gains income may be subject to a lower tax rate than ordinary income (see Chapter 16). Sales of culled breeding livestock or the terminal sale value of the original investment often gives rise to capital gains and should be adjusted by the appropriate marginal tax rate.

Whenever after-tax net cash revenues are used, it is important that an *after-tax discount rate* also be used. This is because interest on borrowed capital is usually tax deductible, and earnings on alternative investments are usually taxable. The after-tax discount rate can be calculated from the equation

$$r = i \times (1 - t)$$

where r is the after-tax discount rate, i is the before-tax discount rate, and t is the marginal tax rate. Marginal tax rate was discussed in Chapter 16. It consists of the amount of added federal income tax, state and local income tax, and self-employment tax due on each added dollar of taxable income.

For example, what is the after-tax discount rate if the before-tax discount rate is 12 percent and the taxpayer is in the 28 percent marginal tax bracket? The answer is

$$12\% \times (1.00 - 0.28) = 8.64\%$$

The after-tax cost of borrowed capital can be calculated from the same equation, with r and i equal to after- and before-tax interest rates. Interest is a tax-deductible business expense, and every dollar of interest expense will reduce taxes by 28 cents for a taxpayer in the 28 percent marginal tax bracket. An investor paying 12 percent interest on borrowed capital would have an 8.64 percent after-tax cost of capital, as computed from the previous equation.

Inflation

Inflation is a general increase in price levels over time, and it affects three factors in investment analysis: net cash revenues, terminal value, and the discount rate. Net cash revenues will change over time due to changes in the prices of inputs and products, even if they increase at different rates. Inflation will also cause terminal values to be higher than would be expected otherwise. Revenues could also change due to changes in basic supply and demand conditions for inputs of products, apart from the effects of general inflation.

When the net cash revenues and terminal value are adjusted for inflation, the discount rate should also include the expected rate of inflation. An interest or discount rate can be thought of as consisting of at least two parts: (1) a real interest rate, or the interest rate that would exist in the absence of inflation, either actual or expected, and (2) an adjustment for inflation, or an inflation premium. During inflationary periods, a dollar received in the future will purchase fewer goods and services than it will at the present time. The inflation premium compensates for this reduced purchasing power through a higher interest or discount rate. An inflation-adjusted discount rate can be estimated in the following manner:

Real discount rate	5%
Expected inflation rate	+ 3%
Adjusted discount rate	= 8%

The inflation-adjusted discount rate is also called the *nominal* discount rate. If the cost of capital is used for the discount rate, that cost generally already includes an expectation of inflation, and the adjustment has been made implicitly.

In summary, there are two basic ways to treat inflation in capital budgeting:

1. Estimate net cash revenues for each year, and the terminal value, using current prices, and then discount them using a *real* discount rate (the nominal rate minus the expected inflation rate).[1] This eliminates inflation from all elements of the equation.

2. Increase the net cash revenues for each year and the terminal value to reflect the expected inflation rate, and then use an inflation-adjusted (nominal) discount rate. This method incorporates inflation into all parts of the equation. If the same inflation rate is used throughout, the NPV will be the same under either method. However, if some revenues or expenses are assumed to inflate at different rates than others, the inflation-adjusted NPV will be different from the real NPV and is the more accurate measure. For example, energy-related costs may be projected to increase faster than the general rate of inflation in the economy, while tax savings from depreciation do not inflate at all once the investment is made.

---

[1] A more precise method for converting a nominal rate to a real rate is $[(1 + i)/(1 + f)] - 1$, where i is the nominal rate and f is the inflation rate. However, the value $(i - f)$ gives a close approximation to the real rate.

Risk

Risk exists in investments because the estimated net cash revenues and the terminal value depend on future production, prices, and costs, which can be highly variable and difficult to predict. Unexpected changes in these values can quickly make a potentially profitable investment unprofitable. The longer the life of the investment, the more difficult it is to estimate future costs and revenues accurately.

One method of incorporating risk into the analysis is to add a risk premium to the discount rate. Investments with greater risk have a higher risk premium. This is based on the concept that the greater the risk associated with a particular investment, the higher the expected return must be before an investor would be willing to accept that risk. In other words, there is a positive correlation between the degree of risk involved and the return an investor would demand before accepting that risk.

The risk-free rate of return can be defined as the rate of return that would exist for an investment with a guaranteed net return. Insured savings accounts and U.S. government securities are typically considered to be nearly risk-free investments. Continuing with the example from the last section, risk can be added to the discount rate as follows:

Risk-free real rate of return	4%
Expected inflation rate	+ 3%
Risk premium	+ 1%
Adjusted discount rate	= 8%

Investments with a greater amount of risk would be assigned a higher risk premium, and vice versa. This procedure is consistent with an earlier statement that the discount rate should be the rate of return expected from an alternative investment with equal risk. However, the risk premium is a subjective estimate, and different individuals may use different estimates for the same investment. Interest rates on borrowed funds typically have some risk premium already built in to reflect the possibility that a borrower will not repay the loan. Borrowers with less financial security will be asked to pay a higher risk premium.

Adding a risk premium to the discount rate does not eliminate risk. It is simply a way to build in a margin for error. The higher the risk premium used, the higher the net cash revenues will have to be to yield a positive NPV.

Sensitivity Analysis

Given the uncertainty that may exist about the future prices and costs used to estimate net cash revenues and the terminal value, it is often useful to look at what would happen to NPV or IRR if the prices and costs were different. Sensitivity analysis is a process of asking several *what if?* questions. What if the net cash revenues were higher or lower? What if the timing of net cash revenues were different? What if the discount rate were higher or lower?

Sensitivity analysis involves changing one or more values and recalculating the NPV or IRR. Recalculating for a number of different values gives the investor an idea of how *sensitive* the results, and therefore the investment decision, are to changes in the values being used. A sensitivity analysis will often give the investor better insight into the probability that the investment will be profitable, the effects of changes in prices or the discount rate, and therefore the amount of risk involved. Decision trees, contingency tables, and other analysis tools presented in Chapter 15 can be used to compare alternative risky investments.

Recalculating the NPV, annual equivalent, and IRR can be tedious and time consuming, particularly for investments with a long life and variable net cash revenues. However, with specialized financial software or electronic spreadsheet programs, the results for many different combinations of net cash revenues, terminal values, and discount rates can be compared quickly and easily.

The Appendix to this chapter presents a complete example of analyzing an investment using the methods discussed. Variable cash flows, tax effects, and inflation are all incorporated.

SUMMARY

The future value of a sum of money is greater than its present value because of the interest it can earn over time. Future values are calculated through a process called compounding. The present value of a sum of money is smaller than its future value, because money invested today at compound interest will grow into the larger future value. Discounting is the process of finding present values for amounts to be received in the future and is the inverse of compounding.

Investments can be analyzed by any of four methods: payback period, simple rate of return, NPV, and IRR. The first two are easy to use but have the disadvantage of not accurately incorporating the time value of money into the analysis. This may cause errors in selecting or ranking alternative investments. The NPV method is widely used, because it properly accounts for the time value of money. An investment with a positive NPV is profitable, because the present value of the cash inflows exceeds the present value of the cash outflows. The IRR method also considers the time value of money. It represents the discount rate at which the NPV of an investment is exactly zero. An IRR higher than the normal discount rate indicates a profitable investment.

All four methods of investment analysis require estimation of net cash revenues over the life of the investment, as well as terminal values. The NPV method also requires selecting a discount rate. Both the net cash revenues and the discount rate should be on an after-tax basis in a practical application of these methods. The cash flows and discount rate can also be adjusted for risk and inflation. A final step in analyzing any investment should be a financial feasibility analysis, particularly when a large amount of borrowed capital is used to finance the investment. Alternative financing methods for the same investment can also be compared by using NPV calculations.

QUESTIONS FOR REVIEW AND FURTHER THOUGHT

1. Put the concepts of future value and present value into your own words. How would you explain these concepts to someone hearing about them for the first time?

2. Explain the difference between compounding and discounting.

3. Assume someone wishes to have $80,000 10 years from now as a college education fund for a child.

   a. How much money would have to be invested today at 6 percent compound interest? At 8 percent?

   b. How much would have to be invested annually at 6 percent compound interest? At 8 percent?

4. If farmland is currently worth $4,150 per acre and is expected to increase in value at a rate of 5 percent annually, what will it be worth in 5 years? In 10 years? In 20 years?

5. If you require a 7 percent rate of return, how much could you afford to pay for an acre of land that has expected annual net cash revenues of $60 per acre for 10 years and an expected selling price of $3,400 per acre at the end of 10 years?

6. Assume you have only $20,000 to invest and must choose between the two investments in the following table. Analyze each one using all four methods discussed in this chapter and an 8 percent opportunity cost for capital (discount rate). Which investment would you select? Why?

	Investment A ($)	Investment B ($)
Initial cost	20,000	20,000
Net cash revenues:		
Year 1	6,000	5,000
Year 2	6,000	5,000
Year 3	6,000	5,000
Year 4	6,000	5,000
Year 5	6,000	5,000
Terminal value	0	8,000

7. Discuss economic profitability and financial feasibility. How are they different? Why should both be considered when analyzing a potential investment?

8. What two approaches can be used to account for the effects of income taxes in investment analysis?

9. What two approaches can be used to account for the effects of inflation in investment analysis?

10. Why would capital budgeting be useful in analyzing an investment of establishing an orchard where the trees would not become productive until six years after planting?

11. What advantages would present value techniques have over partial budgeting for analyzing the orchard investment in question 10?

APPENDIX. AN EXAMPLE OF AN INVESTMENT ANALYSIS

Joe and Sheila Mason have 5 acres of sloping land not suitable for crop production. It is cleared and well fenced, however. They are considering raising Christmas trees on it. Their teenage children would be able to help, and the income would be useful when they are ready for college. However, Joe and Sheila aren't sure if the Christmas tree project would be better than putting money into a savings account each year. With help from a forestry specialist and their farm financial advisor, they put together an investment analysis using net present value techniques.

INITIAL COST

Their first task is to estimate the *initial cost* for 5 acres of Christmas trees. Their budget for establishing the trees is as follows:

	Quantity per acre	Dollars per acre	Total investment
Machinery expense			
Sprayer	3 times over	9.00	$ 45
Mower	3 times over	21.00	105
Tree planter	1 time over	5.00	25
			$ 175
Labor @ $10.00 per hour			
Spraying	2 hours	20.00	$ 100
Mowing	4 hours	40.00	200
Planting	3 hours	30.00	150
			$ 450
Supplies			
Burn-down herbicide		15.00	$ 75
Pre-emergent herbicide		18.00	90
Pesticides		15.00	75
Fall herbicide		15.00	75
Tree seedlings @ $0.70	850	595.00	2,975
			$3,290
Rental of tree planter @ $8.00	3 hours	24.00	120
Total initial cost			$4,035

ESTIMATING CASH EXPENSES AND REVENUES

Joe and Sheila Mason now want to project their cash expenses and revenue for the next eight years for their Christmas tree investment. Each year they will mow around the trees and spray weeds. The first two years they will have to replant some trees by hand. Starting in the third year they will have to shear (prune) the trees each year. They plan to pay a wage of $10 per hour to family members as well as hired workers. No land cost is included because they feel that land costs would not be affected by the investment.

In Years 6, 7, and 8 they expect to begin selling trees. By the end of the eighth year they expect to have sold the last of their trees. Following is a year-by-year summary of their expected production costs.

Year	Machinery operating	Seedlings	Pesticides, herbicides	Labor	Retailing	Total costs
1	$115	$672	$350	$ 500	$0	$1,637
2	115	35	350	1,000	0	1,500
3	115	0	350	1,500	0	1,965
4	115	0	350	1,800	0	2,265
5	115	0	350	2,200	0	2,665
6	115	0	350	2,900	535	3,900
7	115	0	350	4,750	2,000	7,215
8	115	0	350	2,600	850	3,915

They estimate that they can receive an average selling price of $18 per tree, with the largest portion of the sales occurring in Year 7. Following is a year-by-year estimate of their cash income.

Year	Trees sold @ $18	Income
6	535	$ 9,630
7	2,000	36,000
8	850	15,300

THE DISCOUNT RATE

Now that the Masons have estimated their net cash flows from their Christmas tree investment, they must choose a discount rate for calculating the present values. They plan to finance about 60 percent of the costs out of their savings and borrow the rest. They estimate that their savings account will earn interest at an average rate of 5 percent annually. Their lender anticipates that the average interest rate on their borrowed funds will be about 10 percent. Thus, their *weighted cost of capital* is

$$(5\% \times 0.60) + (10\% \times 0.40) = 7.0\%$$

Joe and Sheila did not incorporate any effects from inflation on wages, the prices of inputs, or the selling price of Christmas trees when they estimated their cash inflows and outflows. Thus, their estimates are *real* values. To adjust their discount rate to a real value also, they subtract 2 percent, their anticipated annual rate of inflation in prices over the next eight years, from their weighted cost of capital to get a *real discount rate*:

$$7.0\% - 2.0\% = 5.0\%$$

The Masons are generally in the 28 percent marginal tax bracket for their federal income taxes, and the 5 percent bracket for state income taxes. In addition, they will have to pay self-employment tax at the rate of about 15 percent on their profits. Their total *marginal tax rate* is

$$28\% + 5\% + 15\% = 48\%$$

Until they begin selling trees, they will show a negative taxable income, so their taxes will be decreased by this rate. Their *after-tax real discount rate* becomes

$$5.0\% \times (1.00 - 0.48) = 2.6\%$$

Finally, the Masons realize that several sources of risk are associated with this investment, such as death of trees and fluctuating selling prices. They decide that they would like to earn at least a 3 percent higher return after taxes on this investment compared to their savings account, to compensate for the added risk. Thus, their final *risk-adjusted real discount rate* is

$$2.6\% + 3.0\% = 5.6\%$$

NET CASH REVENUES

Combining the expected cash income for each of the eight years with the expected cash expenses provides the net cash revenue estimates shown in the following table. For tax purposes, the Masons can deduct the initial cost of establishing the trees, $4,035, as a depletion allowance when they begin selling trees. They expect to sell 3,385 trees, so they can deduct ($4,035 ÷ 3,385) = $1.192 for each tree sold in Years 6, 7, and 8. Their total tax deductible expenses (cash expenses plus depletion) are shown in the last column. Negative values represent additional taxable income.

Year	(a) Cash Income	(b) Cash expenses	(c) Net cash revenue (a − b)	(d) Income tax depletion	(e) Total tax deduction (d − c)
1	$ 0	$1,637	$−1,637	$0	$1,637
2	0	1,500	−1,500	0	1,500
3	0	1,965	−1,965	0	1,965
4	0	2,265	−2,265	0	2,265
5	0	2,665	−2,665	0	2,665
6	9,630	3,900	5,730	638	−5,092
7	36,000	7,215	28,785	2,384	−26,401
8	15,300	3,915	11,385	1,013	−10,372

NET PRESENT VALUE

The Masons can deduct the cash operating expenses plus depletion from their business income tax return each year. Given their marginal tax rate of 48 percent, income tax savings are equal to 48 percent of their total tax deductions each year. Once they begin selling trees, their cash income will exceed their deductible costs. They will have to pay additional taxes (negative savings) equal to 48 percent of the difference between income and deductible expenses. The Masons' after-tax, net cash revenue is shown in the following table, along with its present value.

Year	(f) Net cash revenue	(g) Income tax savings (e × 48%)	(h) After-tax net cash revenue (f + g)	(i) Discount factor*	(j) Present value (h × i)
1	$-1,637	$786	$-851	0.947	$-806
2	-1,500	720	-780	0.897	-699
3	-1,965	943	-1,022	0.849	-868
4	-2,265	1,087	-1,178	0.804	-947
5	-2,665	1,279	-1,386	0.761	-1,055
6	5,730	-2,444	3,286	0.721	2,369
7	28,785	-12,672	16,113	0.683	11,003
8	11,385	-4,978	6,407	0.647	4,143
				Present value	$13,140
				Internal rate of return	20.2%

*The discount factor is equal to $(1 + i)^{-n}$, where i is the discount rate and n is the year.

The present value of each year's after-tax net cash revenue is calculated by multiplying by the discount factor for that year, based on the Masons' estimated annual discount rate of 5.6 percent. When the present values for all eight years are summed, they equal $13,140. Subtracting the initial cost of $4,035 from this amount shows a net present value from the Christmas tree enterprise of $9,105. This means that the net value of the revenue generated over the next eight years is equal to receiving $9,105 today, tax free. When they calculate the projected IRR on their investment, it turns out to be over 20 percent, substantially higher than their cost of capital.

The Masons decide that this is more than enough to compensate them for their risk and management and decide to go ahead with the project.

Creatas Images

ENTERPRISE ANALYSIS

CHAPTER OUTLINE

Profit and Cost Centers
The Accounting Period
Types of Enterprises
Land Costs
Verifying Production
Accounting Systems
Summary
Questions for Review and Further Thought

CHAPTER OBJECTIVES

1. Discuss how to analyze individual enterprises by defining profit and cost centers

2. Explain how the accounting period for an enterprise can match either the calendar year or the production cycle

3. Illustrate how various types of revenues and costs can be assigned, including internal transactions

4. Demonstrate how production and inventory values can be verified by comparing sources and uses of crops and livestock

In Chapter 6 procedures for analyzing the profitability and efficiency of the entire ranch or farm business were discussed. However, even when an analysis indicates a problem, the source of the problem may be difficult to identify if there are many different enterprises in the business. Several enterprises may be highly profitable, while others are unprofitable or only marginally profitable.

Enterprise analysis can identify the less-profitable enterprises, so that some type of corrective action can be taken. An enterprise analysis consists of allocating all income and expenses of the farm or ranch among the individual enterprises being carried out. The result is similar to an income statement for each enterprise, showing its gross revenue, expenses, and net income.

At least three major justifications exist for enterprise analysis: (1) Data gathered during an enterprise analysis are extremely useful for developing enterprise budgets for future years. (2) They can be used to calculate the total cost of production per unit, which is helpful for making marketing decisions. (3) When resources such as land, labor, or capital are limited in supply, enterprise analysis can show which activities are generating the highest returns to the scarcest resource.

PROFIT AND COST CENTERS

The first step in enterprise accounting is to define the enterprises being carried out on the farm or ranch. Usually each crop being produced and each species of livestock present is considered to be a separate enterprise. A commonly used accounting term for such enterprises is *profit center*. A profit center has both income and expenses, and is expected to contribute to the overall profitability of the business.

Sometimes a general type of crop can be divided into multiple enterprises that require different production methods or target a special market. An example would be dividing corn production into yellow corn, white corn, sweet corn, waxy corn, and seed corn. Likewise, livestock enterprises can be divided into production phases. A cow-calf ranch may divide its costs and returns into the breeding phase (up to the weaning date of the calves) and a finishing or feedlot phase. A farrow-to-finish hog operation may want to analyze the farrowing, nursery, and finishing phases separately, to find out which one is contributing most to overall profit.

Some resources and activities may contribute to more than one profit center, but not earn any revenue from outside the business. In smaller farms and ranches the costs of such services may simply be divided among enterprises based on the number of acres or number of head of livestock devoted to each enterprise. In larger operations, though, separate *cost centers* can be created in the accounting system. A cost center is not expected to generate any income by itself, but it does incur various costs while providing services to the profit centers. Some examples of cost centers found in agricultural production are

- Farm machinery services
- Farm labor
- A feed mill
- An irrigation system
- Land acquisition activities

All the costs associated with a cost center are accumulated separately in the accounting system and allocated to the various profit centers at the end of the accounting cycle. Some measure of use, such as hours of field time or tons of feed processed, must be kept so that costs can be allocated fairly and easily.

Box 18-1   Guidelines from the Farm Financial Standards Council

Guidelines developed by the Farm Financial Standards Council (FFSC) for constructing financial statements for farms and ranches and analyzing them were introduced in Chapter 3. More recently, the FFSC issued a second report called "Management Accounting Guidelines for Agricultural Producers."

This report explains in detail the various types of centers that can be defined within a farm accounting system, and how the information derived from them can be used to make better management decisions. A sample chart of accounts and a step-by-step process for setting up a system are included. Copies of FFSC publications are available at www.ffsc.org.

Even though cost centers do not generate income, they can be evaluated based on the cost of alternative means of obtaining the same services. For example, in a machinery cost center, the total cost per acre of performing each field operation can be compared to the typical cost for hiring an outside custom operator to perform the same operation. In this case the value of operator labor would need to be included with the other machinery costs to make a fair comparison. A partial budget, such as described in Chapter 12, can be used to determine which costs would actually be reduced by hiring the service done by an outside provider and how they would compare to the added costs. Some fixed costs for the resources being used to perform the service might not change, such as depreciation of machinery that would still be owned.

Some activities can be both a profit and a cost center. A common example would be farmers who do some custom machinery work for other farmers as well as perform all of their own operations. Income earned from the outside work would typically be subtracted from the total machinery costs before they are allocated to the other enterprises on the farm, in order not to overestimate the real cost of providing machinery services to those enterprises.

THE ACCOUNTING PERIOD

The accounting period for analyzing an enterprise must be defined. If the enterprise accounting period is the same as the whole farm's accounting year, it is important to include the value of beginning and ending inventories of crops and livestock in the net income calculation. All the other usual accrual adjustments to income and expenses need to be made as well. For activities that have continuous production, such as dairy, this approach makes sense.

However, other enterprises have definite beginning and ending dates within the year, such as when feeder or stocker livestock are bought and sold. For these enterprises it is often more useful to summarize costs and returns over the production cycle, instead of the calendar or fiscal year. This could be from when the first inputs are purchased until the last bushel of grain or head of livestock is sold. It could include parts of several accounting years, or it can be less than a full year. Enterprise accounting periods may even overlap, such as when seed for the next crop is purchased before the current crop is sold. Accrual adjustments are not necessary, since all costs are paid and all production sold within the enterprise accounting period.

For breeding livestock, the enterprise accounting period should begin when the first expenses toward the next offspring, such as breeding fees, are incurred. Feed expenses should not be allocated toward the next cycle until the last animals have been weaned, however.

For livestock feeding enterprises the accounting period is often less than one year. If multiple groups utilize the facility, the annual ownership costs should be divided among them. For example, if three turns of feeder pigs are finished in one confinement building in the same year, only one-third of the annual depreciation, interest, insurance, and repairs should be charged to each group. If the feeding facilities are used for only one group of animals per year, all of the annual ownership costs should be charged to that group, even if the facility is used for less than 12 months.

Many fixed costs are *period expenses*; that is, they accrue over a period of time and are not directly related to the production of a specific enterprise or the use of a particular resource. Examples are depreciation, property insurance, interest on loans, and salaries and benefits for full-time labor. Care must be taken that all period expenses are ultimately assigned to an enterprise, but that none of them are double-counted. They are often allocated according to how long the production cycle for each enterprise lasts.

For enterprises that have only one cycle per year, an extra digit (such as the last digit of the calendar year) can be added to the account code to indicate the production year to which costs or revenue should be assigned. For example, a 1 would be assigned to all costs for producing the 2016 crop, even if they were incurred in the year 2015 or 2017. Two digits may be needed for livestock or crop enterprises that have more than one production cycle per calendar year. A farm producing three crops of onions in 2016 might assign codes of 61, 62, and 63 to them, for example.

TYPES OF ENTERPRISES

Several examples will be presented to illustrate how an enterprise analysis might look.

Crop Enterprise Analysis

An example of an enterprise analysis for peanuts is shown in Table 18-1. Values are given for the whole farm and per acre. The farm planted 125 acres to peanuts, which was 40 percent of its total crop acres. The first step is to calculate the total income during the accounting period, beginning with sales. In addition, any crop insurance payments or government program benefits received that apply to the peanut crop should be included.

Costs are summarized next. The total costs of items such as seed, fertilizer, and pesticides are relatively easy to calculate from farm accounting records, or they can be estimated based on the quantities actually used.

Costs such as fuel, machinery repairs, depreciation, interest, and wages are more difficult to allocate fairly among enterprises, unless good records are kept on the hours of machinery and labor used by each one. If such records are not available, costs can simply be allocated equally over all planted acres. In the Table 18-1 example, it was estimated that the peanuts took 35 percent of the field time used that year, so 35 percent of the annual machinery and labor costs were charged to the peanut enterprise. Likewise, 40 percent of the cropland was planted to peanuts, so 40 percent of the land charge was assigned to this crop.

Overhead and miscellaneous costs such as property and liability insurance, consultant fees, office expenses, and general building upkeep are also difficult to assign to individual enterprises. They can be allocated in the same proportion as the contribution of each enterprise to gross revenue or to all other expenses, or by some other arbitrary procedure. In the Table 18-1 example, 30 percent of the farm's gross revenue came from peanuts, so 30 percent of the overhead costs were assigned to that crop.

The example data indicate that the peanut enterprise had a total profit above all costs of $8,496 or $67.96 per acre. These values can be compared against similar values for other crop enterprises to determine which ones are contributing the most to overall farm profit. If any crop shows a consistent loss for several years, action should be taken either to improve its profitability or to shift resources to the production of some other more profitable crop. Note that the *gross margin* value can also be used to rank or eliminate enterprises, since fixed costs (by definition) will not be affected by changing the enterprise mix (see Chapter 10).

Livestock Enterprise Analysis

An analysis of a livestock enterprise can be conducted in a manner similar to that used for a crop enterprise. Table 18-2 shows an example for a dairy herd. Several special problems can arise with livestock, however. Since revenue from the sale of culled breeding stock is generally included as a source of cash income, changes in the inventory value of the breeding herd should also be included in income. Otherwise, higher- or lower-than-normal culling rates in some periods could make that group look unfairly profitable or unprofitable. The per-head values for breeding livestock should be held constant from one period to the next to avoid biasing the profit estimates.

Another problem is how to handle farm-raised crops fed to livestock or used as bedding. The amounts of grain, hay, and silage fed or

TABLE 18-1	Example of an Enterprise Analysis for Dryland Peanuts (125 Acres)

	Farm total	Percent peanuts	Peanuts total	Per acre
Peanut production (125 acres)			387,650 lb	3,101
Income:				
Sales of peanuts	$109,375	100%	$109,375	$875.00
Variable costs:				
Seed (peanuts only)	$ 12,188	100%	$ 12,188	$ 97.50
Inoculant (peanuts only)	906	100%	906	7.25
Fertilizer and lime (peanuts only)	7,079	100%	7,079	56.63
Pesticides (peanuts only)	18,395	100%	18,395	147.16
Fuel, lubrication	26,318	35%	9,211	73.69
Machinery repairs	17,357	35%	6,075	48.60
Cleaning and drying (peanuts only)	4,764	100%	4,764	38.11
Insurance, miscellaneous	15,000	35%	5,250	42.00
Labor	12,093	35%	4,233	33.86
Marketing and checkoff	3,743	35%	1,310	10.48
Interest on variable costs	6,324	35%	2,213	17.71
Total variable costs			$ 71,623	$572.99
Gross margin			$ 37,752	$302.01
Fixed costs:				
Machinery ownership	$ 52,664	35%	$ 18,433	$147.46
Land rent	18,125	40%	7,250	58
Farm overhead	11,913	30%	3,574	28.59
Total fixed costs			$ 29,256	$234.05
Total costs			$100,880	$807.04
Profit			$ 8,496	$ 67.96
Average income per pound				$ 0.28
Average cost per pound				$ 0.26
Average profit per pound				$ 0.02

straw used should be measured or estimated. They are then valued according to the market prices that were available during the feeding period, minus potential transportation costs to market and other selling costs. This represents the opportunity cost of not selling the crops to an off-farm buyer.

Another approach can be used when crops are raised solely for the purpose of being used for livestock feed or bedding, and have no realistic

TABLE 18-2	**Example of an Enterprise Analysis for a 250-cow Dairy Herd**

	Farm total	**Per cow**
Income:		
Milk sales	$1,106,651	$4,426.60
Non-breeding livestock sales	49,692	198.77
Breeding livestock sales	60,088	240.35
Change in breeding livestock inventory		
Total breeding stock value, end of year	396,250	
Total breeding stock value, beginning of year	(392,000)	
Change in inventory value of breeding stock	4,250	17.00
Total income	$1,220,681	$4,882.72
Cash expenses		
Breeding fees	$ 10,440	$ 41.76
Employee wages	57,015	228.06
Employee benefits	34,103	136.41
Feed purchased	144,758	579.03
Cost of farm = raised feed		
Corn grain, 25,000 bushels @ $5.35	133,750	535.00
Corn silage, 2,000 tons @ $48.00	96,000	384.00
Alfalfa hay, 1,500 tons @ $128.00	192,000	768.00
Freight and transportation	14,610	58.44
Dairy share of farm insurance (half)	8,175	32.70
Operating interest	7,520	30.08
Building, fence and equipment repairs	15,883	63.53
Livestock supplies	37,310	149.24
Real estate taxes, dairy share (half)	9,340	37.36
Utilities	40,560	162.24
Veterinary and health	31,453	125.81
Marketing fees	13,340	53.36
Purchase of breeding heifers	81,028	324.11
Other livestock expenses	4,850	19.40
Total cash expenses	$ 932,133	$3,728.53
Noncash expenses		
Depreciation on buildings and equipment	$ 94,183	$ 376.73
Change in prepaid expenses	(6,140)	(24.56)
Total noncash expenses	$ 88,043	$ 352.17
Total expenses	$1,020,176	$4,080.70
Net farm income	$ 200,506	$ 802.02
Pounds of milk sold	5,432,750	21,731
Average sale price per hundred-weight of milk		$ 20.37

market of their own. Examples could be corn silage, haylage, corn stover, or pasture. In such cases the crops can be treated as a cost center, and the transfer price would simply be the total cost of production per feed unit (bushel, ton, pound, etc.). This is the approach followed in the example analysis in Table 18-2. If all of a certain crop is used in a single livestock enterprise, it can simply be included as part of the livestock enterprise rather than summarized as a separate enterprise. All of the crop production costs would be included in the feed costs for the livestock.

Livestock manure has value as a replacement for commercial fertilizer products. If a significant amount of the manure produced by a livestock enterprise is applied to crops, it can be valued based on the potential cost of the purchased nutrients that it replaces, minus the costs of handling and applying it. This would appear as revenue for the livestock enterprise and a cost to the crop enterprise that utilizes it.

Enterprise Interactions

One word of caution is needed, however. Enterprise analysis does not identify or value any complementary or detrimental interactions between enterprises. For example, corn may appear to be more profitable than a legume crop such as soybeans or alfalfa. However, growing continuous corn actually may be less profitable than a rotation containing other crops that contribute nitrogen to the soil, break up pest cycles, or spread out peak workloads. Where the presence of one enterprise significantly affects the performance of another, a whole-farm approach must be used in which various crop rotations or even whole-farm plans are compared.

For example, crops grown in rotation or in combination with other crops have different fertility and pest control needs than the same crops grown alone or continuously on the same land. Tillage practices and soil and water runoff also differ. Livestock enterprises can complement crop enterprises by efficiently using feedstuffs with a low market value and returning fertility to the land through manure disposal. There can be interactions among production practices used on a single crop, such as the placement of fertilizer and pesticides and the type of tillage practices followed. These interactions are difficult to quantify and difficult to incorporate into standard enterprise summaries. However, they should be considered in whole-farm planning and budgeting.

Internal Transactions

Table 18-3 shows an example of an annual net income statement using enterprise accounting, with several *internal transactions*. The value of farm-raised feed was entered as a cost to the livestock enterprise that consumed it and as income to the crop enterprise that produced it. Likewise, the value of manure produced and spread on cropland was credited as income to the livestock and an expense to the crops. Machinery costs were recorded separately, and then divided between crops and livestock according to the estimated hours of use for each one. Miscellaneous income and unallocated expenses were summarized in a separate cost center. Their net value was assigned to the cattle and crop enterprises in the same proportion as gross revenue generated.

Other internal transactions that could occur in enterprise accounting include transferring the value of weaned livestock from a breeding to a feeding enterprise, or transferring the value of dairy replacement heifers from a heifer-raising enterprise to the milking herd. Note that internal transactions do not affect the value of total income or expenses for the whole farm. Income to one enterprise is always exactly offset by a cost to one or more other enterprises. Care should be taken to choose a fair market price for valuing internal transactions, so as to not unfairly bias the results toward one enterprise or another.

TABLE 18-3	Income Statement Example with Enterprise Accounting

	Whole farm	Crops	Cattle	Machinery	Overhead
Income					
Cash crop sales	$ 42,644	$42,644			
Cash livestock sales	72,271		72,271		
Government payments	2,100	2,100			
Miscellaneous income	3,369				$3,369
Home consumption	427		427		
Livestock inventory change	(2,870)		(2,870)		
Crop inventory change	13,835	13,835			
Total revenue	$131,776	$58,579	$69,828	$0	$3,369
Expenses					
Crop inputs	$ 16,971	$16,971			
Machine hire	4,693			4,693	
Fuel, lubrication	4,356			4,356	
Machinery repairs	3,780			3,780	
Building repairs	3,224	1,156	2,068		
Purchased feed	6,031		6,031		
Insurance, property taxes	3,462				3,462
Utilities	2,056	456	1,600		
Interest paid	19,433	15,000	3,000		1,433
Livestock health, supplies	1,228		1,228		
Miscellaneous	4,021				4,021
Depreciation	19,058	1,688	3,351	12,944	1,075
Total expenses	$ 88,313	$35,271	$17,278	$25,773	$9,991
Net farm income, unadjusted	$ 43,463	$23,308	$52,550	($25,773)	($6,622)
Internal transactions					
Raised crops fed	0	39,500	(39,500)		
Manure credit	0	(4,500)	4,500		
Machine work (allocated by hours)	0	(23,773)	(2,000)	25,773	
Allocation of net overhead					
(proportional to gross revenue)	0	(2,357)	(4,265)	0	6,622
Net farm income, adjusted	$ 43,463	$32,178	$11,285	$0	$0

Value-Added Enterprises

Some activities are intended to increase the net income received from a commodity after its production cycle has ended. They can be generally described as *value-added enterprises*.

Examples include processing milk, fruits, or vegetables into food products; sorting and packaging products by size or quality; and selling livestock as processed meat. However, most processes that add value to products also add costs. The activities designed to add value can be analyzed as a separate enterprise. The price that the product could have been sold for without further enhancements can be used to transfer it into the value-added enterprise. In the end, the enterprise analysis will tell the manager if the value added to the product is sufficient to pay all the added costs.

Marketing can also be considered a value-added enterprise. The grain enterprise analysis can be terminated at harvest, with the product being valued at its harvest time price and transferred to the marketing enterprise. Items such as broker's fees, options premiums, storage charges, transportation cost, and extra drying costs can be charged to the marketing enterprise. The final selling price, including any gains from hedging or from purchasing options, constitutes the marketing revenue. The net income to marketing shows whether the manager added value to the product with his or her marketing skills or would have been better off simply selling it at harvest.

LAND COSTS

The use of land is typically acquired through ownership, a cash lease, or some type of share lease agreement. These alternatives are described in more detail in Chapter 20. Each has a different type of cost. Owned land generally requires the payment of some real estate taxes and upkeep costs. The cost of cash-rented land is simply the amount of rent paid. Land rented under a typical crop share lease has no direct cost to the tenant. However, only the tenant's share of revenue and production costs should be included in an enterprise summary.

If all the crops produced by the farming unit can be grown on any of the available land, then total land costs can simply be averaged over all acres and the same cost per acre charged to each crop. However, if certain crops can be grown only on certain acres, then a separate cost for each type of land should be estimated and

assigned to the corresponding crop. The same principle applies to pasture land that can be used for any livestock enterprise or only for a particular enterprise.

Comparing Land Units

Some crop producers rent land from multiple owners for different rental rates or under different types of share arrangements. Some of these rented farms may be more productive than others. It is useful to compare the profitability of different land units, especially if their leases can be renewed or terminated yearly. Each crop on each farm can be considered to be a separate profit center. The same rules that were discussed earlier for allocating costs apply. If records allow, costs of inputs can be adjusted for each farm, such as when some units require higher rates of fertilizer application than others. In other cases the total cost for a certain input may simply have to be averaged across all acres. It is very important, though, that the quantity of product harvested from each farm be recorded accurately, in order to fairly assess and compare the profitability of each land unit.

When the quality of the crop harvested varies by farm, this should be reflected in the selling or inventory price assigned to each tract of land. If all the grain is commingled before it is marketed, though, the same average selling price can be assigned to all farms. Finally, a weighted average profit per acre for each land unit can be found based on the income, expenses, and number of acres of each crop grown on that unit.

Farms that consistently show a net loss, or do not at least produce enough income to pay all variable costs plus the land rent, should be dropped from the land base. Some tenants rank all their rented farms by profitability each year, and try to replace the least profitable land units. The cost and income summary can also be used to estimate what a reasonable rental rate would be for each farm. Table 20-3 shows an example of this type of analysis.

Box 18-2   Comparing Land Units

The Sorensen family grows wheat in central North Dakota. They rent land from five different owners, but they were wondering which farms actually contribute the most to their total farm profit. After harvest they separated their direct costs for each farm and developed the comparison below. They did not allocate overhead costs such as machinery depreciation, utilities, or property insurance, because they were not directly affected by the rented land. Likewise, they did not include any expenses on their owned land, such as property taxes.

All the wheat was of similar quality, and was commingled before it was marketed, so they assigned their average marketing price of $6.25 per bushel to all the farms.

Farm name	Turner farm	Richland twp.	Olson estate	Aunt Elizabeth's	Loftsted farm
Acres planted	185	214	144	301	175
Land tenure type	Cash rent	Cash rent	Cash rent	70% share	75% share
Bushels produced per acre	49	64	59	53	55
Gross income @ $6.25 (operator's share)	$306.25	$400.00	$368.75	$231.88	$257.81
Direct expenses per acre (operator's share)					
Seed	$ 13.97	$ 15.90	$ 14.63	$ 9.75	$ 10.95
Fertilizer	71.71	74.27	78.06	51.30	47.05
Pesticides	36.73	33.15	24.14	17.47	18.33
Crop insurance	15.70	21.61	17.03	12.52	13.15
Fuel, oil and repairs	27.52	39.19	34.74	30.04	26.91
Land rent	$ 42.00	$ 64.00	$ 68.00	$ 0.00	$ 0.00
Return over direct expenses and rent	$ 98.62	$151.88	$132.15	$110.79	$141.42

Three of their five farms produced a nice return per acre. The profit from the Turner farm was below average, even though it had the lowest cash rent charge. The Sorensens decided to see if there was other land available in their community that might provide a better return. They also resolved to discuss with Aunt Elizabeth whether a change to a 75 percent share lease might be a more equitable arrangement on her farm.

VERIFYING PRODUCTION

To compute many of the physical efficiency measures used for crop and livestock enterprises, an accurate estimate of the quantity of production is needed. This is usually measured in bushels, tons, pounds, or head. A check on the accuracy of such inventory numbers can be made using the general rule that *sources must equal uses*. Table 18-4 shows the relevant measures of sources and uses for an example crop enterprise, grain sorghum. Most of the quantities and

TABLE 18-4	Verifying Crop Inventories (Grain Sorghum)

Sources	Quantity (cwt.)	Value per cwt.	Value ($)
Beginning inventory	3,100	$5.10	$15,810
Purchased	none		
Produced	13,250		$74,262*
Total	16,350		$90,072

Uses	Quantity (cwt.)	Value per cwt.	Value ($)
Ending inventory	5,300	$5.65	$29,945
Sold	3,470	$5.24	$18,183
Used for seed	None		
Used for feed	7,480	$5.35	$41,944
Spoilage, other losses	100	$0	$ 0
Total	16,350		$90,072

*Equal to total value of crop increase ($90,072 − $15,810).

values can be measured directly or derived from sales receipts or purchase records. However, if all the quantities or values but one are known, the unknown one can be found by calculating the difference between the sum of the sources and the sum of the uses. This is sometimes done to estimate the quantity of feed fed to livestock or the amount of crop produced. However, greater accuracy will be achieved if *all* the physical quantities are measured directly and the equality relation is used to verify their accuracy; that is, the sources total is equal to the uses total.

The total dollar value of production generated during the accounting period can be found by subtracting the value of the beginning inventory and purchases from the total value of all the uses. Note that this value may include a gain or loss on inventories due to price changes during the accounting period, as well as income from sales. If the entire production cycle (through final disposition of the crop) is used as the accounting period, however, no beginning or ending inventories will exist. The value of the crop will simply be the total revenue received from sales and other sources.

Livestock inventories can be verified with a similar procedure. Physical quantities are usually measured in pounds or hundred-weight (100 pounds) as well as number of head. For this reason an extra column is added in Table 18-5. Calves, pigs, or lambs lost to death are assumed to have an ending weight and value of zero. Some animals may also enter or leave the inventory when they are reclassified, such as female offspring selected for breeding stock replacement or calves transferred from the cow herd to the feedlot at weaning.

The numbers for *produced* weight and value can be calculated by subtracting the total of the other sources from the total for uses. The differences are the total pounds of weight gain and the total value of livestock increase, respectively, that occurred during the accounting period.

ACCOUNTING SYSTEMS

Most whole-farm accounting programs also have the ability to perform basic enterprise analysis. Receipts and variable expenses are identified by

TABLE 18-5 | Verifying Livestock Inventories (Cattle)

Sources	Head	Weight (cwt.)	Value per cwt.	Value ($)
Beginning inventory	315	1,890	$160	$302,400
Purchased	265	1,908	$150	$286,200
Produced on the farm	175	2,843*		$388,796**
Total	755	6,641		$977,396

Uses	Head	Weight (cwt.)	Value per cwt.	Value ($)
Ending inventory	296	1,702	$168	$285,936
Sold	415	4,686	$140	$656,040
Death loss	11	—		—
Heifers transferred out	33	253	$140	$ 35,420
Total	755	6,641		$977,396

*Equal to total hundred-weight of gain produced (6,641 – 1,890 – 1,908).
**Equal to total value of livestock increase ($977,396 – $302,400 – $286,200).

enterprise as they are entered, usually by a code number or from a drop-down list. Some programs also have a procedure for automatically allocating overhead costs among enterprises. The computer can quickly sort through all receipts and expenses, collect and organize those that belong to a particular enterprise, and present the results in total dollars per acre or other unit of output.

To do enterprise analysis correctly, the accounting system must track physical quantities of inputs and products as well as monetary values, and allow for internal transactions among enterprises. Considerable care should be taken when setting up the accounting system to clearly define the enterprises to be analyzed and to define the chart of accounts in such a manner as to make it easy to assign income and expenses properly.

SUMMARY

Most farms and ranches produce more than one product. By dividing the business into multiple enterprises, the contribution of each one toward the financial goals of the business can be measured. Profit centers accrue both revenues and expenses, while cost centers only have expenses, and provide services to the other enterprises. The accounting period for an enterprise can be the same as for the whole farm, but it often follows the production cycle of the commodity instead.

Internal transactions can be used to show the value of products produced by one enterprise, such as feed, which are in turn utilized by another enterprise, such as livestock. Keeping careful records of the physical quantities of crops and livestock produced and sold, and using the rules that total sources must equal total uses, can improve the accuracy of enterprise accounting.

QUESTIONS FOR REVIEW AND FURTHER THOUGHT

1. What is the purpose of enterprise analysis?
2. What is the difference between a profit center and a cost center?
3. Think about a farm or ranch with which you are familiar. List all the different enterprises that could be defined for it.
4. What are *internal transactions*? How do they improve the accuracy of enterprise analysis?
5. How can overhead expenses be allocated among enterprises?
6. Given the following inventory, purchase, and sales data for a beef feedlot:

	Head	Weight (lb)	Value ($)
Beginning inventory	850	765,000	918,000
Ending inventory	1,115	936,600	1,095,822
Purchases	1,642	1,018,040	1,298,001
Sales	1,340	1,586,000	1,596,945
Death loss	—	xxx	xxx
Production increase	0	—	—

   a. What was the apparent death loss, in head?
   b. How much beef was produced, in pounds?
   c. What was the value of this production increase?

© Photo by David Nance/USDA

ACQUIRING RESOURCES FOR MANAGEMENT

Although developing farm management skills is the central theme of this book, few people make their living in agriculture from their management skills alone. A large portion of the revenue generated from ranching and farming goes to the providers of the physical, financial, and human resources needed for agricultural production to take place. How much of these resources are available to the manager and how they are obtained may make the difference between operating a business at a profit or a loss.

Net farm income is the return to all resources contributed by the operators. One key to improving it is to increase the quantity or quality of resources owned by the operator over time. Some resources are contributed by the operator and family. Others are obtained through borrowing, renting, or hiring. Some operators contract their management services to other parties. Determining the proper mix of owned and nonowned resources to use is a key management decision. This requires a long-term, strategic plan.

Chapter 19 discusses *capital* as an agricultural production resource. Capital itself does not produce agricultural products, but it can be used to purchase or rent other resources that do. Sources of capital include the operator's equity, borrowing, rented assets, and contributions by partners or investors. The use of credit to acquire capital assets is common in agriculture, but it requires careful planning and control to use it profitably and judiciously.

347

In dollar terms, *land* is the most valuable resource used in agricultural production. Controlling and using farmland is the subject of Chapter 20. Buying and valuing land is discussed, along with the various types of leases and their advantages and disadvantages. Owning and using land for agricultural production requires attention to resource conservation and environmental sustainability as well as profits.

Human resources in agriculture have evolved from doing hard physical labor to carrying out highly skilled tasks using sophisticated equipment and technology. Although labor used in agricultural production has been declining, its productivity has increased greatly. Chapter 21 explains the concepts of planning and managing both hired labor and the operator's own time.

Mechanization has changed the occupation of farming more than any other innovation during the past century. It has caused a rapid increase in productivity per person, which has resulted in less labor in agriculture and larger farms. *Machinery* represents a large capital investment on many farms, and Chapter 22 explores alternatives for acquiring the use of machinery services. Methods for computing and controlling machinery costs and for improving machinery efficiency are also discussed.

Photo by Kirk Withrow

19

CAPITAL AND CREDIT

CHAPTER OUTLINE

Economics of Capital Use

Sources of Capital

Types of Loans

The Cost of Borrowing

Sources of Loan Funds

Establishing and Developing Credit

Liquidity

Solvency

Summary

Questions for Review and Further Thought

CHAPTER OBJECTIVES

1. Point out the importance of capital in agriculture

2. Illustrate the optimal use and allocation of capital

3. Compare different sources of capital and credit in agriculture

4. Describe different types of loans used in ranching and farming

5. Show how to set up various repayment plans for loans

6. Explain how to establish and develop credit worthiness

7. Examine factors that affect the liquidity and solvency of a farm business

Many people think of *capital* as cash, balances in checking and savings accounts, and other types of liquid funds. This is a narrow definition of capital. Capital also includes money invested in livestock, machinery, buildings, land, and any other bought and sold assets.

Agriculture has one of the largest capital investments per worker of any major U.S. industry. This helps make farm operators very productive. Figure 19-1 shows the changes that have taken place in the amount of capital invested in agriculture in the United States,

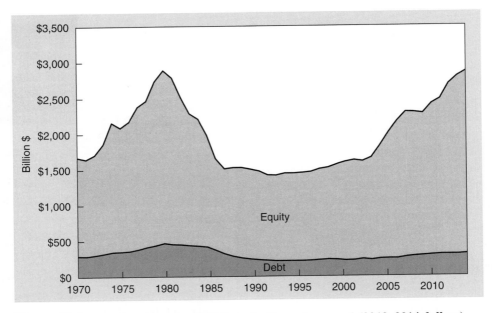

Figure 19-1 **Capital investment in U.S. agriculture, January 1 (2012–2014 dollars).**
Source: U.S. Department of Agriculture, Economic Research Service.

deflated to 2012–2014 dollars. During the 1970s, the value of total agricultural assets grew rapidly. Increases in land prices accounted for much of this increase. Likewise, as land prices declined in many parts of the United States in the early 1980s, the total asset value also declined. Since then, it has grown again, but at a more modest rate.

The capital investment *per farm* has risen even more rapidly than total investment in agriculture, because the number of farms and ranches in the United States has been declining. Some full-time commercial farms and ranches have capital investments over $5,000,000. Land accounts for much of this investment, but buildings, livestock, and machinery are also important. This large capital investment per farm requires a sound understanding of financial management principles to compete in today's international economy.

Credit is important to capital acquisition and use. It is the ability to borrow money with a promise to return the money in the future and pay interest for its use. The use of credit allows farmers and ranchers to acquire productive assets and pay for them later with the income they generate.

ECONOMICS OF CAPITAL USE

Broadly defined, capital is the money invested in the physical inputs used in agricultural production. It is needed to purchase or rent productive assets, pay for labor and other inputs, and finance family living and other personal expenditures. Capital use can be analyzed using the economic principles discussed in Chapters 7 and 8. The basic questions to be answered are

1. How much total capital should be used?
2. How should limited capital be allocated among its many potential uses?

Total Capital Use

When unlimited capital is available, the problem is how much capital the business should use. In Chapter 7, the question of how much input to use was answered by finding the input level where the marginal value product (MVP) was just equal to the marginal input cost (MIC). The same principle can be applied to the use of capital.

Figure 19-2 is a graphical presentation of MVP and MIC, where MVP is declining, as occurs whenever diminishing marginal returns exist. The MVP is the additional net return, before interest payments, that results from an additional capital investment. This added return can be estimated using the partial budgeting techniques described in Chapter 12. The MVP is calculated in a manner similar to the return on assets (ROA), except a factor of 1.00 is added, and ROA measures average rather than marginal return.

Marginal input cost is equal to the additional dollar of capital invested plus the interest that must be paid to use it. Therefore, MIC is equal to $1 + i$, where i is the rate of interest on borrowed funds, or the opportunity cost of investing the farm's own capital. In this example, profit will be maximized by using the amount of capital represented by a, where MVP is equal to MIC; that is, the investment generates just enough revenue to repay the initial capital outlay plus the interest cost.

In some cases, the interest rate increases as more capital is used, such as when a lender classifies a borrower as being in a higher risk category. At the point where this happens, the MIC curve would rise to a higher level. The optimum amount of capital to use would then be less than when the interest rate and MIC are constant.

Allocation of Limited Capital

Many businesses do not have sufficient capital of their own or may not be able to borrow enough to reach the point where MVP is equal to MIC for the total capital being used. In other words, capital is limited to something less than the amount that will maximize total profit. The problem now becomes one of allocating limited capital among its alternative uses. This can be accomplished by using the equal marginal principle, discussed in Chapter 7.

Use of the equal marginal principle results in capital being allocated among alternative uses in such a way that the marginal value product of the last dollar is equal in all uses. Even if no additional capital is available for investment, there may be opportunities for shifting capital between uses to more nearly equate the marginal value products.

This principle is often difficult to apply in an actual farm situation. First, there may be insufficient information available to calculate the marginal value products accurately. Prices and costs are constantly changing. Second, some alternative uses may require large lump-sum investments of an all-or-nothing nature, such as livestock buildings, greenhouses, or irrigation systems. Third, capital invested in assets such as land or buildings cannot easily be shifted to other uses. This makes it difficult to equate the marginal value products for these alternatives with others where capital can be invested dollar by dollar. Nonetheless, difficulties encountered in applying the equal marginal principle should not discourage its use. Whenever limited capital can be reallocated to make the marginal value products more nearly equal, total profit will be increased.

SOURCES OF CAPITAL

Capital consists of cash and assets purchased with cash, so it is relatively easy to intermingle capital from different sources. An important part

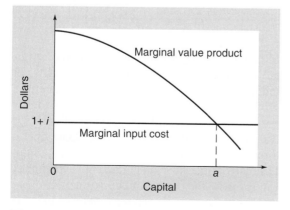

Figure 19-2 Using marginal principles to determine optimal capital use.

Box 19-1   The 1980's Farm Debt Crisis: What Went Wrong?

The 1980s brought financial disaster to many U.S. farmers. Was it caused by irresponsible financial management or by a combination of economic forces over which farmers had no control?

From the end of World War II until the early 1970s, U.S. agriculture increased production by rapidly adopting new technologies. The size of the farms increased as the number of farm units decreased by half. Yet low profit margins and memories of the Great Depression caused farmers to be very conservative in their use of credit.

Suddenly, beginning in 1973, the U.S. farm economy boomed. World demand for agricultural commodities, combined with production shortfalls, caused prices of most farm products to double in less than two years. Farmers saw their net incomes soar and looked to reinvest their windfalls in more land, machinery, and livestock. A plentiful supply of credit and single-digit interest rates fueled the boom (see Figure 19-1). Lenders were only too willing to extend credit when asset values, especially land prices, were increasing at rates of 20 to 30 percent annually (see Figure 20-1).

In the early 1980s, a series of events caused an abrupt turnaround in the farm economy. The Federal Reserve System made a decision to reduce the rate of inflation in the economy (the annual increase in the Consumer Price Index peaked at 13.5 percent in 1980). It restricted the growth of the national money supply, and interest rates suddenly rose to 15 to 20 percent and higher. The subsequent increase in the international value of the dollar caused world demand for U.S. products to decrease, and farm commodity prices plummeted. Many farmers who had borrowed large sums of money through variable interest rate loans now found themselves faced with lower incomes and higher payments.

As the demand for farmland dried up and cash-strapped farmers tried to sell land, real estate values quickly dropped. Lenders moved to foreclose on delinquent loans before collateral values became less than the loan balances. Two severe droughts in the Midwest in 1983 and 1988 further eroded many farmers' financial positions.

An estimated 200,000 to 300,000 farms failed financially during the 1980s. Many rural banks and input suppliers also closed their doors. Many of the farms that survived did so with help from family members, through negotiated write-off of debt by lenders, or because they had been very conservative with their use of credit.

Gradually, interest rates decreased, commodity prices rebounded, and farm asset markets stabilized. However, a whole generation of farmers, ranchers, and agricultural lenders had their attitudes toward capital and credit use profoundly influenced by the Farm Crisis of the 1980s.

of farm financial management is the ability to obtain capital from several sources and combine it in the proper proportions in various uses.

Owner Equity

The farmer's own capital is called owner equity or net worth. It is calculated as the difference between the total assets and the total liabilities of the business, as shown on the balance sheet discussed in Chapter 4. There are several ways the operator can secure or accumulate equity.

Most farmers begin with a contribution of original capital acquired through savings, gifts, or inheritances. As the farm or ranch generates profits in excess of what is withdrawn to pay personal expenses and taxes, retained earnings can be reinvested in the business. Some operators may have outside earnings, such as a non-farm job or other investment income, which they can invest in their farming operation.

Assets already owned may increase in value through inflation or changes in demand. This does not increase the amount or productivity of the physical assets, but additional cash can be obtained by either selling the assets or using them as collateral for a loan.

Outside Equity

Some investors may be willing to contribute capital to a farm or ranch without being the operator. Under some types of share lease agreements, the landowner contributes operating capital to buy seed and fertilizer, or even provides equipment and breeding livestock, as explained in Chapter 20. Larger agricultural operations may include limited or silent partners who contribute capital but do not participate in management. Incorporated farms may sell stock to outside investors. These arrangements increase the pool of capital available to the business but also obligate the business to share earnings with the investors.

Leasing

It is often cheaper to gain the use of capital assets by leasing or renting rather than owning them. Short-term leases make it easier for the operator to change the amount and type of assets used from year to year. However, this also creates more uncertainty about the availability of assets such as land and discourages making long-term improvements. Sometimes assets such as land, buildings, equipment, and breeding livestock can be leased under multiyear agreements. This can reduce annual cash flow commitments, but does not allow the business to build equity in such assets. The advantages and disadvantages of leasing farm assets are discussed in more detail in Chapters 20 and 22.

Contracting

Farmers or ranchers who have restricted access to capital or credit, or who wish to limit their financial risk, may contract their services to agricultural investors. Examples include custom feeding of cattle, finishing pigs on contract, contract broiler or egg production, and custom crop farming. Typically, the operator provides labor and management and some of the equipment or buildings, while the investor pays for the other inputs. The operator receives a fixed payment per unit of production. Any special skills the operator has can be leveraged over more units of production without increasing financial risk. However, potential returns per unit for contract operations may be lower than for a well-managed owner-operated business.

Credit

After owner's equity, capital obtained through credit is the second largest source of farm capital. Borrowed money can provide a means to

- Quickly increase business size,
- Improve the efficiency of other resources,
- Spread out the purchase cost of capital assets over time, and
- Withstand temporary periods of negative cash flow.

Farm debt increased rapidly in the late 1970s and early 1980s (see Figure 19-1). However, in the mid-1980s, many farmers sold assets to reduce debt or had loan payments forgiven when they were unable to pay. Since then, farm liabilities have again grown, but at a slow rate.

Real estate debt, or borrowing for land and buildings, accounts for approximately one-half of the total debt. Borrowing for livestock, machinery, and grain inventories accounts for the other half. Comparing total farm debt against total farm assets indicates that U.S. agriculture is in sound financial condition (see Figure 19-3). However, this does not mean that every individual farmer and rancher is in sound financial condition. There are always individuals and businesses who have more debt than they can easily service.

TYPES OF LOANS

Agricultural loans can be classified by their length of repayment, use of the funds, and type of security pledged. All of them include certain terms used in the credit industry. A prospective borrower needs to be familiar with these terms to communicate effectively with lenders.

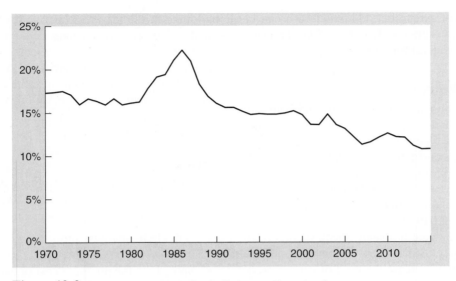

Figure 19-3 **Debt-to-asset ratio for U.S. farms, January 1.**
Source: U.S. Department of Agriculture, Economic Research Service.

Length of Repayment Term

Classifying loans by the length of the repayment period is widely used when preparing balance sheets, as was discussed in Chapter 4. Three time classifications are commonly used in agricultural lending.

Short-Term Loans

Short-term loans are generally used to purchase inputs needed to operate through the current production cycle. Purchases of fertilizer, seed, feeder livestock, and feed are examples. Wages and rents are also financed with short-term credit. Repayment is due when the crops are harvested and sold or when the feeder livestock are sold. Short-term loans are also called *production* or *operating loans* and are listed under current liabilities on the farm balance sheet.

Intermediate-Term Loans

When a loan is repayable over more than one year but less than 10 years, it is classified as an intermediate-term loan. One or more payments are usually due each year. Intermediate-term loans are often used for the purchase of machinery, breeding and dairy livestock, and some buildings. These assets will be used in production for several years and cannot be expected to pay for themselves in one year or less.

Long-Term Loans

A loan with a term of 10 years or longer is classified as a long-term loan. Assets with a long or indefinite life, such as land and buildings, are often purchased with funds from long-term loans. Loans for the purchase of land may be made for a term as long as 20 to 40 years, for example. Annual or semi-annual payments normally are required throughout the term of the loan.

The Farm Financial Standards Council recommends that intermediate- and long-term loans be combined and shown as *noncurrent liabilities* on the balance sheet.

Use

The use or purpose of the funds is another common way to classify loans.

Real Estate Loans

This category includes loans for the purchase of real estate such as land and buildings or where real estate assets serve as security for the loan. Real estate loans are typically long-term loans.

Non–Real Estate Loans

All business loans other than real estate loans are included in this category and are usually short-term or intermediate-term loans. Crops, livestock, machinery, or other non-real estate assets may be pledged as security.

Personal Loans

These are nonbusiness loans used to purchase personal assets such as homes, vehicles, and appliances.

Security

The security for a loan refers to the assets pledged to the lender to ensure loan repayment. If the borrower is unable to make the necessary principal and interest payments on the loan, the lender has the legal right to take possession of the mortgaged assets. These assets can be sold by the lender and the proceeds used to pay off the loan. Assets pledged or mortgaged as security are called loan *collateral*.

Secured Loans

With secured loans, some assets are mortgaged to provide collateral for the loan. Lenders obviously favor secured loans, because they have greater assurance that the loan will be repaid, if not from farm earnings, then from liquidation of assets pledged as collateral. Intermediate- and long-term loans are usually secured by a specific asset, such as a tractor or a parcel of land. Some loans are backed by a blanket security statement, which can even include assets acquired or produced after the loan is obtained, such as growing crops.

Unsecured Loans

A borrower with good credit and a history of prompt loan repayment may be able to borrow some money with only *a promise to repay* or without pledging any specific collateral. This would be an unsecured loan, also called a *signature loan*, as the borrower's signature is the only security provided to the lender. Most lending practices and banking regulations discourage making unsecured loans.

Box 19-2   **Special Loans for Beginning Farmers**

Beginning farmers and ranchers often find it difficult to obtain the credit they need to start their own operations. They generally lack the collateral, net worth, and experience that lenders are looking for in a borrower. However, several types of special programs are available that target new entrants into agriculture.

The Farm Service Agency sets aside a portion of its funds each year to make both farm ownership loans and direct operating loans to beginning farmers and ranchers. FSA can also guarantee a loan from a commercial lender or private party. Farm Credit Services, as well, offers special loan programs to farmers with fewer than 10 years of experience.

Many state departments of agriculture offer beginning farmers loans that are made and serviced through commercial lenders. Often these loans are funded by selling tax-exempt *aggie bonds* that allow a lower interest rate to be charged to the borrower. Typically these loans can be used for the purchase of land, machinery, breeding livestock, and other capital assets.

The definition of a beginning farmer and the eligibility requirements differ for each beginning farmer loan program, but all of them aim to make it easier for the next generation of farmers and ranchers to get established in agriculture.

Repayment Plans

Many types and variations of repayment plans are used for agricultural credit. Lenders try to fit repayment to the purpose of the loan, the type of collateral used to secure the loan, and the borrower's projected cash flow.

When a loan is negotiated, the borrower and the lender should be in agreement about when it is to be repaid. In each case, the total interest paid will increase if the money is borrowed for a longer time. The fundamental equation for calculating interest is

$$I = P \times i \times T$$

where I is the amount of interest to pay, P is the principal or amount of money borrowed or currently owed, i is the interest rate per time period, and T is the number or fraction of time periods over which interest accrues.

Single Payment

A single-payment loan has all the principal payable in one lump sum when the loan is due, plus interest. Operating loans for the purchase of feeder livestock or crop inputs are often of this type. Single-payment loans require good cash flow planning to ensure that sufficient cash will be available when the loan is due.

The interest paid on a loan with a single payment is called *simple interest*. For example, if $80,000 is borrowed for exactly one year at 6 percent annual interest, the single payment would be $84,800, including $80,000 principal and $4,800 interest.

$$\$80,000 \times 6\% \times 1 \text{ year} = \$4,800$$

If the loan is repaid in less than or more than one year, interest would be computed for only the actual time the money was borrowed.

Line of Credit

The use of single-payment loans often means having more money borrowed than the operator really needs at one time or having to take out several individual loans. As an alternative, some lenders allow a borrower to negotiate a *line of credit*. Loan funds are transferred into the farm account as needed, up to an approved maximum amount. When farm income is received, the borrower pays the accumulated interest on the loan first, and then applies the rest of the funds to the principal. There is no fixed repayment schedule or amount.

Table 19-1 contains an example of a line of credit. The amounts borrowed are $80,000 on February 1 and $40,000 on April 1, at 6 percent annual interest. On September 1, a payment of $63,800 is made to the lender. The interest due is calculated as follows:

$$\begin{array}{ll} \$\ 80,000 \times 6\% \times 2/12 = & \$\ \ \ 800 \\ \$120,000 \times 6\% \times 5/12 = & \underline{\$3,000} \\ & \$3,800 \end{array}$$

TABLE 19-1	**Illustration of a Line of Credit**					

Date	Amount borrowed	Interest rate	Amount repaid	Interest paid	Principal paid	Outstanding balance
Feb. 1	$80,000	6%	$ 0	$ 0	$ 0	$ 80,000
Apr. 1	40,000	6%	0	0	0	120,000
Sept. 1	0	6%	63,800	3,800	60,000	60,000
Oct. 1	0	5%	0	0	0	60,000
Dec. 1	0	5%	48,000	800	47,200	12,800

Interest was charged on $80,000 for two months and on $120,000 for five months. The remaining $60,000 is used to reduce the principal balance from $120,000 to $60,000.

On October 1, the interest rate is reduced to 5 percent. Another payment is made on December 1, for $48,000. The interest calculation for this payment is as follows:

$$\$60,000 \times 6\% \times 1/12 = \$300$$
$$\underline{\$60,000 \times 5\% \times 2/12 = \$500}$$
$$\$800$$

Interest accrues at 6 percent for one month and 5 percent for two months. After the interest is paid, the remaining $47,200 goes toward reducing the outstanding principal balance.

A line of credit reduces the borrower's interest costs and results in less time spent in the loan approval process. However, the borrower must exercise more discipline in deciding how to use the loan funds and when and how much to borrow or repay.

Amortized

An *amortized loan* is one that has periodic interest and principal payments. It may also be called an *installment loan*. As the principal is repaid, and the loan balance declines, the interest payments also decline. Assume that $20,000 is borrowed, and the repayment schedule is $10,000 in six months and the remaining $10,000 at the end of one year. The interest calculations would be as follows:

First
payment: $20,000 at 6% for ½ year = $600
Second
payment: $10,000 at 6% for ½ year = $300

Total interest: $900

The total payments would be $10,600 and $10,300.

Interest is paid only on the unpaid loan balance, and then only for the length of time that amount was still borrowed. The total interest is less than if the whole loan was repaid at the end of the year, because only $10,000 was outstanding for the last half of the year.

There are two types of amortization plans: the *equal principal payment* and the *equal total payment*. An amortized loan with equal principal payments has the same amount of principal due on each payment date, plus interest on the unpaid balance. For example, a 10-year, 8 percent loan of $100,000 would have an annual principal payment of $10,000. The loan balance decreases with each principal payment, so the interest payments also decrease, as shown in Figure 19-4.

Borrowers often find the first few loan payments the most difficult to make, because a new or expanded business may take some time to generate its maximum potential cash flow. For this reason, many long-term loans have an amortized repayment schedule with equal total payments, in which all payments are for the same amount. Figure 19-4 also shows the amount of principal and interest paid each year under this plan for a $100,000 loan. A large portion of the total loan payment is interest in the early years, but the interest decreases and the principal increases with each payment, making the last payment mostly principal.

To calculate the total loan payment for an amortized loan with equal total payments, a table of amortization factors can be used. These factors are contained in Appendix Table 1. The annual payment will depend on both the interest rate and the length of the loan. For example, the amortization factor for a 10-year loan at 8 percent interest is 0.14903. This factor is multiplied by the amount of the loan to find the total annual payment. A loan of $100,000 would have an annual payment of (0.14903 × $100,000), or $14,903. This is the same factor and procedure described in Chapter 17 to convert a present sum into an annualized value or annuity. A financial calculator or computer program can be used, instead of the amortization table factors, to compute the size of the payment.

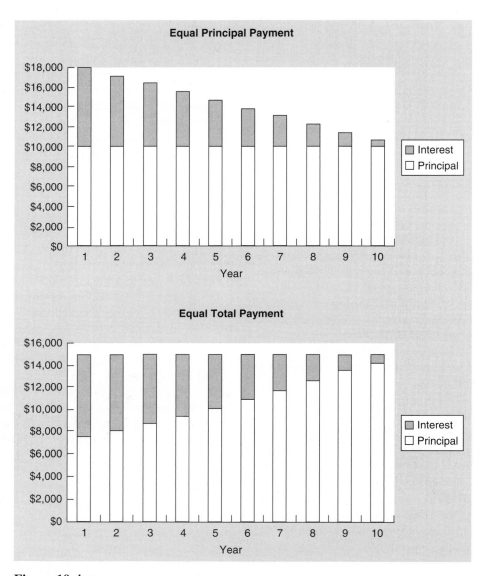

Figure 19-4 Loan repayment under two types of amortization.

Table 19-2 shows the actual principal and interest payments under each plan. The equal total payment method has a smaller total loan payment than the equal principal payment loan in the first four years. However, a total of $49,030 in interest is paid, compared to only $44,000 under the first plan. The advantage of the lower initial payments is partially offset by more total interest being paid over the life of the loan, because the principal is being reduced at a slower rate in the early years.

Balloon Payment Loans

Some amortization schedules are set up with lower periodic payments, such that not all of the principal is repaid by the end of the loan period. For

TABLE 19-2	Amortization of a $100,000 Loan Over 10 Years at 8 Percent Interest							
	Equal principal payments				**Equal total payments**			
Year	Principal paid	Interest paid	Total payment	Principal remaining	Total payment	Interest paid	Principal paid	Principal remaining
1	$ 10,000	$ 8,000	$ 18,000	$90,000	$ 14,903	$ 8,000	$ 6,903	$93,097
2	10,000	7,200	17,200	80,000	14,903	7,448	7,455	85,642
3	10,000	6,400	16,400	70,000	14,903	6,851	8,052	77,590
4	10,000	5,600	15,600	60,000	14,903	6,207	8,696	68,894
5	10,000	4,800	14,800	50,000	14,903	5,512	9,391	59,503
6	10,000	4,000	14,000	40,000	14,903	4,760	10,143	49,360
7	10,000	3,200	13,200	30,000	14,903	3,949	10,954	38,406
8	10,000	2,400	12,400	20,000	14,903	3,073	11,830	26,576
9	10,000	1,600	11,600	10,000	14,903	2,126	12,777	13,799
10	10,000	800	10,800	0	14,903	1,104	13,799	0
	$100,000	$44,000	$144,000		$149,030	$49,030	$100,000	

example, half the principal may be paid with the periodic payments, with the other half due at the end of the loan period. In some cases, the periodic payments may be interest only, with all the principal due at the end of the loan period. These types of loans are called *balloon payment loans*, because the last payment *balloons* in size. They have the advantage of smaller periodic payments but the disadvantage of a large final payment and more total interest cost. Balloon payment loans often require some form of refinancing with another loan when the final payment is due.

The amount of principal that will be paid with the final balloon payment can be set based on the cash flow that is expected to be available to service the loan payments. The periodic payments are calculated by amortizing the principal remaining after subtracting the balloon payment from the original loan using either the equal principal or the equal total payment method previously explained. Then, interest on the principal included in the balloon payment is added to each periodic payment. Table 19-3 shows the annual payments for an equal total payment loan with a 50 percent balloon payment at the end of

TABLE 19-3	Amortization of a $100,000 Loan Over 10 Years at 8 Percent Interest, with a 50 Percent Balloon Payment			
Year	Total payment	Interest paid	Principal paid	Principal remaining
1	$ 11,451	$ 8,000	$ 3,451	$96,549
2	11,451	7,724	3,728	92,821
3	11,451	7,426	4,026	88,795
4	11,451	7,104	4,348	84,447
5	11,451	6,756	4,696	79,752
6	11,451	6,380	5,071	74,680
7	11,451	5,974	5,477	69,203
8	11,451	5,536	5,915	63,288
9	11,451	5,063	6,388	56,900
10	11,451	4,552	6,900	50,000
Balloon	50,000	0	50,000	0
	$164,515	$64,515	$100,000	

the tenth year, amortized with equal total payments in Years 1 through 10.

The general formula for calculating the annual total payment for a balloon payment loan with equal total payments is

$$[(P - B) \times f] + (B \times i)$$

where P is the original principal borrowed, B is the amount of principal in the balloon payment, f is the amortization factor, and i is the interest rate. The value of B can be varied to make the payments match the expected income stream available to service the loan.

An alternative approach to setting a repayment schedule for a balloon payment loan is to calculate an even total payment based on a longer repayment period, but with the requirement that all outstanding principal will be repaid after a shorter period. For example, the same loan assumed in Table 19-3 could be amortized over 20 years instead of 10 years. The annual payment would be equal to

$$0.10185 \times \$100,000 = \$10,185$$

where 0.10185 is the amortization factor from Appendix Table 1 for a 20-year, 8 percent interest loan. The entire repayment schedule is shown in Table 19-4. Note that under this method the regular payments for the first 10 years are smaller, but the balloon principal payment at the end is larger. The total amount of interest paid is also larger, which is the usual trade-off for smaller initial payments.

THE COST OF BORROWING

Lenders use several different methods of charging interest, which makes comparisons difficult. The true interest rate, or *annual percentage rate (APR)*, should be stated in the loan agreement. Some lenders charge loan closing fees, sometimes called *points*, appraisal fees, or other fees for making a loan. These fees, as well as interest rates, affect the total cost of borrowing.

One way to compare the cost of various credit plans is to calculate the dollar amount to

TABLE 19-4	Amortization of a $100,000 Loan Over 20 Years at 8 Percent Interest, with a Balloon Payment after 10 Years

Year	Total payment	Interest paid	Principal paid	Principal remaining
1	$ 10,185	$ 8,000	$ 2,185	$97,815
2	10,185	7,825	2,360	95,455
3	10,185	7,636	2,549	92,906
4	10,185	7,432	2,753	90,153
5	10,185	7,212	2,973	87,180
6	10,185	6,974	3,211	83,969
7	10,185	6,718	3,468	80,502
8	10,185	6,440	3,745	76,757
9	10,185	6,141	4,045	72,712
10	10,185	5,817	4,368	68,344
Balloon	68,344	0	68,344	0
	$170,196	$70,196	$100,000	

be repaid in each time period (principal, interest, and other fees) and find the discounted present value of the series of payments, as described in Chapter 17. If several financing alternatives are being compared, the same discount rate should be used for all of them. Subtracting the original loan from the present value of the payments leaves the net present value, or true cost, of the loan. The true cost can also be expressed in percentage terms, by calculating the *internal rate of return (IRR)* to the lender. These same methods can be used to calculate the true cost of a financial lease under which an operator leases equipment for several years and then purchases it. An example appears in Chapter 22.

Fixed-rate loans have an interest rate that remains the same for the entire length of the loan. However, some lenders do not like to make long-term loans at a fixed interest rate, because the rate they must pay to obtain loan funds may change. Borrowers do not like to borrow long term at a fixed rate if they think interest rates may decrease.

Predicting future interest rates is difficult for both borrowers and lenders.

For this reason, loans with *variable interest rates* have been developed, which allow for adjustment of the interest rate periodically, often annually. There may be limits on how often the rate can be changed, the maximum change on a single adjustment, and maximum and minimum rates. For example, Farm Credit System banks offer a variable rate tied to the average interest rate on the bonds they issue to raise their loan capital. They also make loans with a fixed rate for a period of several years, with the understanding that this rate will be adjusted for the next time period.

Loans with a fixed interest rate typically carry a higher initial interest rate than variable-rate loans, to protect the lender against future increases in rates. The borrower must weigh this higher but known rate against the possibility that the variable rate could eventually go even higher than the fixed rate. The variable rate could also decrease. Variable-rate loans are one way to ensure that the interest rate is always near the current market rate.

SOURCES OF LOAN FUNDS

Farmers and ranchers borrow money from many different sources. Some lending institutions specialize in certain types of loans, and some provide other financial services in addition to lending money. The market shares of the more important sources of funds for both real estate and non-real estate agricultural loans are shown in Figure 19-5.

Commercial Banks

Commercial banks are the single largest source of non–real estate loans for agriculture, and they also provide some real estate loans. Banks limit their long-term loans to maintain the liquidity needed to meet customers' cash requirements and unexpected withdrawals of deposits. However, the use of variable interest rates, balloon payments, and *secondary mortgages* to larger lenders has allowed banks to increase their share of the farm real estate loan market.

The large share of agricultural loans held by banks is due partially to the large number of local banks in rural communities. This proximity to their customers allows bank personnel to become well acquainted with customers and their individual needs. Banks also provide other financial services such as checking and savings accounts, which make it convenient for their farm and ranch customers to take care of all their financial business at one location.

Sometimes smaller banks cannot legally extend enough credit to finance a large-scale

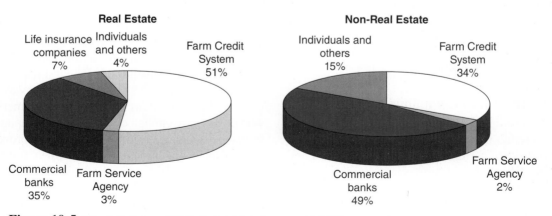

Figure 19-5 **Market share of U.S. farm debt, January 1, 2014.**

Source: U.S. Department of Agriculture, Economic Research Service.

ranch or farm operator. They may arrange for credit to be supplied through a correspondent bank. Banks may arrange for secondary mortgages on real estate loans that are too large or have too long a term for the bank.

Farm Credit System

The Farm Credit System (FCS) was established by the Federal Farm Loan Act passed by the U.S. Congress in 1916 to provide an additional source of funds for agricultural loans. Government funds were used initially to organize and operate the system, but it is now a private cooperative owned by its members/borrowers. The system is supervised, audited, and regulated by the Farm Credit Administration, an independent federal government agency.

The FCS obtains loan funds by selling bonds and notes in the international capital markets. Proceeds from these sales are made available to the four regional Farm Credit Banks and one Agricultural Credit Bank located across the country. These regional banks then provide funds to 81 local associations, which in turn initiate and supervise loans to farmers and ranchers. Loans from the FCS may be used to purchase livestock, machinery, buildings, rural homes, and land. Short-term operating credit is also available.

Because the FCS is organized as a cooperative, borrowers must be members and hold stock equal in value to 2 percent of their outstanding loan balances or $1,000, whichever is less. Likewise, the Farm Credit Banks pay out patronage distributions to borrowers each year from their operating profits.

Life Insurance Companies

Life insurance companies acquire funds from the premiums paid on life insurance policies and from other earnings and reserve funds. They place these funds in various investments, including long-term agricultural real estate loans. The amount of money these companies make available for agricultural loans may vary from year to year, depending on the rate of return from

agricultural loans compared to alternative investments. Life insurance companies generally prefer large farm real estate loans, often over $1 million. Their loan terms are often shorter than those of other lenders, but are amortized with a balloon payment at the end. Insurance companies also make secondary loans to small banks or other lenders who do not have the capacity to finance large purchases.

Farm Service Agency

The Farm Service Agency (FSA) is part of the U.S. Department of Agriculture and has offices in most agricultural counties. A farm loan program division of FSA makes farm ownership and farm-operating loans. Most direct loans made by FSA are now granted to beginning farmers. The FSA also has authority to make emergency loans to qualified farmers and ranchers in officially declared disaster areas. These are temporary loans used to restore normal operations after a natural disaster such as flood or drought. Over the years, FSA has moved away from direct loans made from Congressional appropriations and toward more *guaranteed* loans. With the latter, a private lender provides the loan funds, and FSA guarantees up to 95 percent repayment in case of default by the borrower, and assumes responsibility for servicing the loan.

To be eligible for an FSA loan or guarantee, the borrower must operate a family-size or smaller farm or ranch, receive a substantial portion of total family income from farming or ranching, and be unable to obtain conventional financing from other lending institutions. The last requirement does not mean FSA borrowers are always poor credit risks. Many are beginning farmers who do not have enough equity to borrow capital from other sources. As soon as borrowers improve their financial condition to the point where funds can be obtained from a commercial source, they must switch to another lender.

The FSA also provides short-term *marketing loans* using stored grain or cotton as collateral. Funds come from the Commodity Credit

Corporation (CCC), a subsidiary of the U.S. Department of Agriculture. These loans are made at a fixed rate per bushel or pound and usually carry a below-market interest rate. If the market price for the commodity for which the loan was obtained falls below the loan rate plus interest, the value that the borrower must repay is recalculated using the current marketing price. This in effect offers the farmer a guaranteed minimum price for the commodity. The commodity must be stored until the loan is repaid, but for a maximum of nine months.

The FSA also administers a special loan program to finance the construction of processing and storage facilities for farm commodities. Structures eligible for financing include grain bins, hay sheds, and cold storage rooms.

Individuals and Suppliers

Individuals, farm supply stores, dealers, and others are important sources of agricultural loans, as shown in Figure 19-5. Non–real estate loans can come from friends or relatives, farm supply stores, input manufacturers, or special agricultural credit companies. Many suppliers allow customers 30, 60, or 90 days to pay their accounts before any interest is charged and may finance a purchase for a longer period with interest. This policy is essentially a loan, and the total balances in these accounts can be large at certain times of the year.

Farm equipment and automobile dealers also provide loans by financing purchases themselves or through an affiliated finance company. Such loans are repaid through installment purchase contracts. Producers may also be able to lease equipment for several years with an option to purchase it at the end of that period.

The relatively large portion of real estate debt owed to individuals and others comes mostly from seller-financed land sales. Many land sales use a land purchase installment contract, in which the seller transfers possession of the land and the buyer makes periodic loan payments directly to the seller. This differs from a cash sale, where the buyer borrows from a commercial lender, pays the seller cash for the full purchase price, and then makes periodic payments to the lender. A land purchase contract can have some income tax benefits for the seller, and the buyer may be able to negotiate a lower down payment, lower interest rate, and more flexible repayment terms.

Financial Service Companies

Financial service companies are private financial intermediaries that secure funds from commercial and institutional investors and use them to provide credit to ranches, farms, and other businesses. They often operate on a national or even international scale, and offer large-scale financing packages. They may provide other financial and risk management services to their customers in addition to serving their credit needs.

ESTABLISHING AND DEVELOPING CREDIT

When trying to establish or develop credit, it is useful to look at it from the lender's viewpoint. What does a lender consider when making a decision on a loan application? Why can one business borrow more money than another or receive different interest rates and repayment terms? A borrower should be aware of the need to demonstrate and communicate credit worthiness to lenders. Some of the more important factors that go into making loan decisions are the following:

- Personal character
- Management ability
- Financial position and progress over time
- Repayment capacity
- Purpose of the loan
- Collateral

When using these factors as a guide for establishing and developing credit, a prospective borrower should remember that lenders want to make loans. That is their business. However, they are looking for profitable and safe loans that will be repaid.

Personal Character

Honesty, integrity, judgment, reputation, and other personal characteristics of the loan applicant are always considered by lenders. Credit can be quickly lost by being untruthful in business dealings and slow to meet financial obligations. If the lender is not acquainted with the borrower, character references usually will be requested and checked. To maintain a good credit record, borrowers should promptly inform lenders of any changes in their financial condition or farming operation that might affect loan repayment. An honest and open relationship with lenders is necessary to maintain credit worthiness.

Management Ability

A lender must try to evaluate a borrower's management ability. Established farmers and ranchers will be evaluated on their past records, but beginners can be judged only on their background, education, and training. These factors affect profitability and therefore the ability to repay a loan. Lenders often rate poor management ability as the number one reason for borrowers getting into financial difficulty.

Financial Position

Accurate, well-prepared balance sheets and income statements are needed to document the current financial position of the business and its profitability. Lenders can learn much about a business from these records. A track record of good financial progress over time can be just as important as the current financial position.

Repayment Capacity

Having a profitable business does not guarantee repayment capacity. There must be enough cash income to meet family living expenses and income tax payments, as well as the interest and principal payments on loans. Repayment capacity is best measured by the cash flow generated by the business. A cash flow budget projected for one or more years should be completed before borrowing large amounts and establishing rigid repayment schedules. Too often, money is borrowed for a profitable business only to find that cash flow in the early years is not sufficient to make interest and principal payments. A longer loan term or a more flexible repayment schedule may solve the problem if it is identified in time.

Purpose of the Loan

Self-liquidating loans may be easier to obtain. Self-liquidating loans are for items such as fertilizer, seed, and feeder livestock, where the loan can be repaid from the sale of the crops or livestock. On the other hand, capital asset loans are those used to purchase long-term tangible assets such as land or machinery, which must generate additional revenue without being sold themselves. Capital asset loans may require extra collateral.

Collateral

Land, buildings, livestock, machinery, stored grain, and growing crops can all be used as loan collateral. The amount and type of collateral available are important factors in a loan request. Loans should not be made or requested unless repayment can be projected from farm income. However, lenders still ask for collateral to support a loan request. If the unexpected happens and the loan is in default, it may be the lender's only means of recovering the loan funds. In some cases a lender may ask for personal assets to be pledged as collateral for a business loan or require another operator to co-sign a promissory note.

LIQUIDITY

The ability of a business to meet cash flow obligations as they come due is called *liquidity*. Maintaining liquidity to meet loan payments as they become due is an important part of establishing and improving credit. Several measures of liquidity that can be calculated from the balance sheet were discussed in Chapters 4 and 6.

Factors Affecting Liquidity

A profitable farm or ranch will usually have adequate liquidity in the long run. However, even profitable businesses experience cash flow problems at times, due to several factors.

Business Growth

Holding back ever-increasing inventories of young breeding livestock or feed reduces the volume of production sold in the short run. Construction of new buildings or purchases of land or machinery require large cash outlays upfront, but may not generate additional cash income for months or years. Moreover, when new technology is involved, several production cycles may pass before an efficient level of operation is achieved. All these factors produce temporary cash shortfalls and need to be considered when planning financing and repayment.

Nonbusiness Income and Expenditures

These are especially important for family farms and ranches. Basic family living expenses must be met even when agricultural profits are low. At certain stages of their lives, farm families may have high educational or medical expenses. Generally, nonbusiness expenditures should be postponed until surplus cash is available. However, reinvesting every dollar earned into the farm business may eventually cause family stress and keep personal goals from being reached.

Over the years, farm and ranch families have received more and more of their total income from off-farm employment and investments. A regular and dependable source of outside income may not only stabilize resources for family living expenses but also help support the farm during periods of negative cash flow.

Loan Characteristics

The rates and terms of credit may affect cash flow as much as the amount of debt incurred. Seeking out the lowest available interest rate reduces debt payments. Using longer-term amortizations or balloon payment loans also reduces short-term

financial obligations. Planning for term debt payments to come due during periods that will coincide with product sales or other cash receipts will reduce the need for short-term operating loans.

Debt Structure

Structure refers to the distribution of debt among current, intermediate-term, and long-term liabilities. Generally, loan repayment terms should correspond to the class of assets that they were used to acquire. Financing intermediate- or long-term assets with short-term debt often leads to repayment problems and negative cash flows.

Some balance sheets are *top heavy*; that is, the percent of total liabilities classified as current is greater than the percent of assets classified as current. Refinancing some current debt against longer term assets and amortizing the repayment over several years may improve short-term liquidity.

Financial Contingency Plan

No matter how well the manager budgets or how efficiently the business is managed, there will be periods in which cash flow is negative. Figure 19-3 illustrated how the amount of debt used by farmers and ranchers expanded dramatically from 1981 to 1987. However, interest rates climbed sharply in the early 1980s and reached levels that were not anticipated when many loans were made. This, combined with generally lower agricultural prices, produced what was known as the *farm financial crisis* of the 1980s (see Box 19-1).

The lessons learned during this period made many farmers and lenders more conservative about the use of credit. Every operation should have a *financial contingency plan* to provide for unexpected cash flow shortfalls. In some cases, actions that are not profitable in the long run may have to be taken to cover cash flow obligations in the short run. The following steps are possible financial contingency actions:

1. Maintain savings or stored crops and livestock in a form that can be easily turned into cash and that carries a low risk of loss.

2. Maintain a credit reserve or some unused borrowing capacity for both current and noncurrent borrowing. A long-term credit reserve can be used to refinance excess current liabilities if the need arises.
3. Prepay loans in years when cash income is above average.
4. Reduce nonfarm expenditures or increase nonfarm earnings when farm cash flow is tight. Acquiring additional education or work experience may make obtaining off-farm employment easier.
5. Carry adequate insurance coverage against crop losses, casualties, medical problems, and civil liabilities.
6. Sell off less-productive assets to raise cash. The marginal cost equals marginal revenue principle should be employed to identify assets that will have the least negative effect on total farm profits if sold. In some cases, cash flow can be improved without reducing efficiency by selling assets and then leasing them back, thereby maintaining the size of the operation.
7. Rely on relatives or other personal contacts for emergency financing or for the use of machinery or buildings at little or no cost or in exchange for labor.
8. Declare bankruptcy, and work out a plan to gradually repay creditors while continuing to farm, or conduct an orderly disposal of assets and cancellation of debts.

These actions are not substitutes for operating a profitable business. Some of them may even reduce the farm's profitability in the short run. But any of them can be applied, depending on the severity of the farm's financial condition, as a means to continue operating until profits increase.

SOLVENCY

While liquidity management focuses on cash flow, solvency refers to the amount of debt capital (loans) used relative to equity capital, and the security available to back it up. The debt/asset ratio and other measures of solvency were discussed in Chapter 4.

Leverage and the Use of Credit

Using a combination of equity capital and borrowed capital permits owning a larger business than would be otherwise possible. The degree to which borrowed capital is used to supplement and extend equity capital is called *leverage*. Leverage increases with increases in the debt/asset ratio. A debt/asset ratio of 0.50, or 50 percent, indicates that one-half of the total capital used in the business is borrowed and one-half is equity capital.

When the return on borrowed capital is greater than the interest rate, profits will increase and equity will grow. Higher leverage will increase profits and equity even faster. For example, assume a firm has $100,000 in equity capital, as shown in the upper part of Table 19-5. If $100,000 more is borrowed at 10 percent interest, the debt/asset ratio is 0.50. If the business earns a return on assets (ROA) of 15 percent, or $30,000, and the interest cost of $10,000 is paid, the remaining $20,000 is the return on equity (ROE). The rate of ROE is $20,000 ÷ $100,000, or 20 percent. Increasing the leverage, or debt/asset ratio, to 0.67 by borrowing $200,000 results in an even higher return to equity (25 percent), as shown in the right-hand column of Table 19-5.

However, there is another side to the coin. If the rate of return on total assets is less than the interest rate on borrowed capital, the ROE is decreased by using leverage and can even be negative. This is illustrated in the lower part of Table 19-5, where only a 5 percent ROA is attained, and ROE decreases to a negative 5 percent under high leverage. Thus, high leverage can substantially increase the financial risk of the farm or ranch when interest rates are high or the return on farm assets is low.

Maximum Debt/Asset Ratio

Most lenders use the debt/asset ratio or some variation of it to measure solvency. However, they do not always agree on what constitutes a

TABLE 19-5	**Illustration of the Principle of Increasing Leverage**			
	Debt/asset ratio			
	0.00	0.33	0.50	0.67
Equity capital ($)	100,000	100,000	100,000	100,000
Borrowed capital ($)	0	50,000	100,000	200,000
Total assets ($)	100,000	150,000	200,000	300,000
Good year				
Return on assets (15%)	15,000	22,500	30,000	45,000
Interest paid (10%)	0	5,000	10,000	20,000
Return on equity ($)	15,000	17,500	20,000	25,000
Return on equity (%)	15.0	17.5	20.0	25.0
Poor year				
Return on assets (5%)	5,000	7,500	10,000	15,000
Interest paid (10%)	0	5,000	10,000	20,000
Return on equity ($)	5,000	2,500	0	−5,000
Return on equity (%)	5.0	2.5	0.0	−5.0

safe ratio. Profitability of the business and the cost of borrowed funds also enter into the decision. A simple relation among these factors can be expressed as follows:

$$\text{Maximum debt/asset ratio} = \frac{\text{ROA}}{\text{IR}}$$

where the maximum debt/asset ratio is the highest debt/asset ratio the business can sustain on its own, ROA is the percent return on total assets after deducting the value of unpaid labor, and IR is the average interest rate on the farm's debt.

For example, a farm that earns a 6 percent average ROA and pays an average interest rate of 12 percent on its debt could service a maximum debt/asset ratio of 0.50 without reducing equity. In other words, the return on each $1 of assets pays the interest on $.50 of debt, but no principal. If the ROA dropped to 3 percent though, and

the average interest rate rose to 15 percent, only a 0.20 debt/asset ratio could be sustained.

The *maximum* debt/asset ratio is the level at which ROE is just equal to zero. Debt is serviced without reducing equity or using outside income. Farms that earn a higher ROA or that can borrow money at a lower interest rate can safely carry a higher debt load. The wise manager will maintain a margin of safety, however.

U.S. agriculture as a whole has maintained a fairly conservative debt load. During the 1980s the total debt-to-asset ratio exceeded 0.20, but has generally decreased since then, as shown in Figure 19-3.

Inflation and Capital Gains

In the United States, the rate of inflation has generally been maintained at a low level, although it did surpass 10 percent for several years in the

1970s. Some countries have experienced inflation rates of over 100 percent per year. With high inflation, managers often prefer to hold tangible assets such as land that will increase in value, rather than financial assets such as cash, savings accounts, or bonds. However, tangible assets may also lose value when economic conditions change, as shown in Figure 19-1.

When farmers invest in intermediate- or long-term assets, both their degree of solvency and growth in market-value equity become closely tied to changes in the values of these capital assets. However, changes in asset values alone have little effect on liquidity or cash flow, unless assets are sold. Thus, during periods of rising land values, many farmland owners find their equity rising rapidly without a corresponding increase in cash income. Some managers will sell land and turn the asset appreciation into capital gains, although some cash may be lost to income tax payments. Others will use their new equity as collateral to borrow money, but may have difficulty generating enough cash flow to repay the loans.

In general, the longer the life of a farm asset, the lower its rate of cash return. Many operators have a goal to own land and buildings, with the expectation that they will increase in value over time. The prudent financial manager must carefully balance the need for short-term liquidity against the security and growth potential of long-term investments.

SUMMARY

Capital includes money invested in machinery, livestock, buildings, and other assets, as well as cash and bank account balances. Sources of capital available to farmers include the operator's own equity, equity from outside investors, leased or contracted assets, and borrowed funds. Today's farm managers must be skilled in acquiring, organizing, and using different forms of capital. The economic principles discussed in Chapters 7 and 8 can be used to determine how much total capital can be used profitably and how to allocate limited capital among alternative uses.

Agricultural loans are available for the purchase of real estate and non–real estate assets and for paying operating costs. They can be repaid over periods ranging from less than a year to as long as 40 years or more, in a single payment or with several types of amortized payments. Interest rates, loan terms, and repayment schedules vary from lender to lender and by loan type. Borrowers should compare the annual percentage rate of interest, loan fees, variable rate provisions, and other loan terms when shopping for credit.

Agricultural loans are available from commercial banks, the FCS, the FSA, life insurance companies, individuals and input suppliers, and other sources. Some lenders specialize in certain types of loans, but all are interested in the credit worthiness of a prospective borrower. Borrowers should work at improving their credit by maintaining good personal character, improving management skills, demonstrating adequate financial progress and repayment capacity, and providing sufficient collateral.

Liquidity or cash flow management is affected by business growth, nonbusiness income and expenses, and the characteristics and structure of the debt held. A financial contingency plan should be formulated to provide for unexpected cash flow shortages. Solvency refers to the degree to which farm liabilities are secured by assets. Increased leverage can increase the rate at which equity grows but also increases the risk of losing equity. The amount of debt that a farm can support depends on the rate of return earned on assets and the interest rate paid on debt capital. Inflation increases the market value of assets but does not contribute to cash flow unless the assets are sold.

QUESTIONS FOR REVIEW AND FURTHER THOUGHT

1. What economic principles are used to determine: (a) how much capital to use and (b) how to allocate a limited amount of capital?

2. What is the single largest source of capital used in U.S. agriculture? What other sources are used?

3. Define the following terms:

   a. Secured loan

   b. Amortized loan

   c. Real estate loan

   d. Collateral

   e. Line of credit

   f. Balloon payment

4. How are loans classified as short-term, intermediate-term, and long-term loans? List the types of assets that might serve as collateral for each.

5. Assume that a $200,000 loan will be repaid in 30 annual payments at 9 percent annual interest on the outstanding balance. How much principal and interest will be due in the first payment if the loan is amortized with equal principal payments? If it is amortized with equal total payments? How would these figures change for the second payment in each case? Use Appendix Table 1 to find the amortization factor for the equal total payment case.

6. What are the advantages and disadvantages of a 10-year loan with a 50 percent balloon payment versus a completely amortized loan of the same amount, term, and interest rate?

7. Identify the different sources of agricultural loans in your home town or home county. Which types of loans does each lender specialize in? You might interview several lending institutions to learn more about their lending policies and procedures.

8. Select one agricultural lender and find out the rates and terms currently available for an intermediate- or long-term loan. Are both fixed- and variable-interest rate loans available? What loan closing fees or other charges must be paid?

9. Assume you are a beginning farmer or rancher and need capital to purchase breeding livestock. What information and material would you need to provide a lender to improve your chances of getting a loan? What lenders might be willing to finance you?

10. List several reasons why a request for a loan by one farm operator might be denied while a similar request from another operator is approved.

11. Explain the difference between liquidity and profitability. Give three reasons why a profitable farm could experience liquidity problems.

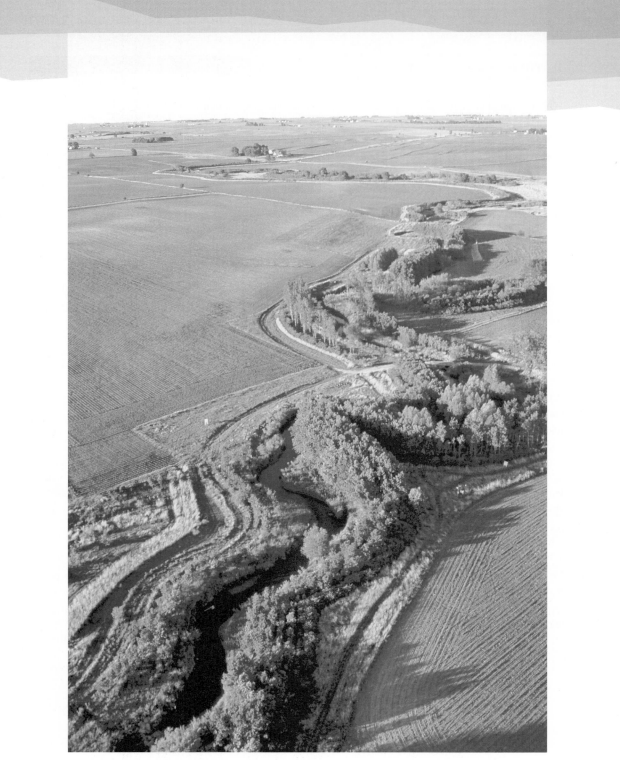

Photo by Lynn Betts, USDA Natural Resources Conservation Service

LAND: CONTROL AND USE

CHAPTER OUTLINE

Factors that Affect Farmland Values
The Economics of Land Use and Management
Controlling Land: Own or Lease?
Buying Land
Leasing Land
Conservation and Environmental Concerns
Summary
Questions for Review and Further Thought

CHAPTER OBJECTIVES

1. Explore the unique characteristics of land and its use in agriculture

2. Compare the advantages and disadvantages of owning and renting land

3. Explain important factors in land purchase decisions, methods of land valuation, and the legal aspects of a land purchase

4. Compare the characteristics of cash, crop share, livestock share, and other leasing arrangements

5. Demonstrate how an equitable share lease arrangement can encourage efficient input use

6. Discuss profitable land management systems that conserve resources and sustain the environment

Agriculture uses large areas of land, distinguishing it from most other industries. Land is the basic resource that supports the production of all agricultural commodities, even livestock, because they depend on land to produce the forage and grain they consume. Land is the single most valuable asset in the balance sheet of U.S. agriculture, accounting for nearly three-fourths of the value of total assets. The index of land values depicted in Figure 20-1 increased sharply in the 1970s, fueled by high grain prices and rapid inflation. In the 1980s, high interest rates, drought, and low

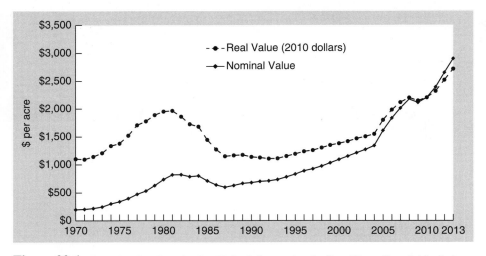

Figure 20-1 **Farmland values in the United States (excluding Hawaii and Alaska).**
Source: U.S. Department of Agriculture.

commodity prices brought financial difficulty for many farmers and ranchers. Those who had high debt loads often could not meet their payments. The result was a sudden decline in land values as farmers tried to liquidate farmland and buyers left the market. Many farmers and ranchers were forced to sell land at a loss or saw their equity decline as the land on their balance sheets decreased in market value. In the late 1980s, land values began to recover, and values rose slowly but steadily until 2004. In recent years farmland values have again risen at double-digit rates, due mostly to increased demand for grain as a feedstock for making ethanol and other biofuels. The dashed line in the figure shows that even real land values (corrected for inflation) rose sharply in the 1970s, fell in the 1980s, then rose again in later decades. Figure 20-2 indicates that land values have increased in most of the years since 1970, except for a decline in the mid-1980s.

FACTORS THAT AFFECT FARMLAND VALUES

Several important factors influence the market value of farmland. The potential profit per acre from production of food, fuel, and fiber is the

most fundamental force behind land values. Increased crop yields made possible by the many technological advances introduced in the past century have contributed to this. In most years, increases in farmland values have kept pace with or exceeded the rate of inflation in the United States, making ownership of land a good hedge against inflation. This factor attracts nonfarm investors, as well as farmers, into the land market. Interest rates and the rates of return available from other investments also affect the demand for and the price of farmland.

In some areas, urbanization has greatly affected land values, causing some agricultural land to be sold to developers for other purposes. Increased urbanization can have negative effects on the overall agricultural economy and on the quality of life in once-rural areas. Urbanization can also lead to conflicts with nonfarm neighbors, with slower moving agricultural vehicles traveling on congested roads, or noise and odor concerns. However, some land owners may welcome the increased land value because it increases the wealth they would have available to use for retirement in the future. Proximity to urban areas may also allow producers to find a profitable niche market, such as a

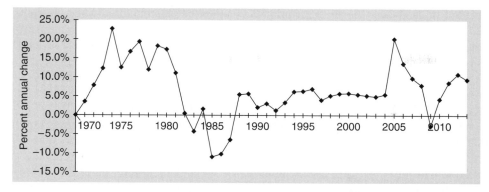

Figure 20-2 **Percent annual change in United States farmland values (excluding Hawaii and Alaska).**

Source: U.S. Department of Agriculture.

pick-your-own produce operation or an agricultural recreation opportunity such as a corn maze. Also, proximity to an urban area may increase the pool of labor available.

In addition to urbanization, some rural land is lost to agricultural production because of a switch to recreational uses. While some recreational uses can co-exist with production agriculture, others cannot. In some areas, producers make extra income leasing hunting rights. Depending on the enterprises, and the time of year, these leases may have little or no impact on the traditional production activities.

While nonfarm investors and developers have been important purchasers of land in some areas, the large majority of land buyers are still farmers and ranchers who desire to expand the size of their operations to increase their incomes and take advantage of larger-scale technologies. They recognize the economies of size they can achieve with a larger operation, the opportunity for greater profits, and the effect of increasing land values on their equity. Although farmland loss to urbanization can be an important local concern, on a national scale it has not had an effect on the productive capacity of U.S. agriculture. Over the entire period from 1945 to 2012, land in farms declined by 20 percent from 1,142 million acres to

915 million acres. Over the same period, agricultural output more than doubled.

THE ECONOMICS OF LAND USE AND MANAGEMENT

Land has a number of unique characteristics not found in other resources. These characteristics greatly influence the economics of land use and management.

Characteristics of Land

Land is a permanent resource that does not depreciate or wear out, provided soil fertility is maintained and appropriate conservation measures are used. Proper management not only will maintain the inherent productivity of land but can even improve it. Land is productive in its native state, producing stands of timber and native grasses, but the management efforts of farmers and ranchers have improved the agricultural productivity of many types of land. This has been accomplished through land clearing, drainage, good conservation practices, irrigation, the introduction of new and improved plant species, and the use of limestone and fertilizer. Land use often changes as a result of these improvements.

Each tract of land has a legal description, which identifies its particular location, size, and shape. Land is immobile and cannot be moved to be combined with other resources. Machinery, seed, fertilizer, water, and other inputs must be transported to the land and combined with it to produce crops and livestock.

Not only is land a unique resource in general, but each farm or specific parcel of land is unique. A piece of land larger than several acres often contains two or more distinct soil types, each with its own set of characteristics. Topography, drainage, organic material, and the existence of natural hazards such as flooding, wind and water erosion, and rock outcrops are other factors that combine to make land resources different from farm to farm.

The supply of land suitable for agricultural production is essentially fixed, although small amounts may be brought into production by clearing and draining or may be lost to nonfarm uses. This makes the price of land very sensitive to changes in the demand for its products. Unlike other agricultural inputs, additional land cannot be manufactured when the demand for it increases. Therefore, changes in the profitability of agricultural production are eventually factored into land prices and rents, and the land owner reaps the economic benefits or losses.

Planning Land Use

The difference in land resources among farms explains why one of the first steps in whole-farm planning is to make a complete inventory of the land, including soil types, drainage, slope, and fertility. Without this information, the most profitable farm plan cannot be developed. The potential livestock and crop enterprises, yields, fertility requirements, and necessary conservation practices are related directly to the nature of the land resources available. Whole-farm planning often involves maximizing the return to the most limited resource. The fixed nature of land in the short run makes it the beginning of most farm planning efforts.

Land use is affected by regional differences in land productivity. Figure 20-3 shows

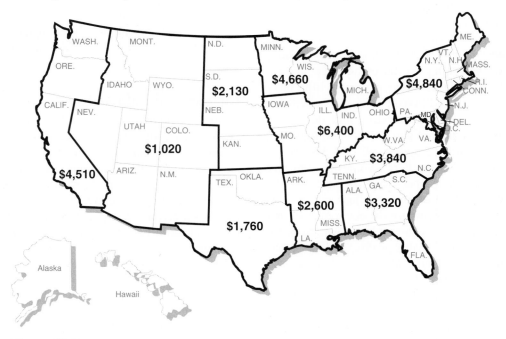

Figure 20-3 **Average value of farmland per acre, by region.**
Source: National Agricultural Statistics Service, USDA (2013).

the wide range in farmland values across regions of the United States. Even within a region, farmland values can vary greatly depending on the quality of the land, its proximity to urban areas, and whether it is irrigated. However, the most profitable use for land also depends on relative commodity prices and production technology. Both can change over time and bring about changes in land use. Cotton production moved westward out of the southeastern United States in the early to mid-twentieth century and was often replaced by pasture and livestock production. Cotton acreage then rebounded in the Southeast during the latter part of the twentieth century following the implementation of a program to reduce or eliminate boll weevil populations. Soybeans were once a minor crop but have become the second most important crop in the Midwest and Mississippi Delta region. The development of irrigation has transformed former livestock grazing areas into important crop production regions. An increase in grain production in the Southern Plains region of Texas and Oklahoma encouraged the development of large-scale cattle feeding operations there. All these changes can be traced to changes in relative prices of products and inputs, new technology, and competing uses for land.

CONTROLLING LAND: OWN OR LEASE?

How much land to control and how to acquire it are two of the most important decisions to be made by any farmer or rancher. Errors made at this point may plague the business for many years. Too little land may mean the business is too small to fully use other resources. At the other extreme, too much land may require borrowing a large amount of money, cause serious cash flow problems, and overextend the operator's management and machinery capacity. Either situation can result in financial stress and eventual failure of the business.

Land acquisition should be thought of in terms of control and not just ownership. Control can be achieved by either ownership or leasing. About 38 percent of U.S. farmland is leased by the operator. Many farmers find a combination of ownership and leasing to be desirable, particularly when capital is limited. They often own some land and buildings to have a permanent *home base* and then lease additional land to attain the desired farm size. These part owners accounted for 25 percent of U.S. farms in 2012. Another 68 percent were full owners, and 7 percent were full tenants who owned no land.

Ownership

Owning land is an important goal for many ranchers and farmers, regardless of the economics involved. A certain amount of pride, satisfaction, and prestige is derived from owning land. It also provides a tangible estate to pass on to one's heirs.

Owning land has the following advantages:

1. *Security of tenure*: Landownership eliminates the uncertainty of losing a lease and having the size of the business reduced unexpectedly. It also ensures that the operator will enjoy the benefits of any long-term improvements made to the land.
2. *Loan collateral*: Accumulated equity in land provides an excellent source of collateral when borrowing money. Increasing land values over time have provided substantial equity for landowners, although some of this equity can be lost with a decline in land values such as occurred during the mid-1980s.
3. *Management independence and freedom*: Landowners are free to make decisions about enterprise combinations, conservation measures, fertilizer levels, and other choices, without consulting with a landlord or professional farm manager.
4. *Hedge against inflation*: Over the long run, land provides an excellent hedge against inflation, because increases in land

values have tended to equal or exceed the rate of inflation in the U.S. economy. However, land values do not necessarily increase every year.

5. *Pride of ownership*: Owning and improving one's property is a source of pride. It assures a future benefit from years of labor and investment.

Ownership of land controlled by the business can also have some disadvantages. These disadvantages are primarily related to the capital position of the business. Possible disadvantages are as follows:

1. *Cash flow*: A large debt load associated with purchasing farmland can create serious cash flow problems. The cash earnings from the land may not be sufficient to meet the required principal and interest payments, as well as other cash obligations of the business.

2. *Lower return on capital*: Where capital is limited, there may be alternative uses for it with a higher return than investing in land. Machinery, livestock, and annual operating inputs, such as fertilizer, seed, and feed, often produce a higher rate of return on investment than land.

3. *Less working capital*: A large investment or heavy debt load on land may restrict the amount of available working capital, severely limiting the volume of production, choice of enterprises, input levels, and profits.

4. *Size limits*: A combination of limited capital and a desire to own all the land to be operated will limit the size of the business. A small size may prevent the use of certain technologies and result in higher average costs.

The disadvantages of landownership are more likely to affect the beginning farmer with limited capital. With the accumulation of capital and borrowing capacity over time, they may become less and less important. Older farmers tend to own more land and have less debt than younger farmers.

Leasing Land

Beginning farmers or ranchers are often advised to lease land. With limited capital, leasing is a means of controlling more acres. Other advantages of leasing land are:

1. *More working capital*: When capital is not tied up in land purchases, more is available to purchase machinery, livestock, and annual operating inputs.

2. *Additional management*: A beginning farmer may be short of management skills. Management assistance can be provided by a knowledgeable landlord or professional farm manager employed by the landlord.

3. *More flexible size*: Lease contracts are often for only one year or, at most, several years. Year-to-year changes in business size or location can be easily accomplished by giving up old leases or leasing additional land.

4. *More flexible financial obligations*: Lease payments are more flexible than mortgage payments, which may be fixed for a long period. The value of share rent will automatically vary with crop yields and prices. Cash rents are less flexible but can be negotiated each time the lease is renewed, taking into account current and projected economic conditions.

Leasing land also has disadvantages, which take on particular significance when all the land operated is leased. These disadvantages are as follows:

1. *Uncertainty*: Given the short term of many leases, the danger always exists that all or part of the land being farmed can be lost on short notice. This possibility discourages long-term investments and contributes to a general feeling of uncertainty about the future of the business.

2. *Poor facilities*: Some landlords are reluctant to invest money in buildings and other improvements. Tenants cannot justify investing in improvements attached

to someone else's property. Thus, family housing, livestock facilities, grain storage, fences, and machinery housing may be obsolete, in poor condition, or nonexistent.

3. *Slow equity accumulation*: Without landownership, equity can be accumulated only in machinery, livestock, or cash savings. In periods of rising land values, tenants may have to pay higher rents without accumulating any equity.

There is no clear advantage for either owning or leasing. Control is still the important factor, as income can be obtained from either owned or leased land. In the final analysis, the proper combination of owned and leased land is the one providing enough land to fully use the available labor, machinery, management, and working capital without creating excessive financial risk.

BUYING LAND

The purchase of a farm or ranch is an important decision that often involves large sums of money. A land purchase will have long-run effects on both the liquidity and the solvency of the business.

The first step in a land purchase decision is to determine the value of the parcel under consideration. Income potential is the most important determinant of land value, but many other factors contribute to it:

1. *Soil, topography, and climate*: These factors combine to affect the crop and livestock production potential and therefore the expected income stream.
2. *Buildings and improvements*: The number, size, condition, and usefulness of buildings, fences, storage structures, and other improvements will affect the value of a parcel of land. A neat, attractive farmstead with a modern house can add many dollars to a farm's value, while rundown, obsolete buildings may detract from it. Farm buildings and improvements are eligible for depreciation expense, creating a potential income tax savings.

3. *Size*: Small and medium-sized farms may sell for a higher price per acre than large farms. A smaller total purchase price puts a farm within the financial reach of a larger number of buyers. Several neighboring farmers may consider the purchase of a small farm a good way to increase the size of their business, and bid up the price.
4. *Markets*: Proximity to a number of markets will reduce transportation costs, increase competition for the farm's products, and possibly raise their net selling prices.
5. *Community*: A farm located in an area of well-managed, well-kept farms or in a community where land seldom changes hands will have a higher selling price.
6. *Location*: Location with respect to schools, churches, towns, recreational facilities, paved roads, and farm input suppliers will also affect value. A prospective purchaser who plans to live on the farm will pay particular attention to the community and services provided in the general area.
7. *Competing uses*: Land close to urban or recreational areas, or that has mineral deposits, may have a higher value than other agricultural land due to its other potential uses. Potential capital gains may outweigh income from agricultural production.
8. *Agricultural program characteristics*: Some parcels of land have historical cropping patterns and yields that affect the level of benefits available for certain USDA commodity programs.
9. *Easements and contracts*: Special easements for conservation programs, irrigation rights, roads, power lines, or gas and oil production can affect potential earnings and, ultimately, land values.

A property under consideration for purchase should be inspected thoroughly to identify any problem areas that might reduce its value. Buildings should be inspected for structural soundness, and soil types properly identified.

Drainage and erosion problems, sand and gravel areas, rock outcroppings, and other soil hazards should be noted and yield expectations adjusted accordingly. Environmental hazards such as old underground fuel storage tanks will also detract from the value of a property.

Land Appraisal

An appraisal is a systematic process for estimating the current market value of a given piece of real estate. There are business firms that provide appraisal services for a fee. They provide their clients with a detailed analysis of the factors that determine the value of the property and conclude the report with an estimate of its current value. Two basic methods used to appraise the value of income-producing property such as farmland are the *income capitalization* method and the *market data* method.

Income Capitalization

This method uses investment analysis tools to estimate the present value of the future income stream from the land. It requires an estimate of the expected annual net income, the selection of a discount rate, and computation of the present value of an annuity, using methods outlined in Chapter 17. It is usually assumed that net income from land has an infinite life. Therefore, the present value equation for a perpetual or infinite stream of income can be used:

$$V = \frac{R}{d}$$

where R is the average annual net return; d is the discount rate, or capitalization rate as it is called in appraisal work; and V is the estimated land value.

The first step is to estimate the net income stream, R, that can be obtained from the property. This requires determining potential long-term yields, selling prices, and production costs for the most profitable enterprise combination. R is the net return to investment in land and is equal to the farm's estimated gross income minus the costs of all resources used to generate it. These costs include land ownership costs such as property taxes, depreciation, and maintenance. The opportunity cost of the capital invested in the land, and any interest cost that might be incurred on a loan used to purchase it, are not subtracted, because the goal is to estimate the returns to the land investment.

The results of such a procedure are shown in the top portion of Table 20-1 for an example 160-acre tract of farmland. It has 150 tillable acres after deducting for roads, ditches, and waterways. The long-run crop plan is assumed to be 90 acres of corn and 60 acres of soybeans, generating an annual gross income of $104,400. Yield and price estimates are important. They strongly influence the final estimate of net income. Both items must be estimated accurately to arrive at an accurate estimate of value. County yield averages and yields based on soil types are useful starting places for yield estimates. Prices should reflect the appraiser's best estimate of long-run prices after a careful review of past price levels. Annual expenses for the crop plan total $70,749, which makes the expected net return equal to $33,651 per year.

The second step in the income capitalization method is to select the discount or capitalization rate. The effect of the capitalization rate on value is readily apparent, so the appropriate rate must be selected carefully. The estimate of the net income does not include any expectation of inflation. Therefore, as explained in Chapter 17, the discount rate should be based on the real interest rate, that is, the nominal or actual rate of interest minus the anticipated rate of inflation.

Actual appraisal practice is to first estimate the average rate of return to land investment in the area for similar farms at recent sale prices. Using this rate in the capitalization procedure will give an appraised value that can be compared to the recent selling prices for other farms. This procedure often results in a capitalization rate of 3 to 6 percent, within the range of historical rates of return to farmland based on current market values. Anticipation of long-term increases in

TABLE 20-1	**Estimated Annual Income and Expenses for Appraisal**
	(160-Acre Tract with 150 Crop Acres)

Income	Acres	Yield (bu/ac)	Price/bu	Total
Corn	90	160	$ 5.00	$ 72,000
Soybeans	60	45	$12.00	32,400
Total income				$104,400

Expenses	
Fertilizer	$16,560
Seed	13,317
Pesticides	6,210
Trucking	932
Drying	2,070
Labor and management	10,800
Machinery ownership costs	9,900
Machinery operating costs	6,660
Depreciation of buildings, tile, fences	1,100
Property taxes and upkeep	3,200
Total expenses	$70,749
Annual net income to land	$33,651

Capitalization of income:

Capitalization rate	Total value	Value per acre
7%	$ 480,729	$3,005
5%	$ 673,020	$4,206
3%	$1,121,700	$7,011

land values helps explain why landowners have historically been willing to accept a current cash rate of return on land that is lower than the rates for other investments that do not have the potential for appreciation in value.

The third step is to divide the expected net return by the capitalization rate. This process is shown in the bottom portion of Table 20-1 for three different capitalization rates. The estimated value of the land in the example, using a 5 percent capitalization rate, is $33,651 divided by 0.05 = $673,020, or about $4,206 per acre.

Market Data

The second appraisal approach compares the land parcels that have been sold recently with the tract of land being appraised. Prices of the comparable sales are adjusted for differences in factors such as soil type, productivity, buildings and improvements, size, proximity to markets, the community, location, and competing uses, as discussed earlier. Three additional factors need to be considered when comparable sales values are being adjusted to reflect the *fair market value* of the farm being appraised.

1. *Financing*: The method and terms of the financing arrangements for the purchase will affect the selling price. Land may be sold with an installment purchase contract where the seller provides the financing. The terms of a contract may include a smaller down payment and/or a lower interest rate than conventional mortgage financing, which will increase the price a buyer is willing to pay for a given piece of land.
2. *Relationships*: If the buyer and seller are closely related, such as parents selling to their children, the selling price is often below the price that would be agreed on by unrelated parties.
3. *Time of sale*: The more time that has passed since the comparable sale took place, the more likely it is that the price must be adjusted to reflect current market conditions. It may be adjusted up or down, depending on whether land values have been rising or falling in recent years.

Financial Feasibility Analysis

A financial feasibility, or cash flow, analysis should be performed on any prospective land purchase that involves repayment of a loan or installment contract. It is not a method for determining land value, but for a given purchase price, it will show if there will be sufficient cash flow to meet both the annual operating expenses and the interest and principal payments. Table 20-2 contains a summarized cash flow analysis for the purchase of the example 160-acre tract of land for $4,500 per acre, or a total of $720,000.

The analysis assumes a 35 percent down payment ($252,000) and a 20-year loan of $468,000, with a 6 percent annual interest rate and equal annual payments. Annual cash receipts and cash expenditures are based on the figures in Table 20-1. Although depreciation is a noncash expense, some cash outflow must be included for long-term machinery replacement. Labor costs should reflect either the cost of hired labor or a portion of family living costs (estimated at $6,000 in this example), but no opportunity cost for unpaid labor. In the example, cash receipts and expenditures are assumed to inflate at a rate of 2 percent annually. Principal payments and the interest rate are assumed to be fixed for the term of the loan, however.

For the first year, the cash operating income is not enough to meet both the cash operating expenses and the required principal and interest

TABLE 20-2	**Cash Flow Analysis of the Purchase of a 160-Acre Tract at $6,000 per Acre***				
Item	**Year 1**	**Year 2**	**Year 3**	**Year 4**	**Year 5**
Cash inflows ($)	$104,400	$106,488	$108,618	$110,790	$113,006
Cash outflows ($)					
Seed, fertilizer, pesticides, trucking, drying	$ 39,089	$ 39,871	$ 40,668	$ 41,482	$ 42,311
Machinery	16,560	16,891	17,229	17,574	17,925
Family living expenses	6,000	6,120	6,242	6,367	6,495
Property taxes and upkeep	3,200	3,264	3,329	3,396	3,464
Annual loan payment	40,802	40,802	40,802	40,802	40,802
Total cash outflows ($)	$105,651	$106,948	$108,270	$109,621	$110,997
Net cash flow ($)	$ (1,251)	$ (460)	$ 348	$ 1,169	$ 2,009

*Assumes a 35% down payment with the balance financed by a 20-year loan of $468,000 at 6% annual interest rate.

payments. Because of increasing receipts each year, net cash flow is projected to become positive in the second year of ownership. In the meantime, the cash flow deficit in the first year must be met with cash from other sources, such as other land being farmed, livestock, or off-farm income.

The situation shown in Table 20-2 is typical of many land purchases. Land prices include expectations of appreciation in value, an infinite income stream, security, and pride of ownership. These factors cause higher prices but do not increase cash returns. Thus, negative cash flows from a debt-financed land purchase tend to be the rule rather than the exception. Earnings often will be insufficient to meet both operating expenses and debt repayment. There are several ways to reduce the cash flow deficit, including a lower purchase price, a lower interest rate, a larger down payment, or a longer term on the loan. An amortized loan with equal total payments or a balloon payment at the end would also help reduce the cash outflow during the first few years, as discussed in Chapter 19. A cash flow projection for the entire business should be done to determine whether cash will be available

from other parts of the business to help make the loan payments on a new land purchase.

Legal Aspects

Purchasing land is a legal as well as a financial transaction. The legal description of the property should be checked for accuracy and the title examined for any potential problems, such as unpaid property taxes or mortgages. The buyer should also be aware of any easements for roads, pipelines, or power lines, which might interfere with the use of the land. Zoning and other local land use restrictions should also be checked.

Any water rights and mineral rights passing to the new owner should be carefully identified and understood. Rights to underground minerals do not automatically transfer with the rights to the land surface. Where oil, gas, coal, or other minerals are important, the fraction of the mineral rights being received can have a large impact on land values. In the irrigated areas of the western United States, the right to use water for irrigation purposes is often limited and allocated on an individual farm basis. The amount, dependability, and duration of these water rights should be

---

Box 20-1   Case Study: A Comparative Sale

Western Land Management Company has been asked to appraise a 750-acre parcel of land near Rio Frio used to grow mostly dryland wheat, milo, and grass hay. Records they have obtained from county courthouses in the area show that several similar properties have sold in the past five years.

The Anderson farm near Habeville was 480 acres, mostly tillable, which sold for $1,000 per acre three years ago. Because it was a smaller farm, bidding on it was more aggressive than is expected for the Rio Frio farm. Thus, Western decides to discount the sale price of the Anderson farm by 15 percent, to reflect a comparable sale.

$1,000 × 85% = $850 per acre

Land values in this part of the state have been increasing in recent years, a total of about 10 percent since the Anderson farm was sold. To put the Anderson sale on a current basis, the Western appraiser adds 10 percent to the adjusted price.

$850 × 110% = $935

The final adjusted price of $935 per acre can now be used as a comparable sale to support the appraised value of the Rio Frio land.

determined carefully. Moreover, an existing but undetected environmental problem could result in anything from difficulty obtaining a loan to a large expense for eliminating the problem.

This is only a partial list of the many ways land buyers can receive less value than they expected because of some unknown restriction or problem. For these and other reasons it is advisable for land buyers to retain the services of an attorney with experience in land transactions before any verbal or written offer is made on a property.

LEASING LAND

Obtaining control of land through leasing has a long history in the United States and other countries. Not all this history has been good. Poor leasing practices have led to exploitation of tenants and sharecroppers and poor land use in some cases. Full landownership has been advocated by some as a means of eliminating these problems. However, it is not likely that we will ever have an agricultural structure where all farmers own all the land they operate. The capital requirements are too large, and improvements in leasing arrangements have reduced or eliminated many of the past problems and inefficiencies.

Leases on agricultural land are strongly influenced by local custom and tradition. The type, terms, and length of leases tend to be fairly uniform within a given area or community. This reliance on custom and tradition results in fairly stable leasing arrangements over time, which is desirable. However, it also means lease terms are slow to change in response to changing economic conditions and new technology. Inefficient land use can result from outdated leasing arrangements.

A lease is a legal contract whereby the landowner gives the tenant the possession and use of an asset such as land for a time in return for a specified payment. The payment may be cash, a share of the production, or a combination of the two. Oral leases are legal in most states but

are not recommended. It is too easy for memories to fail, causing disagreements over the terms of the original agreement. Moreover, documentation may be needed for an estate settlement or a tax audit.

A farm lease should contain at least the following information: (1) the legal description of the land, (2) the term of the lease, (3) the amount of rent to be paid and the time and method of payment, (4) the names of the owner (lessor) and the tenant (lessee), and (5) the signatures of all parties to the lease. These are only the minimum requirements of a lease. A good lease will contain other provisions spelling out the rights and obligations of the landlord and tenant. Many leases also contain a clause that describes the arbitration procedure to be followed in case of unresolved disputes. Dates and procedures for notification of lease renewal and cancellation should also be included, particularly if they differ from state law requirements.

An example of a typical lease form is included at the end of this chapter, but there are many variations. Blank lease forms are available through the Cooperative Extension Service in many states, as well as from attorneys and professional farm managers. It is important to modify the language in a lease form to fit the characteristics of the particular property as well as the laws of the particular state. As with all legal documents, the contracting parties may wish to obtain the advice of an attorney before signing a lease. In the long run, however, a good farm lease is based on mutual trust and communication as well as fair economic terms for both parties.

The three basic types of leases commonly used for renting agricultural land are cash rent, crop share, and livestock share. Each has some advantages and disadvantages to the owner and the tenant.

Cash Rent

About two-thirds of the farm leases in the United States are cash leases. A cash rent lease specifies that the rent will be a cash payment of either a fixed amount per acre or a fixed lump sum. Rent may be due in advance, at the end of the crop season, or as some combination of these. If the rent is set per acre, the number of acres should be specified in the lease. Some cash leases show rent for land and buildings separately. Under a cash lease, the landlord furnishes the land and buildings, while the tenant receives all the income and typically pays all expenses except property taxes, property insurance, and major repairs to buildings and improvements. The lease may contain restrictions on land use and may require the tenant to use certain fertility practices; to control weeds; and to maintain fences, waterways, terraces, and other improvements in their present conditions.

The characteristics of a cash lease create both advantages and disadvantages. Some of the more important are as follows:

1. *Simplicity*: There is less likelihood of disagreement, because the terms can be easily written out and understood by both parties. A cash lease is also easy for a landlord to supervise, because there are few management decisions to be made. For this reason, landlords who live a long distance from their farm or who have little knowledge of agriculture often prefer a cash lease.

2. *Managerial freedom*: Tenants are free to make their own decisions regarding crops, livestock, and other management decisions. Tenants who are above-average managers often favor a cash lease, because they receive the total benefits from their management decisions.

3. *Risk*: A cash lease provides the landlord with a known, steady, and dependable rental income. The tenant stands all risk from yield, price, and cost variability. This avoidance of risk is one reason that cash rents yield lower long-term average returns to the owner than a crop share lease on comparable land.

4. *Capital requirements*: The landlord has fewer capital requirements under a cash lease, because there is no sharing of the annual operating inputs. Conversely, the tenant has a larger capital requirement, including all operating inputs and the cash rent payments.

5. *Land use*: With all income accruing to the tenant, some may be tempted to maximize short-term profits from the land at the expense of long-term productivity, particularly under short-term leases. This can be prevented by including fair and proper land use restrictions in the lease or negotiating long-term contracts.

6. *Improvements*: Landlords may be reluctant to invest in buildings and other improvements when using a cash lease, because they do not share in any additional income they generate. Conversely, some owners may expect to receive rent for buildings for which the tenant has no use.

7. *Rigid terms*: Cash rents tend to be inflexible and slow to change. Unless they are renegotiated each year to reflect changes in prices, land values, and technology, inequities can soon develop.

8. *Tax effects*: Some landlords may favor cash leases for tax purposes. Cash rental income where the landlord is not closely involved in the management of the business may qualify as investment income rather than self-employment income. There is no self-employment tax on investment income. The type of lease agreement used can also affect Social Security payments and estate taxes.

Setting a Fair Cash Rent

The cash rental rates agreed upon for a particular parcel of farmland ultimately depend on the productivity of the land, the value of the crops produced, costs of production, the supply of and

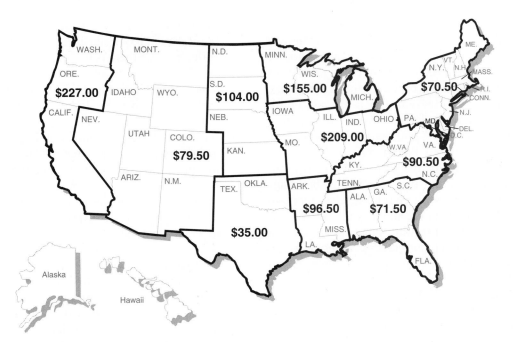

Figure 20-4 **Average cropland cash rental rates per acre, by region.**
Source: National Agricultural Statistics Service, USDA (2013).

demand for farmland in the area, and the bargaining positions of the owner and the operator. Figure 20-4 shows the average cash rent paid per acre for cropland in different regions across the United States.

Several different approaches can be used to estimate a fair rent. These are illustrated in Table 20-3.

1. *Landowner's costs*: In the long run, the landowner wants to receive at least enough rent to cover both the cash and opportunity costs of owning the land. Assume that the owner of the example tract of land in Table 20-1 feels that it has a current market value of $4,500 per acre, or $720,000, and the opportunity cost rate for similar investments is 4 percent. As shown in Table 20-3, the owner's total costs would be 4 percent of $720,000, or $28,800, for opportunity cost

on the investment, $3,200 for property taxes and upkeep, and $1,100 for depreciation, for a total cost of $33,100, or $221 per acre for 150 crop acres. This would be the minimum rent needed to pay all the owner's costs.

2. *Tenant's residual*: A second approach is to estimate how much income the tenant will have left after paying all other costs. Using the same example, the tenant would have gross income of $104,400 and total expenses of $66,449. The difference is $37,951, or $253 per acre. The tenant would not include any land ownership costs such as property taxes. If the rented land could be farmed with no extra machinery investment, then machinery ownership costs could be excluded also. This approach estimates the maximum rent the tenant could pay without allowing for any return to risk or profit.

TABLE 20-3	Setting a Fair Cash Rent (for 150 Crop Acres)

1. Landowner's costs

Opportunity cost on investment (160 acres)	$720,000 × 4.00%	$ 28,800
Property taxes and upkeep		3,200
Depreciation of building, tile, fences		1,100
Total		$ 33,100
Total cost per crop acre		$ 221

2. Tenant's residual

Gross income	104,400
Expenses	
Seed, fertilizer, pesticides, trucking, and drying	39,089
Labor and management	10,800
Machinery	16,560
Total expenses	66,449
Net income available to pay rent	37,951
Net income available per crop acre	253

3. Crop share equivalent

Additional income	$104,400 × 50.00%	$ 52,200
Additional expenses		
Seed, fertilizer, pesticides	$ 36,087 × 50.00%	$ 18,043
Additional net income		$ 34,157
Additional net income per crop acre		$228

4. Share of gross income

Gross income	$104,400
Share of total costs from land	33%
Share of gross income to land	$ 34,452
Estimated rent per crop acre	$ 230

3. *Crop share equivalent*: Another way of finding a fair cash rent is to figure out the income to the owner that would result from a crop share lease. In this example, we used a 50/50 crop share with a 50 percent split in seed, fertilizer, and pesticide expenses. The net gain would be $34,157, or $228 per acre. This is how much cash rent the tenant could pay and still receive the same net return as under a 50/50 crop share lease.

4. *Share of gross income*: The last approach estimates a cash rental rate as a percent of the expected gross income from the land. For example, if land costs generally represent about a third of the total cost of production for particular crops in a region, rent could be estimated at 33 percent of the

expected gross income. In the example, this would be equal to $104,400 \times 0.33 =$ $34,452, or $230 per acre.

These various approaches will not give identical answers. However, they can help define a range within which the owner and operator can negotiate.

Pasture Rent

A fair cash rent for pasture land is more difficult to determine because the potential income is uncertain. Factors such as the quality of the established pasture, water supply, condition of fences, buildings, and location must be considered.

Pasture can be rented for a fixed rate per acre or by the month. One common method is to establish a charge per *animal unit month (AUM)*. An AUM is equivalent to one mature cow grazing for one month. With this method, the rent paid is proportional to the stocking capacity of the pasture.

Crop Share Leases

Crop share leases are popular in areas where cash grain farms are common. These leases specify that the landlord will receive a certain share of the crops produced, with the proceeds from their sale becoming the rent. Where farms are enrolled in certain USDA farm programs, payments received are usually divided in the same proportion as the crop. The tenant typically provides all the labor and machinery. Fertilizer, seed, pesticides, and irrigation costs may be shared, along with harvesting and other costs in some areas. Along with sharing production and expenses, landlords often participate in management decisions about cropping practices.

The landlord's share of the crop will vary depending on the type of crop, local custom, and how operating costs are shared. Soil productivity and climate are important factors, because many of the tenant's costs, such as labor and machinery, will be nearly the same regardless. Therefore, tenants receive a larger share of the

production from less-productive areas, because the value of the land contribution is relatively less important. In the Midwest the landowner's share may be as high as 50 or 60 percent of the crop, but in more arid regions it may be as low as 25 percent. Landlords whose farms contain poorer soils may find they have to take a smaller share and/or pay more of the variable costs to attract a good tenant.

The advantages and disadvantages of crop share leases can be summarized as follows:

1. *Risk*: The crop value will vary with changes in yields and prices, so risk is shared by the tenant and owner. This may be a disadvantage to the owner if the rent is an important part of the owner's total income, which it may be for some retired persons. Variable rent is an advantage to the tenant, as the value of the rental payment varies with the tenant's ability to pay.

2. *Management*: Landlords using a crop share lease usually maintain some direct or indirect control over crop selection and other management decisions. This may be an advantage to an inexperienced tenant and gives the landlord some control over land use.

3. *Capital requirements*: Some crop production expenses are shared and no cash rent must be paid, so the landlord's capital requirements increase, while the tenant's decrease, when compared with a cash lease.

4. *Expense sharing*: One problem with crop share leases is determining a fair and equitable sharing of expenses. As a general rule, variable expenses should be shared in the same proportion as the crop, so as to maintain optimal input levels. The adoption of new technology often creates new problems in determining the proper way to share its costs and benefits. An example of this is presented later in the chapter.

5. *Buildings and pasture*: The landlord receives no direct benefits from buildings (except grain storage) and livestock pasture, so problems can arise concerning a fair rent for these items. A cash rent supplementing the crop share is often paid by the tenant for use of buildings and livestock pasture.

6. *Marketing*: The owner is generally free to sell his/her share of the crops wherever or whenever desired. This requires some additional expertise by the owner. Some owners let the tenant or a professional farm manager make all the marketing decisions.

7. *Material participation*: Landowners are usually more involved in management decisions under a crop share lease. This management participation often qualifies share rent income as self-employment income, which helps build Social Security earnings. Also, it may possibly qualify the farm for a lower valuation for estate tax purposes.

Livestock Share Leases

A livestock share lease is much like a crop share lease, except livestock income and expenses are also shared between landlord and tenant. The tenant typically furnishes all labor and machinery and a share of the livestock and operating inputs, with the landlord furnishing land, buildings, and the remaining share of livestock and operating inputs. Most livestock share leases are 50/50 shares, although other share arrangements are possible depending on the type of livestock and how expenses are shared.

Livestock share leases can be complex, and there is considerable latitude in the number and type of expenses to be shared. In addition to costs such as feed and health expenses, the landlord may share in the cost of machinery and equipment related to livestock production. Examples include milkers, feed grinders, feeders, waterers, and forage harvesting equipment. The lease should contain a full and detailed list of the expenses to be shared and the portion to be paid by each party.

The advantages, disadvantages, and potential problems with a livestock share lease are similar to those of a crop share lease, with the following additional considerations:

1. *Buildings*: A livestock share lease provides the landlord with some return from buildings and pastureland by sharing in the livestock production. Landlords are more likely to furnish and maintain a good set of buildings when they receive part of the income. However, tenants may desire additional building improvements to reduce their labor requirements or improve performance in the shared livestock production.

2. *Records*: The sharing of both crop and livestock income and expenses requires good records to ensure a proper division. There should be a periodic accounting of income and expenses with compensating payments made to balance the accounts.

3. *Lease termination*: Terminating a livestock share lease can be complex and time consuming. All livestock and shared equipment must be divided in a fair and equitable manner. The lease should contain a method for making the division and a procedure for settling any unresolved disputes.

4. *Management*: There is a greater need and opportunity for sharing management decisions under a livestock share lease, and this requires a good working relationship between the landlord and the tenant.

Other Types of Leases

Several other types of leases, and variations of the previously discussed types, are used.

Labor Share Lease

In a labor share lease, the landlord provides all the land, machinery, and other variable inputs. The tenant provides only the labor and receives a share of the production. However, this share would be smaller than with the other share leases. This arrangement works well for a tenant who would like to begin farming but has limited capital and for a landlord who has a complete set of farming resources but is ready to retire.

Variable Cash Lease

Rigidity of cash rents was identified as one of the disadvantages of cash leases. Variable cash leases are sometimes used to overcome this problem and divide some of the risk between the landlord and the tenant. Annual cash rent under a variable cash lease can be tied to the actual yield, or the price received, or both of these factors. For example, the cash rent may increase or decrease by a specified amount for each bushel the actual yield is above or below some base or average yield. The same could be done for price or for gross income. Variable cash leases maintain most of the properties of fixed cash leases but allow the landlord and tenant to share at least part of the price and/or production risk.

One type of variable cash lease sets a rent based on the most likely price and yield, and then adjusts it in proportion to the amount by which the actual price and yield exceed the base values. In the following example, the base rent is $100 per acre for a wheat yield of 60 bushels per acre and a price of $10.00. Assume that the actual yield turns out to be 72 bushels per acre, but the price is only $9.50. The actual rent is then calculated at $114.00 per acre:

$$\text{Actual rent} = \$100 \times \frac{\text{actual yield}}{60} \times \frac{\text{actual price}}{\$10.00}$$

$$= \$100 \times \frac{72}{60} \times \frac{\$9.50}{\$10.00} = \$114.00$$

Another common cash rent formula is to pay a fixed percent of the actual gross income earned from the crops on the rented land. The tenant's ability to pay is affected by price and yield, so a variable cash rent formula that depends on both of these factors provides the most risk reduction. The tenant and landowner should agree in advance on how to determine the actual yield and price to be used for calculating the rent.

Bushel Lease

Another type of lease is the bushel lease, or standing rent lease. With this lease, the landlord typically pays no crop production expenses but receives a specified number of bushels or quantity of the production as rent. Production risk falls entirely on the tenant, because a fixed quantity of the product must be delivered to the landlord regardless of the actual yield obtained. Part of the price risk is shared with the landlord, however, because the actual rent value will depend on the price received for the product.

Custom Farming

The practice of custom farming is not a true leasing agreement but represents another alternative arrangement between the owner and the operator. Typically, the operator provides all machinery and labor for the field operations and hauling in exchange for a fixed payment. There may be a bonus for above-average yields. The owner supplies all the operating inputs, receives all the income, and stands all the price and yield risks. Another variation is for the custom operator to receive a percentage of the production rather than a cash payment.

Table 20-4 summarizes the important characteristics of the lease types discussed in this section.

Efficiency and Equity in Leases

Many leases are based on local custom and tradition, which may result in inefficiencies, poor land use, and a less-than-equitable division of income and expenses. There have been many critics of a land tenure system based on leasing because of the existence of these problems. It may not be possible to write a perfect lease, but improvements can be made. There are two broad areas of concern in improving the efficiency

TABLE 20-4	Comparison of Lease Types				

	Fixed cash	Variable cash	Fixed bushel	Crop or livestock share	Custom farming
Price risk borne by	Tenant	Both	Both	Both	Owner
Production risk borne by	Tenant	Both	Tenant	Both	Owner
Operating capital supplied by	Tenant	Tenant	Tenant	Both	Owner
Management decisions made by	Tenant	Tenant	Tenant	Both	Both
Marketing done by	Tenant	Tenant	Both	Both	Owner
Terms adjust	Slowly	Quickly	Medium	Quickly	Slowly

and equity of a lease. The first is the length of the lease, and the second is the cost-sharing arrangements.

Farm leases are typically written for a term of one year, although some livestock share leases are written for three to five years. Most one-year leases contain a clause providing for automatic renewal on a year-to-year basis if neither party gives notice of termination before a certain date. Many such leases have been in effect for long periods, but the possibility always exists that the lease may be canceled on short notice if the landlord sells the farm or finds a better tenant. This places the tenant in a state of insecurity. It also tempts the tenant to use practices and plant crops that maximize immediate profits rather than conserve or build up the property over time.

Short leases also discourage a tenant from making improvements, because the lease may

---

Box 20-2   Case Study: Negotiating a Lease

Chad and Maria Grabowski started farming five years ago. They own a tractor and some tillage equipment, but share the use of a planter and harvester with Maria's parents. They borrow enough operating capital from their local bank to finance their crop inputs, with a guarantee from the Farm Service Agency (FSA).

The Grabowskis have the opportunity to cash rent 265 acres a few miles away. The owner, whose parents farmed the land for many years, is an attorney in Atlanta. She is asking $90 per acre cash rent, payable in advance.

Chad and Maria visit with their banker and decide that their equity is too small to risk paying the cash rent. They offer to rent the farm on a 60/40 share arrangement, instead. The owner decides that she doesn't want to get involved with paying part of the input bills and selling her share of the crop. However, she is willing to adjust the cash rent each year based on the actual yields obtained and the selling prices available at harvest.

Finally the Grabowskis reach an agreement whereby they will pay $50 per acre on March 1 to rent the farm, plus a bonus equal to 20 percent of the gross income produced, due by December 1. The total rent cannot exceed $100 per acre, however.

be terminated before the cost can be recovered. These problems can be at least partially solved by longer-term leases and agreements to reimburse the tenant for the unrecovered cost. Improvements at the tenant's expense, such as application of limestone and erection of soil conservation structures, can also be covered under an agreement of this type.

Inefficiencies can also arise from poor cost-sharing arrangements if the costs of inputs that directly affect yields are not shared in the same proportion as the income or production. Seed, fertilizer, pesticides, and irrigation water are examples. In Table 20-5, the profit-maximizing level of fertilizer use is where its marginal value product is equal to its marginal input cost, which occurs at 140 pounds of fertilizer per acre. However, if the tenant receives only one-half the crop but pays all the fertilizer cost, the marginal value product for the tenant is only one-half the total. This is shown in the last column of Table 20-5. Under these conditions, the tenant will use only 100 pounds of fertilizer per acre. While this amount will maximize profit for the tenant, it reduces the total profit per acre from what it would be if 140 pounds of fertilizer were applied.

On the other hand, fertilizer has a zero marginal input cost to the landlord. Additional fertilizer use does not increase the landlord's costs but does increase yield, which is shared. The landlord would like to fertilize for maximum yield as the way to maximize profit. This would be at 160 pounds of fertilizer per acre in the example in Table 20-5. These types of conflicts can be eliminated by sharing the cost of yield-determining inputs in the same proportion that the production is shared. If fertilizer costs were shared equally in this example, both parties would pay one-half the marginal input cost and receive one-half the marginal value product. They would both agree on 140 pounds of fertilizer as the profit-maximizing level. Tenants are understandably reluctant to adopt any new yield-increasing technique or technology if they must pay all the additional cost but receive only a share of the yield increase.

Similar problems can arise when changing technology allows the substitution of shared inputs for nonshared inputs. An example is the use of herbicides or pest-resistant seeds in place of mechanical weed control. If the tenant pays only one-half the cost of the seed or herbicide and all the cost of labor and machinery, the profit-maximizing combination of inputs for the tenant will include more herbicide and less labor and machinery than in a cash lease or owner-operator situation.

TABLE 20-5	Example of Inefficient Fertilizer Use Under a Crop Share Lease *(Landlord Receives One-Half of Crop but Pays No Fertilizer Cost)*			
Fertilizer (lb)	Yield (bu)	Marginal input cost at $0.55/lb ($)	Total marginal value product, corn at $5.00 per bu ($)	Tenant's marginal value product ($)
60	97			
80	104	$11.00	$35.00	$17.50
100	110	$11.00	$30.00	$15.00
120	114	$11.00	$20.00	$10.00
140	117	$11.00	$15.00	$ 7.50
160	118	$11.00	$ 5.00	$ 2.50

Determining Lease Shares

The objective of any lease should be to provide a fair and equitable return to both parties for the inputs they contribute to the total farm operation. A fair and equitable sharing arrangement exists when both parties are paid for the use of their inputs according to the contribution those inputs make toward generating income. Application of this principle requires the identification and valuation of all resources contributed by both the landlord and the tenant.

One method for determining the proper shares for a crop share lease is shown in Table 20-6. The example starts from an assumed 50/50 sharing of expenses for fertilizer, pesticides, and seed. Other costs are attributed to the landlord or tenant on the basis of who owns the asset or provides the service.

The most difficult part of this procedure is placing a value on the services of fixed assets such as labor, land, and machinery. The same procedures used for estimating cash and opportunity costs for farm budgets should be applied. Land costs can include cash costs and an opportunity interest charge, or they can be estimated from current cash rental rates. Principal and interest payments should not be used, however, because they represent debt repayment rather than economic costs.

This example has a cost-sharing proportion that is not that far from 50/50. It may be close enough that both parties feel that a 50/50 division of production is fair without any adjustments. If the cost shares are substantially different, one of two methods can be used to make adjustments. First, the party contributing the smaller part of the costs can agree to furnish more of the capital items or pay for more of the variable costs, to make the cost sharing closer to 50/50. The second method would be to change the production shares to 60/40 or some other division to more nearly match the cost shares. This should include changing the cost sharing on the variable inputs to the same ratio, to prevent the inefficiencies discussed earlier. Whichever alternative is used to match the cost and production shares, the result should have the tenant and landlord sharing farm income in the same proportion as they contribute to the total cost of production.

TABLE 20-6	Determining Income Shares Under a Crop Share Lease		
Expenses	**Whole Farm**	**Owner**	**Tenant**
Fertilizer	$16,560	$ 8,280	$ 8,280
Seed	13,317	$ 6,659	$ 6,659
Pesticides	6,210	$ 3,105	$ 3,105
Trucking	932	–	932
Drying	2,070	–	2,070
Labor and management	10,800	–	10,800
Machinery ownership costs	9,900	–	9,900
Machinery operating costs	6,660	–	6,660
Depreciation of buildings, tile, fences	1,100	1,100	–
Property taxes and upkeep	3,200	3,200	–
Opportunity cost for land	28,800	28,800	–
Total expenses	$99,549	$51,144	$48,406
Percent contributed		51%	49%

CONSERVATION AND ENVIRONMENTAL CONCERNS

Conservation can be defined as the use of farming practices that will maximize the present value of the long-run social and economic benefits from land use. This definition does not prevent land from being used but does require following practices that will maintain soil productivity and water quality over time. Farming systems that accomplish these goals are sometimes called *sustainable agriculture*.

Ordinary budgeting techniques are often inadequate for deciding how best to achieve the goals of conservation. Nevertheless, farm and ranch managers need to be conscious of how their decisions affect the long-term quality of life for themselves and society. There are three general areas in which sustainable agriculture considerations go beyond conventional budgeting analysis.

Long-Run Versus Short-Run Consequences

Narrow profit margins and tight cash flows make it tempting to *mine* the soil to maximize short-run returns. Most conservation practices require some extra cash expenditures. They may also reduce crop yields temporarily as soil and cropping patterns are disturbed. This short-run reduction in profit may be necessary to achieve higher profits in the future or to prevent a long-run decline in production if no conservation practices are used. The long-term effects of soil depletion and erosion on productivity are not always well known or understood. Likewise, it takes years of study to determine the long-term effects of continued use of high rates of fertilizer and pesticides on soil, water, wildlife, and humans.

Conservation practices such as terraces, drainage ditches, and diversion structures require large initial investments. The present value methods discussed in Chapter 17 can be used to evaluate the profitability of a particular practice. The opportunity cost of capital and the planning horizon of the landowner become important in the analysis. Higher discount rates reduce the present value of the larger future incomes resulting from conservation. Shorter planning horizons also discourage conservation. For this reason, many farmers look at changes in tillage practices and crop rotations as lower-cost alternatives to achieve conservation goals.

Leasing arrangements also affect the type of conservation practices followed. One-year leases discourage tenants from considering the long-term effects of their farming practices. On the other hand, landlords may be reluctant to make large conservation investments if they feel the tenant will receive all or a large share of the benefits. Share leases should attempt to divide the costs and benefits of such practices whenever possible. Tenants and landlords should fully discuss the farming practices needed to meet long-run conservation and environmental considerations.

Off-Farm Effects

Many of the decisions farmers make with regard to land use and production practices have consequences that go far beyond the boundaries of the farm. Buildup of silt in rivers and lakes, pollution and contamination of groundwater, destruction of wildlife habitat, and the presence of chemical residues in livestock products are a few examples. These effects are difficult to evaluate in monetary terms and are often difficult to relate back to specific agricultural practices and sources. Research into causes and effects is continuing. Agriculture must consider more than farm input costs when making decisions about input use. The total societal costs of using various technologies is quickly becoming an important factor in choosing production practices.

Regulations and Incentives

The state and federal governments have enacted laws to promote and sometimes require land use and livestock production practices that preserve and enhance soil, water, and air resources. As more of these regulations are enacted, future conservation efforts may increasingly become a matter of selecting the least-cost combination of practices to meet the relevant requirements.

Long-term land retirement programs, such as the Conservation Reserve Program, offer guaranteed annual payments in return for taking highly erodible land out of production. The opportunity cost of the lost production must be evaluated against the incentive payments and the long-run conservation benefits. Some programs pay landowners in return for conservation easements that control land use. Other regulations, such as *swampbuster* and *sodbuster* rules, have restricted farmers from cropping certain areas with a history of pasture or wetland use. Farmers are also required to develop and follow an approved conservation plan to participate in some government farm programs.

Restrictions on the application of pesticides, some chemical fertilizers, and animal manure vary among states. Regular soil testing and careful scouting for pest problems will ensure that such products are used only at environmentally safe and economically profitable levels. The principles of marginal cost and marginal revenue can be used to determine the economic threshold at which the potential losses from a pest problem justify the cost of treating it.

Some farms have been found to have environmental hazards such as leaking underground fuel storage tanks or accumulations of discarded pesticide containers. The costs of cleaning up such problems can significantly reduce the value of a farm. Before closing a sale of a farm or ranch property, a prudent buyer will arrange for an environmental audit to identify any potential problems of this type and the estimated costs of correcting them.

Several state and federal laws regulate the discharge of pollutants into streams and lakes. Most agricultural runoff is considered to be *nonpoint source* pollution. However, confined animal feeding operations (CAFOs) are considered to be point sources and are more tightly controlled.

A high opportunity cost of capital, short planning horizons, and limited direct evidence of environmental effects combine to explain why some landowners are reluctant to adopt sound environmental practices. However, society has an interest in conservation to maintain and expand the nation's long-run food production potential, as well as to ensure food safety. Societal planning horizons are typically longer and broader than those for an individual farmer. In recognition of this and the limited capital position of many farmers, society has devised means to encourage more sustainable practices. Technical assistance is available at no cost through local offices of the Natural Resources Conservation Service. Financial assistance to pay for part of the costs of certain conservation structures has been made available through the U.S. Department of Agriculture and some state programs. And, USDA program payments are available to producers who follow certain conservation practices.

Today's farmers and ranchers must think beyond simply complying with regulations and maximizing current profits. The ethic of conservation tells us that owning or using land for agricultural purposes carries with it the responsibility to follow practices that will support society well into the future.

SUMMARY

Land is an essential resource for agricultural production. It is a permanent resource fixed in supply and location. The decision to buy or lease land is an important one that will affect the production capacity and financial condition of the business for many years. Land ownership has many advantages. However, purchasing land requires a strong capital position and adequate cash flow potential if credit is used. Land can be valued based on its net earnings or by comparing sale prices of similar land.

A mixture of owned and leased land exists in many farm and ranch businesses. Cash rent, crop share, and livestock share are the most common types of lease arrangements. Each lease type has advantages and disadvantages to both the tenant and the landlord in terms of the capital contribution required, the amount of price and yield risk to be shared, and the making of management decisions. Share leases should provide for the sharing of income in the same proportion as each party contributes to the total cost of production. Variable inputs should also be shared in this proportion to allocate resources efficiently.

Land use decisions need to consider long-run environmental effects and consequences that occur beyond the borders of the farm to conserve resources and sustain agriculture into the future.

QUESTIONS FOR REVIEW AND FURTHER THOUGHT

1. List as many reasons as you can to explain why people buy farmland. How would your list, or the importance of each reason, be different for an operating farmer than for a nonfarm investor who lives in another state?

2. What are the advantages and disadvantages of owning land compared to renting land?

3. Using the capitalization method of valuation and a 6 percent discount rate, how much could you afford to pay for an acre of land expected to have a net return of $181 per year? What would the answer be with a discount rate of 4 percent?

4. Why is a cash flow analysis important in a land purchase decision?

5. List three advantages and disadvantages of cash and crop share leases for both the tenant and the landlord.

6. What are the typical terms for crop share leases in your community? Analyze them using Table 20-6 as a guide. Are the terms fair? What problems did you encounter in the analysis?

7. Develop a flexible cash lease for your area using price and yield as the variable factors upon which the rent is based. Show how the rent varies for different combinations of yield and price.

8. What environmental concerns are associated with agriculture in your area? Give two examples.

Cash Farm Lease

NCFMEC-01A

For additional information see publication NCFMEC–01 (Fixed and Flexible Cash Rental Arrangements for Your Farm).

This form can provide the landowner and operator with a guide for developing an agreement to fit their individual situation. This form is not intended to take the place of legal advice pertaining to contractual relationships between the two parties. Because of the possibility that an operating agreement may be legally considered a partnership under certain conditions, seeking proper legal advice is recommended when developing such an agreement.

This lease is entered into this _____ day of _____, 20 ___, between

_____, owner, of   _____

_____

Address

_____, spouse, of   _____

_____

Address

hereafter known as "the owner," and

_____, operator, of   _____

_____

Address

_____, spouse, of   _____

_____

Address

hereafter known as "the operator."

I. Property Description

The landowner hereby leases to the operator, to occupy and use for agriculture and related purposes, the following described property:

_____

_____

consisting of approximately _____ acres situated in _____ County (Counties),

_____ (State)

II. General Terms of Lease

A. Time period covered. The provisions of this agreement shall be in effect for _____ year(s), commencing on the _____ day of _____, 20 _____. This lease shall continue in effect from year to year thereafter unless written notice of termination is given by either party to the other at least _____ days prior to expiration of this lease or the end of any year of continuation.

B. Review of lease. A written request is required for general review of the lease or for consideration of proposed changes by either party, at least _____ days prior to the final date for giving notice to terminate the lease as specified in II-A.

C. Amendments and alterations. Amendments and alterations to this lease shall be in writing and shall be signed by both the owner and operator.

D. No partnership intended. It is particularly understood and agreed that this lease shall not be deemed to be, nor intended to give rise to, a partnership relation.

E. Transfer of property. If the owner should sell or otherwise transfer title to the farm, such action will be done subject to the provisions of this lease.

F. Right of entry. The owner, as well as agents and employees of the owner, reserve the right to enter the farm at any reasonable time to a) consult with the operator; b) make repairs, improvements, and inspections; and c) (after notice of termination of the lease is given) do tilling, seeding, fertilizing, and any other customary seasonal work, none of which is to interfere with the operator in carrying out regular operations.

G. No right to sublease. The owner does not convey to the operator the right to lease or sublet any part of the farm or to assign the lease to any person or persons whomsoever, including for purposes of hunting, trapping or other recreational uses.

H. Binding on heirs. The provisions of this lease shall be binding upon the heirs, executors, administrators, and successors of both owner and operator in like manner as upon the original parties, except as provided by mutual written agreement.

Additional agreements regarding terms of lease:

_____

_____

III. Land Use

A. General provisions. The land described in Section I will be used in approximately the following manner. If it is impractical in any year to follow such a land-use plan, appropriate adjustments will be made by mutual written agreement between the parties.

1. Cropland a) Row crops _____ Acres

b) Small grains _____ Acres

c) Hay _____ Acres

d) Rotation pasture _____ Acres

2. Permanent pasture _____ Acres

3. Other: _____ _____ Acres

_____ _____ Acres

Total acres _____ Acres

B. Government Programs. The extent of participation in federal, state or county government programs for purposes of commodity support, conservation enhancement or other objectives will be discussed and decided on an annual basis or when the original contract expires. The course of action agreed upon should be placed in writing and be signed by both parties. A copy of the course of action so agreed upon shall be made available to each party.

IV. Amount and Payment of Rent

If a flexible cash rental arrangement is desired, describe it on the last page of this form and omit section A below.

A. Cash rental rates. The operator agrees to pay as cash rent the amount as calculated in the "Amount of Cash Rent" table for each kind of land; or, one total may be entered for entire farm unit.

Amount of Cash Rent (Part IV-A)

Kind of Land or Improvements	Acres	Rate per Acre	Amount
Row crops	_____	$ _____	$ _____
Small grains	_____	$ _____	$ _____
Hay	_____	$ _____	$ _____
Permanent pasture	_____	$ _____	$ _____
Timber	_____	$ _____	$ _____
Waste	_____	$ _____	$ _____
Farm buildings	_____	$ _____	$ _____
Dwelling	_____	$ _____	$ _____
Other	_____	$ _____	$ _____
Entire farm	_____	$ _____	$ _____

B. Rental payment. The annual cash rent shall be paid as follows:

$_____ on or before _____ day of _____ (Month)

$_____ on or before _____ day of _____ (Month)

$_____ on or before _____ day of _____ (Month)

$_____ on or before _____ day of _____ (Month)

If rent is not paid when due, the operator agrees to pay interest on the amount of unpaid rent at the rate of _____ percent per annum from the due date until paid.

C. Payee information. The rental payments shall be sent to the address of the owner as shown on page 1 of this lease, or to the following address:

_____

D. Liens. The operator acknowledges and agrees that the owner may file and perfect a lien upon the crops grown under this lease to secure the payment of rents or any other amounts due under this lease, and that the operator may execute the same against such crops in accordance with state law.

V. Operation and Maintenance of Farm

In order to operate this farm efficiently and to maintain it in a high state of productivity, the parties agree as follows:

A. The operator agrees:

1. General maintenance: To provide the labor necessary to maintain the farm and its improvements during the rental period in as good condition as they were at the beginning. Normal wear and depreciation and damage from causes beyond the operator's control are excepted.

2. Noxious weeds. To use diligence to prevent noxious weeds from going to seed on the farm. Treatment of the noxious weed infestation and cost thereof shall be handled as follows: _____

_____

3. Conservation. Control soil erosion according to an approved conservation plan; keep in good repair all terraces, open ditches, inlets and outlets of tile drains, and ponds; preserve all established watercourses or ditches including grassed waterways and field borders; and refrain from any operation or practice that will injure such structures.

4. Damage. Upon termination of the lease agreement, to pay the owner reasonable compensation for any damages to the farm for which the operator is responsible. Any decrease in value due to ordinary wear and depreciation or damages outside the control of the operator are excepted.

5. Costs of operation. To pay all costs of operation except those specifically referred to in Sections V-A-4 and V-B.

6. Repairs. Not to buy materials for maintenance and repairs in an amount in excess of $ _____ within a single year without written consent of the owner.

7. Documentation. To provide the owner with yield or production information for harvested crops sufficient to meet requirements for crop insurance documentation and participation in USDA commodity programs.

B. The owner agrees:

1. Loss replacement. To replace or repair as promptly as possible the dwelling or any other building or equipment regularly used by the operator that may be destroyed or damaged by fire, flood, or other cause beyond the control of the operator or to make rental adjustments in lieu of replacements.

2. Materials for repair. To furnish all materials needed for normal maintenance and repairs.

3. Skilled labor. To furnish or pay for any skilled labor tasks that the operator is unable to perform satisfactorily. Additional agreements regarding materials and labor are:

_____

_____

4. Reimbursement. To pay for materials purchased by the operator for purposes of repair and maintenance in an amount not to exceed $ _____ in any one year, except as otherwise agreed upon. Reimbursement shall be made within _____ days after the operator submits the bill.

5. Removable improvements. Let the operator make minor improvements of a temporary or removable nature, which do not mar the condition or appearance of the farm, at the operator's expense. The owner further agrees to let the operator remove such improvements even though they are legally fixtures at any time this lease is in effect or within _____ days thereafter, provided the operator leaves in good condition that part of the farm from which such improvements are removed. The operator shall have no right to compensation for improvements that are not removed except as mutually agreed.

6. Compensation for crop expenses. To reimburse the operator at the termination of this lease for field work done and for other crop costs incurred for crops to be harvested during the following year. Unless otherwise agreed, current custom rates for the operations involved and actual costs for materials applied will be used as a basis of settlement.

C. Both agree:

1. Not to obligate other party. Neither party hereto shall pledge the credit of the other party hereto for any purpose whatsoever without the consent of the other party. Neither party shall be responsible for debts or liabilities incurred, or for damages caused by the other party.

2. Capital improvements. Costs of establishing hay or pasture seedings, new conservation structures, improvements (except as provided in Section V-B-5), or of applying lime and other long-lived fertilizers shall be divided between owner and operator as set forth in the following table. The operator will be reimbursed by the owner either when the improvement is completed, or the operator will be compensated for the share of the depreciated cost of the operator's contribution when the lease ends based on the value of the operator's initial contribution and depreciation rate shown in the "Compensation for Improvements" table. (Cross out the portion of the preceding sentence which does not apply.) Rates for labor, power and machinery contributed by the operator shall be agreed upon before construction is started.

3. Mineral rights and wind/solar development. The landowner shall have the right to enter into agreements for the development of petroleum, wind, solar, or other resources on the property, and may also authorize third parties to enter the property to survey, construct, and/or operate the facilities reasonably necessary to develop those resources. The landowner agrees to reimburse the tenant for any actual damage suffered for crops destroyed by these activities and to release the tenant from obligation to continue farming this property when and if development of such resources interferes materially with the tenant's opportunity to earn a satisfactory return.

4. Environmental issues. The operator shall conduct all operations on the property in a manner consistent with all applicable local, state, and federal environmental codes, regulations, and statutes and shall bear sole responsibility for any violations thereof. The operator shall be solely responsible for securing any permits or approvals necessary for his or her activities on the property. In the event of any legally prohibited release of materials to the environment, the operator will indemnify the landowner for any costs of environmental cleanup and restoration as well as any penalties, fines, judgments or other amounts incurred by the landowner as a result of such release.

5. Arbitration of differences. Any differences between the parties as to their several rights or obligations under this lease that are not settled by mutual agreement after thorough discussion, shall be submitted for arbitration to a committee of three disinterested persons, one selected by each party hereto and to the third by the two thus selected. The committee's decision shall be accepted by both parties.

VI. Amount of Rent to be Paid When Cropland is Rented on a Flexible Basis

Flexible cropland rent. *(Use Method I, II, or III.)*

1. Basic information to be used in Methods I and II

Crop(s)	Base Cash Rent (per acre)	Base Yield (bushel or ton per acre)	Base Price (per bushel or ton)	Base Input Costs (per acre)	Minimum Cash Rent (per acre)	Maximum Cash Rent (per acre)
_____	$ _____	_____	$ _____	$ _____	$ _____	$ _____
_____	$ _____	_____	$ _____	$ _____	$ _____	$ _____
_____	$ _____	_____	$ _____	$ _____	$ _____	$ _____

2. The current price for the current year shall be average price at close of day based on the following time periods(s) and location(s).

Crop(s)_____ ____ Day _____ Month through ____ Day _____ Month at _____ Price Source

Crop(s)_____ ____ Day _____ Month through ____ Day _____ Month at _____ Price Source

Crop(s)_____ ____ Day _____ Month through ____ Day _____ Month at _____ Price Source

3. Base Year Input Costs

Crop(s)	Seed	Fertilizer	Pesticides	Fuel	Total
_____	$ _____	$ _____	$ _____	$ _____	$ _____
_____	$ _____	$ _____	$ _____	$ _____	$ _____
_____	$ _____	$ _____	$ _____	$ _____	$ _____

For each year of this lease, the base cash rent per acre for each crop shall be adjusted at the close of the cropping season by one of the following methods:

Method I - Flexing for Price Only

Crop(s)	Base Rent	×	(Current Price ÷ Base Price)	=	Rent Per Acre†	×	Acres Grown	=	Adjusted Rent for the Year
_____	$ _____	×	_____	=	$ _____	×	_____	=	$ _____
_____	$ _____	×	_____	=	$ _____	×	_____	=	$ _____
_____	$ _____	×	_____	=	$ _____	×	_____	=	$ _____

Method II - Flexing for Price and Yield

Crop(s)	Base Rent	×	(Current Price ÷ Base Price)	×	(Current Yield ÷ Base Yield)[‡]	=	Rent Per Acre[†]	×	Acres Grown	=	Adjusted Rent for the Year
_____	$ _____	×	_____	×	_____	=	$ _____	×	_____	=	$ _____
_____	$ _____	×	_____	×	_____	=	$ _____	×	_____	=	$ _____
_____	$ _____	×	_____	×	_____	=	$ _____	×	_____	=	$ _____

Method III - Flexing for Price, Yield and Input Costs

Crop(s)	Base Rent	×	(Current Price ÷ Base Price)	×	(Current Yield ÷ Base Yield)	×	Base Costs ÷ Current Costs)	×	Acres Grown	=	Flexible Rent
_____	$ _____	×	_____	×	_____	×	_____	×	_____	=	$ _____
_____	$ _____	×	_____	×	_____	×	_____	×	_____	=	$ _____
_____	$ _____	×	_____	×	_____	×	_____	×	_____	=	$ _____

[†]If calculated figure is less than "Minimum Cash Rent" in Part 1, use the set minimum. If calculated figure is more than "Maximum Cash Rent" in Part 1, use the set maximum.
[‡]The current yield shall be the "farm" yield for the current lease year.

Method IV - Work Out and Record Procedure To Be Used

_____

_____

_____

_____

Executed in duplicate on the date first above written:

_____	_____
Operator	Owner
_____	_____
Operator's spouse	Owner's spouse

State of _____

County of _____

On this _____ day of _____ , A.D. 20 _____, before me, the undersigned, a Notary Public in said State, personally appeared _____, _____, _____, and _____ to me known to be the identical persons named in and who executed the foregoing instrument, and acknowledged that they executed the same as their voluntary act and deed.

_____

Notary Public

HUMAN RESOURCE MANAGEMENT

CHAPTER OUTLINE

Characteristics of Agricultural Labor
Planning Farm Labor Resources
Measuring the Efficiency of Labor
Improving Labor Efficiency
Improving Managerial Capacity
Obtaining and Managing Farm Employees
Agricultural Labor Regulations
Summary
Questions for Review and Further Thought

CHAPTER OBJECTIVES

1. Describe trends in the use of human resources in agriculture

2. Illustrate how to plan for the quantity and quality of human resources needed for different farm and ranch situations

3. Outline methods for measuring and improving labor efficiency

4. Suggest ways to improve the management of agricultural employees, including selection, compensation, and motivation

5. Summarize laws regulating agricultural workers and employers

Human labor is one of the few inputs in agriculture whose use has diminished substantially over time. The decline has been especially dramatic since 1950, as shown in Table 21-1. The introduction of mechanization and other labor-saving technology has allowed agricultural production to increase despite the decrease in labor use, however. Energy in the form of electrical and mechanical devices has replaced much of the physical energy provided by humans and draft animals in the past. More of the labor input on today's farm is spent operating, supervising,

TABLE 21-1	Workers on U.S. Farms (1950–2010)		
Year	Operators and unpaid workers	Hired workers	Total workers
1950	7,597,000	2,329,000	9,926,000
1960	5,172,000	1,885,000	7,057,000
1970	3,348,000	1,175,000	4,523,000
1980	2,401,000	1,298,000	3,699,000
1990	1,999,000	892,000	2,891,000
2000	2,062,000	890,000	2,952,000
2010	—	767,500	—

Source: National Agricultural Statistics Service, U.S. Department of Agriculture.

and monitoring these mechanical activities, and less is expended on physical effort. Changes in the tasks performed by agricultural labor have required employees and managers to increase their education, skills, and training.

The mere availability of new technology, such as larger machinery, mechanical feed and manure-handling systems, and computers, does not alone explain its rapid and widespread adoption. There has to be an economic justification before farmers will use any new technology, or it will *sit on the shelf*. Most labor-saving technology has been adopted for one or more of the following reasons:

1. It is less expensive than the labor it replaces.
2. It allows farmers to increase their volume of production and total profit.
3. It makes work easier and more pleasant.
4. It allows certain operations such as planting and harvesting to be completed on time, even when weather is unfavorable or labor is in short supply.
5. It does a better job than could be accomplished manually.

Input substitution occurs because of a change in the marginal physical rate of substitution and/or a change in the relative prices of inputs, as was

discussed in Chapter 8. Both factors have been important in the substitution of capital-intensive technology for labor in agriculture. Marginal rates of substitution have changed as new technology altered the shape of the relevant isoquants, making it profitable to use less labor and more capital.

Interest rates and wage rates have increased since 1950, but wage rates have increased by a higher percentage, making labor relatively more expensive than capital. This change affects the price ratio in the substitution problem, making it profitable to use more capital and less labor. The increased amount of capital per worker in agriculture has caused a substantial increase in the productivity of agricultural labor, making it both feasible and necessary to pay higher wages. This increased productivity has allowed farmers and ranchers to enjoy a standard of living comparable to that of nonfarm families. Table 21-2 shows average wage rates as of April 2014 for farm workers by region of the country.

TABLE 21-2	Wage Rates by Region for Agricultural Workers, 2014
Region	Average hourly wage ($/hour)
Northeast	11.43
Appalachian	10.31
Southeast	10.02
Lake	11.50
Cornbelt	12.32
Delta	10.05
Northern Plains	13.15
Southern Plains	10.10
Mountain	11.13
Pacific	11.19
48 states	11.11

Source: National Agricultural Statistics Service, U.S. Department of Agriculture, 2014.

CHARACTERISTICS OF AGRICULTURAL LABOR

A discussion of agricultural labor must recognize the unique characteristics that affect its use and management on farms. Labor is a continuous-flow input, meaning the service it provides is available hour by hour and day by day. It cannot be stored for later use; it must be used as it becomes available, or it is lost.

Full-time labor is also a *lumpy* input, meaning it is available only in whole, indivisible units. Part-time and hourly labor is often used also, but the majority of agricultural labor is provided by full-time, year-round employees. Table 21-3 shows how farm workers were distributed in April 2014 by number of workers hired. Eight percent of hired farm workers worked on farms where they were the sole hired employee. Nine percent worked on farms with two workers. At the other extreme, 58 percent of farm workers were employed on farms that hired more than 10 workers. If labor is available only on a full-time basis, the addition or loss of an employee constitutes a major change in the labor supply of the farm. For example, a sole proprietor hiring the first employee is increasing the labor supply by 100 percent, and the second employee represents a 50 percent increase. A problem facing a growing business is when and how to acquire the additional resources needed to keep a new worker fully employed. When other resources such as land and machinery also come in *lumpy* units, it becomes difficult to avoid a shortage or excess of one or more resources.

The operator and other family members provide all or a large part of the labor used on most farms and ranches. This labor does not generally receive a direct cash wage, so its cost and value can be easily overlooked or ignored. However, as with all resources, there is an opportunity cost for operator and family labor, which can be a large part of the farm's total noncash fixed costs. Compensation for operator and family labor is received indirectly through expenditures for family living expenses and other cash withdrawals. This indirect wage or salary may vary widely, particularly for non-essential items, as net farm income varies from year to year. Cash farm expenses have a high-priority claim on any cash income, causing nonessential living expenses to fluctuate with farm income.

The human factor is another characteristic that distinguishes labor from other resources. If an individual is treated as an inanimate object, productivity and efficiency suffer. The hopes, fears, ambitions, likes, dislikes, worries, and personal problems of the operator and employees must be considered in any labor management plan.

TABLE 21-3	**Number of Farm Employees on U.S. Farms, 2014, by Number per Farm**

Number of employees per farm	Percent of all employees
1	8
2	9
3–6	18
7–10	7
11–20	13
21–50	13
51 or more workers	32

Source: National Agricultural Statistical Service, U.S. Department of Agriculture, 2014.

PLANNING FARM LABOR RESOURCES

Planning the farm's labor resources carefully will help avoid costly and painful mistakes. Figure 21-1 illustrates the process. The first step is to assess the farm's labor needs, the quantity and quality, and the conditions under which workers will function.

Step 1—Assess your situation

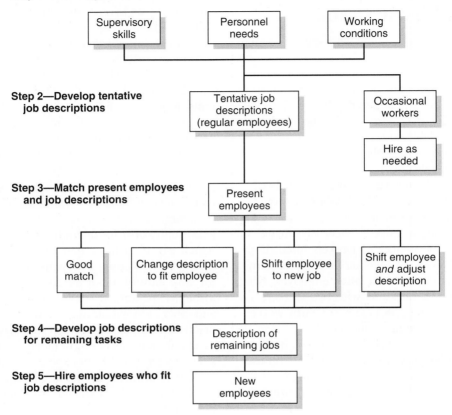

Figure 21-1 Flow chart of the farm labor planning process.

Source: Thomas, Kenneth H., and Bernard L. Erven, *Farm Personnel Management*, North Central Regional Extension Publication 329.

Quantity of Labor Needed

Most farm managers judge the quantity of labor needed by observation and experience. When new enterprises are being introduced, typical labor requirements from published enterprise budgets can be used. A worksheet such as the one illustrated in Table 21-4 is useful for summarizing labor needs.

The seasonality of labor needs also must be considered. For example, labor requirements may exceed available labor during the months when planting, harvesting, calving, and farrowing take place. Figure 21-2 shows an example of

the total monthly labor requirements for all farm enterprises and the monthly labor provided by the farm operator and a full-time employee. The farm operator in this example has a problem common to many farms and ranches. Operator labor will not meet the requirements in some months, but the addition of a full-time employee results in large amounts of excess labor during other months. Longer workdays, temporary help, or hiring a custom operator may be necessary to perform the required tasks on time. A more permanent solution may be to increase the capacity of field machinery or processing equipment or to

TABLE 21-4	Labor Estimate Worksheet

					Total hours for year	December–March	April–June	July–August	September–November
						Distribution of hours			
1	Operator (or Partner No. 1)				2,900	800	750	600	750
2	Partner No. 2								
3									
4	Family labor				1,300	200	300	500	300
5	Hired labor								
6	Custom machine operators								
7	Total labor hours available				4,200	1,000	1,050	1,100	1,050
	Direct labor hours needed by crop and animal enterprises								
	Crop enterprises	Acres	Hours/acre						
8	Wheat	700	1.8		1,260	10	100	750	400
9	Sorghum	300	2.1		630	0	400	0	230
10	Alfalfa	200	6.2		1,240	0	400	700	140
11									
12									
13									
14									
15									
16	Total labor hours needed for crops				3,130	10	900	1,450	770
	Animal enterprises	No. Units	Hours/unit						
17	Beef cows	250	6.0		1,500	600	500	200	200
18	Background	400	0.25		100	50	0	0	50
19									
20									
21	Total labor hours needed for animals				1,600	650	500	200	250
22	Total hours needed for crops and animals				4,730	660	1,400	1,650	1,020
23	Total hours of indirect labor needed				600	200	150	100	150
24	Total labor hours needed				5,330	860	1,550	1,750	1,170
25	Total available (line 7)				4,200	1,000	1,050	1,100	1,050
26	Additional labor hours required (line 24 minus line 25)						500	650	120
27	Excess labor hours available (line 25 minus line 24)					140			

Source: Missouri Farm Planning Handbook, Manual 75: Feb. 1986. Department of Agricultural Economics, University of Missouri, Columbia, MO.

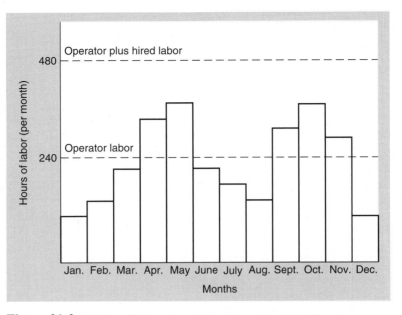

Figure 21-2 Profile of labor requirements and availability.

shift to different enterprises. How much labor should be used to maximize profits depends on its availability, its cost, and whether it is a fixed or variable input.

Labor as a Fixed Cost

The total labor supply of the operator and/or full-time employees may be fixed but not fully used. If this labor is being paid a fixed salary regardless of the hours worked, there is no variable or marginal cost for using another hour. In this situation, labor can be treated as a fixed cost. In an enterprise budget, labor costs do not affect the gross margin, nor the choice of enterprises. In a partial budget, labor costs are not included in added costs nor in reduced costs.

Labor as a Variable Cost

Even permanent labor may have an opportunity cost greater than zero, from either leisure or off-farm employment. This becomes the minimum acceptable earning rate, below which the individual will choose not to work additional hours. Labor has a variable opportunity cost when

increasing or decreasing the size of one enterprise affects the possible scale of other enterprises.

When labor is hired on a part-time or *as-needed* basis, it should always be treated as a variable cost. The cost of such labor, including benefits and payroll taxes, should be included in any budgeting decisions as well as marginal cost, marginal revenue analyses.

Quality of Labor Needed

Not all farm labor is equal in training, ability, and experience. New agricultural technology requires more specialized and sophisticated skills. Some activities, such as application of certain pesticides, may even require special training and certification. An assessment of labor needs requires identification of special capabilities needed, such as operating certain types of machines, performing livestock health chores, balancing feed rations, using computers or electronic control devices, or performing mechanical repairs and maintenance. If the operator or family members do not possess all the special skills needed, then employees must be hired who do, or certain jobs may be contracted to outside consultants, repair shops, or custom operators. Training programs may also be available to help farm workers acquire new skills.

Management Style

Some farm and ranch managers use hired personnel more effectively than others. Many operators are accustomed to working alone and prefer employees who can work independently with a minimum amount of supervision and instruction. Other employers prefer to work closely with workers and give specific instructions about how a job is to be performed. Both management styles can be effective, but a good manager recognizes his or her style and seeks workers who can function effectively under it.

Once the quantity and quality of labor needed by the farm or ranch operation have been analyzed, tentative job descriptions should be developed. Larger operations will allow for much more specialization of duties than smaller ones, of course. Then the skills of the workers currently available should be compared to the job descriptions. Some duties may have to be rearranged to match the skills and interests of certain employees, family members, or partners. Needs that cannot be met by the current workforce will have to be filled by providing job training, securing additional workers, or contracting with outside services.

Developing an organizational chart is especially useful when several employees and managers are involved. In particular, it should be made clear whether some employees are expected to take directions from other employees or from members of the operator's family. Figure 21-3 shows an example of an organizational chart for a medium-size operation. Besides indicating which members of the labor force supervise other members, an organizational chart should show the lines of communication that need to be maintained among managers and employees.

Managing Risks in Hiring Labor

As discussed in detail in the book *Ag Help Wanted: Guidelines for Managing Agricultural Labor* by the Western Farm Management

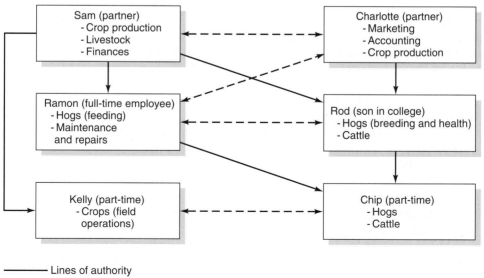

Lines of authority
- - - - Lines of communication

Figure 21-3 Organizational chart for a farm business.

Extension Committee, there are many sources of risk in hiring labor. A major source of risk, one normally well understood in agriculture, is that there will not be enough labor available to perform crucial tasks, meaning that production will suffer and so will profit. Another source of risk is the quality of labor. Poor labor quality can result from poor skills of the employees, poor communication from managers, or improperly set incentives. Indirect costs of labor are another possible source of risk. Excessive turnover and absenteeism can contribute to high indirect costs. Risk can also come from conflict with employees, which can be time consuming and expensive, especially if accompanied by lawsuits. Hiring labor requires compliance with state and federal laws and timely filing of tax forms and other paper work, so improper compliance is another source of risk for employers. Good managers must be aware of all sources of risk in hiring labor and work to minimize the likelihood of problems. Good planning, good communication, and keeping up with relevant laws and regulations are critical components of risk management.

MEASURING THE EFFICIENCY OF LABOR

Labor efficiency depends not only on the skills and training of the labor used but also on the size of the business, enterprises, degree of mechanization, type of organization, and many other factors. Measures of labor efficiency should be used to compare and evaluate results only on farm businesses of approximately the same size and type.

Measures of labor efficiency often use the concept of person-years or full-time equivalents of labor employed. This is a procedure for combining operator, family, and hired labor into a total labor figure, comparable across farms. The example shows 21 months of labor provided from three sources. Dividing this total by 12 converts it into 1.75 person-years, or the equivalent of 1.75 persons working full-time during the year.

Operator labor	12 months
Family labor	4 months
Hired labor	5 months
Total	21 months

$21 \div 12 = 1.75$ person-years equivalent

Measures of labor efficiency convert some physical output, cost, or income total into a value per person-year. The following measures are commonly used.

Value of Farm Production per Person
This measures the total value of agricultural products generated by the farm per person-year equivalent. It is affected by business size, type of enterprise, and the amount of machinery and other labor-saving equipment used.

Labor Cost per Crop Acre
The cost of labor per crop acre is found by dividing the total crop labor cost for the year by the number of acres in crop and fallow. The opportunity cost of operator and family labor is included in total labor cost. Values will be affected by machinery size, type of crops grown, and whether custom operators are used.

Crop Acres per Person
The number of crop acres per person is found by dividing total crop acres by the number of person-years of labor used for crop-related activities.

Cows Milked per Person
The total number of productive cows in the dairy herd is divided by the number of person-years of labor. Other livestock enterprises use similar measures of labor efficiency.

Table 21-5 contains some labor efficiency values for farms in Iowa. The data show that productivity per person usually increases as farm size increases, due mostly to more investment in larger machinery and equipment and spreading labor overhead over more units of production.

TABLE 21-5	Labor Efficiency by Farm Size for Iowa Farms, 2013			

	Farm size, annual sales			
Item	$100,000 to $199,999	$200,000 to $399,999	$400,000 to $799,999	$800,000 and above
Months of labor per farm	7.0	11.0	14.2	27.0
Value of farm production per person	$296,211	$340,495	$465,566	$635,871
Labor cost per crop acre	$98.97	$80.02	$64.60	$72.94
Crop acres per person	283	434	561	516

Source: 2013 Iowa Farm Costs and Returns, Iowa State University Extension Publication, File C1-10.

IMPROVING LABOR EFFICIENCY

Labor efficiency can be improved by more capital investment per worker, the use of larger-scale machinery, or adoption of less-labor-intensive technology. However, if the objective is to maximize profit and not just to increase labor efficiency at any cost, then marginal rates of substitution and the prices of labor and capital must determine the proper combination. Increasing the capital investment per worker will increase profit only if: (1) total cost is reduced, while revenue increases, remains constant, or at least decreases less than costs; or (2) the saved labor can be used to increase output value elsewhere by more than the cost of the investment.

When the labor supply is increased by adding a full-time worker, some additional units of an enterprise may be needed to fully use the available labor. Fig. 21-4 shows how costs per cow may go up when a dairy farm adds employees, if the number of cows is not increased at the same time. This illustrates the *lumpy* nature of full-time labor as an input.

Simplifying working procedures and routines can have a high payoff in increased labor efficiency. Considerable time can be saved by having all necessary tools and other supplies at the work area, not having to stop to open and close gates, keeping equipment well maintained, and having spare parts on hand. Making changes in the farmstead layout, building designs, field

size and shape, and the storage sites of materials relative to where they will be used can also save time and increase labor efficiency. When materials must be moved, consider using conveyers, carts, small vehicles, augers, and other labor-saving devices. Automatic switches and timers can eliminate time spent waiting. Carrying radios or portable telephones in vehicles or tractors makes timely communication possible, allows coordination of activities over a wide geographic area, and reduces trips. Production data can be entered directly into small notebook or hand-held computers, rather than transferred from paper records. As always, the additional cost of any changes must be weighed against the value of the labor saved.

Labor efficiency can also be improved by making sure workers have safe and comfortable working conditions whenever possible. Although most agricultural work is still performed outside, modern machinery cabs and vehicles help reduce fatigue from heat, cold, dirt, and noise. Making sure that machinery guards are in place, livestock buildings are well ventilated, and all safety measures are followed will prevent lost work time from injuries or illness. Workers should be provided with suitable clothing and other safety equipment when working with agricultural chemicals or performing other hazardous jobs.

Planning and scheduling work in advance will help reduce wasted time. Tasks that must be

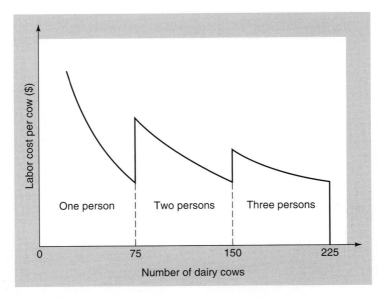

Figure 21-4 **Labor cost per dairy cow for different herd sizes.**

done at a specific time should be scheduled first, and those that are not time specific, such as building repairs or other maintenance, can be planned for months of slack labor. Keep a list of jobs to be done, and assign a priority and deadline to each one. This list should be where all workers can see it, add to it, and cross off tasks that have been completed. From this, a daily work schedule can be planned in a few minutes each morning or evening. Time used to organize the next day's tasks and their order of importance is time well spent.

IMPROVING MANAGERIAL CAPACITY

The most important labor resources in any business are the managers. Although a manager may begin with a wealth of experience and education, this will not be adequate for an entire career. Management skills must continually be reinforced and upgraded.

Agricultural technology is continually evolving. Repeating prescribed techniques is not enough. A successful ranch or farm manager must understand the principles behind the technology, such as animal nutrition, plant physiology, genetics, and farm mechanics. New applications of these principles then will be easier to master. Farm magazines, extension seminars, continuing education courses, and field days are all excellent sources of information about new technology. Ultimately, though, the manager may realize that it is not possible for one person to master everything, and that delegating responsibility to others in the business or hiring outside consultants is a useful way of accessing additional expertise.

Developing an efficient office or business center is also a sign of a good manager. The physical facilities should be comfortable and functional and should allow the use of up-to-date communications and information management technology. The flow of bills, receipts, reports, and correspondence should be quick and frequent. Records should be sorted and filed for easy retrieval.

Farm operators are seldom underemployed. The work to be done always expands to fill the time available, so setting priorities is important. It is easy to look around and do whichever task comes into view first. However, the principles of *immediacy* and *impact* must be applied to decide what jobs will have the most effect on the business and/or need to be completed first. Making monthly and weekly lists helps identify and prioritize tasks to be done.

Many successful farm managers are active in professional and volunteer organizations outside their businesses. Farm policy and commodity organizations, school boards, churches, cooperatives, and marketing clubs offer chances to develop organizational skills and personal relationships. Not only is participation in such groups a way to contribute to a community, but some of the best ideas and philosophies come from interacting with other highly capable people.

Finally, an agricultural manager needs to develop a *world view*. Over 20 percent of U.S. agricultural products are consumed in other countries, so it is important to understand tastes and customs from other cultures. Farmers in other countries also offer significant competition in the marketplace. Traveling abroad, especially with tours designed to learn more about agricultural production, trade, and consumer preferences, is an effective way of learning about the rest of the world. Hosting international visitors and attending programs or reading about other cultures are also valuable ways to increase international understanding.

OBTAINING AND MANAGING FARM EMPLOYEES

Paid employees account for almost 30 percent of the total labor supply on farms and ranches in the United States, as was shown in Table 21-1. Acquisition, training, and retention of hired workers are common subjects of conversation whenever managers of larger farms get together. Farm managers are finding that skills in human relations and personnel management are valuable assets.

Recruiting

The process of hiring an employee starts with recruiting, including announcing the job opening, publicizing it, and receiving applications. Placing newspaper ads; talking with other farmers, relatives, agribusiness people, or farm consultants in the community; or contacting college placement offices and employment agencies are ways to inform individuals about a job opening and identify potential candidates to fill it. In some areas, the employer may have to negotiate with a labor contractor to supply a large number of temporary workers for harvesting or other labor-intensive activities. Recruiting via the Internet has become increasingly common in recent years, and there are several commercial agribusiness recruiting firms that post job openings.

Box 21-1   **Three Ads for the Same Job: Which One Would You Apply For?**

Wanted: Hired Man. Phone 123-456-7890

Wanted: Person for general farm work, full-time, on a beef and crop farm. Requires two years farm experience or equivalent and/or two years ag education beyond high school. Good wage and benefit package. Call: 123-456-7890

Source: Shapley, A. E., *Farm Employer's Handbook*, Michigan State University.

Maple Grove Farms has an opening for a full-time person to assist the owner/operator in management and production of beef and crops. The position offers considerable variety and opportunity for growth. At least two years farm experience and/or ag education beyond high school required. Pay and benefit package based on applicant's experience and training. Write for application to Maple Grove Farms (address).

The job announcement should clearly state the skills and experience desired. In addition, it should make the position sound desirable. Emphasize reasons that the applicant would want to work for this operation instead of the one down the road.

It is generally helpful to provide an application form to each applicant. Basic information should be obtained about the applicant's background, work experience, training, personal goals, references, and other factors.

While employers have great latitude in deciding among employees based on their skill levels and ability to perform a job, state and federal laws prohibit discrimination based on characteristics that do not affect job performance, including race, color, ethnicity, religion, age over 40, sex, marital status, disability and medical status, union activities, arrests that did not lead to convictions, and so on. The Equal Employment Opportunity Act (EEOA) and the Americans with Disabilities Act (ADA) are the two major federal antidiscrimination laws. Employers should be careful when evaluating job applicants not to include questions irrelevant to job performance that could be used to classify employees based on one of the protected categories mentioned. Some states have examples of interview questions to avoid, such as those about children, marital status or marital plans, and social affiliations.

Interviewing and Selecting

Completed application forms can be used to select a small number of candidates for the next step, interviewing. The interview should be planned carefully to efficiently acquire more information. Sufficient time and opportunity should be allowed for the applicant to ask questions about the job, its duties, and responsibilities. Interviewing involves not only obtaining information about applicants but also providing information to them so they can assess their interest in and qualifications for the job. A skills test may be needed for certain technical jobs. A tour of the work location should be provided,

and the candidate should have a chance to visit with other employees.

The information obtained about each job applicant through the application form, the interview, and references must now be evaluated. Many factors must be considered in selecting a candidate, including personal compatibility. Farm employers often work more closely with their employees on a day-to-day basis than other employers do, sometimes under stressful conditions. This close working relationship increases the chance for friction if the individuals are not compatible.

The Employment Agreement

Once a job offer is made and accepted, a written employment agreement should be developed, as shown in the example in Box 21-2. The purpose of the agreement is to clarify the work expectations of both the employer and the employee and to serve as a reference for evaluating performance later on.

The employment agreement should start with the job description, including duties and responsibilities, lines of authority, and the job title. Other important information includes wages and benefits, working hours and days, vacation, sick leave and personal leave policy, safety rules, allowable uses of farm property, training opportunities, bonus or incentive plans, and procedures for evaluation and promotion or termination. A review of the employment agreement should be made once or twice a year as part of the evaluation process.

Compensation

A competitive compensation package is essential in a successful labor-hiring program. The total value of take-home pay, benefits, and bonuses should be compared against comparable figures in other employment.

Wages and Salaries

The cash wage or salary paid is the most important item. Whether to pay a fixed or variable wage is the first decision. Positions in which the duties and hours worked will be fairly constant throughout the year usually receive a fixed weekly or monthly salary. Other positions with highly variable hours, such as those found on a fruit and vegetable or cash grain farm, are often paid by the hour, as are most part-time positions. Workers employed in harvesting activities are sometimes paid on a piece-rate basis.

Salaries depend on the position and the particular skills of the employee filling it. The size of the farming operation, the duties performed, and the number of years the employee has worked on the farm are other factors that influence the level of compensation.

Skill-based pay is an approach that sets compensation based on the level of responsibility each worker has rather than the specific duties. Characteristics such as decision-making authority, supervision of other employees, and specialized skills needed are used to rate each position and assign it a wage range.

Fringe Benefits

Fringe benefits are often a large part of the total compensation for farm employees. Table 21-6 shows benefits reported in a study of employees on swine farms in 2011. In addition to the benefits reported in this table, 96 percent of large hog operations provided medical insurance to employees, as did 68 percent of mid-sized operations. Similarly, a recent survey of agricultural employees in Iowa found that benefits and bonuses accounted for 15 percent of employee compensation.

Prospective employees should be made aware of the value of their benefits so they can fairly evaluate a job offer. Fringe benefits such as retirement plans, insurance, housing, utilities, farm produce, and the use of a vehicle will make a lower cash wage for farm work competitive with nonfarm employment that has a higher cash salary but fewer benefits.

Fringe benefits are most useful when the employer can provide them for less than it would cost the employee to obtain them elsewhere. Examples include the use of existing housing,

TABLE 21-6	Benefits Received by Employees on Swine Farms, 2011

Benefit	Percent Receiving
Paid holidays	63.9
Paid vacation	57.8
Retirement plan	54.4
Fresh or processed pork	49.4
Training and development	43.9
Bonus/Profit-sharing	43.3
Paid sick leave	36.1
Mobile phone	23.9
Vehicle or gas allowance	20.6
Housing	17.8
Continuing education	17.2
Utilities	6.7

Source: National Pork Board, Employee Compensation in Pork Production, 2012.

vehicles, or livestock facilities. Some benefits such as health insurance are tax deductible for the employer but not taxable to the employee.

Incentive Programs and Bonuses

Bonuses are often used to supplement base wages, improve labor productivity, and increase employee retention. According to a 2011 survey conducted by Iowa State University Extension and Outreach, about 65 percent of Iowa farm workers received some sort of bonus or incentive plan. Bonuses help increase labor efficiency only if they are tied closely to performance. If not, employees soon come to expect the bonus and consider it part of their basic cash salary. When the size of the bonus is tied to annual profit, an employer may find it difficult to decrease the bonus in a poor year after employees have experienced several years with a larger bonus.

Most bonus plans are based on one or more factors: volume, performance, longevity, or profitability.

1. *Volume* can be measured by the total number of pigs weaned, calves weaned, or bushels harvested. The employee's wages increase when the work load increases and a modest incentive for efficiency is provided. Care should be taken that higher costs are not incurred to increase production, however.

2. *Performance* can be measured by the number of pigs weaned per sow, calving percentage, milk production per cow, or crop yields per acre. The bonus is often based on how much the actual performance exceeds a certain base level. This type of bonus can be an effective work incentive as long as it is based on factors over which the employee has at least some control.

3. *Longevity* is rewarded by paying a bonus based on the number of years the employee has worked for the business. This recognizes the value of experience and worker continuity to the employer and rewards loyalty.

4. *Profitability* bonuses are usually based on a percent of the gross or net income of the business. They allow the employee to share in the risks and rewards of the business, but may depend on many factors not under the control of the employee. They also require the employer to divulge some financial information to the employee.

Following several basic principles will increase the effectiveness of any incentive program:

- The program should be simple and easily understood by the employee.
- The program should be based on factors largely within the control of the employee.
- The program should reward work that is in the best interest of the employer.
- The program should provide a cash return large enough to motivate improved performance.

- The incentive payment should be made promptly or as soon after the completion of work as possible.
- An example of how the bonus will be computed should be provided in writing, using typical performance levels.
- The incentive payment should not be considered as a substitute for a competitive base wage and good labor relations.

A good bonus plan not only rewards employees financially but also provides recognition for their accomplishments.

Total compensation packages offered to employees should have *internal consistency*; that is, they should be fair among employees of the same business. They should also have *external consistency* when compared to workers doing similar jobs on other farms and ranches.

Training Hired Labor

Farm and ranch managers sometimes hire unskilled workers and then expect them to perform highly skilled tasks in livestock management or with expensive machinery. They may also expect employees to automatically do things exactly the way they themselves would do them. The result is disappointment, frustration, high repair bills, poor labor productivity, and employee dissatisfaction.

Studies of employment practices on ranches and farms show little evidence of formalized training programs for new employees. Even a skilled employee will need some instruction on the practices and routines to be followed for a particular operation and the meaning of specialized terminology. Employees with lesser skills should receive complete instructions and proper supervision during a training period. Employers need to develop the patience, understanding, and time necessary to train and supervise new employees. Unfortunately, in production agriculture, the training period may need to last as long as a year or until the new employee has had a chance to perform all the tasks in a complete cycle of crop and livestock production.

Box 21-2   Employment Agreement Example

FARM EMPLOYER-EMPLOYEE AGREEMENT

Beau Valley Farms (employer) of Greenville, Oregon, agrees to employ _____ (employee), to perform agricultural work in and around Jefferson County, Oregon, beginning (date) _____ and continuing until such time as either wishes to terminate this agreement by a 30-day notice. The employer and the employee agree to comply with the following conditions and actions:

1. To pay the employee $ _____ per hour, from which the employee's income tax and Social Security taxes will be withheld, as required by law. Wages will be paid on the 1st and 15th day of each month.

2. To provide a 3-bedroom house, with utilities to be paid by the employee. Routine maintenance is to be done by the employee and materials paid for by the employer. Any other agreements pertaining to the employee's housing will be noted at the end of the agreement.

3. The normal work week is Monday through Saturday. The normal paid working hours are from 7:00 A.M. to 6:00 P.M. (12:00 noon on Saturdays), with two 30-minute rest breaks and one hour off for lunch. Longer hours may be required to complete jobs in progress. Overtime will be paid for any work done after normal working hours at the rate of 150% of the normal wage rate.

4. Time off shall be given every Sunday and holiday and one Saturday per month. The holidays for purposes of this agreement are New Year's Day, Easter, Memorial Day, Independence Day, Labor Day, Thanksgiving Day, and Christmas Day. Other holidays can be substituted for these by prior agreement.

5. The employee is entitled to 10 days of vacation with pay annually, which shall be taken during the nonheavy work season and agreed upon with the employer 30 days in advance.

6. The employee is entitled to 5 days sick leave with pay annually for the time off due to actual illness, and 1 day of personal emergency leave with pay.

7. The employee is entitled to drive a vehicle provided by Beau Valley Farms for purposes related to employment, including commuting to work and for personal use within Jefferson County.

8. The following insurance plans will be carried on the employee:

   • comprehensive medical and health
   • term life insurance coverage up to $100,000

9. A performance review and written performance evaluation will be carried out once during the first 6 months of employment and once each 12 months thereafter.

10. A bonus or incentive plan (is, is not) included. If included, the provisions are noted on the attached appendix.

11. Other provisions not included above are listed on the appendix to this form.

Employer Representative
  Signature _____ Date _____
Employer Identification No. _____
Employee
  Signature _____ Date _____
Social Security No. _____

Adapted from: Thomas, Kenneth H., and Bernard L. Erven, *Farm Employee Management*.

Periodic retraining may be necessary for even long-time employees. The adoption of new technology in the form of different machinery, new pesticides, feed additives, seed varieties and electronic data gathering, or the introduction of a new enterprise may require additional training for all employees. Extension short courses, bulletins, farm magazines, video tapes, Internet sites, field days, and seminars can be used for employee training. Participation in these activities will improve not only employees' skills but also their self-esteem. The cost of training should be borne partially or wholly by the employer as an investment in future productivity.

Motivation and Communication

Hiring and training new employees is costly in terms of time and money. If labor turnover is high, these costs can become excessive, and labor efficiency will be low. Employers should be aware of the reasons for poor employee retention and take action to improve the situation.

Farm employees often report that they are attracted to farm work by previous farm experience, the chance to work outdoors, and an interest in working with crops and livestock. Disadvantages cited are long hours, little time off, early-morning or late-evening work, uncomfortable working conditions, and poor personal relationships with their employer. Low pay is seldom at the top of the list of disadvantages, indicating that they have personal goals other than obtaining the highest pay.

A set policy for adequate vacations and time off is important to employees, as is the opportunity to work with good buildings, livestock, and equipment. Job titles can also be important. The title *hired hand* contributes little to an employee's personal satisfaction and self-image. Herd superintendent, crop manager, group leader, and machine operator are examples of titles most people associate with a higher status than that of a hired hand.

Good human relations is the most important factor in labor management. This includes such things as a friendly attitude, loyalty, trust, mutual respect, the ability to delegate authority, and the willingness to listen to employee suggestions and complaints. Work instructions should be given in sufficient detail that both parties know what is to be done, when, and how. Everyone responds positively to public praise for a job well done, but criticism and suggestions for improvement should be communicated in private. Unpleasant jobs should be shared by all employees, and everyone should be treated with equal respect. Employers who reserve the larger, air-conditioned tractor for their own use, while employees drive smaller tractors without cabs, for example, are more likely to have labor problems.

As employees grow in their abilities and experience, they should be given added responsibilities, along with the chance to make more decisions. Likewise, the employer must be willing to live with the results of those decisions or tactfully suggest changes.

Working environment also has a major impact on job satisfaction. A national survey of employees on swine farms found that higher levels of gas and dust were associated with lower levels of job satisfaction.

Bridging Cultural Barriers

More agricultural employees are either recent immigrants to the United States or children of immigrants. In 2012, 47 percent of farm laborers and supervisors were foreign born, and 50 percent were of Hispanic ethnicity. When employees and employers do not share a common language and culture, communication and motivation can break down quickly.

Immigrant workers often develop extensive networks of relatives and acquaintances from their native communities. Loyalties are strong and may supersede loyalty to a job or an employer. Such networks can be used effectively for

---

Box 21-3   Case Example: Understanding Cultural Differences

Haitian farm workers in an upstate apple-growing enterprise had been told repeatedly to leave the apples that had fallen on the ground alone and not put them with those picked from the tree. Problems also occurred when different varieties of apples were placed in the same containers. The supervisor had even told a few people not to return for work because of their insubordination. Discussions with the farm workers explained their actions. In Haiti food was in short supply. Leaving what appeared to be good apples on the ground was a waste of food that they could not understand. Mixing varieties was understandable when we consider that only the rich can afford apples in Haiti. The numerous varieties of apples were as unknown to them as the numerous varieties of bananas would be to U.S. born workers.

Source: Embrey, K., *Human Resource Management on the Farm*, Cornell University Cooperative Extension Service.

---

recruiting new employees who will be accepted by the existing workforce. Conversely, mixing workers from different countries or regions may create conflicts, even though the cultural differences are not apparent to the employer.

Training programs are much more critical for employees who have not grown up in the local area. Technical knowledge and habits cannot be taken for granted. Showing nonnative speakers how to perform tasks is much more effective than telling them. Fortunately, many technical manuals, regulations, and employee agreements are available in multiple languages today, especially Spanish.

Cultural attitudes affect employer–employee relationships. In many countries, workplaces have hierarchical structures. Supervisors demand respect from employees based on their position or family ties, not on their abilities or personalities. Workers may be uncomfortable having their employer working side-by-side with them, even though this is common on small farms. They also may feel that asking questions about their duties is a sign of ignorance or disrespect. When something goes wrong, it is often attributed to fate, not to the employee's lack of diligence. Punctuality and timeliness may be less important than personal relationships and family responsibilities.

Religious or cultural holidays observed by some ethnic groups may be different than those of the employer and need to be considered when the employment agreement or work calendar is developed.

Evaluation

Employers are constantly evaluating their employees' performances. Too often, however, communication takes place only when serious problems arise. Regular times for communication and coordination should be scheduled. If meetings are held only when there is a problem, the employee will immediately be put on the defensive when one is called. Some managers take their employees to breakfast once every week or month to discuss plans and problems. The employer should listen carefully to each employee's concerns and ideas, even if not all of them can be acted upon.

Operations with a sizable workforce should use written evaluations, with performance measured against job descriptions. Employees should be warned if their performance is unsatisfactory, first orally and then in writing, and given a chance to improve. If termination is necessary, document the reasons carefully and give written notice in advance. All this takes time, but it can prevent costly complaints or lawsuits later. If an employee quits, hold an exit interview. Discuss the reasons for leaving, and decide whether changes in employee recruitment or management practices are needed.

AGRICULTURAL LABOR REGULATIONS

State and federal regulations that affect the employment of agricultural labor have become an important factor in labor management in the United States. Their thrust is to extend to farm workers much of the same protection enjoyed by nonfarm workers. For the employer, the effects of these regulations may be increased costs for higher wages and more benefits, more labor records to keep, and additional investments for safety and environmental protection, but also a more satisfied and productive workforce.

It is not possible to list and describe all the state and federal regulations pertaining to agricultural labor in a few pages. Only a brief discussion of some of the more important and general regulations will be included here.

Minimum Wage Law

Agricultural employers who used more than 500 person-days of labor in any quarter of the preceding calendar year must pay at least the federal minimum wage to all employees except immediate family members, certain piece-rate hand harvesters, and cowboys and shepherds employed in range production of livestock. The family member exemption does not apply to partnerships and corporations. The minimum wage is subject to increases over time. This law also requires employers to keep detailed payroll records to prove compliance with the minimum wage legislation. Under the federal statute, agricultural employers are not required to pay overtime wages to employees who work more than 40 hours per week. State laws may set more stringent standards for overtime pay and minimum wages, however, and therefore must be checked for compliance.

Social Security

Employers must withhold Social Security, or FICA (Federal Insurance Contributions Act), taxes from an employee's cash wages and match the employee tax with an equal amount. This regulation applies to farmers with one or more employees, if the employee was paid $150 or more in cash wages during the year or the employer paid $2,500 or more in total wages to all employees.

The tax rate, and the maximum earnings subject to this tax, increase from year to year. Employers should contact the local Social Security Administration office or obtain Internal Revenue Service (IRS) Circular E, Employer's Tax Guide, for the current rates. The tax rate and maximum tax figures represent the amounts withheld from the employee's pay. The employer must match them with an equal amount for remittance to the IRS.

Federal Income Tax Withholding

Farmers are required to withhold federal income taxes from wages paid to agricultural workers if the employer and employee meet the same criteria described for Social Security withholding. Circular E or Circular A, Agricultural Employee's Tax Guide, provides details. Records should be kept of each employee's name, age, Social Security number, cash and noncash payments received, and amounts withheld or deducted.

Workers' Compensation

This is an insurance system designed to protect workers who suffer job-related injuries, illnesses, disability, or death. It also frees the employer from additional liability for such injuries. A fixed amount of compensation is paid for each illness or injury. Laws regarding workers' compensation insurance for farm employees vary from state to state, but most states require it of any farm operator with one or more employees. Where required, the employer pays a premium based on a percentage of the total employee payroll.

Unemployment Insurance

The Federal Unemployment Tax Act (FUTA), along with state unemployment systems, replaces part of a person's income lost due to unemployment. Most employers pay both a state and a federal unemployment tax. Benefits are financed by an employer-paid payroll tax. Farm employers must contribute if, during the current or previous calendar year, they: (1) employed 10 or more workers in each of 20 or more weeks or (2) paid $20,000 or more in cash wages in any calendar quarter. State laws vary and need to be checked with the relevant agency.

Child Labor Regulations

Regulations established under the Fair Labor Standards Act require an employee to be at least 16 years of age to be employed in agricultural work during school hours. The minimum age is 14 for employment after school. Children aged 12 or 13 can be employed with written parental consent or when a parent is employed on the same farm, and children under 12 can work on their parents' farm.

Age restrictions also apply to certain jobs that are classified as hazardous. Examples of these jobs are working with certain farm chemicals, operating most farm machinery, working inside storage structures, and working with breeding livestock. Employees under the age of 16 cannot be employed in these hazardous jobs, except that 14- or 15-year-old employees can be certified to operate certain types of farm machinery by taking qualified safety courses. Child Labor Bulletin No. 102 of the U.S. Department of Labor contains details. Employers should also check their state laws and regulations, which may be more stringent than the federal ones.

Occupational Safety and Health Act

The Occupational Safety and Health Act (OSHA) is a federal law passed to ensure the health and safety of employees by providing safe working conditions. Examples are the requirements for slow-moving farm equipment using public roads to display a slow-moving vehicle (SMV) sign on the rear, and for roll bars to be present on most farm tractors. All employers covered by OSHA must report within eight hours any workplace incident resulting in a fatality or a serious injury of three or more employees. Employers with more than 10 employees must also keep additional injury and illness records. Farm employers should check the latest OSHA regulations to be sure they are meeting all current requirements.

Environmental Protection Agency

The Environmental Protection Agency (EPA) regulates pesticide use on farms. The Worker Protection Standard (WPS) for Agricultural Pesticides is designed to reduce the risk of pesticide poisonings and injuries among agricultural workers and pesticide handlers. All farm workers must complete pesticide safety training, while employees designated as pesticide handlers must undergo more extensive training. The 2005 WPS How to Comply (HTC) Manual provides information to employers on these important regulations. State regulatory agencies may impose additional regulations on pesticide use.

Immigration Reform and Control Act

This Act requires employers to check documents that establish the identity and eligibility of all workers they hire. The employer and employee must complete and retain a Form I-9, available from the U.S. Citizenship and Immigration Services, which certifies that the employee is a U.S. citizen, a permanent alien, or an alien with permission to work in the United States.

H-2A Temporary or Seasonal Agricultural Work

Section H-2A of the Immigration Reform and Control Act allows agricultural employers to hire foreign workers for temporary or seasonal work if there is a verifiable shortage of domestic workers available for these tasks. Before hiring foreign workers under this provision, the employer must apply for certification that there are not sufficient qualified workers willing and available to perform the work and that the employment of foreign workers will not adversely affect wages and working conditions of similarly employed workers in the United States. The wage rate for the temporary workers must be set so as not to adversely affect prevailing domestic wages. The program is authorized by the Immigration and Nationality Act (INA) as amended by the Immigration Reform and Control Act (IRCA) of 1986. More details about the program requirements and restrictions are available from the U.S. Department of Labor.

Migrant Labor Laws

The U.S. Migrant and Seasonal Agricultural Worker Protection Act prescribes housing, safety, and record-keeping requirements for employers of migrant or seasonal agricultural workers. State laws may also stipulate minimum housing, safety, and health standards.

Civil Rights

Federal civil rights legislation prohibits employment discrimination on the basis of race, color, religion, sex, national origin, handicap, ancestry, or age. These rules also apply to recruiting and hiring practices.

Disability

The Americans with Disabilities Act of 1990 (ADA) prohibits discrimination against employees or potential employees based on physical or mental handicaps. If there are aspects to a job that could not be carried out by a person with certain disabilities, these should be included in a written job description before applicants are interviewed. The ADA applies to businesses that employ 15 or more workers for 20 or more weeks out of the year. Agricultural employers should check current regulations to be sure they are in compliance.

Affordable Care Act

Under the 2010 Affordable Care Act, employers who hire 50 or more full-time employees are required to provide them with affordable health insurance coverage or they may face a financial penalty if any of the uncovered employees receive a tax credit for health insurance premiums. Seasonal workers employed for 120 days or less in the calendar year are excluded. Additionally, employers must begin withholding an additional 0.9 percent on employee compensation that exceeds $200,000. Other provisions, involving reporting, also apply. Employers with fewer than 50 employees may purchase insurance through the Small Business Health Options Program and those with fewer than 25 employees may be eligible for a Small Business Health Care Tax credit if certain conditions apply. Because this is a new law, provisions should be researched carefully each year.

SUMMARY

Although the number of people employed in agriculture has fallen dramatically in the past 50 years, their productivity and skill level have increased even more rapidly. Planning farm personnel needs involves assessing the quantity, quality, and seasonality of labor needed. Whether the cost of farm labor is considered to be fixed or variable depends on its availability and opportunity costs. Good labor management techniques can improve labor efficiency and satisfaction. Farm managers themselves need to continually upgrade their skills.

Effective management of hired employees starts with the recruiting, interviewing, and selection process. An employment agreement should be drawn up that specifies work rules and compensation, including salary, benefits, and bonuses. Bonuses should be planned carefully to provide the desired work incentives. Farm employees also must be trained, motivated, and evaluated. Where cultural differences exist between managers and employees, special care should be taken to ensure communication is effective.

A manager of farm labor must be familiar with and comply with the various state and federal laws that protect and regulate agricultural workers.

| QUESTIONS FOR REVIEW AND FURTHER THOUGHT |

1. Why has the amount of human resources used in agriculture decreased?

2. Why are the training and skills needed by agricultural workers greater today than in the past?

3. Under what conditions is it profitable to substitute labor for capital?

4. Why are there wide variations in the monthly labor requirements on a crop farm, but less variation on a dairy farm?

5. Why is the labor cost per acre or per head generally less for larger farms than for smaller ones?

6. Observe a routine farm chore such as milking or feeding livestock. What suggestions would you have for simplifying the chore to save labor? Would your suggestions increase costs? Would they be profitable?

7. Write an advertisement to place in a newspaper or magazine for recruiting a livestock manager for a large farm.

8. Write a detailed job description for the same position.

9. Discuss what you would include in a training program for a new farm employee.

10. Discuss some procedures that could be used to improve personal relationships between a farm employer or supervisor and farm employees who grew up in a different culture.

11. What type of incentive program would be most effective on a dairy farm if the objective was to increase milk production per cow? If the objective was to retain the better employees for a longer time?

12. Which labor laws and regulations have to do with withholding taxes? Safety and health? Employing minors?

MACHINERY MANAGEMENT

CHAPTER OUTLINE

Estimating Machinery Costs
Examples of Machinery Cost Calculations
Factors in Machinery Selection
Alternatives for Acquiring Machinery
Improving Machinery Efficiency
Summary
Questions for Review and Further Thought

CHAPTER OBJECTIVES

1. Illustrate the importance of good machinery management on farms and ranches

2. Identify the costs associated with owning and operating agricultural machinery

3. Demonstrate procedures for calculating machinery costs

4. Discuss important factors in machinery selection, including size, total cost, and timely completion of operations

5. Compare owning, renting, leasing, and custom hiring as different means of acquiring the use of machinery

6. Present strategies for increasing the efficiency of machinery use

7. Explain factors that influence when machinery should be replaced

Mechanization has had a dramatic effect on production costs, efficiency levels, energy use, labor requirements, and product quality in agriculture all over the world. The decline in farm labor required and the increase in mechanization over time were discussed in Chapter 21. As tractor sizes have increased, the sizes of other machinery have also increased to keep pace. These increases have contributed to higher machinery investment per farm and more efficient use of labor and have made it possible for an individual operator to farm many acres.

The quality of work performed by farm machinery has also risen dramatically. Field losses during harvesting have been greatly reduced. Improved seed and fertilizer placement has made it possible to reduce the amount of tillage performed. Sensitive yield monitors, more accurate sprayers and applicators, and the use of satellite-positioning technology have spawned a new set of practices known as *precision agriculture.*

Discussions on the increased use of mechanization in agriculture often emphasize tractors and other crop production machinery. Perhaps no less dramatic has been the rise in the use of power and equipment in livestock production and materials handling. Physical labor requirements have been reduced in many areas by the use of small engines and electric motors, conveyors, computers, sensors, and timers. Grain handling, manure collection and disposal, livestock feeding, feed grinding and mixing, hay handling, accounting, and data collection have all been greatly automated to reduce labor requirements and costs and to improve performance.

Good machinery management strives to provide reliable service to the various crop and livestock enterprises at minimum cost. Sometimes machinery is a profit center itself, such as for a custom operator. More often, though, it provides services for other enterprises. This makes it difficult to separate machinery management completely from an analysis of the total business. This chapter will concentrate on the use of economic principles and budgeting in machinery management, as well as the relations among machinery, labor, capital, and the production enterprises.

ESTIMATING MACHINERY COSTS

Machinery is costly to purchase, own, and operate. A tractor can cost well over $200,000, and a cotton picker or combine over $400,000. A manager must be aware of the costs of owning and operating a machine and understand how they are related to machinery use, interest rates, useful life, and other factors. It is easy to underestimate machinery costs, because many of them involve infrequent, though large, cash outlays, or they are noncash costs, such as depreciation.

Machinery costs can be divided into *ownership costs* and *operating costs.* Ownership costs are also called overhead, indirect, or fixed costs, because they are fixed with respect to the amount of annual use. Operating costs are also referred to as variable or direct costs, because they vary directly with the amount of machine use. Fixed and variable costs were discussed in Chapter 9 but will be reviewed in the following sections as they apply to farm machinery.

Ownership Costs

Ownership or fixed costs begin with the purchase of the machine, continue for as long as it is owned, and cannot be avoided by the manager except by selling the machine. For this reason, it is important to estimate how much the resulting ownership costs will be before a machine is purchased.

Ownership costs can be as much as 60 to 80 percent of the total annual costs for a nonpowered implement. As a rule of thumb, yearly ownership costs will be about 10 to 20 percent of the original cost of the machine, depending on the type of machine, its age, expected useful life, and the cost of capital.

Depreciation

Depreciation is a noncash expense that reflects a loss in the value of machinery due to age, wear, and obsolescence. It is also an accounting procedure to recover the initial purchase cost of an asset by spreading this cost over its entire ownership period. Most depreciation is caused by age and obsolescence, so it is considered a fixed cost once the machine is purchased. Research, however, has found that the actual decrease in the market value of machinery can vary by as much as 10 percent, depending on the amount of annual use.

Annual depreciation can be estimated using the straight-line or declining balance methods discussed in Chapter 5. However, if only the *average* annual depreciation is needed, it can be calculated from the same equation as for straight-line depreciation:

$$\text{Depreciation} = \frac{\text{initial cost} - \text{salvage value}}{\text{ownership life}}$$

The salvage value at various ages can be estimated as a percent of the new list price of a similar machine. Table 22-1 shows values that can be used to estimate the salvage value of several common types of farm machinery.

Internal Revenue Service regulations specify certain methods that can be used to calculate machinery depreciation for tax purposes. However, the results may not reflect the actual annual depreciation of the machine. Ownership costs should be based on the concept of economic depreciation, or the actual decline in value, not on income tax depreciation rates. The commonly used tax depreciation methods discussed in Chapter 16 result in high depreciation in the early years of ownership and little or no depreciation in later years. Box 5-3 illustrated another simple approach to estimating the cost of machinery depreciation, for an entire line of machinery.

Interest

Investing in a machine ties up capital and prevents it from being used for an alternative investment. Equity capital has an opportunity cost, and this cost is part of the true cost of machine ownership. The opportunity cost used for machinery capital should reflect the expected return from investing the capital in its next best alternative use. When borrowed capital is used to

TABLE 22-1	Estimated Salvage Value as a Percentage of the New List Price for a Similar Machine					

Age of machine in years	Tractor 80–149 hp	Tractor 150 + hp	Combine	Baler	Tillage	Planter
1	68%	67%	70%	56%	61%	65%
2	62	59	59	50	54	60
3	57	54	51	46	49	56
4	53	49	45	42	45	53
5	49	45	40	39	42	50
6	46	42	36	37	39	48
7	44	39	32	34	36	46
8	41	36	29	32	34	44
9	39	34	26	30	31	42
10	37	32	22	28	30	40
11	35	30	21	27	28	39
12	34	28	18	25	26	38
Assumed hours of annual use	400	400	275	—	—	—

Source: Based on ASABE Standards 2013. American Society of Agricultural and Biological Engineers, St. Joseph, MI, 2013.

purchase machinery, the interest cost is based on the loan interest rate. Depending on the source of capital, the proper interest rate to use is the rate of return on alternative investments, the expected interest rate on borrowed capital, or a weighted average of the two.

The interest component of average annual fixed costs is calculated from the following equations:

$$\text{Average value} = \frac{\text{cost} + \text{salvage value}}{2}$$

$$\text{Interest} = \text{average value} \times \text{interest rate}$$

The first equation gives the *average value* of the machine over its ownership period, or its value at mid-life. The machine is declining in value over time, so its average value is used to determine the average annual interest charge. This procedure assumes that capital tied up in the machinery investment decreases over the life of the machine, just as the balance owed on a loan used to purchase a machine declines as it is paid off.

Taxes

A few states levy property taxes on farm machinery. The charge will depend on the valuation procedure and tax rate in a particular location. Machinery cost studies often use a charge of about 1 percent of the average machine value as an estimate of annual property taxes. A higher rate should be used for pickups and trucks to cover the cost of the license and any other road-use fees.

Some states also levy sales taxes on the purchase of farm machinery. This is a one-time cost, so it should be added to the initial purchase cost.

Insurance

Another ownership cost is the annual charge for insurance to cover damage to the machine from collision, fire, theft, hail, or wind, and for any liability coverage. A charge for insurance should be included in ownership costs, even if the owner carries no formal insurance and personally stands the risk, because some losses can be expected over time. The proper charge for insurance will depend on the amount and type of coverage and insurance rates for a given location. Machinery cost studies often assume an annual charge equal to about 0.5 percent of the machine's average value. This charge should be higher for on-road vehicles because of their higher premiums for property damage, collision, and liability coverage.

Housing

Most machinery cost estimates include an annual cost for housing the machine. Studies have found annual housing costs to be about 0.5 to 1.5 percent of the average value of the machine, so a value of 1.0 percent is often used to estimate machinery housing costs. A housing charge can also be estimated by calculating the annual cost per square foot for the machine shed (possibly including a prorated charge for a shop area) and multiplying this by the number of square feet the machine occupies. Even if a machine is not housed, a charge should be included to reflect additional wear and tear.

Leasing

The use of some machinery is acquired under a long-term leasing agreement. Although the machine is not actually owned, the annual lease payment should be included with other fixed or ownership costs. Typically, the operator also pays the insurance premiums for a leased machine and provides housing for it. Advantages and disadvantages of leasing machinery will be discussed later in this chapter.

Operating Costs

Operating costs are directly related to the level of use of the machinery. They are zero if the machine is not used, but increase directly with the amount of annual use. Unlike ownership costs, they can be controlled by varying the amount of annual use, by improving efficiency, and by following a proper maintenance program.

Repairs

Annual repair costs will vary with use, machine type, age, preventive maintenance programs, and other factors. Table 22-2 shows some average repair costs per 100 hours of use as a percent of the new list price for various types of machines.

TABLE 22-2	Average Repair Costs per 100 Hours of Use, Percent of New List Price

Machine	Percent of list price
Two-wheel-drive tractor up to 150 hp	0.84
Two-wheel-drive tractor over 150 hp	1.12
Four-wheel-drive tractor	0.48
Moldboard plow	5.05
Tandem disk	2.92
Chisel plow	3.70
Rotary tiller	5.40
Field cultivator	3.56
Rotary hoe	3.03
Row crop cultivator	3.91
Planter, grain drill	5.00
Mower, cutterbar	7.47
Mower, rotary	8.80
Mower-conditioner	3.12
Windrower, self-propelled	1.80
Rake	2.45
Small square baler	4.00
Large square baler	1.74
Round baler	5.95
Forage harvester, pull type	2.60
Forage harvester, self-propelled	1.20
Cotton picker, self-propelled	2.65
Sugar beet harvester	7.77
Potato harvester	2.74
Combine, self-propelled	1.33
Grain wagon	2.64
Forage wagon	2.43
Boom sprayer	4.63
Manure-handling equipment	3.09
Fertilizer spreader	6.65
Stalk chopper	3.01

Source: American Society of Agricultural and Biological Engineers Standards, St. Joseph, MI, 2013.

These rates are based on average costs over the useful life of the machine. For example, a four-wheel drive tractor with a new list price of $225,000 would have estimated repair and maintenance costs of 0.48% × $225,000 = $1,080 for every 100 hours of use, or $10.80 per hour.

Machinery repair costs usually increase over time, however. If a new machine is purchased under warranty, repair costs to the operator should be low at first. As more parts wear out, though, repair costs per hour or per year rapidly increase. Repair costs are highly variable, so any rule of thumb for estimating repair costs needs to be used with caution. The best source of information is detailed records of actual repair costs for each machine under the existing level of use, cropping pattern, and maintenance program.

Fuel and Lubrication

Gasoline, diesel fuel, oil and other lubricants, and filters are included in this category. These costs are minor for nonpowered equipment but are important for self-powered machinery, such as tractors, harvesters, and sprayers. Fuel use per hour will depend on engine size, load, speed, and field conditions. Farm records can be used to estimate average fuel use. Data from tractor tests indicate that average fuel consumption in gallons per hour can be estimated from the maximum power-take-off (PTO) horsepower of a tractor, as follows:

$$\frac{\text{Gallons}}{\text{per hour}} = 0.060 \times \text{PTO hp (gasoline)}$$

$$\frac{\text{Gallons}}{\text{per hour}} = 0.044 \times \text{PTO hp (diesel)}$$

Fuel cost per hour then can be estimated by multiplying the estimated fuel use by the purchase price of the fuel.

Costs for lubricants and filters average about 10 to 15 percent of fuel costs for self-powered machines. The cost of lubricants for nonpowered machines is generally small enough that it can be ignored when estimating operating costs.

Box 22-1 — Capital Recovery

An alternative to calculating depreciation and interest is to compute the annual capital recovery charge, as discussed in Chapter 17. The capital recovery charge includes both annual depreciation and interest costs. It can be found using the equation shown below:

The amortization factor is the value from Appendix Table 1 that corresponds to the ownership life and the interest rate for the machine. The capital recovery amount is the annual payment that will recover the initial investment lost through depreciation, plus interest on the investment. This value is generally slightly higher than the sum of the average annual depreciation and interest values calculated from the previous equations, because capital recovery assumes that interest charges are computed on values at the beginning of each year and are compounded annually.

$$\frac{\text{capital}}{\text{recovery}} = [\text{amortization factor} \times (\text{beginning value} - \text{salvage value})] + (\text{interest rate} \times \text{salvage value})$$

Labor

The amount of labor needed for farm machinery depends on the operation being performed, field speed and efficiency, and the size of machinery being used. Labor costs are generally estimated separately from machinery costs but need to be included in any estimate of the total cost of performing a given machine operation. Machinery-related labor costs will be underestimated if only the time spent actually operating the machine in the field is considered. The total labor charge should also include time spent fueling, lubricating, repairing, adjusting, and moving machinery between fields and the farmstead. These activities add as much as 10 to 25 percent to machinery field time.

The value placed on machinery labor should reflect local wage rates. The level of skill needed for various operations should be taken into account, however. A higher hourly labor value should be used for applying pesticides, planting, harvesting, and other highly skilled operations.

Custom Hire or Rental

When a custom operator is hired to perform certain machinery operations, these costs should be included in the machinery variable or operating costs for the farm. Custom machinery rates are usually quoted by the acre, hour, bushel, or ton, and include the value of the custom operator's labor as well as all other ownership and operating costs.

In other cases, a farmer may rent a machine that will be used for only a few days or weeks. The operator supplies the labor and fuel and pays an hourly or daily charge for using the machine. This rental expense should also be included in the farm's total machinery cost.

Other Operating Costs

Some specialized machines have additional operating costs associated with their use. Items such as twine, plastic wraps, and bags may need to be included in the variable costs for certain operations.

Machinery Costs and Use

Annual total ownership or fixed costs are usually assumed to be constant regardless of the amount of machine use during the year. Operating or variable costs, however, increase with the amount of use, generally at a constant rate per acre or per hour. The result is that annual total costs also increase at a constant rate. These relations are shown in Figure 22-1a.

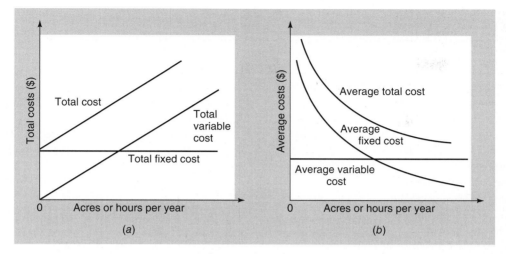

Figure 22-1 **Relations between total and average machinery costs.**

For decision-making purposes, it is often useful to express machinery costs in terms of the average cost per acre, per hour, or per unit of output. Average fixed costs will decline as the acres, hours, or units of output per year increase, while average variable costs will be constant if total variable costs increase at a constant rate. Average total cost is the sum of average fixed and average variable costs, so it also declines with added use. These relations are shown in Figure 22-1*b*, where the vertical distance between the average fixed cost and average total cost curves is constant and equal to the average variable cost.

Machinery cost estimates are only as accurate as the estimate of annual use. As much effort should go into estimating this value as into estimating the various cost components. Use of actual field records and tractor-hour meters will improve the accuracy of annual use estimates.

EXAMPLES OF MACHINERY COST CALCULATIONS

Table 22-3 shows an example of how to calculate total annual costs and average costs per hour and per acre for a new combine.

Ownership Costs

The first step is to list the basic data, such as list price, purchase cost, salvage value, ownership life, estimated annual use, and interest rate on the capital invested. Next, total ownership costs are calculated and converted into average ownership cost per hour of use. For the combine in the example, average ownership costs are estimated to be $30,250 per year over 10 years. Based on the assumed 300 hours of annual use, the average ownership cost per hour is estimated to be $100.40. Remember that this value will change if the hours of use are more or less than 300.

If the capital recovery method had been used, the estimate of fixed costs would have been slightly higher. The capital recovery factor from Appendix Table 1 for a 10-year life and 6 percent interest rate is 0.13587, so the capital recovery amount would be

$$[0.13587 \times (250{,}000 - 66{,}000)] + (0.06 \times 66{,}000) = \$28{,}960$$

Total fixed costs would be $28,960 + 2,370 (taxes, insurance, housing) = $31,330 per year, or $104.43 per hour, using this method compared to $30,250 computing depreciation and interest separately (Table 22-3).

TABLE 22-3	Calculating Machinery Costs for a New Combine

Step 1: List basic data

New combine, 24-foot header, 240-hp diesel engine:

List price	$300,000
Purchase cost	$250,000
Salvage value (22% of new list price, Table 22-1)	$ 66,000
Average value ($250,000 + $66,000) ÷ 2	$158,000
Ownership life	10 years
Estimated annual use	300 hours
Interest rate (cost of capital)	6%
Price of fuel per gallon	$3.50
Labor cost ($/hour)	$ 15.00
Performance rate (acres/hour)	8

Step 2: Calculate ownership costs

Depreciation ($250,000 − $66,000) ÷ 10 years	$ 18,400
Interest (6% × $158,000)	$ 9,480
Taxes, insurance, and housing (1.5% × $158,000)	$ 2,370
Total annual ownership costs	$ 30,250
Ownership costs per hour ($30,250 ÷ 300 hours)	$ 100.83

Step 3: Calculate operating costs

Repairs (1.34%, from Table 22-2, × $300,000 list price × 300 hours ÷ 100)	$ 11,970
Diesel fuel (240 horsepower × 0.044 gallon/hp-hour × $3.50 per gallon × 300 hours)	$ 11,088
Lubrication and filters (15% of fuel costs)	$ 1,663
Labor (300 hours × $15.00 per hour × 1.20)*	$ 5,400
Total annual operating costs	$ 30,121
Operating cost per hour ($30,121 ÷ 300 hours)	$ 100.40

Step 4: Calculate total cost per hour

Ownership cost per hour	$ 100.83
Operating cost per hour	$ 100.40
Total cost per hour	$ 201.23

Step 5: Calculate cost per acre

Total cost per acre ($200.23 ÷ 8 acres/hour)	$ 25.15

*Labor requirements are increased by 20% to account for time spent servicing, adjusting, repairing, and transporting machinery.

The list price for a new machine comparable to the one for which costs are being estimated is used as a benchmark for estimating the salvage value and repair costs. In some cases the purchase cost may be the same as the list price, but often some discount is available. This is almost always true when a used machine is being purchased. If costs are being calculated for a machine that is already owned, use an estimate of its current market value instead of the purchase price. However, the list price should still be for a new, comparable model.

Operating Costs

Total operating costs, including repairs, fuel, lubrication, and labor, are calculated in Step 3 of Table 22-3. The labor requirement was increased by 20 percent over the estimated machine use, to account for time required to service, adjust, and transport the combine. Average operating cost per hour is estimated to be $100.40.

The total of ownership and operating costs per hour is estimated at $201.23 (Step 4). This figure can be converted to a cost per acre or per unit of output if the field capacity per hour is known. In Step 5 of the example, the performance rate of the combine is assumed to be 8 acres per hour, resulting in a total cost per acre of $25.15.

When the same tractor is used to pull several implements, or more than one harvesting head is used on the same self-propelled unit, ownership and operating costs must be calculated separately for the power unit and for the attachment. They can then be added together to find the combined cost of performing the operation. In Table 22-4, ownership costs and operating costs per hour have been estimated for a 185-horsepower tractor and a 25-foot chisel plow, using the methods previously explained.

Fuel and lubrication costs and labor charge were assigned only to the tractor. The costs *per hour* for the tractor and chisel plow are combined to find the total cost of chiseling per hour, and are then divided by the field capacity to find the cost per acre. It is not correct to combine the *annual* costs of two machines when only part of the annual use of the power unit occurs with this implement. Instead, the *hourly* costs should be combined. For a self-propelled machine or an implement alone, however, the cost per acre can be found by dividing the annual cost by the annual *acres* of use, without calculating hourly costs first.

TABLE 22-4	Combined Cost of a Tractor and Implement

	185-hp tractor	25-ft chisel plow
Annual ownership costs	$12,800	$2,900
Annual hours of use	400	120
Ownership cost per hour	$ 32.00	$24.17
Operating costs per hour:		
Fuel and lubrication	26.45	
Repairs	7.85	14.15
Labor	14.00	
Total cost per hour	$ 80.30	$38.32
Combined cost per hour	$118.62	
Field capacity, acres per hour	12.0	
Combined cost per acre	$ 9.88	

FACTORS IN MACHINERY SELECTION

One of the more difficult problems in farm management is selecting machinery with the proper capacity. This process is complicated not only by the wide range of types and sizes available, but also capital availability, labor requirements, the particular crop and livestock enterprises in the farm plan, tillage practices, and climatic factors. The objective in selecting machinery is to purchase the machine that will satisfactorily perform the required task within the time available for the lowest possible total cost. This does not necessarily result in the purchase of the smallest machine available, however. Labor and timeliness costs must also be considered.

Machinery Size

The first step in machinery selection is to determine the field capacity for each size available. The formula for finding capacity in acres per hour is

$$\text{Field capacity} = \frac{\text{speed (mph)} \times \text{width (feet)} \times \text{field efficiency (\%)}}{8.25}$$

For example, a 12-foot-wide windrower operated at 8 miles per hour with a field efficiency of 82 percent would have an effective field capacity of

$$\frac{8 \text{ mph} \times 12 \text{ ft.} \times 82\%}{8.25} = 9.5 \text{ acres/hour}$$

Field efficiency is included in the equation to recognize that a machine is not always used at 100 percent of its maximum capacity because of work overlap and time spent turning, adjusting, lubricating, and handling seed and other materials. Operations such as planting, which require frequent stops to refill seed, pesticide, and fertilizer containers, may have field efficiencies as low as 50 or 60 percent. On the other hand, field efficiencies as high as 85 to 90 percent are possible for some tillage operations, particularly in large fields when turning time and work overlap are minimized. Larger machines typically have higher field efficiencies due to less overlapping and less time needed for adjustments relative to the area covered.

The next step in selecting machinery is to compute the minimum field capacity (acres per hour) needed to complete the task in the time available. This value is found by dividing the number of acres the machine must cover by the number of field hours available to complete the operation. In turn, the number of field hours available depends on how many days the weather will be suitable for performing the operation and the number of labor or field hours available each day. The formula for finding the minimum field capacity in acres per hour is

$$\text{Minimum field capacity} = \frac{\text{acres to cover}}{\text{hours per day} \times \text{days available}}$$

Suppose the operator wants to be able to windrow 180 acres in two days and can operate the windrower 10 hours per day. The minimum capacity needed is

$$[180 \text{ acres} \div (10 \text{ hours} \times 2 \text{ days})] = 9.0 \text{ acres per hour.}$$

The 12-foot windrower, with a capacity of 9.5 acres per hour, should be large enough.

When several field operations must be completed within a certain number of days, it may be more convenient to calculate the days needed for each operation and sum them to test whether an entire set of machinery has sufficient capacity. The formula for each operation is

$$\text{Field days needed} = \frac{\text{acres to cover}}{\text{hours per day} \times \text{acres completed per hour}}$$

Using the same example, assume the operator also owns a baler with an effective field capacity of 5 acres per hour. If baling can be performed only 8 hours per day, then 180 acres requires

$$\frac{180 \text{ acres}}{8 \text{ hours} \times 5 \text{ acres/hour}} = 4.5 \text{ days}$$

Thus, a total of 2.0 + 4.5 = 6.5 suitable field days are required each time the 180 acres are windrowed and baled. The operator must decide if typical weather patterns will permit these many days to be available without a risk of significant crop losses.

The size of machinery required to complete field operations in a timely manner can be reduced by: (1) increasing the labor supply so that machinery can be operated more hours per day or two operations can be performed at the same time, (2) reducing the number of field operations performed, or (3) producing several crops with different critical planting and harvesting dates, instead of just one crop. Often these adjustments are less costly than acquiring larger machinery.

Machinery selection may also involve a choice between one large machine or two smaller ones. The purchase cost and annual fixed costs will be higher for two machines, because the same capacity can usually be obtained at a lower cost in one large machine. Two tractors and two operators will also be needed if tractor-drawn equipment is involved. The primary advantage of owning two machines is an increase in reliability. If one machine breaks down, work is not completely stopped but can continue at half speed using the remaining machine.

Timeliness

Some field operations do not have to be completed within a fixed period, but the later they are performed, the lower the harvested yield is likely to be. The yield reduction may be in terms of quality, such as for fruits and vegetables that become overly mature, or in quantity, such as for grain that suffers from too short a growing season when planting is late or has excessive field losses during harvesting. Figure 22-2 shows a typical relation between planting time and potential yield for corn in southern Illinois, where planting too early or too late reduces yield.

A yield decrease reduces profit and should be included as part of the cost of using a smaller machine. This cost is referred to as *timeliness cost*. The dollar cost of poor timeliness is difficult to estimate, because it will vary from year to year depending on weather conditions and prices. However, some estimate should be included when comparing the cost of owning machines of different sizes. Figure 22-3 is a hypothetical example of how timeliness and other machinery costs change as larger machinery is used to perform the same amount of work. At first, larger machinery reduces timeliness and labor costs and lowers total costs. After some point, no more gains in timeliness are available, and higher ownership costs cause total costs to rise. Timeliness concerns and labor costs must be balanced against higher ownership costs to select the least-cost machine size.

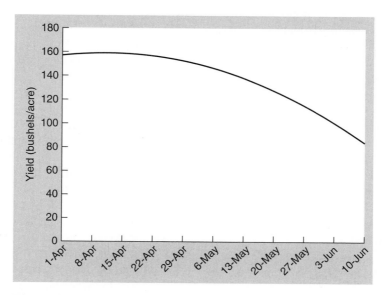

Figure 22-2 Estimated corn yields as a function of planting date, southern Illinois.

Source: Planting Decisions Model, Farmdoc, University of Illinois.

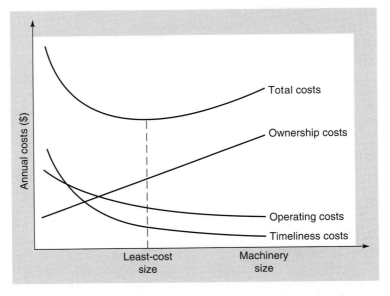

Figure 22-3 **Hypothetical effect of timeliness and machine size on cost.**

Partial budgeting is a useful tool to use when making decisions about machinery sizes where timeliness is important. Table 22-5 contains an example that emphasizes the importance of considering all costs. Changing from a 20-foot grain drill to a 30-foot drill will double annual ownership costs. However, the manager estimates that improved timeliness of planting will result in an average yield increase of 0.5 bushel of wheat per acre each year on 1,000 acres. The value of this added yield, combined with slightly lower labor costs, results in a projected increase in net income of over $1,600 per year from purchasing the larger drill. The date of planting also depends on how long it takes to complete tillage and other pre-plant operations. Thus, the total machinery complement should be analyzed when timeliness factors are considered.

ALTERNATIVES FOR ACQUIRING MACHINERY

Efficient machinery management means having the correct size and type of machine available to use at the right time at a minimum cost. Once the size and type have been selected, there are several common alternatives for acquiring the use of farm equipment.

Ownership

Most farm and ranch managers prefer to own their machinery. Ownership gives them complete control over the use and disposal of each machine. However, machinery and vehicles represent a large investment on commercial farms and ranches, as much as $200 to $300 per acre on many cash grain farms and over $2,000 per acre on some horticultural farms. Managers must be careful to control the size of this investment and the related operating costs. Machinery investment can be reduced by: (1) using smaller machinery, (2) increasing annual machine use to lower the average ownership cost per unit of output, (3) keeping machinery longer before trading, (4) purchasing used machinery, and (5) using alternatives to ownership such as rental, leasing, and custom hiring.

Trading machinery frequently for new models can result in higher-than-average ownership costs. However, some managers like to have their

TABLE 22-5	**Example of a Partial Budget for Selecting the Most Profitable Machine**

Adjustment: Sell a 3-year-old 20-foot grain drill worth $18,000 ($25,000 new list price) and buy a 30-foot drill ($40,000 purchase cost and list price). Assumes $10,000 salvage value after 10 years of ownership for the old drill, a $16,000 salvage value for the new drill, and a 7 percent interest rate. The 20-foot machine requires 137 hours to drill 1,000 acres. The 30-foot machine requires 82 hours. Fuel and lubrication costs are assumed to be the same for both drills.

Additional revenue:

0.5 bushel per acre increase in yield × 1,000 acres × $6.25 per bushel

Total additional revenue		$3,125

Reduced costs:	Ownership	Operating	
Depreciation ($18,000 − $10,000) ÷ 7	$1,143		
Interest ($18,000 + $10,000) ÷ 2 × 7%	980		
Taxes and insurance ($18,000 + $10,000) ÷ 2 × 1%	140		
Repairs ($25,000 × 5% × 137 hours ÷ 100)		$1,713	
Labor (55 hours less at $15/hour)		825	
Subtotal	$2,263	$2,538	
Total reduced costs			$4,801
Total of additional revenue and reduced costs			$7,926

Additional costs:	Ownership	Operating	
Depreciation ($40,000 − $16,000) ÷ 10	$2,400		
Interest ($40,000 + $16,000) ÷ 2 × 7%	1,960		
Taxes and insurance ($40,000 + $16,000) ÷ 2 × 1%	280		
Repairs ($40,000 × 5% × 82 hours ÷ 100)		$1,640	
Subtotal	$4,640	$1,640	
Total additional costs			$6,280

Reduced revenue:		
		0

Total additional costs and reduced revenue		$6,280
Net change in net farm income		$1,646

machinery under a manufacturer's warranty to avoid high repair costs and prefer to always have the latest technology available.

Some machinery dealers offer *rollover* ownership plans. The operator purchases a new machine, and then trades it for another new machine each year. The cost to trade may depend on the number of hours that the machine to be traded was used. Rollover plans allow for the use of a new machine each year, usually under warranty, with a known cost.

Rental

When investment capital is limited or interest rates are high, renting a machine may be preferable to owning it. Short-term rental arrangements usually cover a few days to a whole season. The operator pays a rental fee plus the cost of insurance and daily maintenance, but not major repairs. Machinery rental is especially attractive when: (1) a specialized machine is needed for relatively low use, (2) extra capacity or a replacement machine is needed for a short time, or (3) the operator wants to experiment with a new machine or production practice without making a long-term capital investment.

Sometimes large, expensive machines such as combines are rented to two or more farms in the same year, often in different states. This works best when the harvesting season is different for each renter. The rental company is usually responsible for servicing and transporting the machine, but the renter must relinquish it by a set date.

Leasing

A lease is a long-term contract whereby the machine owner (the lessor), often a machinery dealer or leasing company, grants control and use of the machine to the user (the lessee) for a specified period for a monthly, semiannual, or annual lease payment. Most machinery leases are for three to five years or longer, with the first payment usually due at the start. As with any formal agreement, the lease should be in writing and should cover such items as payment rates and dates; excess use penalties; payment of repairs, taxes, and insurance; responsibilities for loss or damage; and provisions for early cancellation.

Some leases allow the lessee to purchase the machine at the end of the lease period for a specified price. These are called *operating leases*. To maintain the income tax deductibility of lease payments, the purchase at the end of the lease must be optional and for a value approximately equal to the market value of the machine at that time. Otherwise, the lease may be considered to be a *capital lease*, and depreciation and interest become tax deductible for the lessee instead of the lease payments, as if the operator owned the machine.

Leasing machinery may help operators reduce the amount of capital they have tied up in noncurrent assets. Although lease payments represent a cash flow obligation just like a loan payment, they are generally smaller. Leasing reduces the risk of obsolescence, because the lessee is not obligated to keep the machine beyond the term of the lease agreement. Moreover, operators who prefer the reliability and performance of a newer machine often lease it for a few years and then exchange it for a new one. Some operators who have little taxable farm income and cannot use depreciation deductions, such as section 179 expensing, may find leasing to have a lower after-tax cost than owning.

There are also some disadvantages to leasing farm machinery. The practice is not well established in many areas, and the desired model may not be available for lease. Lease payments are operating expenses, and a late payment may cause cancellation of the lease. Further, the lessee may not be allowed to cancel the lease early without paying a substantial penalty. A lease does not allow the operator to build equity value in the machine. Unless the purchase option is exercised when the lease expires, the machine reverts to the owner and the operator has no financial interest in it.

The decision to lease or own equipment should be studied carefully, as shown in the example in Box 22-2. The economic cost (net present value of all payments) and the cash flow requirements should be analyzed and matched to the financial situation of the business.

Custom Hire

Custom hiring is an important practice in some areas for operations such as applying chemicals and harvesting grain or forages. The decision of whether to own a machine or custom hire the service depends on the costs involved, the skills needed, and the amount of work to be done. For machines that will be used very little, it is often more economical to hire the work done on a custom basis. However, the availability and dependability of custom operators must be considered. A manager may not want to rely on a custom operator for a task such as planting, where timeliness is often critical.

Total costs per acre or per unit of output should be compared when deciding between

Box 22-2   Buy Versus Lease Comparison Example

The Struthers family farm needs to replace one of its medium-sized tillage tractors. They have located the model they want at a local dealership. The dealer offers them two alternatives:

1. Purchase the tractor for $65,000, minus $11,000 allowed for the trade-in value of their old tractor that has a tax basis of $0. They can pay off the $54,000 difference in five equal annual payments of $13,883 starting in 12 months. The annual interest rate is 9 percent.
2. Lease the tractor for five years. The annual lease payments would be $11,000 per year. The first one is due when the lease is signed, but can be paid by the trade-in allowance.

After five years they can purchase the tractor for $32,000.

The Struthers have a cost of capital of 7 percent and a marginal income tax rate of 40 percent, so their after-tax discount rate is $0.07 \times (1.00 - 0.40) = 0.042$, or 4.2 percent. Purchasing the tractor would allow them to take advantage of the Section 179 depreciation expensing option discussed in Chapter 16, up to the full purchase cost. They could also expense the purchase cost at the end of the lease period.

By using the investment analysis techniques discussed in Chapter 17, they are able to calculate the net present value of the cost of acquiring the tractor under either option, as shown here.

Purchase with loan	Year 1	Year 2	Year 3	Year 4	Year 5	Total
a) Loan principal	$ 9,023	$ 9,835	$10,720	$11,685	$12,737	$54,000
b) Loan interest	4,860	4,048	3,163	2,198	1,146	15,415
c) Tax depreciation	54,000	0	0	0	0	54,000
d) Tax savings (b + c) × .40	23,544	1,619	1,265	879	459	27,766
e) Net cash outflow (a + b − d)	−9,661	12,264	12,618	13,004	13,424	41,649
f) Discount factor	.960	.921	.884	.848	.814	
g) Present value (e × f)	−9,275	11,295	11,154	11,027	10,928	35,129

Lease and purchase later	Year 1	Year 2	Year 3	Year 4	Year 5	Total
a) Lease payment	$11,000	$11,000	$11,000	$11,000		$44,000
b) Purchase cost					$32,000	32,000
c) Tax depreciation					32,000	32,000
d) Tax savings (a + c) × .40	4,400	4,400	4,400	4,400	12,800	30,400
e) Net cash outflow (a + b − d)	6,600	6,600	6,600	6,600	19,200	45,600
f) Discount factor	.960	.921	.884	.848	.814	
g) Present value (e × f)	6,336	6,079	5,834	5,597	15,629	39,475

The net present value of purchasing the tractor and paying off the loan is $35,129 after taxes, less than the $39,475 cost of leasing the tractor and then exercising the purchase option. This means that in this example purchasing is cheaper in the long run. However, in Years 2, 3, and 4 the net cash outflow is considerably higher for purchasing than leasing. If cash flow is expected to be tight during these years, the leasing option might still be preferred.

machine ownership and custom hiring. Custom charges are typically a fixed rate per acre, hour, or ton, while ownership costs per unit will decline with increased use. These relations are shown in Figure 22-4. At low levels of use, hiring a custom operator is less expensive, while for higher usage, the cost is lower if the machine is owned. The point where the cost advantage changes, or the break-even point, is shown as output level *a* in Figure 22-4.

When the necessary cost data are available, the break-even point can be found from the following equation:

$$\text{Break-even units} =$$
$$\frac{\text{total annual fixed costs}}{\text{custom rate} - \text{variable costs per unit}}$$

For example, the ownership or fixed costs for the combine in Table 22-3 were $30,250 per year. The variable costs for operating it were $12.55 per acre for a performance rate of 8 acres per hour ($100.40 per hour divided by 8 acres per hour). If the custom hire rate for a similar combine is $35.00 per acre, the break-even point would be

$$\frac{\$30,250}{\$35.00 - 12.55} = 1,347 \text{ acres}$$

If the machine would be used on less than 1,347 acres, it would be less costly to custom hire the work done, while above 1,347 acres, it would be less expensive to own the machine. A determination of the break-even point provides a useful guide to help managers choose between machine ownership and custom hiring.

Labor use is another consideration in custom hiring. The custom operator typically provides the labor necessary to operate the machine, which frees the farm operator's labor for other uses. This can be an advantage if it reduces the amount of hired labor needed or if the owner's labor has a high opportunity cost at the time the custom work is being performed. It can also be an advantage for operations that take special skills to perform, such as the application of pesticides.

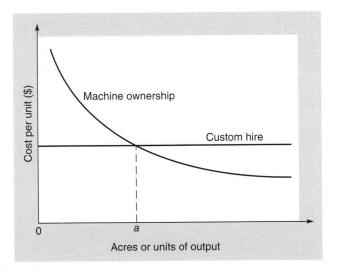

Figure 22-4 Cost per unit of output for machine ownership and custom hiring.

Many operators who own machinery find it advantageous to do custom work for other farmers. This helps spread their fixed ownership costs over more acres. It is important for custom operators to accurately estimate their costs so that they can arrive at a fair and profitable charge, though.

IMPROVING MACHINERY EFFICIENCY

Several values can be used to measure the efficiency of machinery use. One is *machinery investment per crop acre*, calculated by dividing the current value of all crop machinery by the number of crop acres on the farm. The current value of all machinery for a given year can be found by taking the average of the beginning and ending machinery inventory market values for the year.

A second measure of machinery efficiency is *machinery cost per crop acre*. It is found by dividing the total annual machinery costs, both ownership and operating, by the number of crop acres. Some farm record analyses also include pickup and truck expenses, machinery lease payments, and custom hiring expenses in machinery costs. When possible, the cost of machinery used for livestock purposes should be excluded to make a fair comparison among farms.

These values should be used with caution. Numerous studies have shown that investment per acre and cost per acre decline with increases in farm size and also vary by farm type. Compare values only to those calculated in the same manner for farms of the same approximate size and type. Table 22-6 shows recent machinery costs per acre, excluding labor, for a group of grain producers in Kentucky. Low-profit producers incurred as much as $50 per acre higher machinery costs than high-profit producers in the same size group.

Several techniques can be employed to improve machinery efficiency. Choosing the optimum machine size and the least-cost alternative

TABLE 22-6	Total Machinery Costs for Kentucky Grain Producers (per Crop Acre)	
	High-profit third	**Low-profit third**
0 to 999 acres	$159.95	$217.93
1,000 to 1,999 acres	$149.97	$154.86
Over 2,000 acres	$136.25	$187.42

Source: Kentucky Farm Business Management Program, 2012 Annual Summary. University of Kentucky Cooperative Extension Service.

for acquiring machinery services has already been discussed. There are four other areas that have a large impact on efficiency: maintenance and operation, machinery use, new versus used equipment, and replacement decisions.

Maintenance and Operation

Repairs are a large part of machinery variable costs, but they are a cost that can be controlled by proper use and maintenance. Agricultural engineers report that excessive repair costs can generally be traced to: (1) overloading or exceeding the rated machine capacity, (2) excessive speed, (3) poor daily and periodic maintenance, and (4) improper adjustment. These items can be corrected by constant attention and proper training of machine operators. Modern farm machines have more monitoring systems and automatic adjustment controls that help maintain efficient operating levels.

A system of scheduling and recording repairs and maintenance is essential for controlling repair costs. Adherence to the manufacturer's recommended maintenance schedule will keep warranties in effect, prevent unnecessary breakdowns, and reduce lifetime repair costs. Complete records of repairs on individual machines will help identify machines with higher than average repair costs. These machines should be considered for early replacement.

The manner in which a machine is operated affects repair costs and field efficiency. Speed should be adjusted to load a machine to capacity without overloading it or lowering the quality of the work being done. Practices that improve field efficiency reduce costs by allowing more work to be done in a given period, either by permitting the same work to be done in less time or by allowing a smaller machine to be used. Small and irregular-shaped fields that require frequent turns, frequent stops, and work overlap reduce field efficiency. For example, a 30-foot disk operated with a 3-foot overlap loses 10 percent of its potential efficiency from this factor alone.

Machinery Use

Using an expensive specialized machine on only a few acres also contributes to high machinery costs. Because of this, some farmers purchase low-use machines jointly with other operators. Some producers have even formed machinery cooperatives with 5 to 10 members owning all their machinery in common. This not only decreases the fixed costs per unit but also decreases the investment required from each individual. It is important that the joint owners are compatible and can agree on the details of the machine's use. Whose work will be done first and the division of expenses such as repairs, insurance, and taxes should be agreed to before the machine is purchased. Normally, operators provide their own fuel and labor, then pay a fixed rate per acre into a fund used to pay ownership costs and repairs.

Some owners exchange the use of specialized machinery. If the use or value of the machines is not equal, some payment may be exchanged as well. Many operations such as harvesting are accomplished more efficiently when two or more people work together, anyway.

Machinery costs per unit of output can also be reduced by performing custom work for other operators. If custom work does not interfere with timely completion of the owner's work, it will provide additional income to help pay ownership costs. The custom rate charged should reflect the opportunity cost of labor in the operator's own business as well as the costs of owning and operating the machine. Some skilled operators do custom machinery work as a part-time or full-time alternative to farming their own or rented land, with less financial risk.

New Versus Used Machines

Used farm machinery is readily available from both dealers and private sellers. There are several Internet sites, as well, that list used machinery items available for sale.

Farm machines, particularly tractors and other self-powered machines, decline in market value most rapidly during the first few years of their useful lives, so buying used machinery is an economical way to lower machinery investment and ownership costs. Offsetting some of the lower ownership costs may be higher repair costs and decreases in reliability and timeliness. A used machine may also become obsolete sooner than a new one. The owner's ability as a mechanic and availability of the facilities and time to do major repair work at home are often crucial factors for owning used machinery. Used machinery should be considered when capital is limited, interest rates are high, the machine will have a relatively small annual use, and reliability and timeliness are not critical.

Replacement Decisions

When to replace or trade a machine is one of the more difficult decisions in machinery management. There is no easy rule that applies to all types of machines and conditions. Besides costs and reliability, replacement decisions must also consider the effects of a purchase or trade on income taxes and cash flow.

The decision to replace a machine can be made for any of the following reasons:

1. *The present machine is worn out*: Its age and accumulated use are such that it is no longer capable of performing the required task reliably.

2. *The machine is obsolete*: New developments in machinery technology, or changes in farming practices, allow a newer machine to perform the job better, or with greater safety and comfort.

3. *Costs are increasing with the present machine*: Repair, fuel, and timeliness costs are increasing rapidly, in both total and per unit of output.

4. *The capacity is too small*: The area in production has increased, or timeliness has become so critical that the old machine cannot complete the job on time.

5. *Income taxes*: In a high-profit year, machines may be replaced to take advantage of the tax-reducing benefits of fast depreciation deductions or because the owner has a higher marginal tax rate that year. However, replacement decisions should not be made on the basis of tax savings alone.

6. *Cash flow*: Many machines are replaced in years of above-average cash income to avoid borrowing funds later. Likewise, replacement of machinery is often postponed in years when cash flow is tight.

7. *Pride and prestige*: There is a certain pride of ownership involved in the purchase of new and larger machinery. While this may be important to some individuals, it can be a costly reason and one that is difficult to justify from a purely economical point of view.

These reasons can be used individually or in combination to determine the replacement age for a specific machine. Annual cost and repair records on each machine are useful for making the replacement decision. In Figure 22-5 the annual costs for a 165-horsepower tractor purchased new are estimated for each year of a 20-year ownership period. At first, total costs decline because depreciation and interest are decreasing. Eventually, though, increasing repair costs more than offset the declining ownership costs. Total annual costs are lowest around Year 10, but this depends on how many hours the tractor is used annually. The higher the annual use, the faster annual repair costs will increase.

Most farm and ranch managers have an overall strategy for replacing machinery. The

Figure 22-5 **Estimated annual total cost of a 165-horsepower tractor.**

appropriate strategy for each operation depends on the financial resources available, the mechanical skills of the work force, and the particular priorities and objectives of the manager.

Keep and Repair

One strategy is to keep and repair equipment as long as possible. This is usually the least-cost strategy in the long run, especially if most of the repair work can be done on the farm. The risk of suffering a breakdown at a critical time is higher, though.

Trade Often

Operations that emphasize maximum use of machinery usually prefer newer and more reliable equipment. They may even find that leasing machinery and exchanging it for a new model every year or two is the strategy that best meets their objectives.

Trade When Income Is High

Managers who want to avoid the use of credit may wait until a year when above-average cash income occurs to trade machinery. This strategy also helps reduce taxable income in a high-income year.

Invest Some Each Year

Finally, some operators prefer to upgrade part of their machinery line each year by trading or disposing of their least reliable units. This avoids having to invest large amounts of capital in any one year and works especially well for businesses that experience only small year-to-year variations in cash net income.

Although machinery investment decisions are made infrequently, they often involve many thousands of dollars. Taking time to carefully analyze each decision can have a large impact on the long-run profitability of the farm or ranch.

SUMMARY

Machinery investment is the second largest investment on most farms and ranches after real estate. Annual machinery costs are a large part of a farm's total annual costs. Ownership costs include depreciation, interest, taxes, insurance, housing, and lease payments. Repairs, fuel, lubrication, labor, custom hire, and rental payments are included in operating costs.

Selection of the optimum machine size to own should consider total costs and the effects of timeliness in completing operations. Rental, leasing, and custom hiring are other alternatives for acquiring the use of farm machinery, particularly for specialized machines with low annual use. Operators short of capital or skilled in machinery repair can benefit from investing in used rather than new machinery.

Machinery efficiency can be improved by proper maintenance and operation, by owning equipment jointly with other operators, or by exchanging the use of individually owned machines. The proper time to trade machinery depends on repair costs, reliability, obsolescence, cash flow, income tax considerations, and personal pride.

QUESTIONS FOR REVIEW AND FURTHER THOUGHT

1. What is the total annual fixed cost for a $90,000 tractor with a 12-year life and a $18,000 salvage value, when insurance and taxes are 2 percent of average value and there is a 9 percent opportunity cost on capital? What is the average fixed cost per hour if the tractor is used 400 hours per year? If it is used 700 hours per year?

2. What is the field capacity in acres per hour for a 28-foot-wide tandem disk operated at 5 miles per hour with 80 percent field efficiency? How much would it change if the field efficiency could be increased to 90 percent? How much time would be saved on 800 acres?

3. Assume that owning and operating a certain machine to till 850 acres has an ownership cost of $10,000 per year, plus operating costs of $10.00 per acre, including labor. Leasing a machine with a field capacity of 5.0 acres per hour to do the same work would cost $80 per hour, plus the same operating costs. Hiring the work done on a custom basis would cost $24 per acre. Which alternative has the lowest total cost?

4. Assume that a self-propelled windrower has annual fixed costs of $7,650 and variable costs of $4.25 per acre. A custom operator charges $12.60 per acre. What is the break-even point in acres per year?

5. List ways to improve the field efficiency of machinery operations such as planting, disking, and combining grain.

6. What are the advantages and disadvantages of owning, leasing, and custom hiring farm machinery?

7. If you decide to use your forage harvester to do custom work for others, what would happen to total ownership cost? Average ownership cost per acre harvested? Total operating cost? Average operating cost per acre harvested?

8. What factors are important in machinery replacement decisions? How would you rank them in order of importance? Would your ranking be different for different types of machines?

9. Do you think farmers and ranchers should invest in used machinery? What are the advantages and disadvantages?

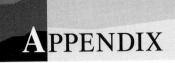

APPENDIX

TABLE 1 | Amortization Factors for Equal Total Payments

Interest Rate

Years	4%	5%	6%	7%	8%	9%	10%	11%	12%	13%	14%	15%	16%
1	1.04000	1.05000	1.06000	1.07000	1.08000	1.09000	1.10000	1.11000	1.12000	1.13000	1.14000	1.15000	1.16000
2	0.53020	0.53780	0.54544	0.55309	0.56077	0.56847	0.57619	0.58393	0.59170	0.59948	0.60729	0.61512	0.62296
3	0.36035	0.36721	0.37411	0.38105	0.38803	0.39505	0.40211	0.40921	0.41635	0.42352	0.43073	0.43798	0.44526
4	0.27549	0.28201	0.28859	0.29523	0.30192	0.30867	0.31547	0.32233	0.32923	0.33619	0.34320	0.35027	0.35738
5	0.22463	0.23097	0.23740	0.24389	0.25046	0.25709	0.26380	0.27057	0.27741	0.28431	0.29128	0.29832	0.30541
6	0.19076	0.19702	0.20336	0.20980	0.21632	0.22292	0.22961	0.23638	0.24323	0.25015	0.25716	0.26424	0.27139
7	0.16661	0.17282	0.17914	0.18555	0.19207	0.19869	0.20541	0.21222	0.21912	0.22611	0.23319	0.24036	0.24761
8	0.14853	0.15472	0.16104	0.16747	0.17401	0.18067	0.18744	0.19432	0.20130	0.20839	0.21557	0.22285	0.23022
9	0.13449	0.14069	0.14702	0.15349	0.16008	0.16680	0.17364	0.18060	0.18768	0.19487	0.20217	0.20957	0.21708
10	0.12329	0.12950	0.13587	0.14238	0.14903	0.15582	0.16275	0.16980	0.17698	0.18429	0.19171	0.19925	0.20690
11	0.11415	0.12039	0.12679	0.13336	0.14008	0.14695	0.15396	0.16112	0.16842	0.17584	0.18339	0.19107	0.19886
12	0.10655	0.11283	0.11928	0.12590	0.13270	0.13965	0.14676	0.15403	0.16144	0.16899	0.17667	0.18448	0.19241
13	0.10014	0.10646	0.11296	0.11965	0.12652	0.13357	0.14078	0.14815	0.15568	0.16335	0.17116	0.17911	0.18718
14	0.09467	0.10102	0.10758	0.11434	0.12130	0.12843	0.13575	0.14323	0.15087	0.15867	0.16661	0.17469	0.18290
15	0.08994	0.09634	0.10296	0.10979	0.11683	0.12406	0.13147	0.13907	0.14682	0.15474	0.16281	0.17102	0.17936
16	0.08582	0.09227	0.09895	0.10586	0.11298	0.12030	0.12782	0.13552	0.14339	0.15143	0.15962	0.16795	0.17641
17	0.08220	0.08870	0.09544	0.10243	0.10963	0.11705	0.12466	0.13247	0.14046	0.14861	0.15692	0.16537	0.17395
18	0.07899	0.08555	0.09236	0.09941	0.10670	0.11421	0.12193	0.12984	0.13794	0.14620	0.15462	0.16319	0.17188
19	0.07614	0.08275	0.08962	0.09675	0.10413	0.11173	0.11955	0.12756	0.13576	0.14413	0.15266	0.16134	0.17014
20	0.07358	0.08024	0.08718	0.09439	0.10185	0.10955	0.11746	0.12558	0.13388	0.14235	0.15099	0.15976	0.16867
21	0.07128	0.07800	0.08500	0.09229	0.09983	0.10762	0.11562	0.12384	0.13224	0.14081	0.14954	0.15842	0.16742
22	0.06920	0.07597	0.08305	0.09041	0.09803	0.10590	0.11401	0.12231	0.13081	0.13948	0.14830	0.15727	0.16635
23	0.06731	0.07414	0.08128	0.08871	0.09642	0.10438	0.11257	0.12097	0.12956	0.13832	0.14723	0.15628	0.16545
24	0.06559	0.07247	0.07968	0.08719	0.09498	0.10302	0.11130	0.11979	0.12846	0.13731	0.14630	0.15543	0.16467
25	0.06401	0.07095	0.07823	0.08581	0.09368	0.10181	0.11017	0.11874	0.12750	0.13643	0.14550	0.15470	0.16401
30	0.05783	0.06505	0.07265	0.08059	0.08883	0.09734	0.10608	0.11502	0.12414	0.13341	0.14280	0.15230	0.16189
35	0.05358	0.06107	0.06897	0.07723	0.08580	0.09464	0.10369	0.11293	0.12232	0.13183	0.14144	0.15113	0.16089
40	0.05052	0.05828	0.06646	0.07501	0.08386	0.09296	0.10226	0.11172	0.12130	0.13099	0.14075	0.15056	0.16042

TABLE 2 | **Future Value of a $1 Investment**

Interest Rate

Years	4%	5%	6%	7%	8%	9%	10%	11%	12%	13%	14%	15%	16%
1	1.0400	1.0500	1.0600	1.0700	1.0800	1.0900	1.1000	1.1100	1.1200	1.1300	1.1400	1.1500	1.1600
2	1.0816	1.1025	1.1236	1.1449	1.1664	1.1881	1.2100	1.2321	1.2544	1.2769	1.2996	1.3225	1.3456
3	1.1249	1.1576	1.1910	1.2250	1.2597	1.2950	1.3310	1.3676	1.4049	1.4429	1.4815	1.5209	1.5609
4	1.1699	1.2155	1.2625	1.3108	1.3605	1.4116	1.4641	1.5181	1.5735	1.6305	1.6890	1.7490	1.8106
5	1.2167	1.2763	1.3382	1.4026	1.4693	1.5386	1.6105	1.6851	1.7623	1.8424	1.9254	2.0114	2.1003
6	1.2653	1.3401	1.4185	1.5007	1.5869	1.6771	1.7716	1.8704	1.9738	2.0820	2.1950	2.3131	2.4364
7	1.3159	1.4071	1.5036	1.6058	1.7138	1.8280	1.9487	2.0762	2.2107	2.3526	2.5023	2.6600	2.8262
8	1.3686	1.4775	1.5938	1.7182	1.8509	1.9926	2.1436	2.3045	2.4760	2.6584	2.8526	3.0590	3.2784
9	1.4233	1.5513	1.6895	1.8385	1.9990	2.1719	2.3579	2.5580	2.7731	3.0040	3.2519	3.5179	3.8030
10	1.4802	1.6289	1.7908	1.9672	2.1589	2.3674	2.5937	2.8394	3.1058	3.3946	3.7072	4.0456	4.4114
11	1.5395	1.7103	1.8983	2.1049	2.3316	2.5804	2.8531	3.1518	3.4785	3.8359	4.2262	4.6524	5.1173
12	1.6010	1.7959	2.0122	2.2522	2.5182	2.8127	3.1384	3.4985	3.8960	4.3345	4.8179	5.3503	5.9360
13	1.6651	1.8856	2.1329	2.4098	2.7196	3.0658	3.4523	3.8833	4.3635	4.8980	5.4924	6.1528	6.8858
14	1.7317	1.9799	2.2609	2.5785	2.9372	3.3417	3.7975	4.3104	4.8871	5.5348	6.2613	7.0757	7.9875
15	1.8009	2.0789	2.3966	2.7590	3.1722	3.6425	4.1772	4.7846	5.4736	6.2543	7.1379	8.1371	9.2655
16	1.8730	2.1829	2.5404	2.9522	3.4259	3.9703	4.5950	5.3109	6.1304	7.0673	8.1372	9.3576	10.7480
17	1.9479	2.2920	2.6928	3.1588	3.7000	4.3276	5.0545	5.8951	6.8660	7.9861	9.2765	10.7613	12.4677
18	2.0258	2.4066	2.8543	3.3799	3.9960	4.7171	5.5599	6.5436	7.6900	9.0243	10.5752	12.3755	14.4625
19	2.1068	2.5270	3.0256	3.6165	4.3157	5.1417	6.1159	7.2633	8.6128	10.1974	12.0557	14.2318	16.7765
20	2.1911	2.6533	3.2071	3.8697	4.6610	5.6044	6.7275	8.0623	9.6463	11.5231	13.7435	16.3665	19.4608
21	2.2788	2.7860	3.3996	4.1406	5.0338	6.1088	7.4002	8.9492	10.8038	13.0211	15.6676	18.8215	22.5745
22	2.3699	2.9253	3.6035	4.4304	5.4365	6.6586	8.1403	9.9336	12.1003	14.7138	17.8610	21.6447	26.1864
23	2.4647	3.0715	3.8197	4.7405	5.8715	7.2579	8.9543	11.0263	13.5523	16.6266	20.3616	24.8915	30.3762
24	2.5633	3.2251	4.0489	5.0724	6.3412	7.9111	9.8497	12.2392	15.1786	18.7881	23.2122	28.6252	35.2364
25	2.6658	3.3864	4.2919	5.4274	6.8485	8.6231	10.8347	13.5855	17.0001	21.2305	26.4619	32.9190	40.8742
30	3.2434	4.3219	5.7435	7.6123	10.0627	13.2677	17.4494	22.8923	29.9599	39.1159	50.9502	66.2118	85.8499
35	3.9461	5.5160	7.6861	10.6766	14.7853	20.4140	28.1024	38.5749	52.7996	72.0685	98.1002	133.1755	180.3141
40	4.8010	7.0400	10.2857	14.9745	21.7245	31.4094	45.2593	65.0009	93.0510	132.7816	188.8835	267.8635	378.7212

TABLE 3 Future Value of a $1 Annuity

Interest Rate

Years	4%	5%	6%	7%	8%	9%	10%	11%	12%	13%	14%	15%	16%
1	1.0000	1.0000	1.0000	1.0000	1.0000	1.0000	1.0000	1.0000	1.0000	1.0000	1.0000	1.0000	1.0000
2	2.0400	2.0500	2.0600	2.0700	2.0800	2.0900	2.1000	2.1100	2.1200	2.1300	2.1400	2.1500	2.1600
3	3.1216	3.1525	3.1836	3.2149	3.2464	3.2781	3.3100	3.3421	3.3744	3.4069	3.4396	3.4725	3.5056
4	4.2465	4.3101	4.3746	4.4399	4.5061	4.5731	4.6410	4.7097	4.7793	4.8498	4.9211	4.9934	5.0665
5	5.4163	5.5256	5.6371	5.7507	5.8666	5.9847	6.1051	6.2278	6.3528	6.4803	6.6101	6.7424	6.8771
6	6.6330	6.8019	6.9753	7.1533	7.3359	7.5233	7.7156	7.9129	8.1152	8.3227	8.5355	8.7537	8.9775
7	7.8983	8.1420	8.3938	8.6540	8.9228	9.2004	9.4872	9.7833	10.0890	10.4047	10.7305	11.0668	11.4139
8	9.2142	9.5491	9.8975	10.2598	10.6366	11.0285	11.4359	11.8594	12.2997	12.7573	13.2328	13.7268	14.2401
9	10.5828	11.0266	11.4913	11.9780	12.4876	13.0210	13.5795	14.1640	14.7757	15.4157	16.0853	16.7858	17.5185
10	12.0061	12.5779	13.1808	13.8164	14.4866	15.1929	15.9374	16.7220	17.5487	18.4197	19.3373	20.3037	21.3215
11	13.4864	14.2068	14.9716	15.7836	16.6455	17.5603	18.5312	19.5614	20.6546	21.8143	23.0445	24.3493	25.7329
12	15.0258	15.9171	16.8699	17.8885	18.9771	20.1407	21.3843	22.7132	24.1331	25.6502	27.2707	29.0017	30.8502
13	16.6268	17.7130	18.8821	20.1406	21.4953	22.9534	24.5227	26.2116	28.0291	29.9847	32.0887	34.3519	36.7862
14	18.2919	19.5986	21.0151	22.5505	24.2149	26.0192	27.9750	30.0949	32.3926	34.8827	37.5811	40.5047	43.6720
15	20.0236	21.5786	23.2760	25.1290	27.1521	29.3609	31.7725	34.4054	37.2797	40.4175	43.8424	47.5804	51.6595
16	21.8245	23.6575	25.6725	27.8881	30.3243	33.0034	35.9497	39.1899	42.7533	46.6717	50.9804	55.7175	60.9250
17	23.6975	25.8404	28.2129	30.8402	33.7502	36.9737	40.5447	44.5008	48.8837	53.7391	59.1176	65.0751	71.6730
18	25.6454	28.1324	30.9057	33.9990	37.4502	41.3013	45.5992	50.3959	55.7497	61.7251	68.3941	75.8364	84.1407
19	27.6712	30.5390	33.7600	37.3790	41.4463	46.0185	51.1591	56.9395	63.4397	70.7494	78.9692	88.2118	98.6032
20	29.7781	33.0660	36.7856	40.9955	45.7620	51.1601	57.2750	64.2028	72.0524	80.9468	91.0249	102.4436	115.3797
21	31.9692	35.7193	39.9927	44.8652	50.4229	56.7645	64.0025	72.2651	81.6987	92.4699	104.7684	118.8101	134.8405
22	34.2480	38.5052	43.3923	49.0057	55.4568	62.8733	71.4027	81.2143	92.5026	105.4910	120.4360	137.6316	157.4150
23	36.6179	41.4305	46.9958	53.4361	60.8933	69.5319	79.5430	91.1479	104.6029	120.2048	138.2970	159.2764	183.6014
24	39.0826	44.5020	50.8156	58.1767	66.7648	76.7898	88.4973	102.1742	118.1552	136.8315	158.6586	184.1678	213.9776
25	41.6459	47.7271	54.8645	63.2490	73.1059	84.7009	98.3471	114.4133	133.3339	155.6196	181.8708	212.7930	249.2140
30	56.0849	66.4388	79.0582	94.4608	113.2832	136.3075	164.4940	199.0209	241.3327	293.1992	356.7868	434.7451	530.3117
35	73.6522	90.3203	111.4348	138.2369	172.3168	215.7108	271.0244	341.5896	431.6635	546.6808	693.5727	881.1702	1120.7130
40	95.0255	120.7998	154.7620	199.6351	259.0565	337.8824	442.5926	581.8261	767.0914	1013.7042	1342.0251	1779.0903	2360.7572

TABLE 4 | Present Value of a $1 Lump Sum

Interest Rate

Years	4%	5%	6%	7%	8%	9%	10%	11%	12%	13%	14%	15%	16%
1	0.96154	0.95238	0.94340	0.93458	0.92593	0.91743	0.90909	0.90090	0.89286	0.88496	0.87719	0.86957	0.86207
2	0.92456	0.90703	0.89000	0.87344	0.85734	0.84168	0.82645	0.81162	0.79719	0.78315	0.76947	0.75614	0.74316
3	0.88900	0.86384	0.83962	0.81630	0.79383	0.77218	0.75131	0.73119	0.71178	0.69305	0.67497	0.65752	0.64066
4	0.85480	0.82270	0.79209	0.76290	0.73503	0.70843	0.68301	0.65873	0.63552	0.61332	0.59208	0.57175	0.55229
5	0.82193	0.78353	0.74726	0.71299	0.68058	0.64993	0.62092	0.59345	0.56743	0.54276	0.51937	0.49718	0.47611
6	0.79031	0.74622	0.70496	0.66634	0.63017	0.59627	0.56447	0.53464	0.50663	0.48032	0.45559	0.43233	0.41044
7	0.75992	0.71068	0.66506	0.62275	0.58349	0.54703	0.51316	0.48166	0.45235	0.42506	0.39964	0.37594	0.35383
8	0.73069	0.67684	0.62741	0.58201	0.54027	0.50187	0.46651	0.43393	0.40388	0.37616	0.35056	0.32690	0.30503
9	0.70259	0.64461	0.59190	0.54393	0.50025	0.46043	0.42410	0.39092	0.36061	0.33288	0.30751	0.28426	0.26295
10	0.67556	0.61391	0.55839	0.50835	0.46319	0.42241	0.38554	0.35218	0.32197	0.29459	0.26974	0.24718	0.22668
11	0.64958	0.58468	0.52679	0.47509	0.42888	0.38753	0.35049	0.31728	0.28748	0.26070	0.23662	0.21494	0.19542
12	0.62460	0.55684	0.49697	0.44401	0.39711	0.35553	0.31863	0.28584	0.25668	0.23071	0.20756	0.18691	0.16846
13	0.60057	0.53032	0.46884	0.41496	0.36770	0.32618	0.28966	0.25751	0.22917	0.20416	0.18207	0.16253	0.14523
14	0.57748	0.50507	0.44230	0.38782	0.34046	0.29925	0.26333	0.23199	0.20462	0.18068	0.15971	0.14133	0.12520
15	0.55526	0.48102	0.41727	0.36245	0.31524	0.27454	0.23939	0.20900	0.18270	0.15989	0.14010	0.12289	0.10793
16	0.53391	0.45811	0.39365	0.33873	0.29189	0.25187	0.21763	0.18829	0.16312	0.14150	0.12289	0.10686	0.09304
17	0.51337	0.43630	0.37136	0.31657	0.27027	0.23107	0.19784	0.16963	0.14564	0.12522	0.10780	0.09293	0.08021
18	0.49363	0.41552	0.35034	0.29586	0.25025	0.21199	0.17986	0.15282	0.13004	0.11081	0.09456	0.08081	0.06914
19	0.47464	0.39573	0.33051	0.27651	0.23171	0.19449	0.16351	0.13768	0.11611	0.09806	0.08295	0.07027	0.05961
20	0.45639	0.37689	0.31180	0.25842	0.21455	0.17843	0.14864	0.12403	0.10367	0.08678	0.07276	0.06110	0.05139
21	0.43883	0.35894	0.29416	0.24151	0.19866	0.16370	0.13513	0.11174	0.09256	0.07680	0.06383	0.05313	0.04430
22	0.42196	0.34185	0.27751	0.22571	0.18394	0.15018	0.12285	0.10067	0.08264	0.06796	0.05599	0.04620	0.03819
23	0.40573	0.32557	0.26180	0.21095	0.17032	0.13778	0.11168	0.09069	0.07379	0.06014	0.04911	0.04017	0.03292
24	0.39012	0.31007	0.24698	0.19715	0.15770	0.12640	0.10153	0.08170	0.06588	0.05323	0.04308	0.03493	0.02838
25	0.37512	0.29530	0.23300	0.18425	0.14602	0.11597	0.09230	0.07361	0.05882	0.04710	0.03779	0.03038	0.02447
30	0.30832	0.23138	0.17411	0.13137	0.09938	0.07537	0.05731	0.04368	0.03338	0.02557	0.01963	0.01510	0.01165
35	0.25342	0.18129	0.13011	0.09366	0.06763	0.04899	0.03558	0.02592	0.01894	0.01388	0.01019	0.00751	0.00555
40	0.20829	0.14205	0.09722	0.06678	0.04603	0.03184	0.02209	0.01538	0.01075	0.00753	0.00529	0.00373	0.00264

TABLE 5 | Present Value of a $1 Annuity

Interest Rate

Years	4%	5%	6%	7%	8%	9%	10%	11%	12%	13%	14%	15%	16%
1	0.9615	0.9524	0.9434	0.9346	0.9259	0.9174	0.9091	0.9009	0.8929	0.8850	0.8772	0.8696	0.8621
2	1.8861	1.8594	1.8334	1.8080	1.7833	1.7591	1.7355	1.7125	1.6901	1.6681	1.6467	1.6257	1.6052
3	2.7751	2.7232	2.6730	2.6243	2.5771	2.5313	2.4869	2.4437	2.4018	2.3612	2.3216	2.2832	2.2459
4	3.6299	3.5460	3.4651	3.3872	3.3121	3.2397	3.1699	3.1024	3.0373	2.9745	2.9137	2.8550	2.7982
5	4.4518	4.3295	4.2124	4.1002	3.9927	3.8897	3.7908	3.6959	3.6048	3.5172	3.4331	3.3522	3.2743
6	5.2421	5.0757	4.9173	4.7665	4.6229	4.4859	4.3553	4.2305	4.1114	3.9975	3.8887	3.7845	3.6847
7	6.0021	5.7864	5.5824	5.3893	5.2064	5.0330	4.8684	4.7122	4.5638	4.4226	4.2883	4.1604	4.0386
8	6.7327	6.4632	6.2098	5.9713	5.7466	5.5348	5.3349	5.1461	4.9676	4.7988	4.6389	4.4873	4.3436
9	7.4353	7.1078	6.8017	6.5152	6.2469	5.9952	5.7590	5.5370	5.3282	5.1317	4.9464	4.7716	4.6065
10	8.1109	7.7217	7.3601	7.0236	6.7101	6.4177	6.1446	5.8892	5.6502	5.4262	5.2161	5.0188	4.8332
11	8.7605	8.3064	7.8869	7.4987	7.1390	6.8052	6.4951	6.2065	5.9377	5.6869	5.4527	5.2337	5.0286
12	9.3851	8.8633	8.3838	7.9427	7.5361	7.1607	6.8137	6.4924	6.1944	5.9176	5.6603	5.4206	5.1971
13	9.9856	9.3936	8.8527	8.3577	7.9038	7.4869	7.1034	6.7499	6.4235	6.1218	5.8424	5.5831	5.3423
14	10.5631	9.8986	9.2950	8.7455	8.2442	7.7862	7.3667	6.9819	6.6282	6.3025	6.0021	5.7245	5.4675
15	11.1184	10.3797	9.7122	9.1079	8.5595	8.0607	7.6061	7.1909	6.8109	6.4624	6.1422	5.8474	5.5755
16	11.6523	10.8378	10.1059	9.4466	8.8514	8.3126	7.8237	7.3792	6.9740	6.6039	6.2651	5.9542	5.6685
17	12.1657	11.2741	10.4773	9.7632	9.1216	8.5436	8.0216	7.5488	7.1196	6.7291	6.3729	6.0472	5.7487
18	12.6593	11.6896	10.8276	10.0591	9.3719	8.7556	8.2014	7.7016	7.2497	6.8399	6.4674	6.1280	5.8178
19	13.1339	12.0853	11.1581	10.3356	9.6036	8.9501	8.3649	7.8393	7.3658	6.9380	6.5504	6.1982	5.8775
20	13.5903	12.4622	11.4699	10.5940	9.8181	9.1285	8.5136	7.9633	7.4694	7.0248	6.6231	6.2593	5.9288
21	14.0292	12.8212	11.7641	10.8355	10.0168	9.2922	8.6487	8.0751	7.5620	7.1016	6.6870	6.3125	5.9731
22	14.4511	13.1630	12.0416	11.0612	10.2007	9.4424	8.7715	8.1757	7.6446	7.1695	6.7429	6.3587	6.0113
23	14.8568	13.4886	12.3034	11.2722	10.3711	9.5802	8.8832	8.2664	7.7184	7.2297	6.7921	6.3988	6.0442
24	15.2470	13.7986	12.5504	11.4693	10.5288	9.7066	8.9847	8.3481	7.7843	7.2829	6.8351	6.4338	6.0726
25	15.6221	14.0939	12.7834	11.6536	10.6748	9.8226	9.0770	8.4217	7.8431	7.3300	6.8729	6.4641	6.0971
30	17.2920	15.3725	13.7648	12.4090	11.2578	10.2737	9.4269	8.6938	8.0552	7.4957	7.0027	6.5660	6.1772
35	18.6646	16.3742	14.4982	12.9477	11.6546	10.5668	9.6442	8.8552	8.1755	7.5856	7.0700	6.6166	6.2153
40	19.7928	17.1591	15.0463	13.3317	11.9246	10.7574	9.7791	8.9511	8.2438	7.6344	7.1050	6.6418	6.2335

GLOSSARY

A

Account payable An expense that has been incurred but not yet paid.

Account receivable Income that has been earned but for which no cash payment has been received.

Accounting A comprehensive system for recording and summarizing business transactions.

Accounting period The period over which accounting transactions are summarized.

Accrual accounting An accounting system that recognizes income when it is earned and expenses when they are incurred.

Accrued expense An expense that has been incurred, sometimes accumulating over time, but has not been paid.

Accrued liability A liability that has been incurred but not yet paid, such as accrued interest.

Accumulated depreciation The sum of all depreciation taken on an asset from time of purchase to the present.

Adjusted basis The income tax basis of an asset, equal to the original basis reduced by the amount of depreciation expense claimed and/or increased by the cost of any improvements made.

Amortized loan A loan scheduled to be repaid in a series of periodic payments.

Animal unit month (AUM) A unit used for renting pasture, equal to one mature beef cow or equivalent grazing for one month.

Annual percentage rate (APR) The true annual rate at which interest is charged on a loan.

Annuity A series of equal periodic payments.

Appraisal The process of estimating the market value of an asset.

Appreciation An increase in the market value of an asset.

Asset Physical or financial property that has value and is owned by a business or individual.

Asset turnover ratio Total revenue divided by total farm assets. A measure of financial efficiency.

Average fixed cost (AFC) Total fixed cost divided by total output; average fixed cost per unit of output.

Average physical product (APP) The average amount of physical output produced for each unit of input used; total output divided by total input.

Average total cost (ATC) Total cost divided by total output; average cost per unit of output.

Average variable cost (AVC) Total variable cost divided by total output; average variable cost per unit of output.

B

Balance sheet A financial report summarizing the assets, liabilities, and equity of a business at a point in time. Also called a net worth statement.

Balloon payment loan A loan amortization method in which a large portion of the principal is due with the final payment.

Bankruptcy A legal action that a business can take when it no longer has the financial resources to pay its debts and must reorganize or go out of business.

Basis (marketing) The difference between the local cash price and the futures contract price of the same commodity at a point in time.

Basis (tax) The beginning value of an asset for income tax purposes.

Bonus (wage) A payment made to an employee, in addition to the normal salary, based on superior performance or other criteria.

Book value The original cost of an asset minus the total accumulated depreciation expense taken to date.

Boot When trading a used asset for a new one, the cash paid to make up for the difference in value.

Borrowing capacity The maximum amount an individual or business can borrow based on ability to repay and other factors.

Break-even price The selling price for which total income will just equal total expenses for a given level of production.

Break-even yield The yield level at which total income will just equal total expenses at a given selling price.

Breeding livestock Livestock owned for the primary purpose of producing offspring.

Budget An estimate of future income, expenses, and/or cash flows.

Bushel lease A leasing arrangement in which the rent is paid as a specified number of bushels of grain delivered to the owner.

Business plan   A detailed description of a farm's intended business activities and the strategies and resources available to carry them out.

C

C corporation   A "regular" corporation that files its own income tax return. (See also S corporation.)

Call option   A contract that gives the buyer the right to buy a futures contract for an agricultural commodity at a specified price. It is used to set a maximum purchase price in advance.

Capital   A collection of physical and financial assets that have a market value.

Capital asset   An asset expected to last through more than one production cycle that can be used to produce other saleable assets or services.

Capital budgeting   A process for determining the profitability of a capital investment.

Capital gain   The amount by which the sale value of an asset exceeds its cost or original tax basis.

Capital lease   A contract that allows the lessee to purchase the leased equipment over a period of time.

Capital loss   The amount by which the sale value of an asset is below its cost or adjusted tax basis.

Capital recovery   The annualized equivalent value of the initial investment cost of a capital asset.

Capitalization method   A procedure for estimating the value of an asset by dividing the expected annual net returns by an annual discount rate.

Cash accounting   An accounting system that recognizes income when it is actually received and expenses when they are actually paid.

Cash expenses   Expenses that require the expenditure of cash.

Cash flow   The movement of cash funds into and out of a business.

Cash flow budget   A projection of the expected cash inflows and cash outflows for a business over an accounting period.

Cash rent lease   A rental arrangement in which the operator makes a cash payment to the owner for the use of certain property, pays all production costs, and keeps all the income generated.

Chart of accounts   An organized list of the names and code numbers for all the asset, liability, income, expense, and equity items in a farm accounting system.

Closed cooperative   A farmer cooperative to which members agree to sell a fixed amount of production on a regular schedule.

Coefficient of variation   A measure of the variability of the outcomes of a particular event; equal to the standard deviation divided by the mean.

Collateral   Assets pledged as security for a loan.

Commodity Credit Corporation (CCC)   A corporation owned by the U.S. Department of Agriculture. Its primary purpose is to support agricultural prices through the use of commodity loans.

Comparable sale   An actual land sale used in an appraisal to help estimate the market value of a similar piece of land.

Comparative advantage   The ability of a firm, region or country to produce a good or service at a relatively lower cost than another firm, region, or country.

Comparative analysis   The comparison of the performance level of a farm business to the performance level of other similar farms in the same area or to other established standards.

Competitive enterprises   Enterprises for which the output level of one can be increased only by decreasing the output level of the other.

Complementary enterprises   Enterprises for which increasing the output level of one also increases the output level of the other.

Compounding   The process of determining the future value of an investment or loan, in which interest is charged on the accumulated interest as well as on the original capital.

Compound interest   The reinvestment of each interest payment so that it becomes part of the principal that earns interest in future periods.

Contingent liabilities   Liabilities that will come into existence only if some specific event should occur. An example would be income taxes due should an asset such as land be sold.

Contributed capital   Capital invested in a business by its owner(s), other than earnings produced by and retained in that business.

Control   The process of monitoring the progress of a farm business and taking corrective action when desired performance levels are not being met.

Cooperative   A form of business organization in which profits are distributed as patronage refunds and all members have a single vote.

Corporation   A form of business organization in which the owners have shares in a separate legal entity that itself can own assets and borrow money.

Cost basis   A balance sheet in which assets are valued at their cost value.

Cost center   An accounting unit of a farm business that incurs costs but does not produce revenue.

Cost recovery   The system or method used to compute depreciation for income tax purposes.

Cost value   The value at which an asset was originally acquired, minus accumulated depreciation, plus the cost of improvements made.

Credit (accounting)   An entry in the right hand side of an account ledger that causes a decrease in assets or an increase in liabilities and equity.

Credit (financial)   The capacity or ability to borrow money.

Creditor   Someone to whom a debt is owed, such as a lender.

Crop share lease   A lease agreement in which crop production and certain input costs are divided between the operator and the land owner.

Cumulative distribution function (CDF)   A graph of all the possible outcomes for a certain event and the probability that each outcome, or one with a lower value, will occur.

Current assets   Assets normally used up or sold within a year.

Current liabilities   Liabilities normally paid within a year.

Current ratio   The ratio of current assets to current liabilities; a measure of liquidity.

Custom farming   An arrangement in which the land owner pays the operator a fixed cash amount to perform all the labor and machinery operations needed to produce and harvest a crop.

Custom hire   An arrangement in which an operator performs one or more machinery operations for someone else for a fixed charge.

Cwt   An abbreviation for hundredweight, equal to 100 pounds. Many livestock products and some crops are priced by this unit.

D

Debit   In accounting, an entry on the left-hand side of a ledger that will increase assets or decrease liabilities and equity.

Debt   An obligation to pay, such as a loan or account payable.

Debt/asset ratio   The ratio of total liabilities to total assets; a measure of solvency.

Debt/equity ratio   The ratio of total liabilities to owner's equity; a measure of solvency.

Debt service   The payment of debts according to a specified schedule.

Decision tree   A diagram that traces all the possible strategies and outcomes for a particular decision or sequence of related decisions.

Declining balance method   A depreciation method that results in high depreciation in the early years of life and smaller amounts in the later years.

Deferred taxes   The amount by which income taxes will increase or decrease at some future time when assets and liabilities shown on a current balance sheet are sold or paid.

Deflation   A general decrease in the level of all prices.

Depreciation   An annual, noncash expense to recognize the amount by which an asset loses value due to use, age, and obsolescence. It also spreads the original cost of the asset over its useful life.

Depreciation recapture   Taxable income that results from selling a depreciable asset for more than its adjusted tax basis.

Diminishing returns   A decline in the rate at which total output increases as more inputs are used; a declining marginal physical product.

Direct costs   Costs that occur as a direct result of producing. Also called variable costs.

Discount rate   The interest rate used to find the present value of an amount to be paid or received in the future.

Discounted cash flow   The present value of a series of net cash flows to be received over time. Often used in investment analysis.

Discounting   The process of reducing the value of a sum to be paid or received in the future by the amount of interest that would be accumulated on it to that point in time.

Diseconomies of size   A production relationship in which the average total cost per unit of output increases as more output is produced.

Diversification   The production of two or more commodities for which production levels and/or prices are not closely correlated.

Double-entry accounting   An accounting system in which changes in assets, liabilities and equity, as well as income and expenses, are recorded for each transaction.

Down payment   The portion of the cost of purchasing a capital asset financed from owner's equity, usually in the form of cash.

Dual values   Values produced by a linear programming solution; the value of one more unit of a resource (shadow price), or the penalty from forcing a unit of a non-selected activity into the solution (reduced cost).

E

Economic efficiency   The ratio of the value of output per physical unit of input or per unit cost of the input.

Economic profit   Total income minus total costs, including opportunity costs, for an enterprise or whole farm.

Economies of size   A production relation in which average total cost per unit of output decreases as output increases.

Efficiency   A ratio showing the number of units or value of production generated per unit of resource utilized.

Enterprise   An individual crop or type of livestock, such as wheat, dairy, or lettuce. A farm's production plan will often consist of several enterprises.

Enterprise analysis   An analysis of one or more individual enterprises, in which a portion of the whole-farm income and expenses is allocated to each enterprise.

Enterprise budget   A projection of all the costs and returns for a single enterprise.

Environmental audit   A thorough inspection of a tract of land to determine whether any environmental hazards exist.

Equal marginal principle   The principle that a limited resource should be allocated among competing uses in such a way that the marginal value products from the last unit in each use are equal.

Equity   The amount by which the value of total assets exceeds total liabilities; the amount of the owner's capital invested in the business.

Equity/asset ratio   The ratio of owner's equity to total assets; a measure of solvency.

Expected value   The weighted average outcome from an uncertain event, based on its possible outcomes and their respective probabilities.

Expenditure   An outlay of cash for operating or investment purposes.

Expense   Cost incurred in the process of producing a commodity. May be cash or noncash.

Expensing   An option allowed by the IRS by which much or all of the initial cost of a depreciable asset can be deducted in the year it is purchased. Also called Section 179 expensing.

Extension Service   An educational service for farmers and others provided jointly by the U.S. Department of Agriculture, state land grant universities, and county governments.

External scanning   A critical analysis of the business and economic environment in which a farm or ranch operates.

F

Farm (includes ranch)   The definition used by the U. S. Department of Agriculture is any business entity that sold, or would have sold in a normal year, $1,000 or more of agricultural products.

Farm Credit System (FCS)   A borrower-owned cooperative established by the authority of the U. S. Congress that makes loans to farmers and ranchers.

Farm Financial Standards Council (FFSC)   A committee of agricultural financial experts that developed a set of guidelines for uniform financial reporting and analysis of farm businesses.

Farm management   The process of making decisions about the allocation of scarce resources in agricultural production for the purpose of meeting certain management goals.

Farm Service Agency (FSA)   An agency of the U.S. Department of Agriculture that administers farm commodity and conservation programs and provides direct and guaranteed loans to farmers and ranchers.

Feasibility analysis   An analysis of the cash inflows generated by an investment compared to the cash outflows required.

Federal Insurance Contributions Act (FICA)   A federal law that created a retirement and disability program commonly called Social Security.

Feeder livestock   Young livestock purchased for the purpose of being fed until they reach slaughter weight.

Field efficiency   The actual accomplishment rate for a field implement as a percent of the theoretical accomplishment rate if no time were lost due to overlapping, turning, and adjusting the machine.

Financial statements   Often used as another term for a balance sheet but also used as a general term for other documents relating to the financial condition of a business such as an income statement, statement of cash flows, and statement of owner equity.

Financing   The acquisition of funds to meet the cash flow requirements of an investment or production activity.

Fiscal year   An annual accounting period that does not correspond to the calendar year.

Fixed assets   Assets expected to have a long or indefinite productive life. They are included in noncurrent assets on a two-category balance sheet.

Fixed costs   Costs that will not change in the short run even if no production takes place.

Foreclosure   Legal action taken by a creditor to obtain possession of collateral whenever a borrower is unable to make loan payments.

Forward price contract   A contract between a buyer and seller that fixes the price of a commodity before it is delivered, possibly many months before delivery.

Fringe benefits   Compensation provided to employees in addition to cash wages and salary.

Future value (FV)   The value that a payment or set of payments will have at some time in the future, when interest is compounded.

Futures market   A central market where contracts for future sales of agricultural commodities are bought and sold.

G

General partnership   A partnership in which all partners are general partners; they all participate in management and have unlimited financial liability for partnership actions.

Globalization   A trend toward more integration of consumer tastes, agricultural production, and trade among nations.

Gross income   The total income, cash and noncash, received from an enterprise or business, before any expenses are paid.

Gross margin   The difference between gross income and variable costs; also called income above variable costs.

Gross revenue   The total of all the revenue received by a business over a period; same as gross income.

H

Half-year convention   A provision of the income tax depreciation system (MACRS) that allows one-half year of depreciation in the year an asset is purchased regardless of the date of purchase. May not apply if too many assets are purchased in the last quarter of the year.

Hedging   A strategy for reducing the risk of a decline in prices by selling a commodity futures contract in advance of when the actual commodity is sold.

I

Implementation   The process of carrying out management decisions.

Improvements   Renovations or additions to capital assets that improve their productivity and/or extend their useful lives.

Incentive program   Provisions in an employment contract that pay the employee a bonus for achieving certain performance levels.

Income   Economic gain resulting from the production of goods and services, including receipts from the sale of commodities, other cash payments, increases in inventories, and accounts receivable.

Income statement   A report that summarizes the income and expenses and computes the resulting profit of a business over an accounting period.

Indirect costs   Costs arising from owning an asset, which are affected little by the asset's degree of use (also called fixed costs and ownership costs).

Inflation   A general increase in the level of all prices over time.

Input   A resource used in the production of an output.

Intangible assets   Assets that have financial value but are not in a physical form.

Interest   The amount paid to a lender for the use of borrowed money, or the opportunity cost of investing equity capital in an alternative use.

Intermediate asset   An asset with a useful life greater than one year but less than 10 years. Included as a noncurrent asset when following FFSC recommendations.

Intermediate liability   A liability with an intermediate asset as collateral and payments spread over 2–10 years. Included as a noncurrent liability when following FFSC recommendations.

Internal rate of return (IRR) The discount or interest rate at which the net present value of an investment is just equal to zero.

Internal scanning A critical analysis of the physical, financial, and human resources that a business has available to meet its goals.

Internal transaction A noncash accounting transaction carried out between two enterprises within the same business.

Inventory A complete listing of the number, type, and value of assets owned at a point in time.

Isoquant A line on a graph connecting points that represent all the possible combinations of inputs that can produce the same output.

J

Joint venture Any of several forms of business operation in which more than one person is involved in ownership and management.

L

Labor share lease A leasing agreement in which the operator receives a share of the production in exchange for contributing only labor.

Land contract An agreement by which a land buyer makes principal and interest payments to the seller on a regular schedule.

Law of diminishing returns A relation observed in many physical and biological production processes, in which the marginal physical product declines as more units of a variable input are used in combination with one or more fixed inputs.

Lease An agreement that allows a person to use and/or possess someone else's property in exchange for a rental payment.

Lessee A person who leases property from the owner; a tenant.

Lessor A person who leases owned property to a lessee; a landlord.

Leverage The practice of using credit to increase the total capital managed beyond the amount of owner equity.

Liabilities Financial obligations (debts) that must be paid at some future time.

Limited liability company (LLC) A form of business organization similar to a partnership but offering its owners the advantage of limited financial liability.

Limited partnership A form of business in which more than one person has ownership, but some (the limited partners) do not participate in management and have liability limited to the amount of their investment.

Line of credit An arrangement by which a lender transfers loan funds to a borrower as they are needed, up to a maximum amount.

Linear programming A mathematical technique used to find a set of economic activities that maximizes or minimizes a certain objective, given a set of limited resources and/or other constraints.

Liquidate To convert an asset into cash.

Liquidity The ability of a business to meet its cash financial obligations as they come due.

Livestock share lease A lease agreement in which both the owner and operator contribute capital and share the production of livestock.

Loan repayment capacity The ability to repay loans based on collateral and revenues. Often used to describe the maximum principal and interest that can be paid in a year.

Long run That period long enough for the manager to change the amounts of all inputs or resources available for use.

Long-term assets Assets that have an expected useful life beyond 10 years, usually land and buildings. Also called fixed assets.

Long-term liabilities Liabilities scheduled to be repaid over a period of 10 years or longer.

Loss Financial result that occurs when expenses exceed revenue, which causes a decrease in equity.

Lumpy input A resource that can be obtained only in certain indivisible sizes, such as a tractor or a full-time employee.

M

Marginal cost (MC) The additional cost incurred from producing an additional unit of output.

Marginal input cost (MIC) The additional cost incurred by using an additional unit of input.

Marginal physical product (MPP) The additional physical product resulting from the use of an additional unit of input.

Marginal revenue (MR) The additional income received from selling one additional unit of output.

Marginal tax rate The additional tax that results from an additional dollar of taxable income at a given income level.

Marginal value product (MVP) The additional income received from using an additional unit of input.

Market basis A balance sheet in which assets are valued at their market values.

Market livestock Animals fed for eventual slaughter, not for the production of offspring.

Market value The value for which an asset would be sold in an open-market transaction.

Marketable securities Stocks, bonds, and other financial instruments that can be readily and easily converted into cash.

Marketing loan A loan that can be obtained from the Farm Service Agency using grain or cotton as collateral. The amount of the loan is a fixed rate per bushel or ton, and the commodity must be stored until the loan is repaid.

Minimum price contract A forward price contract that guarantees the seller a minimum price but allows a higher price if the market is above the minimum when the commodity is delivered.

Mission statement A short, descriptive statement of why the farm or ranch business exists and its goals.

Modified Accelerated Cost Recovery System (MACRS) A system for calculating income tax depreciation, as specified by IRS regulations.

Mortgage A legal agreement by which a lender receives the right to acquire a borrower's property to satisfy a debt if the repayment schedule is not met.

N

Natural Resources Conservation Service (NRCS) An agency of the U.S. Department of Agriculture that provides technical and financial assistance for carrying out soil and water conservation practices.

Net farm income The difference between gross revenue and total expenses, including gain or loss on the sale of all capital assets; also the return to owner equity, unpaid labor, and management.

Net farm income (from operations) The difference between total revenue and gross expenses, not including gain or loss on the sale of certain capital assets.

Net operating loss (NOL) A negative net farm profit for income tax purposes, which can be used to offset past and/ or future taxable income.

Net present value (NPV) The present value of the net cash flows that will result from an investment, minus the amount of the original investment.

Net worth The difference between the value of the assets owned by a business and the value of its liabilities. Also called owner equity.

Net worth statement A summary of all the assets, liabilities and net worth of a business. Also called a balance sheet.

Noncash expense An expense that does not involve the expenditure of cash, such as depreciation.

Noncurrent asset An asset that will normally be owned or used up over a period longer than a year.

Noncurrent liability A liability that will normally be paid over a period longer than a year.

Non–real estate All assets other than land and items attached to land, such as buildings and fences.

O

Operating agreement An arrangement between two or more individuals whereby they perform some of their business activities jointly while maintaining individual ownership of the resources being used.

Operating costs Costs for the purchase of inputs and services used up relatively quickly, usually in one production cycle.

Operating lease A lease agreement that allows the lessee to use an asset but does not obligate the lessee to purchase or own the asset at the end of the lease period.

Operating profit margin ratio The value represented by net farm income from operations, plus interest expense, minus opportunity cost of operator labor and management, expressed as a percentage of total revenue.

Opportunity cost The income that could be received by employing a resource in its most profitable alternative use.

Option A marketing transaction in which a buyer pays a seller a premium to acquire the right to sell or buy a futures contract at a specified price.

Ordinary income For income tax purposes, any taxable income that is not capital gain income.

Organizational chart A diagram that shows the supervisors and workers involved in a business and the lines of authority and communication among them.

Output The result or yield from a production process, such as raising crops and livestock.

Overhead costs Costs not directly related to the type and quantity of products produced; a type of fixed cost.

Owner equity The difference between the total value of the assets of a business and the total value of its liabilities; also called net worth.

Owner withdrawals Business assets, generally cash, transferred to the owner(s) for their personal use.

Ownership costs Costs that result from owning assets, regardless of how much they are used; fixed costs.

P

Partial budget An estimate of the changes in income and expenses that would result from carrying out a proposed change in the current farm plan.

Partnership A form of business organization in which more than one operator owns the resources and/or provides management. (See general and limited partnerships.)

Payback period The length of time it takes for the accumulated net returns earned from an investment to equal the original investment.

Payoff matrix A contingency table that illustrates the possible outcomes for a particular occurrence and their respective probabilities.

Period expenses Expenses that accrue over time, but are not directly related to the level of production of specific commodities.

Person-year equivalent A total of 12 months of labor contributed by one or more persons.

Physical efficiency The ratio of output produced per unit of input used, all in physical units.

Precision agriculture A production system using global positioning equipment to precisely apply different levels of inputs to different locations in a field according to their individual requirements.

Prepaid expense A payment made for an input or service prior to the accounting period in which it will be used.

Present value (PV) The current value of a set of payments to be received or paid out over a period.

Price ratio (input) The ratio of the price of the input being added to the price of the input being replaced.

Price ratio (output) The ratio of the price of the output being gained to the price of the output being lost.

Principal The amount borrowed, or the part of the original loan that has not yet been repaid.

Probability distribution A set of possible outcomes to a particular event and the probability of each occurring.

Production function A physical or biological relation showing how much output results from using certain quantities of inputs.

Production possibility curve (PPC) A line on a graph that connects points representing all the possible combinations of outputs that can be produced from a fixed set of resources.

Profit Total revenue minus total expenses, including opportunity costs of labor and capital.

Profit center   An accounting unit within a farm business that both incurs costs and produces revenue.

Profitability   The degree or extent to which the value of the income derived from a set of resources exceeds their cost.

Progressive tax rates   A tax structure that imposes a higher marginal tax rate on higher levels of taxable income.

Promissory note   A legal agreement that obligates a borrower to repay a loan.

Put option   A contract that gives the buyer the right to sell a futures contract for an agricultural commodity at a specified price. It is used to set a minimum selling price in advance.

R

Ranch   An agricultural business that engages in extensive livestock production.

Real estate   Land or assets permanently attached to land.

Reduced cost   A value from the solution to a linear programming problem that shows how much the gross margin would be reduced by forcing into the solution one unit of an enterprise not included in the optimal farm plan.

Repayment capacity   A measurement of the ability of a borrower to repay loans.

Retained farm earnings   Net income generated by a farm business used to increase owner equity rather than being withdrawn to pay for living expenses, taxes, or dividends.

Return on assets (ROA)   The value represented by net farm income from operations, plus interest expense, minus the opportunity cost of operator labor and management. It is usually expressed as a percentage of the average value of total assets.

Return on equity (ROE)   The net return generated by the business before gains or losses on capital assets are realized, but after the value of unpaid labor and management is subtracted. Usually expressed as a percent of the average value of owner's equity.

Return to management   The net return generated by a business after all expenses have been paid and the opportunity costs for owner's equity and unpaid labor have been subtracted.

Revenue   Payments received from the sale of products and services, or from miscellaneous sources of income.

Revenue insurance   An insurance policy that guarantees crop producers a minimum level of gross income per acre. It protects against combinations of low prices and low yields.

Risk   A situation in which more than one possible outcome exists, some of which may be unfavorable.

Rule of 72   A relation used to estimate the time it will take for an investment to double in value; found by dividing 72 by the percent rate of return earned on the investment.

S

S corporation   A corporation that is taxed like a partnership; that is, all income, expenses, and capital gains are passed pro rata to the stockholders to include with their other taxable income.

Salvage value   The market value of a depreciable asset at the time it will be sold or removed from service.

Secondary mortgage   A legal agreement by which a lender receives the right to acquire a borrower's property to satisfy a debt only after the primary lender's debt has been satisfied.

Secured loan   A loan for which the borrower agrees to let the lender take possession of and sell certain assets if the repayment terms are not met.

Self-employment tax   A tax paid on profits earned by self-employed individuals, used to fund the Social Security and Medicare programs.

Self-liquidating loan   A loan that will be repaid from the sale of the assets originally purchased with the loan funds.

Sensitivity analysis   A procedure for assessing the riskiness of a decision by using several possible price and/or production outcomes to budget the results, and then comparing them.

Shadow price   A value obtained from a linear programming solution that shows the amount by which total gross margin would be increased if one more unit of a limiting input were available.

Short run   That period for which at least one production input is available only in a fixed quantity.

Short-term loan   A loan scheduled to be repaid in less than a year.

Signature loan   A loan for which no collateral is pledged.

Single-entry accounting   An accounting system in which income and expenses are recorded but changes in assets and liabilities are not.

Skill-based pay   An approach to setting worker compensation based on levels of responsibility rather than specific duties.

Social Security   A tax on wages and self-employment income to provide retirement and disability income for individuals. (See Federal Insurance Contributions Act.)

Sole proprietorship   A form of business organization in which one operator or family owns the resources and provides the management.

Solvency   The degree to which the liabilities of a business are backed up by assets; the relationship between debt and equity capital.

Standard deviation   A measure of the variability of possible outcomes for a particular event; equal to the square root of the variance.

Statement of cash flows   A summary of the actual cash inflows and cash outflows experienced by a business during an accounting period.

Statement of owner equity   A financial statement showing the causes and amounts of change in owner equity during an accounting period.

Straight-line depreciation   A depreciation method that results in an equal amount of depreciation for each year of an asset's useful life.

Strategic alliance   An agreement among multiple individuals or businesses for the purpose of obtaining economic advantages that would not be available to them individually.

Strategic management   The process of charting the overall long-term course of the farm or ranch.

Subjective probability   A probability based only on individual judgment and past experiences.

Substitution ratio   The ratio of the amount of one input replaced to the amount of another input added, or the amount of one output lost to the amount of another output gained.

Sunk cost   A cost that can no longer be reversed, changed, or avoided; a fixed cost.

Supplementary enterprises   Enterprises for which the level of production of one can be increased without affecting the level of production of the other.

Sustainable agriculture   Agricultural production practices that maximize the long-run social and economic benefits from the use of land and other agricultural resources.

Systems analysis   An evaluation of individual enterprises and technologies that takes into account their interactions with other enterprises and technologies.

T

Tableau   An array of values showing activities, constraints, technical coefficients, and gross margins; used to solve a linear programming problem.

Tactical management   The process of making and implementing short-term decisions that keep the farm or ranch moving toward its long-term goals.

Tangible asset   Any asset that has a physical presence such as land, buildings, machinery, and livestock.

Tax-free exchange   A trade of one piece of farm property for another similar piece of property, such that any taxable gain is reduced or postponed.

Tax credit   An amount by which a taxpayer can reduce the amount of income tax owed if certain conditions are met.

Technical coefficient   The rate at which units of input are transformed into output.

Technology   A particular system of inputs and production practices.

Tenant   A farm operator who rents land, buildings, or other assets from their owner; a lessee.

Tenure   The manner by which an operator gains control and use of real estate assets, such as renting or owning them.

Tillable acres   Land that is or could be cultivated.

Timeliness cost   Loss of revenue resulting from a lower quality or quantity of crop harvested due to planting, harvesting, or other field operations not being completed on time.

Total cost (TC)   The sum of total fixed cost and total variable cost.

Total fixed cost (TFC)   The sum of all fixed costs.

Total physical product (TPP)   The quantity of output produced by a given quantity of inputs.

Total revenue (TR)   The income received from the total physical product; same as total value product.

Total value product (TVP)   Total physical product multiplied by the selling price of the product.

Total variable cost (TVC)   The sum of all variable costs.

Trend analysis   Comparison of the performance level of a farm business to the past performance of the same business.

U

Uncertainty   A situation in which neither the possible outcomes of an event nor their probabilities of occurring are known.

Unsecured loan   A loan for which the borrower does not give the lender the right to possess certain assets if the repayment terms are not met; there is no collateral.

USDA   The U.S. Department of Agriculture, which oversees many federal programs and policies related to agriculture including farm programs, extension, research, and food distribution programs.

Useful life   The number of years an asset is expected to be used in a business.

V

Value of farm production   The market value of all crops, livestock, and other income generated by a farm business, as measured by accrual accounting, after subtracting the value of purchased livestock and feed.

Variable cash lease   A leasing arrangement in which a cash payment is made in return for the use of the owner's property, but the amount of the payment depends on the actual production and/or price received by the tenant.

Variable costs   Costs that will occur only if production takes place and that tend to vary directly with the level of production.

Variable interest rate   An interest rate that can change during the repayment period of a loan.

Variance   A measure of the variability in the possible outcomes of a particular event.

Vertical integration   A contractual or other business arrangement involving two or more stages in the production of a commodity.

W

Weighted average   A long-run expected outcome from an event, found by multiplying each possible outcome by its respective probability and summing the results. Also called "expected value."

Whole-farm budget   A projection of the total production, income, and expenses of a farm business for a given whole-farm plan.

Whole-farm plan   A summary of all the intended types and size of enterprises to be carried on by a farm business.

Workers' compensation insurance   An insurance plan required by law in most states that protects employees from job-related accidents or illnesses and sets maximum compensation limits for such occurrences.

Working capital   The difference in value between current assets and current liabilities; a measure of liquidity.

INDEX

A

Accounting, 43–54
  debits and credits, 46
  double-entry, 46–47
  equation, 46
  single-entry, 46–47
  terms, defined, 43
  whole farm *vs.* enterprise, 47
Accounting period, 44
Accounting system
  basic *vs.* complete, 47–48
  double-entry, 46–47
  options in, 43–44, 46–48
  output from, 52–54
  single-entry, 46–47
Accrual accounting, 49–52
Accrual adjusted net farm income, 87–89
Affordable Care Act, 422
Americans with Disabilities Act, 422
Amortized loan, 359–360
Annual equivalent, 321
Annual percentage rate (APR), 362
Annuity, 313
Appraising land, 380–382
Asset turnover ratio, 111–112
Asset valuation, 62–63
Assets, 59
  current, 59
  intermediate, 62
  liquid, 59
  noncurrent, 59
  return on (ROA), 90–92
  valuation of, 62–63
Average
  simple, 274
  weighted, 274
Average fixed cost (AFC), 157
Average physical product (APP), 125
Average total cost (ATC), 159
Average variable cost (AVC), 158

B

Balance sheet
  analysis, 69–72
  cost basis, 63–64
  example, 65–69
  format, 58–62
  market basis, 64
  notes to, 68
  purpose and use, 58
Balloon payment loan, 360–362
Banks, commercial, 363–364
Basis, tax, 302–303
Biohazards, 165
Book value, 63
Borrowing, cost of, 362–363
Break-even analysis, 189–190
Break-even price, 189
Break-even yield, 189
Budgets, 175
  cash flow, 228–229
  enterprise, 177–179
  long-run *vs.* short-run, 207
  partial, 215
  whole-farm, 193–194
Bushel lease, 390
Business plan, 26

C

Capital, 351
  allocation of limited, 353
  economics of use, 352–353
  sources of, 353–355
Capital budgeting, 317
Capital debt repayment margin, 117
Capital gain, 306–308
  livestock and, 297, 308
  taxation of, 260
Capital recovery, 184, 321, 430
Cash accounting, 48

Cash flow budget
constructing a, 230–231
example, 234–238
features of, 228–230
form, 231–234
investment analysis and, 239–242
structure of, 229–230
uses for, 238–239
Cash flows, monitoring, 239
Cash rent lease, 384–385
Cash/noncash expenses, 155
Child labor regulations, 421
Civil rights, 422
Coefficient of variation, 275–276
Collateral, 357, 366
Competitive enterprises, 144–145
Complementary enterprises, 148–149
Compounding interest, 313
Conservation, 394–395
Consumer demands, 14–15
Contingent taxes, 66
Contracting, 15, 285–286, 355
Cooperatives, 263–264
Corporation, 258
advantages, 260–261
C corporation, 259
disadvantages, 261
example, 262
operation of, 261
organization and characteristics, 258–260
S corporation, 259–260
Cost center, 334
Cost curves, 167–168
Cost of production, 188–189
Costs
application of, 159–163
average fixed, 157
average total, 159
average variable, 158
curves, 167–168
direct, 179
fixed, 156–158
indirect, 179
machinery, 426–431
marginal, 159
operating, 178
opportunity, 154–155
ownership, 179
total, 159
total fixed, 157
total variable, 158
variable, 158
Credit, 355
establishing and developing, 365–366
leverage and use of, 368
line of, 358–359
management ability and, 366
personal character and, 366
reserve, 288
Crop share lease, 388–389
Cultural barriers, 418–419
Cumulative distribution function, 276–277
Current ratio, 69–70, 115
Custom farming, 284, 390

D

Debt, repayment capacity, 366
Debt structure ratio, 72
Debt/asset ratio, 70, 114–115, 368–369
Debt/equity ratio, 71
Decision making
environment in agriculture, 30–31
steps in, 26–29
tactical, 26
under risk, 278–281
Decision tree, 279–280
Decisions
characteristics of, 29–30
tactical, 26
Deferred taxes, 66
Depreciation, 81–82
defined, 81
expensing, Section 179, 304–305
machinery, 426–427
MACRS, 303–304
partial year, 85
recapture for taxes, 306
Depreciation expense ratio, 112
Depreciation methods, 82–84
comparing, 84–85
declining balance, 83–84
straight line, 82–83
Depreciation recapture, 306
Diagnosing a farm income problem, 104–105
Diminishing returns, law of, 126–127
Discount rate, 315
Discounted cash flow, 319
Discounting, 315
Diversification, 282–283

E

Earnings before interest, taxes, depreciation
and amortization (EBITDA), 94
Economic efficiency, measures of, 111–113
Economies of size. *See* Size

Efficiency
    economic, 111–113
    labor, 410–411
    machinery, 441–444
    measures of, 110–114
    physical, 113
Employment agreement, 414, 417
Enterprise accounting, 47
Enterprise analysis, 333–345
    crop, 336
    livestock, 336–339
Enterprise budget, 177–179
    breakeven analysis, 189–190
    computers and, 184
    constructing a crop, 180–184
    constructing a livestock, 185–187
    expenses on, 180–184
    interpreting and analyzing, 188–190
    revenue on, 181
    third party, 188
Enterprises
    combinations, 144–149
    competitive, 144–145
    complementary, 148–149
    profit ratio, 145–148
    substitution ratio, 145–148
    supplementary, 148
    value-added, 340
Environmental concerns, 15–16, 394–395, 421
Environmental Protection Agency, 421
Equal marginal principle, 133–135
Equity/asset ratio, 71
Expectations, forming, 273–275
Expected value, 274
Expenses, on income statement, 80–81
Expensing, Section, 179, 304–305
External scanning, 24

F

Farm business
    activities, 42–43
    environment, 30–31
    transferring a, 264–267
Farm business analysis
    diagnosing a problem, 104–105
    efficiency, measures of, 110–114
    financial, measures of, 114–116
    profitability, measures of, 105–109
    standards of comparison, 103–104
    types of, 102–103
Farm Credit System, 364
Farm Financial Standards Council (FFSC), 40, 52, 229
Farm Service Agency, 357, 364–365
Farm size, measures of, 109–110
Farming systems, 208, 394

Farms
    business strategies, 9–10
    number of, 8
    structure of, 8–9
Federal Insurance Contributions Act. *See* Social Security tax
Financial contingency plan, 367–368
Financial feasibility analysis, 322–323
    cash flows and, 239–242
    of land purchase, 382–383
Fiscal year, 44
Fuel, machinery, 429
Functions of management, 20
Future value (FV), 313
    of a present value, 313–314
    of an annuity, 314–315

G

Gain or loss
    income tax, 302–303
    on income statement, 79–80
Global positioning system (GPS), 11, 130
Globalization, 16–17
Goals, formulating, 22–23
Gross margin, 179, 182, 197

H

H-2A temporary or seasonal agricultural work, 421
Hedging, 286
Human resources, 13–14. *See also* Labor

I

Identity preservation, 12
Immigration Reform and Control Act, 421
Income statement
    accrual adjustments to cash, 87–89
    analyzing, 89–95
    definition, 77
    expenses on, 80–81
    format, 85–87
    gains and losses on, 79–80
    revenue, 78–79
Income taxes
    accounting methods, 296–298
    capital gain and, 306–308
    depreciation and, 303–304
    expensing, Section 179, 301, 304–305
    income averaging, 300
    management strategies, 299–302
    net operating loss, 301
    objectives of tax management, 295

record requirements, 297–298
tax basis, 302–303
tax free exchanges, 301–302
tax rates, 298–299
types of, 294–295
Inflation, 324–325
Information age, 12–13
Initial cost, of investment, 317
Input
    combination, least cost, 144
    combinations, 140–144
    price ratio, 131-132
    substitution ratio, 140
Input level, how much to use, 127–128
Input substitution ratio, 140
Insurance
    managing risk with, 283–284
    types of, 283–284
Interest
    annual percentage rate (APR), 362
    compounding, 313
    fixed cost as a, 157, 428
Interest expense ratio, 112
Internal rate of return, 321–322
Internal scanning, 23
Internal transaction, 339
Inventories
    valuing, 59, 63
    verifying, 342–344
Investment analysis, 317–322
    example, 328–331
    income taxes and, 323–324
    inflation and, 324–325
    information needed for, 317–318
    internal rate of return, 321–322
    net present value, 319–321
    payback period, 318–319
    risk and, 325–326
    simple rate of return, 319
    using cash flow budget, 239–242
Isoquant, 140

J

Joint ventures, 252–253

L

Labor
    characteristics of agricultural, 405
    communication, 418
    compensation, 414–416
    cultural barriers, bridging, 418–419
    efficiency, 410

employment agreement, 414
    evaluation, 419–420
    interviewing and selecting, 414
    motivation, 418
    planning farm use, 405–410
    recruiting, 413–414
    regulations, government, 420–422
    training, 416–418
Labor efficiency
    improving, 411–412
    measuring, 410–411
Land
    appraisal of, 380–382
    buying, 379–384
    characteristics of, 375–376
    controlling, 377–379
    economics of use, 375–377
    financial feasibility of buying, 382–383
    leasing, 378–379, 384–393
    ownership, 377–378
    planning use of, 376–377
    values, 374–375
Law of diminishing marginal returns, 126–127
Lease shares, determining, 393
Leases
    advantages of, 378–379
    bushel lease, 390
    cash rent, 384–388
    content, 384
    crop share, 388–389
    efficiency and equity in, 390–392
    labor share, 389–390
    livestock share, 389
    variable cash, 390
Leasing, 378–379, 384–393
Least cost input combination, 140–141
Leverage, 368
Liabilities, 59
    current, 60
    intermediate, 62
    noncurrent, 60
Liability insurance, 289
Life cycle, 250–251
Life insurance, 289
Limited liability companies, 261–263
Line of credit, 358–359
Linear programming, 200–204
    additional features, 202–203
    basics, 200
    dual values, 203–204
    graphical example of, 210–212
    reduced costs, 203–204
    sensitivity analysis, 205
    shadow prices, 203–204
    tableau, 200–202

Liquidity, 367
    analyzing, 69–70, 207
    defined, 58
    factors affecting, 367
    tests for problems with, 116
Livestock share lease, 389
Loans
    amortized, 359–360
    balloon, 360–362
    collateral for, 366
    intermediate-term, 356
    length of, 356
    long-term, 356
    non-real estate, 357
    personal, 357
    real estate, 357
    repayment plans, 356, 358–362
    secured, 357
    self-liquidating, 288, 366
    short-term, 356
    signature, 357
    single payment, 358
    sources of, 363–365
    types of, 355–362
    unsecured, 357
Long run
    average cost curve, 167–168
    defined, 156
Lubrication, machinery, 429

M

Machinery
    acquiring, alternatives for, 436–441
    costs, estimating, 426–431
    custom hire, 430, 438–441
    efficiency, improving, 441–444
    fuel, 429
    housing, 428
    insurance, 428
    leasing, 428, 438
    lubrication, 429
    maintenance and operation, 441–442
    new *vs.* used, 442
    operating costs, 428–430
    ownership, 436–437
    ownership costs, 179
    rental, 437–438
    repairs, 429
    replacement, 442–444
    selection, factors in, 433–436
    size, 433–435
    taxes on, 428
    timeliness, 435–436
Machinery cost per crop acre, 441

Management
    capacity, improving, 412–413
    functions of, 20–21
    strategic, 21–26
    style, 409
    tactical, 21, 26
Marginal analysis, 125
Marginal cost (MC), 159
Marginal input cost (MIC), 132–133
Marginal physical product (MPP), 125–126
Marginal principle, an example, 128–132
Marginal revenue (MR), 128–129
Marginal value product (MVP), 132–133
Marketing year price, 287
Mean, 274
Migrant labor laws, 422
Minimum wage law, 420
Mission statement, 21–22
Modified Accelerated Cost Recovery
        System (MACRS), 303–304

N

Natural Resources Conservation Service, 395
Net cash revenues, of investment, 317–318
Net farm income, 89
    analysis of, 89–95
    from income statement, 90
Net farm income from operations ratio, 112
Net operating loss (NOL), 301
Net present value, 319–321
Net worth, 61
Net worth statement. *See* Balance sheet

O

Occupational Safety and Health Act (OSHA), 421
Operating agreements, 253–255
Operating costs, 178, 426
Operating expense ratio, 112
Operating profit margin ratio (OPMR), 93–94, 108
Opportunity cost
    defined, 154
    of capital, 154–155
    of labor, 154
    of management, 154
Options, commodity, 286–287
Output level, profit maximizing, 131, 159–160
Owner equity, 354–355
    balance sheet and, 61, 68–69
    change in, 94–96
    sources, 61
    statement of, 54, 72–73
Ownership costs, 179, 426

P

Partial budget, 215
  computing changes, 222
  examples, 219–221
  format, 217–219
  limitations, 223–224
  procedure, 216–217
  sensitivity analysis of, 222–223
  uses of, 216
Partnership
  advantages, 257
  decision rules for, 280–281
  disadvantages, 257–258
  example, 256
  general, 255
  income taxes, 257
  limited, 255
  operation, 258
  organization and characteristics, 255–257
Payback period, 318–319
Payoff matrix, 280
Physical efficiency, measures of, 113
Precision agriculture, 426
Present value (PV), 315–317
  of an annuity, 313–314
  of future value, 316
Price ratio, input, 131-132
Price ratio, output, 131-132
Production function, 124–125
Production possibility curve (PPC), 144–145
Profit center, 334
Profitability, measures of, 105–109

R

Records, purpose and use, 40–42
Repairs, machinery, 429
Repayment capacity, debt, 366
Resources, taking inventory, 194–196
Retained farm earnings, 95
Return on assets (ROA), 90–92
Return on equity (ROE), 92–93
Return to labor, 94
Return to labor and management, 94–95, 108
Return to management, 94, 108
Revenue, and income statement, 78–79
Risk
  ability and willingness to bear, 272–273
  decision making under, 278–281
  financial, 271–272
  investment analysis and, 325–326
  legal, 272
  personal, 272
  price and market, 271

  production and technical, 270–271
  sources of, 270–272
Risk management tools
  legal, 289
  market, 285–287
  personal, 289
  production, 282–285
Rule of 72, 314

S

Salvage value, 82, 427–428
Seasonal agricultural work, 421–422
Self-employment tax, 299
Sensitivity analysis, 205, 222–223, 324
Short run
  defined, 156
  production rules for, 162–163
Simple rate of return, 319
Size
  constant returns to, 164
  decreasing returns to, 164
  diseconomies of, 165
  economies of, 163–167
  increasing returns to, 164
  long run, 164–165
  measures of, 109–110
  short run, 156, 163–164
Small Business Health Options Program, 422
Social Security tax, 420
Sole proprietorship
  advantages, 252
  disadvantages of, 252
  income taxes and, 251–252
  organization and characteristics, 251
Solvency, 368–370
  analyzing, 70–71
  defined, 58
Standard deviation, 275
Statement of owner equity, 54, 72–73
Strategic alliances, 11, 25
Substitution
  ratio, constant, 140–141
  ratio, input, 140
Sunk cost, 158
Supplementary enterprises, 148
Sustainable agriculture, 394
Systems analysis, 208

T

Tax basis, 302–303
Tax free exchange, 301–302
Technology, 11, 130

Term debt and capital lease coverage ratio, 117
Terminal value, 318
Time value of money, 312
Total cost (TC), 128
Total fixed cost (TFC), 157
Total physical product (TPP), 124, 125
Total variable cost (TVC), 158
Transferring the farm business, 264–267
    key areas to transfer, 265
    stages in, 265–267

U

Uncertainty, 270
Unemployment Insurance, 421
Useful life, 82

V

Valuation methods, 62–63
Value of farm production, 110, 410
Variable cash lease, 390

Variability, measures of, 275–277
Vertical integration, 15

W

Whole farm accounting, 47
Whole farm budget, 193–194
Whole farm plan, 193–194
    definition of, 193–194
    example of, 198–205
    farming systems and, 208
    planning procedure, 194–198
    sensitivity analysis, 205
Workers' compensation insurance, 420
Working capital, 70, 115–116